THE
UPPER PALAEOZOIC
AND
POST-PALAEOZOIC
ROCKS OF WALES

THE
UPPER PALAEOZOIC
AND
POST-PALAEOZOIC
ROCKS OF WALES

PUBLISHED IN MEMORY OF
SIR ARTHUR TRUEMAN

AND ALSO TO CELEBRATE THE
FIFTIETH ANNIVERSARY
(1920–1970)

OF THE
DEPARTMENT OF GEOLOGY
UNIVERSITY COLLEGE OF SWANSEA

Edited by
T. R. OWEN

CARDIFF
UNIVERSITY OF WALES PRESS
1974

PRINTED IN GREAT BRITAIN BY
CSP LTD, FAIRWATER, CARDIFF

PREFACE

In 1969, the University of Wales Press published "The Pre-Cambrian and Lower Palaeozoic Rocks of Wales", edited by Professor Alan Wood, in honour of two of its most distinguished geologists—the late Professor O. T. Jones and Sir William Pugh. The Department of Geology at the University College of Swansea has recently celebrated its fiftieth year and it has been thought fitting to similarly mark this milepost in its history and, at the same time to pay tribute to Swansea's first Professor of Geology— A. E. Trueman. The name of Trueman, like those of his successors to the Chair of Geology at Swansea, has been closely associated with studies (both stratigraphical and palaeontological) of the Upper Palaeozoic and post-Palaeozoic rocks of Wales. Each author in this volume has close associations with the Geology Department at Swansea. Professor T. Neville George was one of the first of Trueman's students and succeeded him to the chairs at Swansea and Glasgow. It is particularly fitting that he should write about Sir Arthur Trueman's work. Professor D. V. Ager is the present holder of the Swansea chair. Dr. Bloxam, Dr. Kelling, Dr. Walmsley and the editor are members of the Swansea department whilst Dr. Jones and Dr. Thomas are past students. Professor Allen, Dr. Bowen, Dr. Calver, Dr. Davies, Dr. Ivimey-Cook, Dr. Ramsbottom and Mr. Smith are all good friends of the department, besides being experts in their own particular fields. The editor sincerely thanks the authors for their generous contributions and also thanks Dr. R. Brinley Jones (the Director) and his staff at the University of Wales Press for their great assistance in the production of the book. The editor particularly wishes to thank the University College of Swansea for their generous contribution towards the cost of publishing this book and the Director of the Institute of Geological Sciences for his permission to publish chapters 7, 8, 9 and 14.

I wish to acknowledge help given by Professor Alan Wood at various stages during the preparation of this work.

CONTENTS

page

vii

FOSSIL MOLLUSCS AND MOLLUSCS IN STRATIGRAPHY: THE GEOLOGICAL WORK OF A. E. TRUEMAN

T. Neville George

I. BACKGROUND

A. E. TRUEMAN (1894–1956) was the first Head of the Department and Professor of Geology in the University College of Swansea (1920–33), where also he founded the Department of Geography. He moved as Professor of Geology to the University of Bristol in 1933, and to the University of Glasgow in 1937. He resigned from Glasgow in 1946 to become Chairman of the University Grants Committee, a post he occupied until his death. (See Pugh 1958.)

Trueman was a man of many parts. Beyond his more strictly geological interests he wrote on population changes, on the iron industry, and on regional geography in South Wales. He promoted a popular concern for geology in books some elementary and introductory, some linking geology and scenery. He entered the field of formal geomorphology in studies of the landforms and erosion levels of South Wales and the British district that, in a correlation with more widely developed surfaces in southern England, led him to infer late-Cenozoic eustatic uplift of marine benches that truncated indifferently the variety of rocks and structures to be found in Armorican and Alpine Britain.

These, however, were interests peripheral in relation to his main activities in geological research, which lay in the morphology, systematics, and evolution of Jurassic and Carboniferous molluscs, and in the application of molluscan palaeontology to refinements in the stratigraphy of the Lias and the Coal Measures. His work deeply penetrated palaeontological theory, in which it posed and partly answered problems that continue to be stimulants to research; and it radically transformed the manner of interpreting Upper Carboniferous sequence not merely by adding vastly to factual knowledge, but even more by reorganising method (and by pointing the way to further reorganisation) to give exactitude in stratigraphy and palaeontology.

Trueman's work needs to be put into its contemporary setting. He matured in a period of major discovery and enlightment prompted by the application to fossils of a Darwinian analysis of phylogeny and comparative morphology, as in Rowe's heart-urchins and Carruthers's corals, that gave a meaning to lineage hitherto unrealised. His mentors were H. H. Swinnerton (1877–1968), a thoughtful zoologist turned palaeontologist who put life into dead fossils and saw them as the stuff of an evolutionary natural history, and S. S. Buckman (1860–1929), an amateur brilliant in his essays on the sequence and the zones of the Jurassic rocks and deeply discerning (if with a speculative eye) in his interpretations of ammonite and brachiopod phylogeny and evolutionary process. Trueman was only to a degree less influenced by F. A. Bather (1863–1934) and W. D. Lang (1878–1966), both of whom in

1

complementary ways prompted him to theorise on the significance of variants and transients and on the alinements of evolutionary sequence. In the first half-century of Darwinian enthusiasm, when genetics was a little-understood science and a synthetic integration of genetic control and adaptive response had not been merged into comprehensive theory, discussion by these men was exceptionally stimulating to Trueman's receptive and critical mind; and although the discussion may now seem antiquated in its terms and in its assumptive and suppositious bases it had a firmness and a solidity of content that greatly encouraged him to build his own superstructure of systematised palaeonotology and stratigraphy on partly derivative, mainly original, lines.

His earlier papers on Jurassic rocks and fossils were perhaps the less independent parts of his achievement; but when he transferred his interests to the Coal Measures he was able to carry method and instrument into an almost untouched field in which crude lithostratigraphy dominated classification and correlation, and fossils were at best uncertain and at worst thoroughly misleading guides to age. In the transfer he elaborated a taxonomic system of describing Westphalian bivalves that enlarged greatly and in part changed the meanings of classificatory categories; and he devised kinds of stratal zones that differed significantly from the zones long recognised in the Jurassic rocks.

His attention in all his work was concentrated on molluscs—ammonites, oysters, and some gastropods in his Liassic studies, mussels in his Westphalian— but inevitably interest in biostratigraphical sequence led him to facies sequence, to lithology and palaeogeography, and to sedimentation and sedimentary environments, and he was induced to consider tectonism as an influence on deposition, and to explain stratigraphical sequence as in part a response to earth movement.

II. ONTOGENY AND PHYLOGENY

(a) Recapitulation

The identification of evolutionary relationships, notably in the work of Buckman, rested partly in a time-sequence of graded forms, but also (in analytical mode perhaps more than in blinkered advocacy) in a Haeckelian dependence on recapitulation. The deficiencies or limitations of ontogenetic evidence were of course well recognised, and at the turn of the century no reputable palaeontologist naively thought that, without qualification, 'ontogeny recapitulates phylogeny'; but despite the acknowledged need to allow for diversionary modifications of ontogeny it was taken for granted that an ideal recapitulation could be discerned or restored or postulated by the expert when allowance was made for the diversions. It was then possible to scan the course of evolution, to reconstruct the lineage, from a few specimens whose comparative stages of growth allowed a plotting of morphogenetic series as standards of reference.

At the time of Trueman's writing, a typical attitude to recapitulation was expressed by Bather (1920, p. 67). He conceded that how growth came about was mysterious, that genetic controls on the growth of fossils were unknown, and that even amongst living animals, to which genetic tests could be applied, the detailed characterisation of growth was almost equally unknown, so that 'fluctuations' and 'temporary modifications' must be supposed to be nongenetic. He also conceded that departures from a recapitulatory rectilinear ontogeny were to be recognised in larval adaptations; and he thought that deviations in ontogeny that disrupted a simple linear progress from infant to adult were signs of a correspondingly complex phylogeny. But in essence, minor unresolved elements put aside, he expressed the central assumption of a fundamental 'biogenetic law'—a 'law' usually linked with Haeckel's name, but having antecedents long pre-Darwinian—when he regarded ontogeny in the later members of a lineage as additive, as incorporating by a process of hypermorphosis a condensed version of the successive evolutionary stages passed through by the earlier members: 'however they may be explained, the instances of recapitulation afford convincing proof of descent'.

Trueman in his work on ammonites accepted, did not question, the assumption. Without discussing the nature of the genetic instructions given in ontogeny (about which fossils could say nothing), he took for granted that descendants inherit from ancestors potentialities of growth that it is in the nature of the ancestors to transmit; that ontogeny is narrowly determined by genetic controls in the lineage if the individual organism is to retain its specific identity; and that in more or less degree the sequence of ancestors (the more recent of them the more assertive) is, in the nature of heredity, to be discerned in the ontogeny of the descendants. But in his assumption of a summarised linear course of ancestral evolution in an ideal or conceptual descendant ontogeny, he was no less aware of the many possible ways in which an actual ontogenetic sequence may obfuscate or may cease to reflect phylogeny.

Already in his studies of liparoceratid lineages (1919, p. 254) he recognised in some characters a relative retardation of development (a bradygenic or 'catagenetic' feature), which offset a tachygenic or 'anagenetic' acceleration in other characters brought about by the hypermorphic addition and telescoping of terminal ontogenetic stages (notably seen in the ammonites by the encroachment of changing pattern of ornament or coiling angle or whorl profile onto progressively earlier whorls in successive descendants). He emphasised the interplay of tachygenic and bradygenic influences in modifying the relative rates of bioserial change and thus in promoting the association of contrasted character-grades in ontogeny, to the point of wide departure between any ontogenetic and any ancestral stage. Following Buckman and Lang, he drew attention to the possibilities arising in condensed ontogenetic sequence when intermediate stages deviant from simple linear development are 'skipped' as a speeded-up ephebic stage is reached in a 'straightening-out' of the ontogeny, the recapitulatory reference to some ancestral forms being suppressed in such truncated palingenesis (see Trueman 1922a).

Similarly in his description of growth in Liassic gastropods he noted the contrasting emphasis, to be seen in the ontogeny of different species, on an abbreviated or a prolonged stage of smooth initial whorls, on alternating developments of axial and radial ribs and on consequent variant patterns of reticulate ornament, and on recurrent kinds of homoeomorphs arising from permutative ontogenetic relations of bioserial change (McDonald & Trueman 1921, p. 303).

The fullest exposition of his views is to be found in his account of echioceratid evolution (Trueman & Williams 1925, p. 700), in which there are distinguished in ontogeny caenogenetic characters, incidentally adapted to larval needs but not of great significance in ammonite evolution, palingenetic characters following 'a direct line of development from the protoconch to the adult', and palingenetic characters highly informative in being deviant from a direct ontogenetic line but then recapitulating 'recent' phylogenetic departures from a direct evolutionary line. A particular sign of process is given in the nepionic stage of all ammonites, for while deviant characters are most likely to be 'skipped' in later members of a lineage, a stout protoconch and depressed first whorls are retained throughout the evolution of the group, no matter what adult whorl shapes may evolve, 'because they are of value in fitting the young individual to its environment'. (See also Trueman 1929, p. 135; Swinnerton 1932, p. 333; George 1933.)

Contemporary criticism, notably by Spath (1933, p. 445; 1936, p. 177) and indirectly by Schindewolf (1929), drew attention to the imbalance in the way Trueman used recapitulatory evidence, mainly in his neglect of paedomorphosis or protero-genesis; for he discounted as a rarity or as unproved the evolutionary appearance, first as a 'larval' or nepionic insertion in ancestral ontogeny, of a character that in successive descendants encroached on later and later stages of growth until it persisted into adult life. Spath, perhaps in opposed imbalance, went so far as to reject recapitulation as a consistently reliable sign of phylogenetic relationships, and thought its evidence might as much mislead as inform. But while it is true that Trueman generally disregarded the possibility of a nepionic expression of a newly mutant gene, as he also undervalued the allometric correlation of character-expression with size, the prolongation of a paedomorphic character into the later ontogenetic stages of phylogenetic descendants is in one aspect but a particular instance of what (descriptively) Trueman understood as a combination of caeno-genesis and a sustained bradygenic retardation. (See Swinnerton 1938; Lang 1940, p. 110; Gould 1966.)

(b) The ammonite septum

An analogue of recapitulation, which Trueman continued to invoke for a number of years, was provided by his early studies, under Swinnerton's guidance, of the pattern of the ammonite septum (Swinnerton and Trueman 1918; Trueman and Williams 1925, p. 701). In particular, he used 'septal sections' (transects of the contoured surface) of an adult septum, plotted as concentric profiles from mid-septum to periphery and showing progressive complication towards the shell wall, in demonstration of the close parallelism (the innermost sections and the first suture excepted) of the profile sequence with ontogenetic suture sequence; and he then concluded that the close similarity, repeated in variously shaped ammonites, con-firmed an ontogenetic sequence: 'the sutural development of an ammonite [as shown by the adult septum] represents the most direct series of changes from the suture of the protoconch to the adult suture'.

The successive profiles were explicitly distinguished from 'localised stages of development' (elements of adult structure that to R. T. Jackson recalled nepionic or neanic stages or palingenetic adult ancestral stages, as in the latest-formed stalk plates of a crinoid, or apical plates of a sea-urchin showing ambulacral 'crushing'). The single septum, deposited very rapidly (if not continuously at a moment), was moulded in its contoured form by the dorsal mantle; and in ontogenetic sequence it usually shows progressive elaboration of lobes and saddles to maturity. It is obvious that with strong bilateral symmetry in whorl cross-section, which in many evolute ammonites approaches a radial symmetry, the frills of the marginal suture (profile) must diminish in amplitude and abundance as they are traced towards the septal centre, so that the notional mid-point of the septum is both flat and 'horizontal'. An ontogenetic septal sequence to the adult septum, accompanying a progressively increasing radius of whorl cross-section, may thus be expected *a priori* to parallel in increasing elaboration the changing concentric profiles of an adult septum with increasing radial distance from the septal centre, until sequential septa and centrifugal profiles converge in an identical (coincident) transect-profile at the marginal suture—provided ontogeny is (in Trueman's sense) direct, displaying sustained anagenesis. If however there is catagenesis of septal corrugations and sutural frilling in later after earlier anagenetic whorls, ontogeny not then being direct, Trueman's analogy breaks down, the concentric profiles of an adult septum then displaying a simpler sequence than the ontogenetic sutural sequence, and the one (like a lipopalingenetic ontogeny) reflecting a truncated series in the other.

(c) Evolutionary process

Trueman was much influenced in his analysis of evolutionary change by the method of morphological dissection. As illustrated for instance in the structural units of Swinnerton (1921; 1932) and Lang (1923a; 1923b), an organism may conceptually be regarded as an additive association of characters—in a broad sense the biocharacters of Osborn—that have some independence one from another in both ontogeny and phylogeny. Evolution may then be synthesised as the phylogenetic summation of character changes along anagenetic (and catagenetic) lines or bioseries. The phylogenetic independence of the bioseries may be recognised in the manner of combination of bioserial stages in the single organism, in reaching whose phenetic status the rates of bioserial change show a variability and a differential emphasis characteristic of the lineage.

Thus the liparoceratid ammonites (Trueman 1919, p. 257) in their phylogenesis show progressive bioserial change in (conveniently segregated) whorl shape, involution, sutural complexity, and ornament. In Trueman's evolutionary reconstruction, the more 'primitive' members have (in the adult) subrounded whorls and are evolute, and they are strongly costate (capricorn) in ornament; the more 'advanced' members have more or less depressed sphaeroconic whorls and are tightly involute, and they are relatively finely costate (the ribs often bifurcate) and tuberculate. In evolution some lineages are 'progressive' in one character, 'conservative' in another; other lineages are the converse, the various trends then being repeated again and again but at various rates of change. Similarly, in the procerithiid and loxonematid gastropods of the Lias (McDonald & Trueman 1921, pp. 301 ff.) bioserial changes

include modifications in the apical angle, in the persistence of the smooth 'embryonic' whorls, in the onset of radial and concentric ornament and in its catagenetic
loss, and in the permutative combinations of various bioserial stages reached in
different lineages. Ontogenetic criteria, the association of stages expressed in the
unitary organism, persistently emerge as bases for identifying the course of evolution,
with consequent implications for classification (see p. 8).

The analytical method, penetratingly developed by Trueman, illustrates means
of distinction between grade (a term he used) and clade (a term more recently
introduced); and although his method was foreshadowed in the work of such other
contemporary palaeontologists as Swinnerton and Lang, he is to be counted
amongst the pioneers in a sustained application of a principle that has become
commonplace in palaeontological theory (see George 1971, p. 205 ff.). The elucidation of evolutionary process that followed from his analysis also led him to build
upon Buckman's views on homoeomorphy when in a comparative ontogeny of
his gastropods he showed that the permutations of bioserial anagenesis and catagenesis were likely to engender both cyclical and transversal homoeomorphs, with
consequent danger in an undiscerning taxonomics (McDonald & Trueman 1921
p. 305).

At a later date Trueman (1940) returned to the problems that beset evolutionary
explanation when repetitive bioserial trends are seen in divergent and heterochron-
ous lineages as signs on the one hand of a directed evolution (Lang 1923a) and on
the other of a repeated adaptive response to repeated environmental stimulus
(Bather 1920, pp. 74 ff.). His attitude was perhaps ambivalent, for on the one hand he
thought that parallel series expressing 'trends' might as much be attributable to
environmental as to internal factors, and on the other that lineages may reveal
evidence of a lack of 'harmony' with their environment, 'progressing so far in a
given direction that extinction automatically supervenes'. Signs of the latter process
he professed to see in the excessive development of a calcareous skeleton in some
Cretaceous bryozoans—'excessive' being a postulate that originated with Lang
(1919, p. 102; 1923a, p. 12) but was roundly refuted by Bather (1920, p. 78)—and in
the excessive incurvature of the left valve of the shell in some gryphaeoid oysters of
which the umbo 'pressed against the right valve and so prevented it from opening,
thus perhaps leading to the extinction of the gens' (an end that in fact was never
achieved in recognisable umbonal pressure in any actual gryphaea). At the same
time, in a milder interpretation of orthogenesis, he discouraged the use of the term
programme evolution' as it carried overtones of causation and predetermination'
that he thought might be misleading.

(d) Ammonite form

In a highly original discussion of growth and development in relation to mode
of life in ammonites, Trueman (1941) demonstrated his constant interest in fossils
not as dead forms to be accorded only passive morphological or morphogenetic
description but as once-living organisms actively responsive to their physical
environment, whose shape was to be seen in the light of that relationship. In an
ingeniously devised dissection of the morphic elements of a variety of shells, with

which he combined well-founded inferences on specific gravity, he determined the centre of buoyancy and the centre of gravity of the living ammonites; and in imaginative reconstruction he demonstrated convincingly the kinds of adaptive responses to be read into ammonite form explicated in a hydrodynamic context.

Thus length of body chamber is generally to be correlated with magnitude of spiral angle and not with closeness of involution; but although body form in some ammonites was comparatively short and stout (in *Ludwigia* for instance the body occupied a chamber spanning little more than half a whorl), and in others was long and vermiform (in *Dactylioceras* spanning a whorl, in some echiocerates nearly a whorl and a half), the volume ratio of body-chamber to air chambers—a major factor in shell attitude—is commonly about 2 or 3 to 1. A combination of proportionate length of body chamber, spiral angle of coiling (but not the apical angle of the spiral cone), and degree of involution appears to have determined the position of rest of the different kinds of ammonites. Many evolute serpenticones, like *Dactylioceras* and *Caloceras*, were presumably benthic with aperture on the underside of the shell in the resting position, but, the centres of buoyancy and gravity being almost coincident, the animals had little difficulty in rotating their shells and swimming in any chosen direction. *Ludwigia* and *Oxynoticeras*, in which the centre of buoyancy is relatively far removed from the centre of gravity, on the other hand appear to have rested with their apertures opening upwards; and while doubtless they had no difficulty in swimming into whatever position they wished to assume, it may be supposed that they were less consistently benthic than the serpenticones. Their hydrodynamical centres being still more widely separated, the scaphitids and some of the criocerates, with reflexed apertures not appropriate to a benthic habit, contrast with the baculitids (and the analogous orthocerate nautiloids) and the turrilitids, more or less 'vertical' when at rest, characteristic benthos, perhaps ambulatory in gastropod fashion. The very short body of *Sigaloceras*, in whose shell the volume ratio of body-chamber to air-chambers is as low as 0·84 to 1, suggests that the net specific gravity of shell plus soft parts was significantly less than that of sea-water, and the animal may well have spent much of its time floating or swimming near the sea surface.

These interpretations of shell attitude of the living ammonite make their comment no less on ontogenetic change. The 'larval' protoconch and perhaps the first chambers (which together Trueman thought to be the 'embryonic' initial shell) may well represent a planktonic stage (and therefore in their stout sphaeroconic form be conservatively persistent throughout the group); but, with an increasing maturity and an adaptive radiation to a variety of nektonic and benthic habitats, the ammonites in their neanic whorls indicate their divergent modes of life. When ontogeny is direct, adaptation to the adult habitat was also presumably direct; but when it is indirect, when there are major neanic interpolations and diversions, it is clear that the ammonite adopted radically new hydrodynamic postures on becoming adult.

Thus a capricorn stage in the ontogeny of sphaeroconic liparoceratids implies a neanic form of comparatively low specific gravity, evolute and easily manageable in swimming, and an aphebic form far less buoyant, involute and more persistently benthic. The criocerates and scaphitids, close-coiled in nepionic and early neanic stages, progressively uncoiled in well-represented form series and in later neanic stages, reflexed in the ephebic stages of advanced members, record rest attitudes from down-facing to up-facing apertural positions, and perhaps an abandonment of a freely swimming habit. The heteromorphs—*Anaklinoceras* for instance turreted, *Heteroceras* turbinate, in early whorls, both crioconic in the adult—show even wider departures from a direct anagenesis in adaptive response and await further analysis along Trueman's lines.

III. TAXONOMIC THEORY

(a) Differential and progressive characters

Attempts to co-ordinate palaeontological and genetical theory fifty years ago were in part frustrated by an over-simple distinction between variation postulated as mutational, and variation postulated as environmental; and such terms as 'modification', 'fluctuation', 'somatic', 'germinal', 'substantive', 'continuous', 'discontinuous', all intended or supposed to reflect the elements of a crudely understood Mendelian kind of analysis, were the common coin of discussion. Most of the terms were not directly applicable to fossils, but allusively they were often invoked in explanation *a posteriori* of variant or transient differences considered to be either inherent and important or superficial and unimportant. They also lent themselves, presumptively if not explicitly, to supporting a Linnaean system of defined taxonomic categories; and indirectly they penetrated classificactory modes in a distinction between differential and progressive characters, the one kind considered or felt to be substantive and discontinuous, the other modificational and continuous.

Trueman was a man of his time, and absorbed contemporary attitude with little explanatory gloss in his selection of definitive and diagnostic criteria for his ammonite and gastropod taxa. His exposition of phylogenetic or morphogenetic progression in the liparoceratid series was of anagenesis in ornament of the adult shell from smooth (not seen in any known member but indicated in ontogeny) through coarsely ribbed capricorn to finely ribbed and imparinode tuberculate forms (extreme end members sometimes showing catagenesis in a reversion to parinode tuberculation); of change in whorl shape from stout involute sphaeroconic (also not known in any adult member—unless in a distantly ancestral *Cymbites*—but indicated in ontogeny) through evolute capricorn to involute sphaeroconic; and of increasing sutural complexity. These bioserial characters he therefore called progressive, in contrast to relatively static differential characters identified in the pattern of sutural elements (lobes and saddles), and the position of tubercles in relation to the sutural elements. The differential characters helped to distinguish his genera, the progressive grades his species. On this taxonomic grid he then demonstrated the recurrence of parallel homoeomorphs, and the modification of ontogeny in a particular member that arose from bradygenesis, tachygenesis, and lipopalingenesis, factors that at once were defined by the taxonomics and contributed to a justification of the taxonomics.

Similarly the cerithiid gastropods (McDonald & Trueman 1921, p. 309) possess characters that Trueman sharply distinguished between progressive (whorl shape, kind of concrescent ornament, strength and character of ornament) and differential (primary axial or spiral ornament, holostome or siphonstome aperture, obliquity of the columella, differential acceleration of character development); and on the distinction he was able to demonstrate comparable grades of structure in the occurrence of transversal homoeomorphy between members of the postulated lineages.

Such a segregation of differential and progressive characters is a relic of an Aristotelian classificatory system. It is still a major element in some kinds of systematics, and is likely to remain so until the bonds of Linnaean techniques are broken; but in most instances of its continued use the arbitrary choice of characters is obvious and is correspondingly unconvincing both in practice and in principle. Trueman was criticised by Spath (see particularly 1938), less on theoretical than on observational grounds, for his gratuitous selection of liparoceratid differentiae: Spath pointed out that capricorn ammonites are relatively late arrivals in the Lias sequence, the earliest liparoceratids known being tuberculate sphaerocones, and he then reversed Trueman's evolutionary order to match the stratigraphy, and rejected the palingenetic significance of a direct ontogeny. He regarded ammonite phylogenesis commonly to be one of simplification of structure, the evolute capricorns being descended from involute sphaerocones; and he derived an appropriate ontogeny by a proterogenetic extension in lineage sequence of a caenogenetic capricorn stage.

No comparable critique of cerithiid classification has yet been made; but it is not easy to discern the basic signs of differentiae in such morphic elements as ornament tachygenesis or columellar obliquity, unless differentiae are initially presupposed to exist (and therefore to be discovered), or, if indeed they have 'objective' existence, unless they can be 'objectively' identified. (See George 1971, pp. 222, 224.)

It is significant that in his account of the echiocerates, forms that display much greater superficial uniformity than the liparoceratids, Trueman made no attempt to distinguish between progressive and differential characters but separated his genera on comparatively minor differences of rib-pattern, involution, and ventral form, the lineages he recognised being essentially clades and the species grades (Trueman & Williams 1925, pp. 703, 706; 1926, p. 248). Even so, Arkell, a 'lumper', in subjective disagreement, suppressed all of Trueman's genera as synonyms (one of them objective), not thinking the cladal branches to be of generic magnitude (Arkell and others 1957, p. 243).

(b) Liassic oysters

Few ammonite lineages are preserved in an abundance of local sequential demes to allow a study of chronoclines in biometrical detail: constrained by the small number of individual specimens available to him, Trueman was formal and orthodox in his taxonomics of the liparoceratids and the echioceratids and in the manner of his selecting differentiae to isolate his taxa. It is all the more surprising that at precisely the same time he should have written a short paper on oyster evolution (Trueman 1922b), radically different in method and approach, that was to become of immense importance in transforming concepts of lineage and in giving new criteria to taxonomics.

The modesty of his original intention in studying the oysters was inherent in the primary purpose of the paper, which was to see how far species of *Gryphaea* could be used to supplement ammonites in a zoning of the Liassic rocks; and in fact Trueman did not apply himself to further substantial study of oysters for nearly twenty years (McLennan and Trueman 1942); but the wider implications of his discoveries were not lost upon him. Almost immediately he recognised that the essence of his work in its bearing on evolution was the demonstration of the significance of the 'interbreeding' community (the palaeodeme in modern terms) as the constituent member of the lineage at every geological moment. He found he could segregate (by analytical dissection) three or four biocharacters in his oysters— coiling of the left valve, area of attachment, thickness of the left valve, sulcus of the left valve—which, combined with absolute size, showed various degrees of expression in the members both of any one community and of communities in chronoclinal sequence. Moreover, his analysis showed him that there was not always close correlation between the biocharacters as they were comparatively expressed in neighbouring individuals, nor was there close correlation between them amongst members of communities in succession. He demonstrated (what is now commonplace) that, with a sufficiently large number of specimens available for statistical manipulation, Bather's variants and transients were subsumed in fields of variation in chronoclinal sequence, and that evolution could be measured in terms of variation fields migrating through time.

The holomorph continuity of the *Gryphaea* lineage as Trueman traced it in its abundant members through the Lower Liassic rocks became pointed comment on the nature of species and genus, for differentiae of the kind convenient in ammonite and gastropod taxonomics were not to be identified. In three or four pages of a collateral paper Trueman (1924) summarised the implications of his work by placing the biocharacters of variants and transients in a plexus of interweaving strands of bioserial change, which (in an analogy not now to be stressed) he compared with the cross-bred strands of Mendelian inheritance—an analogy that had also occurred to Swinnerton. The intricacies of individual relationships in such a lineage gave significance to a notional anagenesis (or catagenesis) not in one individual's being more 'advanced' or more 'primitive' than another, but in modal differences between one biometrically defined community and another; and it became difficult to allocate an individual to its proper species without violating the notion of a morphotypic standard on the one hand or the unity of a deme on the other. (See George 1956, p 134.)

Strangely, Trueman did not follow his own guiding hand to a logical conclusion. Trammelled by a Linnaean net and the conventions of nomenclatural practice, he accepted the need to retain binomials, despite his demonstrations of the continuity of the holomorphic lineage: he not only continued to refer to the species *irregularis*, *dumortieri*, *obliquata*, and *incurva* as stages in the evolution of the lineage, but also placed *irregularis* in the genus *Ostrea*, *dumortieri* in *Gryphaea*, without a suggestion of a differentia to separate the genera. Moreover, he also continued to accept the morphotypic standard and to subscribe to the need for a holotype as the sign of a species, and he even went so far as to say that he was 'in favour of restricting the use of the specific name to specimens identical with the holotype'—although the basis of his work lay in the bioserial independence of the biocharacters he used, so that no second individual was likely to be identical with the holotype, and although the unity of the deme, of which the holotype was but an incidental member, would be denied by such a practice.

An extremely narrowly conceived morphospecies thus emerged as a taxonomic category. All individuals not identical with the holotype but falling into the same 'species-group' Trueman proposed to identify by the use of 'cf.' or 'aff.'—adjuncts intrinsically not applicable to members of holomorph lineages, for an individual not belonging to a morphospecies cannot in its nature have 'affinity' with (although it may have some resemblance to) the morphotype, and conversely every individual in its lineage has in its nature some (if not close) resemblance to the specimen called the morphotype. It is even stranger that, having committed himself so far, Trueman recognised that his system 'must lead to enormous increases in the number of specific names', when a central part of his theme was the breaking down of barriers between species; and stranger still that he acknowledged the defects of his proposals in 'It may be urged, however, that the Linnaean system of binomial nomenclature is in many cases inadequate to meet the needs of palaeontology'.

The importance of the work on *Gryphaea* cannot be over-emphasised. It threw a brilliant light on the processes of evolution as revealed by fossils. It gave a new meaning to lineage in its integration of variants and transients. It altogether re-alined method and purpose in much subsequent palaeontological research. Nothing can now take away the value of its contributions to theory, even if the foundations on which it was built should prove to be unsound.

Recent criticism, despite support for and defence of Trueman by Swinnerton (1939; 1940; 1964) and Philip (1962), asserts that the foundations are unsound. In mutually conflicting argument on technique, but in combined attack on Trueman's biometrics, Hallam (1959; 1968, pp. 96–113), Joysey (1959), and Burnaby (1965), working on larger, more carefully collected, and more precisely located samples than Trueman, and adding a stringency to their analysis, regarded Trueman's statistical methods to be without rigour, and to fail altogether to support his conclusions. Joysey was prepared to concede that the conclusions were merely not proved and might be true. Hallam and Burnaby, emphasising Trueman's unawareness of the allometric effects of increase in the size of shell, rejected an assumption of steadily increasing tightness of shell incurvature as Lias times progressed—they even suggested that on their biometrical data there were signs of incipient uncoiling. Hallam concluded that the later thick-shelled coiled forms (*Gryphaea*) were genetically (and so generically) sharply distinct at an evolutionary step from the earlier thin-shelled flat forms (*Liostrea*), without there being graded continuity between them.

IV. LIAS STRATIGRAPHY

(a) *Zonal sequence*

Trueman's interests in ammonites and oysters were directly linked with, and were partly prompted by, his interests in the Jurassic rocks. In his first publications (1915; 1918) he described the Liassic rock and fossil succession of the Midlands in orthodox fashion, in the manner established notably by Buckman and Lang, accepting (and corroborating in a finely resolved fossil distribution) the zonal and hemeral divisions as rock and time units at variance with the lithological units of earlier workers. His meticulously refined stratigraphy enabled him to show that marked variations in zonal thicknesses occurred between Grantham and Lincoln (with oolites developing where the thinning was due to slow deposition), and that some of the variation was to be attributed to nonsequence.

A comparable study of the Lower Lias in east Glamorgan (Trueman 1920), dominated by a correlation of the ammonite sequence with the lithology, included a brief discussion of zonal attributes, in which incidental phrases anticipated the defence of ammonite zones he was to elaborate later. When the study was continued into mid Glamorgan (Trueman 1922c; 1930), further refinement of the ammonite sequence, but with changes in zonal nomenclature that indicated his underlying concept of 'zone' to be biostratigraphical, added much to precise correlation; and in a close analysis of the highly fossiliferous rocks of the Radstock shelf (Tutcher & Trueman 1925) the zonal scheme was amplified (on a basis laid by Buckman and others) by the inclusion not of 'subdivisions' of the major zones but of 'units' some of which were called 'horizons' and some 'faunas', and some of which look suspiciously like subzones.

The Jurassic work was carried out at a time of wide divergence of views on the validity of ammonite zones and subzones in correlation: Stamp (1925) summarised the opinions of many critics when he contrasted the limited range of many living organisms, controlled in their habitat by a multiplicity of environmental factors, with the extended (to the sceptic unlimitedly extended) range bestowed by Jurassic stratigraphers on their ammonites. Trueman (1923) participated in the controversy, giving clarity and precision to the meaning of terms, and justifying their use. He emphasised that zones were rock-units (not 'assemblages of fossils') spanning a prescribed time-interval, and that at any locality the rock-sequence might be discontinuous but the time-interval was not; he offered theoretical justification for identifying interrupted deposition (nonsequence) when expected Jurassic zones were absent; he disagreed with Buckman's restriction of 'zone' to the deposits of a single hemera, and in his acceptance of an expanded (traditional) meaning to 'zone' offered the new term 'epibole' (the subzone of some authors, equivalent to Buckman's small-scale zone) as the rock equivalent of a hemeral time-span.

These improvements, and Trueman's part in promoting them, are now built into the historical corpus of Jurassic stratigraphy, and have formed part of the basis of an evolving stratigraphical system to the present day. At the same time, the limitations of method (some of them still surviving) to be read into Trueman's theme and practice indicate an overlay of ammonite dominance in what is basically a biostratigraphy. In his exposition, zones (and subzones and epiboles) were not defined in relation to a particular rock-suite but rested on ammonite occurrence (the entry, sometimes the disappearance, of species or 'faunas'); and the quirks of ammonite sequence, notably in 'faunal repetition', made zonal terms only 'approximate', and zonal limits 'could not be precisely defined'. Some of his zones (or at least the epiboles) were acme zones, some weighted assemblage zones, some little more than marker-bed zones; and he recognised some of the zones to rest on ammonite misidentification and to be invalid. Trueman, oppressed by the many practical shortcomings of zones as defined rock-units—they could 'have only general value'—proposed to replace them in systematic correlation by chronostratigraphical terms ('ages') which were 'more precisely defined' and reflected 'the ammonite sequence quite as well as the zonal terms': that is, he promoted a conceptual stratigraphy based on conceptual fossil time-range.

He did not pursue his ideas, and in the papers on Radstock (Tutcher & Trueman 1925, p. 597) and Nash Point (Trueman 1930, p. 149)—his last substantial contributions to Jurassic stratigraphy—he used 'zone' and 'subzone' in a conventional sense with ammonite 'horizons' dominating method.

(b) Palaeogeography

The littoral facies of the Blue Lias in mid-Glamorgan gave Trueman (1922c; see also Hallam 1960) the opportunity of collating fauna with rock type as the sediments were traced from the off-shore shales and muddy limestones of the Cardiff district into the massive pebbly and conglomeratic rocks of the Southerndown and Sutton Beds, and then of making a close palaeogeographical restoration of a landscape drowning beneath an advancing sea. When he studied the area the Mesozoic topography had been broadly plotted by members of the Geological Survey, notably in a tracing of the unconformable base of the Mesozoic sediments (of whatever age—Triassic, Rhaetic, Liassic) on the Palaeozoic foundation, as an archipelago of islands flanking an inferred mainland to the north. Trueman was able to add much further detail both of zonal extension and of lithological diachronism in his elucidation of the stratigraphy. He showed that the rapid passage, often in less than a mile, from shales to boulder beds was accompanied by biofacies changes from an offshore ammonite fauna of relatively small forms, through an intermediate fauna with swarms of gryphaeas and with large ammonites, to an inshore fauna of thick-shelled bivalves, turreted gastropods, isastraeoid corals, but few ammonites. The fossils giving him a zonal sequence in whatever kinds of rocks, he traced the stages in the drowning of the islands from Keuper through Rhaetian and Hettangian to Sinemurian times, when the last remnant Palaeozoic summits of the Vale of Glamorgan were finally submerged.

Throughout the transgression of the Lower Liassic sea Glamorgan appears not to have been appreciably deformed by even the gentlest folding; but the contemporary deposits in the Midlands in their nonsequences, despite their being fine-grained without sign of a neighbouring coast, point to a tectonic restlessness that repeatedly caused the rocks to be warped into subdued folds and the anticlinal crests to be eroded. In their dissection of the pattern of the folds, Cox & Trueman (1920), prompted by the earlier work of Buckman in the Cotswolds, saw the posthumous influence of 'basement' structures (and reciprocally could postulate 'basement' controls) in explanation of the nonsequences. They showed that the intra-Jurassic folding, strongly marked in the Lias, probably died down by the end of Middle Jurassic times, much as it did farther west.

The inspiration of Buckman (see Davies 1930) was even stronger in the exploration by Tutcher and Trueman (1925) of the Lias of the Radstock shelf. The systematics of the exceptionally abundant and varied fossils found in the rocks was an immense task, mainly undertaken by Trueman. The interpretation of the bewildering changes of sequence between neighbouring outcrops was an even greater task, and demanded a penetrating eye in the identification of faunal and stratal gaps as signs of sedimentary environments in an unusual kind of palaeogeographical context. In result, the Radstock shelf was shown to be a region of clear waters, not often recognisably coastal but always shallow and often in retreat, that nourished a wealth of organisms whose skeletons contributed to the formation of limestones, only a few metres thick, ranging through Rhaetian, Hettangian, Sinemurian, and lower Pliensbachian stages. Relatively great variations in thickness and in limestone lithology, and the recurrence of diastems, contributed to the difficulties of interpreting a sequence characterised less by deposition than by nondeposition, in which zones are often no more than a few centimetres thick and fossils are often derived. Integrating the evidence, and emphasising a major nonsequence, Tutcher and Trueman suggested that there were uplift and a widespread retreat of the sea between Hettangian and Sinemurian times, a basal Sinemurian limestone (the *bucklandi* Bed) resting transgressively on a floor of gently folded Hettangian and Rhaetian sediments, the fold axes having an armoricanoid trend in posthumous reflection of Palaeozoic 'basement'. Some parts of the integration may be challenged, but the generally convincing picture that emerged threw a flood of light on Liassic conditions bordering the Mendips, and vindicated in detail the value of Buckmanian methods in Jurassic stratigraphy.

V. BIVALVES OF THE COAL MEASURES

(a) Biometrics

It is not easy to convey, more than forty years later, the excitement evoked at the time by Trueman's (and similar) work in Jurassic palaeontology and stratigraphy, when the substance of it is now submerged in the general field of geological knowledge. His discoveries in the Coal Measures and his organising of them into coherent theory, although begun fifty years ago, were, however, sustained until his death, and in the hands of his colleague John Weir were continued until 1968, when the last part of their joint monograph on the bivalves appeared. The impact of his work in Westphalian geology is as direct now as it was when he first began his revolutionary studies.

It is not without interest to recall that in 1921 accidents of circumstance turned his mind to the Coal Measures—his moving to Swansea College on the Lower Coal Series, his developing acquaintance with J. H. Davies who knew intimately the Coal Measures from the inside, and his appreciation in his *Gryphaea* work of the place of biometrics in palaeontological technique. The first beginnings of studies in the Westphalian bivalves (Trueman & Davies 1923) included a naive supposition that the mytiliform anisomyarian *Naiadites* was evolved through '*Anthracomya*' from the dimyarian *Carbonicola*, in conformity with notions (encouraged by Swinnerton) of functional adaptation in phylogenesis; but they also included an awareness that 'at any horizon the various members of a single species-group [of mussels] show different degrees of advancement, but the proportions of the different members of that species-group are constant everywhere at that horizon' —an observation that was directly transferred from the gryphaeas, and that contained a biometrical element. Moreover, the proportions of various 'species-groups' changed in successive mussel bands, and allowed the identification of 'horizons' not on index specimens but on assemblages—a theme also illustrated by Dix & Trueman (1924) in a comparison of mussel occurrence between north and south crops of the South Wales coalfield.

The first experimental probings were rapidly integrated into what was to become a foundation work in a modern Westphalian geology (Davies & Trueman 1927). A great part of the integration was the successful management of the bewildering abundance of many kinds of mussels found in the recurrent fossiliferous bands of the Coal Measures. Earlier workers, notably Wheelton Hind, had recognised the analogy between the variability of living forms in different ecological environments and the morphotypic variability of the Westphalian bivalves; but the uncertainty of their analysis of the variation, and the limited ways in which the analysis was applied, failed to provide full support for the species they established, or for the criteria used in classification; and the crudities of specific identification combined with the fewness of the species recognised prevented their going more than a short way towards a comprehensive systematisation of the whole mussel fauna, or towards a use of the mussels in stratigraphy. In Davies & Trueman's deeper studies what immediately became clear was the need for a less generalised concept of a taxonomic group (and a less vaguely extended taxonym) and, if the random meaninglessness of incidental names was to be avoided in the maze of thousands of morphic variants, for the bulk manipulation of the variants by biometrical means.

The illustrative examples they used included a demonstration that forms of the same age but from different localities displayed significant differences in variant range—inferentially an effect of ecological control and of topoclinal gradation— that were hidden under a single taxonym; and that the measurable biocharacters, while displaying correlated variation in more or less degree (as in the ratio of thickness to height), commonly led to a near-random scatter of variants whose interrelations were permutative, to make the individual only arbitrarily important.

Essentially both method and conclusion, derived from the oyster studies, followed theoretical lines that were to be developed over the next forty years not only by Trueman himself but also by a number of colleagues and students who were instructed or inspired by him. They particularly offered a critique of Linnaean method, Trueman reiterating in 1927 what he had already said of oysters in 1924— 'the Linnaean or binomial system of nomenclature is scarcely adequate to meet the needs of the palaeontologist in such cases [as those of the mussels]'.

There followed a stream of papers many of them collaborative (notably Trueman 1929, 1930, 1938; Clift & Trueman 1929; Dix & Trueman 1931, 1932; Wray & Trueman 1931, 1934; Trueman & Ware 1932; Moore & Trueman 1937) that culminated in the monographed description, in great part on the basis of the biometrics, of all the known kinds of Upper Carboniferous nonmarine mussels of Britain (Trueman & Weir 1946–68)—a work of immense labour not merely abundantly rich in morphological detail, but also fully explanatory of the techniques of classification that followed from the biometrics, and the problems of nomenclature that consequently arose.

(b) Species and genus

The problems that Trueman (and with him Weir) faced in classification were those of the oysters on a much wider scale. At any one locality and at any one horizon a 'community' (what would now be called a palaeodeme) of many individuals, found in a proximity suggesting the members of an inter-breeding biospecies, usually displays a wide range of variants any one of which departs in more or less degree from the norm in a number of biocharacters. Contemporaneous morphs of what appears to be the same kind of mussel in a community at a distance commonly show a shift in the norm—the expression of a topoclinal environmental gradient—and there are similar shifts in the norms of communities in chronoclinal sequence.

An identification of such communities, and of individuals in them, is not easily recorded when there are many communities and vast numbers of variants to be classified. Community variance may, for two or three biocharacters, be comparatively easily summarised in simple graphs or scatter-diagrams (see for instance Davies & Trueman 1927, pp. 213 ff.). It may, much more tediously, be expressed in an elaborated statistical study of covariance, with a precision that in effect allows a close definition of the community dependent only on the adequacy of the data.

Commonly, however, the variance is more immediately and directly illustrated in the pictograph, a device, much used by Trueman's associates but not by Trueman himself, intended to show the distribution of variants in a community by placing them in a parametral frame of character-axes (see for instance Leitch 1936, 1940, 1942; Eagar 1947, 1953, 1972; Bennison 1954; Broadhurst 1959). The distribution of variants from two communities, arranged on a standard frame, may also be used as an instrument of comparison to demonstrate clinal (ecological, facies) gradients. While however the pictographic method is intended to embody the concept of trends of variation, and to give visual evidence of the differences between the variants, it is in its present development subjectively constructed both in its frame and in its scatter of variants, it links parameters that are not commensurable, and it has a central defect of construction that invalidates its purpose when individual members are located each on only a single parameter when in fact they each should be placed on all (see George 1954, p. 222; 1971, pp. 200 ff.).

However the variation is assessed, the actualities of variant differences, especially of demes in sequence, demand some means of identification. Trueman chose the same technique, developing it much further, as he adopted in the naming of the Liassic oysters: he retained a Linnaean-style binomial nomenclature, although he fully realised its impropriety in application to variants and transients, and he continued to conform to the Rules of Nomenclature by designating a 'holotype'. He defined the specimens to which the specific name should be given as being 'identical' or with slight relaxation 'practically identical' with the holotype, or 'forming a narrow ring around the holotype', regardless of whether they fell into the biometrical unity of a deme or not (Davies & Trueman 1927, p. 219; Trueman & Weir 1946, p. xxi). All other specimens, like the variant oysters, were labelled as having some affinity with ('aff.') the holotype, or as being morphologically comparable with ('cf.') 'but not known to be related to' the holotype: 'aff.' being applicable to all those demic variants of whatever shape not quite so nearly identical as (subjectively) to be regarded as conspecific with the holotype.

Such practices were justified partly because they subscribed to the formalities of the binomial system, mainly because they were found to be of the greatest convenience in recording mussel occurrence in stratigraphical studies. Taxonomy being a contrived 'science', the pragmatic uses to which its instruments are put are the only measures of its methods; and the 'proof' of the validity of Trueman's species lies in the high success of their application to Trueman's purpose.

Nevertheless, the taxa as defined by Trueman and Weir, while given orthodox binomial labels, are of a kind remote from any based on orthodox assumptions of the comprehensiveness of a class on the one hand, and the exclusiveness of a class on the other. Thus in defining the range of a species narrowly about a holotype, Trueman and Weir excluded all remoter variants—the vast majority of variants—from any recognised species; and in defining a species in terms of a close likeness to the holotype they created the narrowest of morphospecies that nevertheless incorporated shells of no matter what community (in time or location) having the appropriate shape. The two practices are factors that at once separate forms that, in the single community, have the closest relationship, and unite forms that, in separate communities, may belong to very different biometrical sets. (See Fig. 1.)

When the morphospecific criterion is set up by the 'competent systematist' (that is, the systematist who consistently applies his own principles), there is moreover the common establishment of two or more species, or the identification of two or more holotypes, in a single unitary community (for the general case see Trueman and Weir 1946, fig. iii, p. xxi; for an instance see Eagar 1953, fig. 3, p. 152), variants outside the inner rings being labelled as having an affinity with ('aff.') any of the

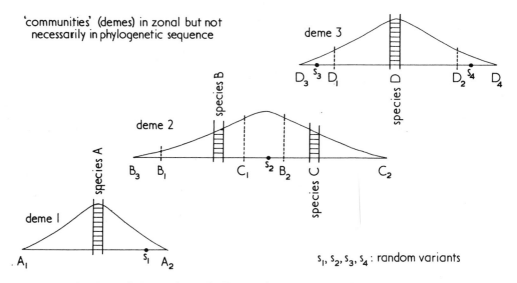

Fig. 1.   Species and nomenclature in Trueman's taxonomics. The model is of three demes, not contemporaneous, graphically plotted as fields of variants. Variants of a single deme fall into an arbitrary species when they are closely similar to the selected holotype in form (as in the lined bands of the diagram). They are regarded as not strictly conspecific though having affinity ('aff.'), if they range (but not too far) beyond the narrow specific limits (between A_1 and A_2, B_1 and B_2, C_1 and C_2, D_1 and D_2). Peripheral variants outside the penumbra of 'affinity' (B_1–B_3, D_1–D_3, D_2–D_4) are in limbo, each nameless in the context of its own deme. If the demic range of variants is 'unusually' wide, two or more species may be designated conveniently to accommodate the morphs (species B and C of deme 2): random variants (s_2) may then indifferently be labelled as having 'affinity' with ('aff.') more than one species. When demes of different loci or times are compared nameless peripheral members may acquire names: s_1, 'aff.' species A, would be called 'cf.' species A if its morph were found in the B_1–B_3 group of variants; s_3 as a morph falls precisely into species C, and can be said not to do so (being called 'cf.' C) only because it is found in its context of deme 3; s_3 acquires a name through the extraneous taxonomics of deme 2, but s_4 remains nameless if no other overlapping deme is found.

species indifferently. Despite sustained defence of Trueman's systematics by Weir (1968, pp. xxxviii), such method is retrograde in accepting (perhaps perforce) a particulate binomial Linnaean system, in ignoring phylogeny and palaeodemes in chronoclinal sequence while emphasising ecophenes, and in resting on the highly subjective judgement of the systematist in the definition of species and the spread of species-range. (See George 1949, pp. 23 ff.; 1956, p. 134; 1971, pp. 211 ff.).

   Classification of the mussel genera, going beyond the initial simplicities of a form-evolution from *Carbonicola* through '*Anthracomya*' to *Naiadites*, was developep on the basis of major morphological contrasts by Trueman & Weir (1946, p. xiv; 1968, pp. xxxv; see also Dix & Trueman 1932; Trueman 1954, pp. 62 ff.; Weir in Arkell and others, 1969, pp. 291, 406). The genera were distinguished notably on the form of the hinge and in mode of growth, between myalinids (*Anthraconaia, Anthraconauta, Curvirimula, Naiadites*) heteromyarian with duplivincular ligament, and anthracosiids (*Carbonicola, Anthracosia, Anthracosphaerium*) more or less isomyarian with parivincular ligament. Within the families minor generic differentiae were variously selected—the tilt of the growth lines, the form of the umbones, the shell profile in the flatness or roundness of the valves and the sharpness of definition of the anterior muscle scar, the strength of the hinge plate, the development of teeth. The basis of classification, bringing a refinement of dissection to the shedding of homoeomorphs, was thus very different from the anagenetic separation of *Gryphaea* from *Ostrea*, and in principle reverted to the use of differential characters as in the classification of the Jurassic liparoceratids and cerithiids (the 'progressive' characters, as represented in the abundance of mussel variants, being the criteria for specific distinction).

(c) Ecology

Not least of the difficulties that Trueman met in inventing an appropriate pragmatic classification for his mussels arose out of the circumstances of fossilisation in the Coal Measures. As nonmarine forms, some attached by a byssus, some perhaps burrowing, the mussels were sensitive to the rapid changes in facies of the Westphalian sediments, preferring always shales and fine silts and almost never being found in the coarser sandstones, especially the many thick sandstones of the upper Coal Measures. Even in the shales they are usually found concentrated in a comparatively few and thin mussel bands, suggesting a response to non-lithological factors in the environment. The advent of an invading sea (now represented in the recurrent marine beds) caused marked changes in the assemblage of shells to be found at successive horizons, the myalinids being more tolerant of a slight increase in salinity than the anthrocosiids, and in full development caused all the mussels to emigrate (or perhaps to be extinguished). At the other extreme of a Westphalian rhythm, during the slight emergence that allowed the widespread growth of coal-forming peat, there was a similar inhospitable blanket of sediment from which the mussels were excluded. Even in a narrow sequence of fossiliferous beds 'large species of *Carbonicola* are rarely if ever associated in one bed with smaller species of *Carbonicola*, or with species of *Anthracomya*', and 'horizons with large species of *Carbonicola* alternate with horizons yielding *Anthracomya, Naiadites*, and small *Carbonicola*'. Comparable lateral changes in facies at a single horizon are similarly to be recognised: 'On the south crop [in contrast to the north crop of the South Wales coalfield] there are fewer horizons with large *Carbonicola*, while *Anthracomya* and *Naiadites* are more common'; and minor swells between basins were influential in restricting the ranges of some species and in changing the proportions of species or variants in assemblages at a distance. (See Davies & Trueman 1927, p. 221; Trueman 1946, p. lxii.)

The sporadic distribution of the mussels in the Westphalian rocks (despite their great abundance in individual beds) and the repeated signs of the subleties of ecological control on the kinds of mussels briefly and locally encouraged, point to exodus and replacement in stratigraphical sequence; and at each cycle, however brief, of emigration (or extinction) and immigration, there is never an expectation that at any one locality an incoming fauna will be composed of the descendents of an emigrant earlier fauna, or even will have the same aspect as an assemblage

(Trueman & Weir 1946, p. xxiv).

The large-scale distribution of the mussels in the Coal Measures—the combined result of evolution, migration, and extinction—contributed to the founding of the two major stages of the Westphalian series by Dix & Trueman (1937, p. 197): the Ammanian with abundant *Carbonicola*, *Anthracosia*, *Anthracosphaerium*, and *Naiadites*, but with few *Anthraconauta*, the Morganian with abundant *Anthraconauta* but with few other kinds except *Anthraconaia*. Within the Ammanian further division is possible, *Carbonicola* being virtually limited to the lower part, *Anthracosia* to the upper part (with most occurrences of *Naiadites* also) (see Trueman & Weir 1946, p. xxix).

VI. CLASSIFICATION OF THE COAL MEASURES

(a) Lithology and zones

When Trueman first began his studies systematic stratigraphical knowledge of the Coal Measures was very imperfect. In most British coalfields it was dominated by a formational recognition of rock units, notoriously a poor guide in sediments showing high variability in lithological sequence, especially in the lenticular form and lateral inosculation of the many sandstones. Over comparatively small areas coal seams, some having an individuality that made them readily identifiable, were used in correlation and in structural interpretation; and occasionally a widespread marker bed—the Gin Mine of the Midlands, Skipsey's Marine Band in Scotland— was a useful datum; but classification was for the most part local, generalised, and inexact. (See Trueman 1924a, p. 292, for a revealing insight into the state of knowledge of the Coal Measures of South Wales, at a time when the first stages in the determination of bivalve sequence were being tentatively applied to the litho-stratigraphical divisions of Lower Coal Series, Pennant Series, and Upper Coal Series.)

In the Midlands Wheelton Hind had begun to use mussel bands, not merely as lithological datum-horizons but as units each characterised by species assemblages potentially useful in zoning. His work however was not extended into other coal-fields; indeed it was neglected even in the Midlands, partly because statistical methods of bulk analysis had not then evolved, and partly because palaeontological refinements necessary to distinguish his species were then beyond the patience of most stratigraphers, who tended to ignore the bivalves in a belief that they were useless in correlation. Greater faith was put in the plants of the Coal Measures, a British version of the continental system being represented in the four Upper Carboniferous stages of Lanarkian, Yorkian, Staffordian, and Radstockian—stages that in the event were to prove increasingly difficult to apply, and (as the mussels duly showed) were erroneously founded in what were supposed to be their definitive floras.

The first tentative approaches to a zonal scheme (Trueman & Davies 1923, pp. 377, 380) are to be seen in Trueman's awareness, from Davies's collecting, that although there were notable differences in the species assemblages in successive mussel bands in the lower part of the sequence in South Wales, *Carbonicola* was the dominantly persistent genus in the lower part until at a well-marked horizon *Anthraconaia* ('*Anthracomya*') suddenly appeared and remained represented by a number of species in later measures. In a few years (Davies & Trueman 1927, pp. 237 ff.) the main zones were established: in upward sequence the Zone of *Carbonicola ovalis*, changed in 1946 to the Zone of *Carbonicola communis* for nomenclatural reasons, characterised by the presence of large specimens of *Carbonicola*, with only rare specimens of other mussels except *Naiadites;* the Zone of *Anthraconaia modiolaris* with only small and relatively few specimens of *Carbonicola*, but with abundant specimens of *Anthracosia* and large specimens of *Anthraconaia;* the Zone of *Anthraconaia similis*, with abundant specimens of *Anthracosia* and *Anthracosphaerium*, but comparatively few of *Anthraconaia;* the Zone of *Anthraconaia pulchra*, with a return of common specimens of *Anthraconaia* but almost without *Anthracosia;* the Zone of *Anthraconauta phillipsii*, the index species almost the only common member; and the Zone of *Anthraconauta tenuis*, the index species the only common member but with an increasing number of specimens of *Anthraconaia*. The impoverishment of the mussel fauna at the top of the *similis-pulchra* Zone and the sudden incoming of abundant specimens of *Anthraconauta* were the bases for distinguishing the Ammanian stage beneath from the Morganian stage above.

The nature of the zones emerges in the naming. The very rapid alternations in the Westphalian environment from hospitable to inhospitable, with correspondingly rapid migrations of species and genera, are reflected in the fluctuations in proportions and kinds of forms seen in successive zones. The forms of each mussel band commonly constitute a unique assemblage for the horizon; and although a number of morphospecies recur in upward sequence the zonal succession for the most part reflects not evolutionary or phylogenetic lineage segments but replacement faunas having only indirect relationships with faunas of earlier or later date.

The zones were first established in South Wales. In a few years they were found to be generally applicable over much of Britain (Clift & Trueman 1929; Wray & Trueman 1931, 1934; Weir & Leitch 1936; Moore & Trueman 1937; Stubblefield & Trueman 1947); but signs of their nature, as zones whose assemblages were patterned in palaeoecological and regional response to their environments, is shown by adjustments Trueman made to the zonal system. In Yorkshire the lowest measures, much richer than the equivalent beds in South Wales, proved to contain a fauna characterised by *Anthraconaia lenisulcata*, the index species of a newly created zone (Wray & Trueman 1931, p. 71); and in Somerset strata above the last occurrence of *Anthraconauta tenuis* were referred to a newly created Zone of *Anthraconaia prolifera* on a return of that genus (Trueman 1941, p. 72). In contrast, the *similis* and *pulchra* zones, sufficiently distinguished in South Wales, proved to have a widely overlapping range of species of *Anthraconaia* in Nottinghamshire (Clift & Trueman, 1929, p. 100) and the two zones were then united as the *similis-pulchra* Zone.

The most effective cause of the disruption of the molluscan sequence in the Coal Measures was the accumulation of the many coals, widespread in range (some of them extending from Britain into the Continental coalfields) and deposited under conditions wholly inhospitable to mussels. Trueman in recognising this influence identified his zonal limits at or immediately beneath coal seams, for, as might be expected, he found the greatest faunal changes, or the locally abrupt ending of a faunal facies, to be marked by the more prominent coals. Zonal range was thus in a sense negative, the gaps in the fossiliferous sequence as much reflecting local circumstances of deposition as synchronous time-intervals. A small change Trueman made in a zonal boundary has high significance as a sign of environmental influence: initially the top of the *modiolaris* Zone (negatively coincident with but not identified in the base of the *similis-pulchra* Zone) in South Wales was placed immediately beneath the Stanllyd Vein and its equivalents; but the finding, over a small area immediately above the coal, of a shelly band with a *Carbonicola-Anthraconaia* fauna typical of rocks placed in the *modiolaris* Zone of northern England caused Trueman to revise his identification of the top of the zone and to place it immediately above the Stanllyd Vein (see Trueman & Ware 1932, p. 78; Clift & Trueman 1929, p. 87; Wray & Trueman 1931, p. 83).

The change is also a sign of Trueman's view of the nature of his zones, for although he defined his zones at specified horizons that were determined by the rhythmic sequence in the measures, he recognised the zones by reference to fossil occurrence; and where fossil range (in upward continuity) was locally greater than he expected he was prepared to change the zonal boundaries—indeed he took for granted that he should change the zonal boundaries—to accommodate the new discoveries. That is, his zones were conceived to be biostratigraphical on a basis of assemblages, not in the first instance stratotyped. They were of different kind, both in their criteria and in their definition, from his ammonite zones (acme zones) and from his potential *Gryphaea* zones (lineage zones) in the Lias.

The high success of his zonal system, in bringing order to Westphalian correlation over the whole of Britain in less than ten years, had direct and destructive effect on the series of plant divisions erected mainly by Kidston at the beginning of the century and later revised by Crookall—divisions which, like the mussel zones, were essentially assemblage zones, the Lanarkian having a relatively 'impoverished' assemblage, the Yorkian an enriched assemblage. The refinements of correlation made possible by the use of Trueman's zones included a demonstration that the inferred junction between Lanarkian and Yorkian fluctuated widely between Namurian horizons and horizons in the *modiolaris* and *similis-pulchra* zones, the Lanarkian sometimes being regarded as absent in an unbroken sequence of Ammanian strata. It became evident that 'Lanarkian' and 'Yorkian' were terms embracing facies floras, not validly applied in correlation (Dix, Pringle, and Trueman 1930). Floral (assemblage) zones collating with the mussel zones (and in part anchored by them) were not established in Britain until Dix revised the plant sequence in 1934.

(b) Marine beds

The close study of sequence stimulated by Biast's work on the Namurian goniatites and Trueman's on the Westphalian mussels led to increased recognition of the importance of shelly horizons (in addition to mussel bands) in the Upper Carboniferous succession. When Trueman began his work the local occurrence of marine beds in the Coal Measures had long been known, even in South Wales, but they found only an incidental place in his early papers, including the 1927 paper in which his mussel zones were founded and in which there is a single allusion to a *Myalina* band in the *pulchra* Zone above the Red Vein at Cwmgors, correlated (as it happens inexactly) with the Bay Coal Marine Band of Staffordshire and the Rimbert horizon in France (Davies & Trueman 1927, pp. 244 ff.). The Cwmgors boreholes transformed knowledge of the recurrence of marine beds, a dozen or more (including *Lingula* bands) being found in a thickness of less than a hundred metres in the uppermost part of the *pulchra* Zone (Davies, Dix, and Trueman 1928, p. 82); and in a year or two most of the more important marine bands of several coalfields became recognised and correlated—the Wernffrwd (Halifax Hard Bed, Bullion Mine) in the *lenisulcata* Zone, the Amman (Clay Cross, Pennystone, Katharina of the Continent) in the middle of the *modiolaris* Zone, the Cefn Coed (Mansfield, Gin Mine, Dukinfield, Aegir of the Continent) in the *similis-pulchra* Zone, and the Upper Cwmgors (Top, Shafton) towards the top of the *similis-pulchra* Zone. (See Dix & Trueman 1928; Ware 1930, pp. 455 ff.; Jones 1933, p. 416; Jones 1934, p. 320.)

The finding of the marine beds, and the realisation of their widespread occurrence (the more prominent of them with a range exceeding that of most coal seams) had an insidious effect on the classification of the Coal Measures. As datum horizons, found to be more readily and more precisely identifiable than the boundaries of the mussel zones, and as sediments, characteristic of transgressive seas that had radical effect on the viability of the nonmarine bivalves, were considered to be even more significant indicators than the zonal boundaries of palaeoecological facies changes in the Coal Measures; and they came to be regarded as having greater claims to being classificatory signposts than the mussel zones.

Thus the Ammanian-Morganian junction, between the *similis-pulchra* and *phillipsii* zones, was originally placed at a horizon immediately beneath the Lower Pinchin coal and its correlatives in South Wales, but increasingly it came to be identified at the top of the Upper Cwmgors (*Anthracoceras cambriense*) Marine Band, at a lower horizon. The *similis-pulchra* Zone became formally divided into an upper and a lower part not at the original junction between the *pulchra* and the *similis* zones but at the Cefn Coed (Aegir, *Anthracoceras aegiranum*) Marine Band not because of any significant differences—there are none—between the faunas of the two parts but because the marine band could be widely recognised. With the formal establishment by international decree in 1935 of the *Gastrioceras subcrenatum* Marine Band as the definitive boundary between the Namurian and the Westphalian series, the base of the *lenisulcata* Zone was made to coincide with the marine band, whereas in faunal continuity it had earlier been placed at an uncertain horizon lower in the 'Millstone Grit'. (See for instance Trueman 1954, pp. 106, 111, 139, 179, 180, for evidence of the changes and of the varying application of the zonal terms). A final stage in the dominance of the marine beds in classification is found in the reorganised lithostratigraphy of the Coal Measures by Stubblefield and Trotter

(1957), whose allusions to the mussel zones were fleeting as they made the boundaries of their rock groups coincide with the major marine bands. (See also Calver 1969, p. 235.) (See Fig. 2.)

In the success of his zonal system, out of which emerged the importance as widely traceable horizons of the marine bands in the British coalfields, Trueman may thus be said to have prepared the ground for the desuetude if not the demise of his zones—they are for instance almost wholly ignored as a stratigraphical frame in the most recent Survey memoirs on the South Wales coalfield, the home of their invention (Archer 1969, pp. 22 ff.; Squirrell & Downing 1969, pp. 90 ff.). Moreover, in the change of method, zonal technique has gone. The definition of the divisions in the Stubblefield-Trotter stratigraphy includes no reference to their fossil contents, either as fossil assemblages or as index fossils (indeed it may be said to include no reference to specific contents of any kind). It rests entirely on an identification of marker-bed horizons whose relationship with the mussel zones is collated by a coercive parallelism neither supplementary nor complementary. It (rightly) uses the term 'zone' not at all.

In a comparable lithostratigraphical system, Woodland and his colleagues (1957) divided the Upper Coal Measures of South Wales (the base at the Upper Cwmgors Marine Band) into rock groups each defined by a coal seam used as a marker band. This was a sophisticated reversion to nineteenth-century method, refined and securely founded on a sustained tracing of each identifiable seam across the coalfield: but it subsumed the *phillipsii* Zone under the Lower Pennant Measures, and the *tenuis* Zone under the Upper Pennant; and with the supersession of the zones the local use of a Morganian Stage vanished also.

(c) *Facies and earth-movements*

A conspectus of the wide stratigraphical implications of his work allowed Trueman (1946, 1947) to relate sequence to environment, and to create a palaeogeography of Westphalian times that complemented his classification of the measures into zones and stages. The rhythms of sedimentation, and of controlling earth-movement, to be seen in the thickening or the splitting of a coal seam, the rapid lateral changes in lenses of delta and channel sands, the contrasted influence of compaction on shales and silts at neighbouring localities, were to him the incidental signs of environmental controls that in major degree were to be seen in the recurrent marine bands, the spacing of coals in vertical sequence, and the variation (topoclinal and sequential) in the non-marine faunas.

The marine bands attracted his close attention partly for the evidence they provided of a singular uniformity of conditions over very wide areas of north-western Europe, and of the corresponding uniformity in the paralic environment in which the non-marine measures were also deposited. At the same time, he noted the contrasts in the development of any one marine band—in thickness, in faunal assemblage, in bed-by-bed composition—as it is traced from one locus to another; and he particularly contrasted the range (as he knew it) of the Cwmgors marine beds, found only in parts of Britain, with the widespread development of the marine beds at lower horizons (notably the Aegir and Katharina horizons). In his regional

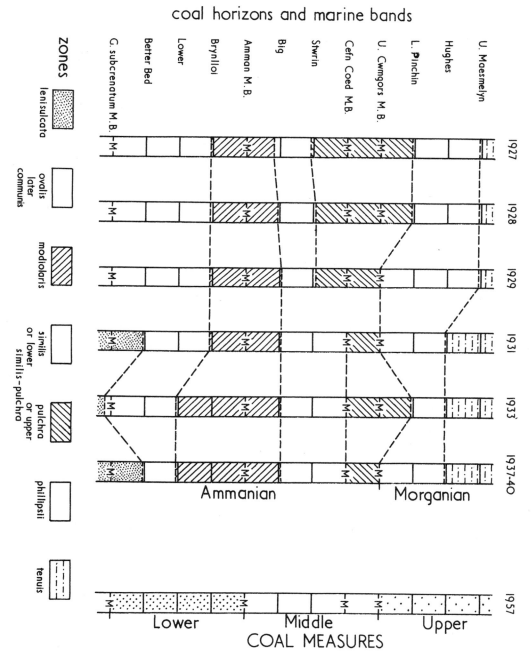

Fig. 2. The mussel zones of the Coal Measures. The changing position of the zonal boundaries adopted by Trueman reflect adjustments arising from biostratigraphical discoveries, mainly of assemblage faunas, and upon increasing weight put upon marine bands as marker beds. The 1957 classification is that of Stubblefield & Trotter, the mussel zones subordinated to the marine bands. The diagram is slightly generalised.

tracing of the marine beds he initiated a study of their facies changes that, particularly in the hands of Calver (see 1969, pp. 246 ff.), has since become an elaborate analysis of palaeoecological relations disclosing the subtleties of the controls on sedimentation of the Coal Measures.

The distribution of the nonmarine bivalves in the paralic measures also enabled Trueman to invoke factors of distribution that went beyond the obvious relations of shells to entombing sediments (sands and coarse silts being unfossiliferous), and beyond the obliteration or exclusion of faunas by a widespread peat or by a saline incursion. Thus the Aegir Marine Band not only closed a long period of nonmarine sedimentation, rich in coals and mussel bands, it also marked virtually the extinction of a whole long-persistent bivalve fauna of anthracosiids and naiaditids in north-western Europe; and the Upper Cwmgors Marine Band, although of more limited range, similarly recorded a large-scale geographical change in dominant rock types, seen in the incoming of thick and massive Pennant-like sandstones and red beds, and in the great diminution in the number and the quality of the coals, and in the replacement of the *Anthraconaia* fauna of the upper *similis-pulchra* Zone of the Ammanian by the much impoverished *Anthraconauta* fauna of the *phillipsii* and *tenuis* zones of the Morganian (see also Trueman, 1941a).

Trueman placed these secular changes of Westphalian times in local settings when he traced his zones over Britain and into Continental Europe. The southward overstep of the Coal Measures onto the Mercian Highlands (part of the 'Wales–Brabant ridge') had long been known when he showed that it continued over their crest into the 'Variscan foredeep' of the South Wales, Forest of Dean, and Bristol coalfields, unconformities developing at the base of the Ammanian stage and especially at the base of the Morganian (whose contrasts with the Ammanian in rocks and faunas are thus in part explained) (see Trueman 1947, figs. 1, 2, and pl. A). He also showed that to the north of the Mercian Highlands tectonic controls similarly affected sequence, not only in the occurrence of a basal Ammanian unconformity (and of breaks at higher horizons) over much of the Midlands, but also in differential sagging of the floor as the Westphalian rocks accumulated, the Ammanian basin of the southern Pennine area being followed to the north by the minor Northumbrian, south Ayrshire, and Midland Valley basins, in subdued reflection of an earlier Carboniferous palaeogeographical frame. In these more northern areas, beyond the Mercian Highlands, the Morganian rocks are commonly red beds of great thickness in which fossils of any kind, including bivalves, are rare; and in applying his classification Trueman was able to give yet further significance to a delimited Morganian Stage.

On the Continent, the Coal Measures to the north of the Variscan front were traced by Trueman (1946, pl. A) from France (Nord) through Belgium and Germany to the Donetz basin, where he suggested links with the partly marine Moscovian rocks. Within the fold belt, the isolation of the intramontane limnic basins and the peculiarities of their sediments (many of them deltaic and coarse-grained, of local provenance, often affected by earth-movements and broken by non-sequence or unconformity, the Ammanian members without marine beds) resulted in an absence of the nonmarine mussels on which Trueman's zones were based. He attributed the absence to difficulties of transport of bivalve eggs or larvae from paralic regions to the inland basins at a time when birds or other migrant agents had not evolved; but the absence may possibly be attributable to a preference for a paralic or brackish-water environment by anthracosiids and myalinids.

The Pennsylvanian rocks of North America provided him only with vague and generalised grounds for correlation with Britain, but the occurrence of *Carbonicola*, *Anthracosia*, and *Naiadites* gave him strong ground for recognising Ammanian rocks in the southern Appalachian coalfields, records of *Anthraconaia modiolaris* and *Anthraconaia pulchra* were even more precise in their zonal implications, and the wide range of species of *Anthraconauta*, including *phillipsii*, in the Acadian province and in Pennsylvania was also a sign of his Morganian Stage. Although he himself did not study the American rocks and fossils at first hand, his conjectures (Trueman 1946, fig. 10) were remarkably prophetic: he restored a North-Atlantic palaeogeographical province (fore-shortened to fit du Toit's distribution of continental massifs in Carboniferous times) in which he could link the Acadian measures some with the paralic rocks of north-western Europe, some to the south-east with the intramontane limnic basins of central France. In doing so he provided a synthesis of knowledge of fossils, rocks, and structures that for the first time gave an inter-continental coherence to the Westphalian Series.

The range of his work was enormous. It combined excursions into theory and principle with discovery directly applicable to coalfield economics; it directed research along paths that were at once academic and practical; it rendered meaningless a contrast between a 'pure' and an 'applied' geology; and it stamped him as a pioneer both in what he did and in the manner of his doing it. It was distinguished in its quality, and it brought distinction to him and to the institutions in which he worked.

VI REFERENCES

Only those of Trueman's papers are included that are referred to in the text. For a full list see Pugh (1958).

ARKELL, W. J., and others. 1957. Mollusca: Cephalopoda Ammonoidea. *Treat. invert. Paleontol.*, N.

BATHER, F. A. 1920. Fossils and life. *Rep. Brit. Assoc.*, 61–86.

BROADHURST, F. M. 1959. *Anthraconaia pulchella* sp. nov. and a study of palaeoecology in the Coal Measures of the Oldham area of Lancashire. *Quart. J. geol. Soc. Lond.*, **114**, 522–45.

BURNABY, T. P. 1965. Reversed coiling trend in *Gryphaea arcuata*. *Geol. J.*, **4**, 257–78.

CALVER, M. A. 1969. Westphalian of Britain. *C. R. 6me Congr. intern. Strat. Géol. Carb.* (*Sheffield 1967*), **1**, 233–54.

COX, A. H., and TRUEMAN, A. E. 1920. Intra-Jurassic movements and the underground structure of the southern Midlands. *Geol. Mag.*, **57**, 198–208.

DAVIES, A. M. 1930. The geological work of Sidney Savory Buckman. *Proc. Geol. Ass.*, **41**, 221–40.

DAVIES, D. F., DIX, E., and TRUEMAN, A. E. 1928. Boreholes in Cwmgorse valley. *Proc. S. Wales Inst. Engin.*, **44**, 37–136.

DAVIES, J. H. and TRUEMAN, A. E. 1927. A revision of the non-marine lamellibranchs of the Coal Measures. *Quart. J. geol. Soc. Lond.*, **83**, 210–59.

DIX, E., PRINGLE, J., and TRUEMAN, A. E. 1931. The significance of the term 'Lanarkian

Series'. *Naturalist*, 321–26.

—— and TRUEMAN, A. E. 1924. The correlation of the Coal Measures in the western portion of the South Wales coalfield. Part II: The Coal Measures of north Gower. *Proc. S. Wales Inst. Engin.*, **40**, 353–83.

—— and ——. 1928. Marine horizons in the Coal Measures of South Wales. *Geol. Mag.*, **65**, 356–63.

—— and ——. 1931. Some non-marine lamellibranchs from the upper part of the Coal Measures. *Quart. J. geol. Soc. Lond.*, **87**, 180–211.

—— and ——. 1932. Some observations on the genus *Naiadites*. *Ann. Mag. nat. Hist.*, (10) **9**, 1–20.

—— and ——. 1937. The value of nonmarine lamellibranchs for the correlation of the Upper Carboniferous. *C. R. 2me Congr. Strat. Carb.* (*Heerlen 1935*), 185–201.

EAGAR, R. M. C. 1947. A study of a non-marine lamellibranch succession in the *Anthraconaia lenisulcata* Zone of the Yorkshire Coal Measures. *Phil. Trans. roy. Soc.*, **B233**, 1–54.

——. 1948. Variation in shape of shell with respect to ecological station. A review dealing with Recent Unionidae and certain species of the Anthracosiidae in Upper Carboniferous times. *Proc. roy. Soc. Edinburgh*, **63**, 130–48.

——. 1953a. Relative growth in shells of the fossil family Anthracosiidae in Upper Carboniferous times. *Proc. Linn. Soc.*, **164**, 148–73.

——. 1953b. Variation with respect to petrological differences in a thin band of Upper Carboniferous nonmarine lamellibranchs. *Lpool and Manchr geol. J.*, **1**, 161–90.

——. 1954. New species of Anthracosiidae in the Lower Coal Measures of the Pennine region. *Mem. Proc. Manchr. lit. phil. Soc.*, **95**, 1–26.

——. 1972. Use of the pictograph. *Palaeontology*, **15**, 378–80.

GEORGE, T. N. 1933. Palingenesis and palaeontology. *Biol. Rev.*, **8**, 107–35.

——. 1949. Evolution in fossil communities. *Proc. roy. phil. Soc. Glasgow*, **73**, 23–42.

——. 1954. Fossil species. *Sci. Progr.*, **42**, 220–8.

——. 1956. Biospecies, chronospecies, and morphospecies. *Publ. Syst. Assoc.*, **2**, 123–37.

——. 1960. Fossils in evolutionary perspective. *Sci. Progr.*, **48**, 1–30.

——. 1962. The concept of homoeomorphy. *Proc. Geol. Ass.*, **73**, 9–64.

——. 1971. Systematics in palaeontology. *J. geol. Soc.*, **127**, 197–245.

GOULD, S. J. 1966. Allometry and size in ontogeny and phylogeny. *Biol. Rev.*, **41**, 587–640.

HALLAM, A. 1960. A sedimentary and faunal study of the Blue Lias of Dorset and Glamorgan. *Phil. Trans. roy. Soc.*, **B243**, 1–44.

——. 1968. Morphology, palaeoecology, and evolution of the genus *Gryphaea* in the British Lias. *Phil. Trans. Roy. Soc.*, **B254**, 91–128.

JONES, S. H. 1934. The correlation of the coal seams around Ammanford. *Proc. S. Wales Inst. Eng.*, **23**, 131–56.

——. 1935. The Lower Coal Series of north-western Gower. *Proc. S. Wales Inst. Eng.*, **26**, 172–254.

JOYSEY, K. A. 1959. The evolution of the Liassic oysters *Ostrea–Gryphaea*. *Biol. Rev.*, **33**, 297–332.

LANG, W. D. 1919. Old age and extinction in fossils. *Proc. Geol. Ass.*, **30**, 102–13.

——. 1923a. Evolution: a resultant. *Proc. Geol. Ass.*, **34**, 7–20.

——. 1923b. Trends in British Carboniferous corals. *Proc. Geol. Ass.*, **34**, 120–36.

——. 1940. Ammonites. *Proc. Dorset nat. Hist. Arch. Soc.*, **61**, 98–116.

LEITCH, D. 1936. The *Carbonicola* fauna of the Midlothian Fifteen-Foot Coal: a study in variation. *Trans. geol. Soc.* Glasgow, **19**, 390–403.

——. 1940. A statistical investigation of the anthracomyas of the basal *similis-pulchra* Zone in Scotland. *Quart. J. geol. Soc. Lond.*, **96**, 13–37.

——. 1942. *Naidites* from the Lower Carboniferous of Scotland: a variation study. *Trans. geol. Soc. Glasgow*, **20**, 208–22.

——. 1951. Biometrics and systematics in relation to palaeontology. *Proc. Linn. Soc.*, **162**, 159–70.

MACLENNAN, R. M., and TRUEMAN, A. E. 1942. Variation in *Gryphaea incurva* (Sow.) from the Lower Lias of Loch Aline, Argyll. *Proc. roy. Soc. Edinburgh*, **B61**, 211–32.

MOORE, L. R., and TRUEMAN, A. E. 1937. The Coal Measures of Bristol and Somerset. *Quart. J. geol. Soc. Lond.*, **93**, 195–240.

—— and ——. 1939. The structure of the Bristol and Somerset coalfields. *Proc. Geol. Ass.*, **50**, 46–67.

PHILIP, G. M. 1962. The evolution of *Gryphaea. Geol. Mag.*, **99**, 327–44.

PUGH, W. J. 1958. Arthur Elijah Trueman. *Biogr. Mem. Fellows roy. Soc.*, **4**, 291–315.

SCHINDEWOLF, O. H. 1929. Ontogenie and Phylogenie. *Paläontol. Zeitschr.*, **11**, 54–67.

SPATH, L. F. 1933. The evolution of the Cephalopoda. *Biol. Rev.*, **8**, 418–62.

——. 1936. The phylogeny of the Cephalopoda. *Paläontol. Zeitschr.*, **18**, 156–81.

——. 1938. A catalogue of the ammonites of the Liassic family Liparoceratidae. *Brit. Mus. (nat. Hist.).*

SQUIRRELL, H. C., and DOWNING, R. H. 1969. Geology of the country around Newport (Mon.). *Mem. geol. Surv. Gt. Britain.*

STAMP, L. D. 1925. Some practical aspects of correlation: a criticism. *Proc. Geol. Ass.*, **36**, 11–25.

STUBBLEFIELD, C. J., and TROTTER, F. M. 1957. Divisions of the Coal Measures on Geological Survey maps of England and Wales. *Bull. geol. Surv. G.B.*, **13**, 1–5.

——. and TRUEMAN, A. E. 1947. The faunal sequence in the Kent Coalfield. *Geol. Mag.*, **83**, 266–79.

SWINNERTON, H. H. 1921. The use of graphs in palaeontology. *Geol. Mag.*, **58**, 357–64, 397–408.

——. 1932. Unit characters in fossils. *Biol. Rev.*, **7**, 321–35.

——. 1938. Development and evolution. *Rep. Brit. Assoc.*, 57–84.

——. 1939. Palaeontology and the mechanics of evolution. *Quart. J. geol. Soc. Lond.*, **95**, xxxiii–lxx.

——. 1940. On the study of variation in fossils. *Quart. J. geol. Soc. Lond.*, **96**, lxxvii–cxviii.

——. 1964. The early development of *Gryphaea. Geol. Mag.*, **101**, 409–420.

——. and TRUEMAN, A. E. 1918. The morphology and development of the ammonite septum. *Quart. J. geol. Soc. Lond.*, **73**, 26–58.

TRUEMAN, A. E. 1915. The fauna of the Hydraulic Limestones in south Nottinghamshire. *Geol. Mag.*, **6, 2**, 150–66.

——. 1918. The Lias of south Lincolnshire. *Geol. Mag.*, **6, 5**, 64–73, 102–11.

——. 1919. The evolution of the Liparoceratidae. *Quart. J. geol. Soc. Lond.*, **74**, 247–98.

——. 1920. The Liassic rocks of the Cardiff district. *Proc. Geol. Ass.*, **31**, 93–107.

——. 1922a. Aspects of ontogeny in the study of ammonite evolution. *J. Geol.*, **30**, 140–3.

——. 1922b. The use of *Gryphaea* in the correlation of the Lower Lias. *Geol. Mag.*, **59**, 256–68.

——. 1922c. The Liassic rocks of Glamorgan. *Proc. Geol. Ass.*, **33**, 245–84.

——. 1923. Some theoretical aspects of correlation. *Proc. Geol. Ass.*, **34**, 193–206.

——. 1924. The species-concept in palaeontology. *Geol. Mag.*, **61**, 355–60.

——. 1929. Fossil shells and some problems of evolution. *Proc. Bristol Nat. Soc.*, **4, 7**, 131–7.

——. 1930a. The Lower Lias (*bucklandi* Zone) of Nash Point, Glamorgan. *Proc. Geol. Ass.*, **41**, 148–59.

——. 1930b. Results of some recent statistical investigations of invertebrate evolution. *Biol. Rev.*, **5**, 296–308.

——. 1933. A suggested correlation of the Coal Measures of England and Wales. *Proc. S. Wales Inst. Engin.*, **49**, 63–94.

——. 1940. The meaning of orthogenesis. *Trans. geol. Soc. Glasgow*, **20**, 77–95.

——. 1941a. The periods of coal formation represented in the British Coal Measures. *Geol. Mag.*, **78**, 71–6.

——. 1941b. The ammonite body-chamber, with special reference to the buoyancy and mode of life of the living ammonite. *Quart. J. geol. Soc. Lond.*, **96**, 330–83.

——. 1946. Stratigraphical problems in the Coal Measures of Europe and North America. *Quart. J. geol. Soc. Lond.*, **102**, xlix–xcii.

——. 1947. Stratigraphical problems in the coalfields of Great Britain. *Quart. J. geol. Soc. Lond.*, **103**, lxv–civ.

—— (editor) 1954. *The coalfields of Great Britain*. London.

—— and DAVIES, J. H. 1923. The correlation of the Coal Measures in the western portion of the South-Wales coalfield. Part I: the Mollusca of the anthracite area. *Proc. S. Wales Inst. Engin.*, **39**, 367–91.

—— and WARE, W. D. 1932. Additions to the fauna of the Coal Measures of South Wales. *Proc. S. Wales Inst. Engin.*, **48**, 67–85.

—— and WEIR, J. 1946–68. The Carboniferous non-marine Lamellibranchia. *Mem. palaeontogr. Soc.*

—— and WILLIAMS, D. M. 1925. Studies in the ammonites of the family Echioceratidae. *Trans. roy. Soc. Edinburgh*, **53**, 699–739.

—— and WILLIAMS, D. M. 1926. Notes on some Lias ammonites from the Cheltenham district. *Proc. Cotteswold Nat. F. C.*, **22**, 239–53.

TUTCHER, J. W., and TRUEMAN, A. E. 1925. The Liassic rocks of the Radstock district. *Quart. J. geol. Soc. Lond.*, **81**, 595–666.

WOODLAND, A. W., and others. 1957. Classification of the Coal Measures of South Wales with special reference to the Upper Coal Measures. *Bull. geol. Surv. G.B.*, **13**, 6–13.

WRAY, D. A., and TRUEMAN, A. E. 1931. The non-marine lamellibranchs of the Upper Carboniferous of Yorkshire and their zonal sequence. *Summ. Progr. geol. Surv. G.B.*, 1930, **3**, 70–92.

THE BASE OF THE UPPER PALAEOZOIC

V. G. Walmsley

I. INTRODUCTION

IN common usage the Upper Palaeozoic has long been taken to begin with the Devonian System. If there were clear agreement as to the base of the Devonian System there would be no problem. However, the boundary separating this system from the underlying Silurian System has provoked uncertainty, argument and a large volume of geological communication extending over a century.

In recent years the matter has become of international concern and specialists from many countries have met in symposia and field studies in order to arrive at an agreed boundary. To some non-specialists this has perhaps been seen as an arid exercise and to them and to other more sympathetic but equally bewildered non-specialists, some explanation is now due of what has been achieved and why it was important to achieve it. International agreement and presumably, therefore, stability, is within sight, and may well be established at the International Geological Congress of 1972 in Montreal (See addendum). The events leading to this were a mixture of mistakes, misunderstandings and hard work, a record of which provides a case history of an important stratigraphic procedure.

II. THE PROBLEM

Generally speaking, when the various stratigraphical systems were erected during the last century, little attention was paid to the precise boundaries between them. They were to be recognised by their gross and typical characteristics, and indeed it was accepted that transitional series linked adjacent systems. In the twentieth century the succession of phanerozoic systems previously established has been broadly maintained, but in response to a growing need for precision, attempts have been made to define boundaries between them. Lack of precision leads inevitably to faulty correlation, which in turn leads to inaccurate pictures of palaeogeography, and of which environments, faunas, floras, etc. were synchronous.

Moreover, the makers of geological maps in various parts of the world need to work to a common datum in colouring the areal distribution of each system, if comparison of their results is to be meaningful.

Murchison's first detailed account of his Silurian System in 1839 gave no precise limits to it. The popular belief that he used the Ludlow Bone Bed as an upper boundary is unfounded and a careful reading of pages 181, 198 of his Silurian System, 1839, indicates that he included the 'Downton Castle Building stone' as the upper strata of the Silurian. His 1839 boundary, therefore, would be some 16 to 19 feet (5 to 6 metres) above the Ludlow Bone Bed, but he was clearly aware of the transitional nature of the junction. In 1857, p. 290, he further stated that 'the Tilestones of Shropshire and Herefordshire which connect the Silurian and Devonian rocks may, according to the predominance of certain fossils, be classed either with the inferior or the superior system. Their maximum thickness may be considered to be about 40 or 50 feet'.

Subsequent authors dealing with the Silurian and Devonian rocks of the Welsh Borderland selected a variety of boundary horizons for a variety of reasons, and by 1950, White (pp. 58–65) in an analysis of the 'vexed question of the Silurian—Old Red boundary', produced a chart in which the plot of these changes resembles a seismogram, with the proposed boundary ranging between the Ludlow Bone Bed and the top of the Dittonian. The problem then being tackled was the somewhat parochial one of deciding where best to place an arbitrary but recognisable datum horizon in an apparently continuous series of strata in the Welsh Borderland. White, supporting the claims of the law of priority 'having regard to the demands of practicability' suggested accepting the slight adjustment involved in using the Ludlow Bone Bed as the 'datum line from which to mark the boundary in other areas'.

In their account of the Silurian rocks of the Ludlow district, Holland, Lawson and Walmsley (1963, p. 146) accepted the temporary stability created by White's suggestion and defined the base of the Ludlow Bone Bed as the base of the Downtonian Stage and the Lower Old Red Sandstone Series.

It should be remembered that the Old Red Sandstone is not a formal system but is a generally non-marine facies of the Devonian System. The real difficulty in determining the boundary between the Silurian and Devonian systems in Britain arose from the facts that in the type area of the Silurian, there is no upward passage into marine Devonian, and in the type area of the Devonian (in Devon) the base of the marine Devonian and the passage down into Silurian is not available. In short no means existed for an exact correlation of the basal Old Red Sandstone with its marine equivalent. The lowest Devonian of Devon (the Dartmouth Slates) are now correlated with the Siegenian below which on the continent, a succession is known down to the Gedinnian. Boucot (1960, pp. 286 and 288) showed that the Gedinnian brachiopod fauna is post-Ludlow and directed attention to a post-Ludlow, pre-Gedinnian interval in which were deposited the Köbbinghauser Schichten of the Sauerland and their equivalents in other areas.

Unfortunately, the equivalent deposits in the Welsh Borderland cannot be readily determined and their systematic assignment was therefore in doubt.

III. THE SOLUTION

Although it has always been possible for individual authors to advocate a particular solution and there have been many attempts made, it is now clear that three things are essential to a widely satisfactory and lasting solution.

These are, firstly, agreement on the philosophical approach, secondly, the accumulation of sufficient scientific data, and thirdly, international agreement.

(a) The Philosophical Approach

Boundaries, as pointed out by McLaren (1970, p. 802) have often been established on some physical event commonly assumed to be tectonic and of world-wide significance. The 'event' was commonly supposed to be chronologically significant over a wide area.

A minority view regards biological criteria, i.e. the occurrence of particular species, as the only practical way of defining a boundary, despite the possibility that such a biological 'event' may occur at different times at different places.

Lawson (1962 p. 139) summarised the defects of these approaches by his phrase 'the better the boundary—the worse it is', meaning that if a boundary were obvious then continuity of record of geological time was suspect, and therefore the horizon or site was a bad one.

In recent years the concept of what is desirable in a boundary has swung towards the view that boundaries should be defined at stratotype sections and that these should be in sequences of continuous sedimentation. Drastic changes in lithology and/or fauna should be avoided as possibly indicating gaps in the record.

This 'golden spike' concept is based on an argument which may be summarised as follows. The continuum of geological time is represented by the deposition of sediment *somewhere* on the earth. In an apparently unbroken sequence of sediments at a particular place, a particular bedding plane represents a moment of geological time—so far as the present degree of resolution permits. A spike marking this plane at this place uniquely defines a moment of geological time which may then arbitrarily be chosen as a boundary between successive units of geological time.

It is of course accepted that the *recognition* of this 'time plane' away from such a stratotype will depend upon whatever means of correlation are available and will rarely result in a completely reliable equation of all synchronous events. It is essential, therefore, that in choosing the particular bedding plane and location of the stratotype section, as many aids to correlation as possible are available. Thus the horizon may be *defined* at one point, but *recognised* (as near as practicable) in other areas by the occurrence of, and preferably the evolutionary stage of, as many fossil groups as possible.

The important point about this approach is that no matter how imperfect the correlation or how the 'tools' of correlation may change in the future, there is a unique reference point which will remain; even if the account of the distribution of fossils about this point later turned out to be inadequate or even inaccurate, the reference point would remain.

Since the data of lithology, fauna, etc. about this point would become of basic importance in subsequent correlations, the stratotype should be chosen with certain additional criteria in mind. Ideally it should be accessible in both the physical and political senses, and should be in an area where its protection is assured and from which specialists would be permitted when necessary to make comparative collections. Above all, it should be internationally agreed, if it is to have any real stability.

In the case of the Silurian/Devonian boundary, this approach has predominated and has proceeded in two phases. Firstly, to agree internationally on a particular horizon, and subsequently to aim to precisely define that horizon at a selected stratotype section.

How then was the particular horizon chosen? Briefly the answer is that when correlations of the various Silurian/Lower Devonian sequences were at last aligned in a manner which received wide agreement, it became clear that one horizon, the base of the *M. uniformis* Zone, was more widely recognisable than most, and as this horizon already coincided with local boundaries already drawn in various areas it became the natural choice for most countries concerned. The question then becomes, how did this widely acceptable correlation come about?

(b) The Scientific Data

The successions of the Welsh Borderland and of Bohemia have always been at the heart of this problem. The Bohemian succession has long been established as:—

Lochkov Beds
Přídolí Beds
Kopanina Beds
Liteň Beds.

In Britain no graptoloids have been found younger than Ludlow (Leintwardinian). In Bohemia monograptids were known from the Lochkov and as the shelly faunas of the two areas appear to have little in common it was for a long time assumed that the last occurrences of monograptids in the two areas were more or less synchronous. Hence at the 1958 Prague symposium the main reason for advocating the upper boundary of the Lochkov as the Silurian/Devonian boundary was that 'the extent of the Silurian ought to be preserved equivalent to the extent of the youngest graptolite zone', (*M. hercynicus* was then regarded as the youngest). Hence, correlation tables produced in Czechoslovakia as recently as 1958 equated the Lochkov, Přídoli and Kopanina together as equivalents of the Ludlow (see Bouček 1960 and Horný 1960).

On present views this was a false correlation to the extent that both the Přídolí and Lochkov are now regarded as post-Ludlow. The successions of Britain and Bohemia (and other areas more directly correlated with Bohemia), were for almost a century locked in a miscorrelation. The 'mistake' was discovered when the higher graptoloid-bearing beds of Bohemia were shown to have brachiopod faunas which in terms of the standard Rhineland/Ardennes Devonian successions were clearly Lower Devonian in age.

Bouček, Horný and Chlupáč (1966 p. 51) in discussing developments after the Prague symposium, outlined the steps by which the correction was gradually made. The revision of the type Ludlow area (Holland, Lawson and Walmsley, 1959, 1963) provided a basis for the comparison with the Bohemian area. The comparative study of brachiopods from the beds close to the Silurian/Devonian boundary by Boucot 1960 and Boucot and Pankiwskyj 1962 showed the highest strata of the Bohemian 'Silurian' to be equivalent to the Rhineland Gedinnian, a view supported by conodont studies of Walliser 1962 and 1964, who went further in assigning the youngest Lochkov Beds to the Siegenian.

Jaeger, 1962, Table 3, was already equating the post-*M. ultimus* graptolite zones (Přídolí and younger) with post-Ludlow and in 1963, Alberti related the base of the Gedinnian to the base of the Lochkov by using sub-species of the trilobite *Warburgella rugulosa* (Alth). Solle (1963) recorded mid-Siegenian spiriferids from the *M. hercynicus* Zone of Thuringia thus confirming Jaeger's (1962) assignment of this zone to mid-Siegenian; and the discovery of faunas of mixed Rhineland and Bohemian character in Morocco, (Holland (1962, 1963) Legrand (1965)) and in Poland,

(Tomczyk (1962) and Teller (1964)), added confirmation to these correlations.

Graptoloids, although absent from the topmost Ludlow and younger beds in Britain, were at last seen to have survived into the early Devonian in central Europe. The initial feeling that this must be an isolated survival rapidly changed as results from many areas including N. Africa, the Yukon, central Asia and Australia showed that, far from Bohemia being an exception in having post-Ludlow graptoloids, Britain was indeed the exception in not having them.

Jaeger, 1970, has summarised the record of Lower Devonian graptoloid discoveries and has shown that in addition to the six post-Ludlow, (Přídolí) zones not only Lochkov (Gedinnian-Siegenian) but also the younger Praguian (Siegenian-Lower Emsian) graptoloids are now known from many localities on all continents except Antarctica and S. America. At least a further 5 to 6 graptolite zones may be recognised even above and including the zone of *M. uniformis* i.e. even in the Lower Devonian as newly defined. His fig. 1 shows monograptid species extending in range almost to the base of the Middle Devonian.

Once the mental block (that monograptid meant Silurian) had been lifted, a whole series of observations of various fossil groups and rock sections rapidly converged on the same result. In 1965 Jaeger, in summarising and discussing the 34 papers from some 80 participants from 14 countries at the Bonn/Brussels symposium of 1960 and subsequent discoveries, was able to publish a correlation chart of late Silurian to early Devonian from England to Podolia (reproduced here, fig 3.) which was convincing and which rapidly became accepted.

Perhaps the most important aspect of this chart was the clear indication of some 9 graptolite zones of post-Ludlow age and extending up to Siegen age. The fact that the Gedinnian, as basal Devonian, does not extend down to Ludlow in the Ardennes/Rhineland areas was only then (1965) becoming slowly realised, despite Shirley having stated this as early as 1938 and Boucot in 1957 and 1960 having confirmed Shirley's view.

Given the situation that the marine facies of the Welsh Borderland failed to extend above the Ludlow Bone Bed and that the Devonian marine sequence of the Ardennes/Rhineland area was regarded as starting with the Gedinnian, there were only three possibilities. Either the Gedinnian immediately succeeded the Ludlow, or the two overlapped, or there were intermediate strata.

It gradually became clear that the latter view was the correct one, and such units as the Calcaire de Liévin of Artois and the Köbbinghauser Schichten of the Rhineland (both previously regarded as Ludlow) fell into place as post-Ludlow— pre-Gedinnian age deposits with equivalents in the Kellerwald, Hartz and Thuringian successions which correlated with the Přídolí of Bohemia, the Rzepin beds of Poland and the Skala of Podolia.

The previous miscorrelations had been largely the result of misidentification of fossils. For example, in the Calcaire de Liévin a species of *Dayia* was assumed to be *Dayia navicula* which was further assumed to indicate a mid-Ludlow age. In fact the species is not *D. navicula* and even if it were, this species in the type Ludlow ranges from early Ludlow (Eltonian) to late Ludlow (Lower Whitcliffian) (see Shirley, 1939, p. 360). In the Köbbinghauser Schichten, the dalmanelloid brachiopod assigned to *Dalmanella orbicularis* (a Ludlow species) is in fact *Protocortezorthis fornicatumcurvata*, a descendant post-Ludlow species, (Walmsley 1965, p. 465, 467 and Johnson & Talent, 1967, p. 157).

In the non-marine Old Red Sandstone of Britain the approximate equivalent of these beds was seen as the Downtonian, as restricted by Allen and Tarlo in 1963. The Ludlow Bone Bed thus appeared to correspond roughly with the base of these units and the base of the Gedinnian with the top of them. It was this latter horizon marked by the base of the zone of *M. uniformis* and conveniently separating the Přídolí from the Lochkov in Bohemia and the Skala from the Borszczow in Podolia which gradually came to be recognised as a widely correlatable horizon. Not only is it well marked in the graptolite sequence, but Alberti (1963) had shown how the trilobite *Warburgella rugulosa rugosa* appeared immediately above it in the Rhineland, Bohemia and in Poland, whilst immediately below it occurred the widespread *Scyphocrinus elegans* in the Rhineland, Kellerwald, Hartz mountains, Thuringia, Carnic Alps and Poland. As Jaeger well realised, the exact position of the Downtonian/Dittonian boundary in terms of this European correlation cannot be stated with certainty but until more data is available it provided a rough approximation to the horizon (based on the base of the *M. uniformis* Zone).

Not only is this horizon well recognised in Europe, but Berdan *et al* 1969 show how it approximates to the base of the Helderberg in N. America, traditionally used in N. American stratigraphy as the base of the Lower Devonian. They regard this horizon (*M. uniformis* boundary) as coincident with the base of the Gedinnian and the base of the conodont zone of *Icriodus woschmidti*, 'it is recognisable in the brachiopod-coral-trilobite succession by the disappearance of the pentamerids, *Atrypella* and *Gracianella*, halysitids and *Encrinurus* and by the incoming of terebratulids, *Cyrtina* and common *Schizophoria*'.

Although the Přídolí/Lochkov boundary is well defined and easily recognised in Bohemia, its position in terms of conodonts is not yet explicitly stated (but see Walliser 1964). The writer has collected material from many horizons and localities in Bohemia in order to study the brachiopods. After extraction of brachiopod material the debris has been processed for conodonts and an account of them is in preparation jointly with R. Aldridge and R. Austin.

(c) *International Agreement*

Even though it was generally recognised that the horizon of the Ludlow Bone Bed approximated fairly closely to the zone of *M. ultimus*, in selecting the horizon to be used as the Silurian/Devonian boundary, opinion in most countries concerned, converged on the use of the zone of *M. uniformis*.

This was claimed to have more widespread recognition and it must be remembered that the solution sought was one of maximum global utility rather than a solution based on previous usage, stability or history.

Many British geologists, realising that the horizon of *M. uniformis* in Britain is presumably represented by some unknown and as yet unrecognisable horizon in the Lower Old Red Sandstone, tended to prefer a solution which would approximate to the Ludlow Bone Bed—even though it may be expressed in terms of the graptolite zonal scheme. It was felt by many that modern work on microfossils held out hope that a correlation between the Ludlow area and the eastern Baltic region might still be achieved—whence correlation by graptolites would permit the expression of this horizon (the Ludlow Bone Bed) in terms of the central European graptolite sequence, presumably about the horizon of the zone of *M. ultimus*. Shaw (1969, p. 70) considered that his study of Downtonian Beyrichiacean ostracodes confirmed a correlation of the base of the Downtonian with the graptolite sequence 'in the vicinity of the *M. ultimus* zone.' Martinsson (1967 p. 380) had already indicated the position of the Ludlow Bone Bed horizon in the eastern Baltic, based on an ostracode correlation. Moreover Richardson & Lister (1969) emphasised the potential usefulness of spores and microplankton for linking vertebrate zoned continental strata with invertebrate zoned marine sequences. Unfortunately nothing comparable with the distinctive Lower Downtonian spore assemblages had been described from elsewhere. They were of the opinion that the striking change in spore assemblage composition at the Ludlow-Downtonian boundary was essentially time controlled rather than an environmental effect.

However, it must be remembered that international agreement was also necessary for a satisfactory solution and international opinion converged strongly in favour of the *M. uniformis* horizon. The manner in which this opinion made itself felt and the formal arrangements by which it was recognised and recorded are therefore of some interest.

The first international symposium on the Silurian/Devonian boundary was held in Prague in 1958. A more representative second symposium in 1960, with some 80 participants from 14 countries, was held in Bonn and Brussels, with field excursions in the Rhineland, the Hartz Mountains and the Ardennes, under the auspices of the Geological Societies of Germany, France and Belgium. Erben (1961 and 1962) and Jaeger (1965) have provided accounts of the proceedings and scientific results of this symposium. During this meeting a 'committee on the Silurian/Devonian boundary and stratigraphy', was set up under the International Commission on Stratigraphy of the I.U.G.S. Professor Dr. H. Erben of Bonn, the first Chairman of this committee, was succeeded in early 1968 by Dr. D. J. McLaren of Calgary.

During the period 1960–1967, little of a formal nature took place but a Devonian Symposium in Rennes in 1964, was followed by the much larger Calgary Devonian Symposium of 1967. During the latter a meeting of the Silurian/Devonian boundary committee took place and a report of this was published by Professor Dr. Erben in the Geological Newsletter 1967, No. 4.

It was, however, during this period that much of the scientific progress, prompted by the 1960 symposium was made. In the absence of formal international meetings, the Ludlow Research Group Bulletins maintained contact between numerous specialists and the widespread annual reporting of preliminary results probably helped to bring about and have recognised, an agreed revised correlation much sooner than might otherwise have been likely. An indication of the growth of interest in such problems is given by the rise in numbers of specialists listed in the Directory.

The first Directory issued in 1962 with L.R.G. Bulletin No. 10, included some 130 interested parties. The 1970 figure was close to 300.

In 1968 the 3rd International Symposium on 'The Silurian/Devonian Boundary and Stratigraphy of the Lower and Middle Devonian', was arranged by Soviet geologists with meetings in Leningrad and field excursions to the Devonian of Salair, (Siberia) and the Silurian/Devonian of Podolia (Ukraine). The Ministry of Geology of the U.S.S.R., the All Union Geological Scientific Research Institute (VSEGEI), the Academy of Sciences of the U.S.S.R., the Western Siberian Geological Dept., the Institute of Geology and Geophysics of the Siberian branch of the Academy of Sciences of the U.S.S.R., the Academy of Sciences of the Ukrainian S.S.R. and the Ministry of Geology of the Ukrainian S.S.R. all contributed to make this an outstanding symposium under the Chairmanship of Academician D. V. Nalivkin. Some 50 visiting participants from 18 countries were joined by at least as many Soviet participants. A brief account of this symposium was published by Holland, (1969). During the symposium six meetings of the Silurian/Devonian Boundary Committee took place. At the first, attended only by Committee members, it was agreed that a decision on the boundary should no longer be delayed and should be reached at this symposium. Subsequent meetings in Leningrad, Novosibirsk and Lvov were open to all interested participants. During these meetings the proposal that 'the boundary between the Silurian and Devonian systems shall be drawn in relation to the base of the *M. uniformis* zone' was carried by a large majority of the Committee. So also was a second proposal that the Committee 'shall now proceed to study localities where a suitable boundary stratotype may be defined in accordance with proposal 1, and to make further recommendations to the Commission of Stratigraphy on an agreed locality and horizon.' A report on these proposals together with dissenting opinions was submitted by the Committee Chairman (Dr. McLaren) to the Commission of Stratigraphy at Prague on August 22nd, 1968, (see Geological Newsletter, 1969, No. 1, in which Dr. D. J. McLaren has published a record of the voting by committee members). Committee members and others interested were then invited to submit suggestions on suitable localities for the Boundary Stratotype to the Chairman, who subsequently indicated the criteria both scientific and political/physical which should be taken into account. The aim was to produce a shortlist of nominated areas, which if necessary, would be examined by some of the Committee before making a final recommendation. In arriving at the shortlist, however, submissions were encouraged from as many areas as possible in the hope that these carefully documented submissions would serve to demonstrate hypotheses of correlation between major cratonic regions on several continents at the horizon chosen for the boundary. They would serve as 'paraboundary strato-types', local, regional or even continental, reliably correlated with the type boundary.

Of the submissions so far made (and summarised in the following Geological Newsletters) those from Czechoslovakia by I. Chlupáč (1969, No. 4, pp. 322–336), Northern Yukon by A. C. Lenz (1969, No. 4, pp. 337–344), Podolia (Ukraine) by O. I. Nikiforova & M. Predchentsky (1970, No. 2, pp. 150–153), Roberts Mountains area, Nevada by M. A. Murphy (1970, No. 4, pp. 342–360) and northwest Morocco by G. K. B. Alberti (*in press*) all argue for the selection of the boundary stratotype in that particular area. The Podolian section was studied during the 1968 symposium, some Committee members visited the Roberts Mountains sections in May 1970, the Bohemian Czechoslovak sections in September, 1970, and sections in Morocco in September, 1971. Other submissions which will serve as regional paratypes have been submitted (and summarised in Geological Newsletters as follows): Carnic Alps (Austria/Italy) by G. B. Vai (1970, No. 2, pp. 157–162), Kazakhstan by H. L. Bublichenko & G. A. Stukalina (1970, No. 2, pp. 163–170), Tien Shan by V. Gorianov (1970, No. 2, pp. 171–175), the Holy Cross Mountains (Poland) by M. Pachlowa, E. Tomczykowa and H. Tomczyk (1970, No. 3, pp. 245–250), Algerian Sahara by Ph. Legrand (1970, No. 3, pp. 250–260), Aragon (Spain) by P. Carls (1970, No. 4, p. 341), Gaspé (Quebec) by P. A. Bourque *et al.* (1970, No. 4, pp. 361–375), Australia by G. M. Philip (1971 No. 1 p. 12) and Canadian Arctic Archipelago by J. W. Kerr and R. Thorsteinsson (*in press*). After a study of all these submissions and visits to some of the areas concerned, the Committee expect to make their final recommendation to the International Stratigraphic Commission at the International Geological Congress in Montreal in 1972. Assuming that these recommendations are accepted, the matter will then be formally settled and the Committee will disband (See addendum).

IV. CONSEQUENCES FOR BRITAIN

In the submissions for stratotypes listed on pp. 38-39, Britain (so long the home of the boundary) is noticeably absent. The reason is not far to seek. The zone of *M. uniformis* has not been recognised in Britain and is most unlikely to occur, since in the Welsh Borderland, open marine sedimentation appears to have ceased about 120 feet (40 metres) above the zone of *M. leintwardinensis*. In these 40 metres of Whitcliffe Beds, no scraps of graptolite have yet been found (even by micro-palaeontologists) and it therefore seems most unlikely that some further six graptolite zones will be shown to be represented by the Whitcliffian.

The succeeding Lower Old Red Sandstone beds of Shropshire have been subjected to a number of subdivisions and classifications of which the latest is that proposed by Allen & Tarlo (1963).

These authors revised the boundary between the Downtonian and Dittonian stages to coincide with 'the major palaeontological break in the region' which they claim approximates to a major lithological break between the Downton and Ditton Series. This lithological break is 'where brackish intertidal sediments give way to fluviatile beds whether these begin with limestone or sandstones.' (p. 147).

They point out that, although White and Toombs (1948) proposed a revision of King's Downtonian/Dittonian boundary to a position between the *Traquairaspis symondsi* and *Pteraspis leathensis* zones (regarding the replacement of *Traquairaspis* by *Pteraspis* as significant), nevertheless no sharp boundary can be drawn between the two zones as in a number of cases *T. symondsi* and *P. leathensis* occur together.

Although Allen & Tarlo (1963, p. 151) suggest that the marine and brackish water vertebrates of the Downtonian include a large proportion of animals of Silurian type whilst the freshwater Dittonian vertebrates are typically Devonian, they did not then specifically discuss the Silurian/Devonian boundary and certainly made no claim that their newly proposed Downtonian/Dittonian boundary could be correlated closely with the *M. uniformis* boundary. Tarlo later (1965) argued that the Silurian/Devonian boundary should be drawn at the revised Downtonian/Dittonian boundary.

Jaeger (1965) and Martinsson (1969) have extended their correlations of the *M. uniformis* horizon into Britain by neatly equating it with the Downtonian/Dittonian boundary 'sensu Allen and Tarlo'.

It may well be that the Downtonian/Dittonian boundary of Allen and Tarlo represents the position of the base of *M. uniformis* zone in Britain but there is as yet no evidence for this and until we have a means of correlation we should resist the temptation of too readily accepting it as fact merely because it is convenient. This warning was made by the writer in the discussion of the Silurian Correlation Charts presented by Cocks, Holland, Rickards and Strachan at the Geological Society of London in February 1970 although no account of this discussion accompanies their paper published in 1971.

That the proposed boundary between Downtonian/Dittonian using the onset of fluviatile deposition is not universally acceptable, let alone equatable with the recommended Silurian/Devonian boundary, is revealed by the recent paper by Westoll *et al.* (1971) who point out (p. 287) that this must be a diachronous boundary.

These latter authors also express the view that the correlation of the Downtonian/Dittonian junction, 'a most difficult biostratigraphical and lithostratigraphical one to define and map', with an horizon in the graptolitic facies, is neither easier nor more acceptable to them than the correlation of the base of the Ludlow Bone Bed with successions in Europe which can now be dated in terms of graptolite zones.

Their preference for a Silurian/Devonian boundary expressed in graptolite terms but using an horizon (e.g. the zone of *M. ultimus*) as close as possible to the Ludlow Bone Bed is precisely the view expressed by the present writer at the open meeting of the Silurian/Devonian Boundary Committee in Leningrad in 1968. It is also expressed in the writer's paper submitted to the Leningrad Symposium (Walmsley, 1971). The majority of British geologists then present held similar views, (see Lawson 1971, p. 629) but these unfortunately were not expressed in the voting of the British committee members of whom two of the three voted in favour of the *M. uniformis* boundary. However, as the total voting in favour of the *M. uniformis* boundary resulted in 28 in favour, 2 against, with 1 abstention, it seems that at least one requirement, a large measure of international agreement, has now been met, and even if all three British votes had been against the *M. uniformis* horizon it could hardly have altered the result.

The problem that does remain for Britain, however, is to find a means of recognising the new Silurian/Devonian boundary in the Old Red Sandstone facies. This means devising a chain of correlation (presumably using vertebrates and plants) between the graptolitic facies and the Shropshire succession.

There are other related problems, e.g. the naming of the post-Ludlow pre-Gedinnian interval. In a formal nomenclature it can hardly be called Downtonian (see Cocks *et al.* 1971 p. 105). Recent usage (e.g. Walmsley, Boucot and Harper

1969 and Walmsley & Boucot 1971, Berry and Boucot 1970) has tended to use Přídolí for this post-Ludlovian Silurian in the absence of a formally agreed term. It should be remembered however that although the correlation of Kopanina with Ludlow is generally accepted, there is no *precise* correlation available of the topmost Ludlow with successions elsewhere, and that therefore the equation of Kopanina/ Přídolí boundary with the base of the Ludlow Bone Bed is also a convenient but as yet unproved, device.

Where then is the base of the Upper Palaeozoic in South Wales and the Welsh Borderland? In the minds of many British stratigraphers it probably still rests at the base of the Ludlow Bone Bed. Others possibly believe that it has now been established at the new Downtonian/Dittonian boundary.

If we accept that the international majority view should prevail, then it rests at the base of the zone of *M. uniformis* (at an as yet undefined stratotype), (see addendum) and its chronostratigraphic equivalents in other areas.

It is clear therefore, that before we can put a finger on the boundary in our area certain events must take place.

Firstly, the formal acceptance by the International Stratigraphic Commission of the proposed horizon and stratotype recommended to them by the Silurian/ Devonian Boundary Committee. Secondly, a convincing correlation must be made between the designated stratotype and some appropriate section in Britain. Thirdly, even if the boundary should then happily coincide with or even be capable of being expressed in terms of, the new Downtonian/Dittonian boundary, it will still be necessary to locate this latter 'boundary' in much of the area. Westoll *et al.* (1971, p. 287) argue that it is inherent in the analysis of Allen & Tarlo 1963 that their boundary must be diachronous.

Finally, we may therefore need a standard section *defining* the Downtonian/ Dittonian boundary at an appropriate site from which correlation must be made by whatever means possible (and inevitably only approximately) to other British areas

Meanwhile, as Lawson (1971, p. 629) has indicated, we in Britain must live with a Silurian/Devonian boundary as vague almost as our Permian/Triassic boundary.

V. ACKNOWLEDGEMENTS

I am indebted to the publishers of 'Geologie' and to Dr. Hermann Jaeger for permission to reproduce his 1965 correlation chart (see p. 45). Grants from the Royal Society (1968) and the University College of Swansea (1960 and 1968) enabled me to participate in the International Symposia in Bonn and Leningrad. They are gratefully acknowledged.

Addendum

At the International Geological Congress held in Montreal in August 1972, the Commission on Stratigraphy accepted the recommendation of the Silurian/Devonian Boundary Committee that:—

the boundary between the Silurian and Devonian systems be defined in the Barrandian area at the Klonk section, in the upper part of bed 20, immediately below the first occurrence of *M. uniformis* (see Chlupač, Geological Newsletter, 1969, No. 4, p. 133).

VI. REFERENCES

ALBERTI, G. 1963. Zur Kenntnis rheinisch-herzynischer Mischfaunen (Trilobiten) in Unterdevon. *Mitt. Geol. Staatsinst.* Hamburg, **32**, pp. 148–59.

ALLEN, J. R. L., and TARLO, L. B. 1963. The Downtonian and Dittonian Facies of the Welsh Borderland. *Geol. Mag.*, **100**, 2, pp. 129–55.

BERDAN, J. M., BERRY, W. B. N., BOUCOT, A. J., COOPER, A., JACKSON, D. E., JOHNSON, J. G., KLAPPER, G., LENZ, A. C., MARTINSSON, A., OLIVER, W. A., RICKARD, L. V., and THORSTEINSSON, R. 1969. Siluro-Devonian Boundary in North America. *Geol. Soc. America Bull.*, **80**, pp. 2165–74.

BERRY, W. B. N., and BOUCOT, A J. (Eds.). 1970. Correlation of the North American Silurian Rocks. *Geol. Soc. America*, Spec. Paper 102.

BOUČEK, B. 1960. Einige Bemerkungen zur Entwicklung der Graptolithenfaunen in Mitteldeutschland und Böhmen. *Geologie*, **9**, 5, Berlin.

——, HORNÝ, R. and CHLUPÁČ, I. 1966. Silurian versus Devonian. *Acta mus. Nat. Prague*, **22B**, 2, pp. 49–66.

BOUCOT, A. J. 1957. Position of the North Atlantic Silurian-Devonian Boundary (Abstract). *Bull. Geol. Soc. Amer.*, **68**, 2, p. 1702.

——. 1960. Lower Gedinnian Brachiopods of Belgium. *Mém. L'Inst. Geol. L'Univ. Louvain*, **21**, pp. 283–324.

—— and PANKIWSKYJ, K. 1962. Llandoverian to Gedinnian Stratigraphy of Podolia and adjacent Moldovia. *Symposiums Band Silur/Devon*. Grenze, pp. 1–11 Stuttgart.

COCKS, L. R. M., HOLLAND, C. H., RICKARDS, R. B., and STRACHAN, I. 1971. A correlation of Silurian rocks in the British Isles—*Jl. geol. Soc.*, **127**, pp. 103–36.

ERBEN, H. K. 1961. Ergebnisse der 2. Arbeitstagung über die Silur/Devon-Grenze und die Stratigraphie von Silur und Devon. Bonn und Brüsel 1960—*Zeitschr. deutsch geol. Ges.*, **113**.

——. 1962. *Symposiums-Band der 2 Internationalen Arbeitstagung über die Silur/Devon-Grenze und die Stratigraphie von Silur und Devon, Bonn-Bruxelles* 1960. Stuttgart. (the title of this volume has been abbreviated in other references in this list).

HOLLAND, C. H. 1969. Third International Symposium on the Silurian-Devonian Boundary and Stratigraphy of the Lower and Middle Devonian. Leningrad, U.S.S.R., 18 July–5 August 1968. *Geological Newsletter*, No. 1, pp. 20–4.

——, LAWSON, J. D., and WALMSLEY, V. G. 1959. A revised classification of the Ludlovian succession at Ludlow. *Nature*, London, **184**, pp. 1037–39.

——, —— and ——. 1963. The Silurian Rocks of the Ludlow District, Shropshire. *Bull. Brit. Mus. (Nat. Hist.) Geol.* **8**, pp. 93–171.

HOLLARD, H. 1962. Etat des recherches sur la limite siluro-dévonienne dans le Sud du Maroc. *Symposiums Band-Silur/Devon Grenze*. Stuttgart.

——. 1963. Les *Acastella* et quelques autres Dalmanitacea du Maroc, présaharien. Leur distribution verticale et ses conséquences pour l'étude de la limite Silurien-Dévonien. *Notes et mém. Serv. géol.*, **176**, 60p., Rabat.

HORNÝ, R. 1960. Notes on the Correlation of the Bohemian and British Silurian. *Věstník Ú.Ú.G.*, **35**, pp. 251–54. Praha.

JAEGER, H. 1962. Das Silur (Gotlandium) in Thüringen und am Ostrand des Rheinischen Schiefergebirges (Kellerwald, Marburg, Giessen)—*Symposiums Band-Silur/Devon-Grenze*. Stuttgart.

——. 1965. Referat: Symposium Band der 2 internationalen Arbeitstagung über die Silur/Devon-

Grenze und die Stratigraphie von Silur und Devon, Bonn-Bruxelles. 1960. *Geologie*, **14**, pp. 348–64. Berlin.

——. 1970. Remarks on the stratigraphy and morphology of the Praguian and probably younger monograptids. *Lethaia*, **3**, pp. 173–82. Oslo.

JOHNSON, J.G., and TALENT, J. A. 1967. Cortezorthinae, a new subfamily of Siluro-Devonian dalmanellid brachiopods. *Palaeontology*, **10**, 1, pp. 142–70.

LAWSON, J. D. 1962. Stratigraphical Boundaries. *Symposium Band-Silur/Devon-Grenze*. Stuttgart.

——. 1971. The Silurian-Devonian Boundary. *Jl. geol. Soc.*, **127**, pp. 629–30.

LEGRAND, P. 1965. Nouvelles connaissances acquises sur les limites des systemes Silurian et Dévonien du Sahara Algérien—Coll. Dév. inf. Rennes, 1964. (Résumés). *Bur. Rech. Géol.*, etc., **33**, pp. 50–2. Paris.

MARTINSSON, A. 1967. The succession and correlation of ostracode faunas in the Silurian of Gotland. *Geol. Fören. Stockholm Förh.*, **89**, Stockholm.

——. 1969. The series of the redefined Silurian System. *Lethaia*, **2**, pp. 153–61. Oslo.

McLAREN, D. J. 1969. Report from the Committee on the Silurian-Devonian Boundary and Stratigraphy to the President of the Commission on Stratigraphy. Prague, August 9, 1968. *Geological Newsletter*, No. 1, pp. 24–34.

——. 1970. Presidential Address: Time, Life and Boundaries. *J. Paleont.*, **44**, pp. 801–15.

MURCHISON, R. I. 1839. The Silurian System. 768 p. London.

——. 1857. Note on the relative position of the strata near Ludlow containing the Ichthyolites described by P. Egerton. *Quart. J. geol. Soc. Lond.*, **13**, pp. 290–91.

RICHARDSON, J. B., and LISTER, T. R. 1969. Upper Silurian and Lower Devonian spore assemblages from the Welsh Borderland and South Wales. *Palaeontology*, **12**, 2, pp. 201–52.

SHAW, R. W. L. 1969. Beyrichiacean ostracodes from the Downtonian of Shropshire. *Geol. Fören. Stockholm Förh.*, **91**, pp. 52–72.

SHIRLEY, J. 1938. Some aspects of the Siluro-Devonian Boundary problem. *Geol. Mag.*, **75**, pp. 353–62.

——. 1939. Note on the occurrence of *Dayia navicula* (J. de C. Sowerby) in the Lower Ludlow Rocks of Shropshire. *Geol. Mag.*, **76**, pp. 360–61.

SOLLE, G. 1963. *Hysterolites hystericus* (Schlotheim). Brachiopoda; Unterdevon, die Einstufung der oberen Graptolithen-Schiefer in Thuringen und die stratigraphische Stellung der Zone des *Monograptus hercynicus*. *Geol. Jb.*, **81**, pp. 171–220, Hannover.

TARLO, L. B. H. 1965. Psammosteiformes (Agnatha)—a review with descriptions of new material from the Lower Devonian of Poland. I General Part. *Palaeont. pol.*, **13**, pp. 1–135 (1964).

TELLER, L. 1964a. Graptolite Fauna and Stratigraphy of the Ludlovian Deposits from the Chelm Borehole (Eastern Poland). *Studia Geol. Pol.*, **13**, Warsaw.

——. 1964b. On the stratigraphy of Beds younger than Ludlovian and the Silurian-Devonian Boundary in Poland and Europe. *Acta Geol. Pol.*, **14**, No. 2, Warsaw.

TOMCZYK, H. 1962. Stratigraphic problems of the Ordovician and Silurian in Poland in the light of recent studies. *Trav. Serv. geol. Pologne*, **35**, pp. 1–134.

WALLISER, O. H. 1962. Conodontenchronologie des Silurs (=Gotlandiums) und des tieferen Devons mit besonderer Berücksichtigung der Formationsgrenze. *Symposiums Band-Silur/Devon-Grenze*. Stuttgart.

——. 1964, Conodonten des Silurs. *Abh. Hess. Landesamt. Bodenf.*, **41**, 106 p. Wiesbaden.

WALMSLEY, V. G. 1971. Silurian dalmanellid brachiopods of Britain—Transactions 3rd International Symposium (The Silurian/Devonian Boundary and Silurian biostratigraphy), Leningrad 1968. Vol. 1, Leningrad, Izdatelstvo 'Nauka' pp. 65–78 [Russian]. (Granitsa Silura i Devona i biostratigrafiya Silura. Trudy III Mezhdunarodnogo Simpoziuma, Tom. 1. Leningrad 1968).

——. 1965. *Isorthis* and *Salopina* (Brachiopoda) in the Ludlovian of the Welsh Borderland. *Palaeontology*, **8**, 3, pp. 454–77.

WALMSLEY, V. G., and BOUCOT, A. J. 1971. The Resserellinae—a new subfamily of Late Ordovician to Early Devonian dalmanellid brachiopods. *Palaeontology*, **14**, pp. 487–531.

——, —— and HARPER, C. W. 1969. Silurian and Lower Devonian salopinid brachiopods. *J. Palaeo.*, **43,** pp. 492–516.

WESTOLL, T. S., SHIRLEY, J., DINELEY, D. L., and BALL, H. W. 1971. The Silurian Devonian Boundary—Written contribution to a paper 'A correlation of Silurian rocks in the British Isles'. *Jl. geol. Soc.*, **127,** pp. 285–88.

WHITE, E. I. 1950. The vertebrate faunas of the Lower Old Red Sandstone of the Welsh Borders. *Bull. Brit. Mus.* (*Nat. Hist.*) *Geol.*, **1,** pp. 51–67.

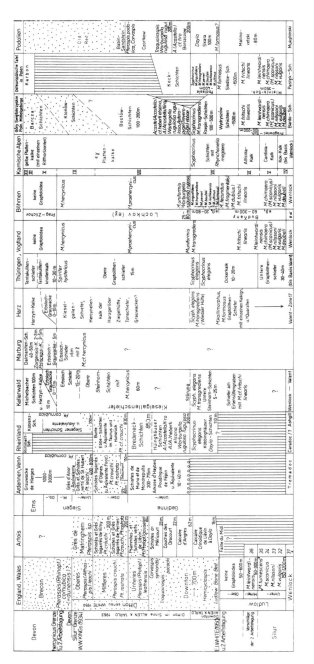

Fig. 3. Correlation of the Siluro-Devonian boundary beds from England to Podolia.

This correlation chart is taken from Jaeger's 1965 paper, the References in which give the sources of much of the data. It is an expanded version of the correlation chart given as Tab. 3 of Jaeger's 1962 Symposium paper in which the correlations are developed in detail. Detailed correlation of the British-Rhineland O.R.S. facies was made possible by the work of White, (1950 and later papers) and W. Schmidt (1954 and 1959) using the succession of *Pteraspis* species. The classification of the Podolian sections was made on the same basis. We are indebted mainly to Shirley and Boucot for the knowledge that the rich invertebrate faunas underlying the fish bearing Gedinnian of the Rhineland-Ardennes area are post-Ludlow in age.

In this chart, new columns have been added for the Hartz Mountains, Carnic Alps (after Walliser) and Poland after E. & H. Tomczyk. An important contribution was the addition of the Gedinnian index trilobites after Alberti (1962–3) and E. & H. Tomczyk. The work of these authors is based on the fundamental monographic studies of R. & E. Richter, (1954). As far as possible the most recent information available in 1965 (not only from the 1960 Symposium) had been considered and the author (H. Jaeger) had selected the most convincing view from divergent opinions.

THE DEVONIAN ROCKS OF WALES AND THE WELSH BORDERLAND [1]

J. R. L. Allen

I. GENERAL INTRODUCTION

IN Lower Palaeozoic times Wales and its surroundings was part of a geosynclinal complex in which marine sediments formed, partly in deep-water basins to great thicknesses and partly on shallow shelves, with at times an accompanying vulcanicity. Although the development of this region is interpreted in different ways (Jones, 1938, 1956; Dewey, 1969; Rast & Crimes, 1969; Ziegler, 1970), it is clear that between late Silurian and mid-Devonian times there occurred orogenic deformations of such a profound influence as to drive out the sea and cause continental conditions of sedimentation to replace the former marine ones. The deposits of this continental episode constitute the Old Red Sandstone, which crops out widely in South Wales and the Welsh Borderland, in Anglesey, and around the Mouth of the Severn. The Old Red Sandstone is of Devonian age and its equivalents in southwest England are mainly marine sediments, representing a new geosynclinal sea that lay across the southern parts of the British Isles and persisted into Carboniferous times. Thus these Devonian rocks give evidence of a time of profound change when one configuration of continent and ocean was replaced by another.

The basis of knowledge of the Old Red Sandstone of Wales and the Welsh Borderland was laid long ago by Murchison (1839), De la Beche (1846), Salter (1863) and Symonds (1872). Around the turn of the century, and in subsequent years, valuable stratigraphical data was accumulated by the Geological Survey during the mapping of the South Wales Coalfield. Of this period, Dixon's (1921) work in Pembrokeshire stands out as the first detailed stratigraphical scheme for the Old Red Sandstone, based on carefully observed lithological characteristics. Equally a milestone was the establishment by the amateur W. W. King (1925, 1934) of a detailed sequence in the Welsh Borderland. The long-known vertebrate faunas of the Old Red Sandstone were put to stratigraphical use by White and Toombs (1948), White (1950a) and Ball and others (1961), particularly as regards the succession in the West Midlands. Dixon (1921) cited pertinent observations on the conditions of deposition of the Old Red Sandstone, and these sedimentological investigations were continued by Ball and others (1961), Allen (1962, 1964a, 1965a), Pick (1964), and Allen & Tarlo (1963).

Our knowledge of the Old Red Sandstone is nevertheless meagre, for the greater part of the outcrop has never been adequately mapped or explored palaeontologically. Consequently any attempt at stratigraphical and environmental synthesis is bedevilled by ill-defined lithological units, a confused stratigraphical nomenclature, and a general lack of faunal control. The following presentation is not excepted and is best treated sceptically, as regards both certain of the supposed facts concerning the Old Red Sandstone and the interpretations placed on them. The course of future work may, however, become a little clearer.

II. OUTLINE OF STRATIGRAPHY

With the necessary international agreements and correlations relating to the Siluro-Devonian boundary at present incomplete (but see Ch. 3), it is here convenient to retain the Ludlow Bone Bed as the local base of the Devonian and the Downtonian as the first division of the System applicable in the present region (figures 4, 5). Following White (1956) and Schmidt (1959), the local divisions of the Old Red, Sandstone are compared with the standard divisions of the Devonian in Fig. 6 showing the probable relationship of the Old Red Sandstone to the marine Devonian in the south. Evidently the formations which compose the Old Red Sandstone can be divided between ai least three episodes of selimentation, each episode being preceded by a period of erosion caused in at least two of the three cases by a substantial crustal deformation. The first episode is represented by the Lower Old Red Sandstone, and the second and third together by the Upper Old Red Sandstone (Fig. 6).

The vertebrates, as summarized by White (1950a, 1951), Ball and others (1961), and Allen & Tarlo (1963), are the organisms mainly used to define the local divisions of the Old Red Sandstone. Macroscopic plant remains are not sufficiently abundant to be stratigraphically useful, but the plant spores are fairly widely dispersed and show marked temporal changes (Mortimel, 1967; Richardson, 1967; Chaloner & Streel, 1968; Richardson & Lister, 1969), which may in future make them important in stratigraphy.

The chief Downtonian vertebrates are *Didymaspis*, *Hemicyclaspis*, *Kallostrakon*, *Sclerodus* and *Thyestes*. *Cyathaspis* persists from the Ludlovian but the Lower Palaeozoic form *Archegonaspis* dies out. *Hemicyclaspis* occurs in the early Downtonian, whereas *Kallostrakon* and *Didymaspis* appear at later stages.

The Dittonian, as defined by Allen & Tarlo (1963), is characterized by a different vertebrate fauna. The base of the division is marked by the incoming of *Tesseraspis* and *Traquairaspis*, shortly followed by a series of pteraspids, largely on the basis of which the Dittonian is divided into zones. *Anglaspis* and *Corvaspis* are members of the early Dittonian fauna, together with *Didymaspis* and *Kallostrakon* surviving from the Downtonian. Cephalaspids extend through the Dittonian as well as the Downtonian.

The Downtonian and Dittonian faunas are known from very many localities and the lower boundaries of the two divisions are perhaps adequately defined. The same cannot be said of the boundary between the Dittonian and Breconian divisions, as Allen and others (1968) have pointed out. In the Clee Hills, the type area of the Dittonian, the upper limit of the *Althaspis leachi* zone has not yet been discovered in terms of the appearance of *Rhinopteraspis cornubica*, the Breconian zonal fossil. Moreover, *R. cornubica* occurs at only one locality in the whole of South Wales and the Welsh Borderland, at a level substantially above the highest appearance in the Clee Hills of *A. leachi*. The *R. cornubica* reported from the Forest of Dean has proved to be *A. leachi* (Allen and others, 1968). The upper limit to the Breconian is even more unsatisfactory, since the rocks embraced by the division are essentially unfossiliferous, of substantial thickness, and truncated by an unconformity. The dubious status of this division can only be resolved by faunal studies in an area

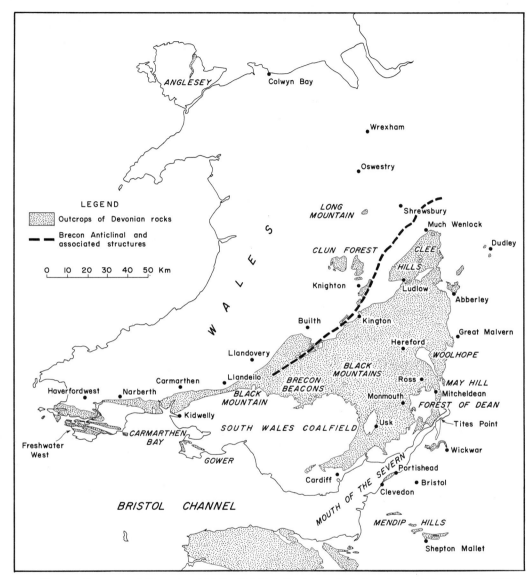

Fig. 4. Outcrops of Devonian (chiefly Old Red Sandstone facies) rocks in Wales and the Welsh Borderland, together with the location of the principal places referred to in the text.

where the Dittonian pteraspids and *R. cornubica* can all be found, or by choosing other organisms for zonal purposes.

The Farlovian division is the only one designated in the Upper Old Red Sandstone of South Wales and the Welsh Borderland. The vertebrate fauna includes *Bothriolepis*, *Eusthenopteron*, *Holoptychius* and *Sauripterus*. *Coccosteus* is found locally, suggesting the presence of Middle Devonian elements. Again locally, the

Farlovian yields a shallow-marine fauna, including the brachiopod *Cyrtospirifer verneuili*. The boundaries of the Farlovian are also unsatisfactorily defined. The division is limited below by an unconformity and above by the incoming of facies-controlled marine faunas of Carboniferous type.

III. BASE OF THE LOWER OLD RED SANDSTONE

(a) *West Midlands and the Mouth of the Severn*

Here the base of the Lower Old Red Sandstone generally rests at outcrop on late Ludlovian (Whitcliffian) beds in a calcareous shelly siltstone facies (Holland and Lawson, 1963). The contact is marked by an abrupt change of facies, usually at a bed (Ludlow Bone Bed) whose base is erosional and locally down-channelling. The facies change does not, however, depart from a shallow-marine depositional motif.

Near Dudley the Ludlow Bone Bed is a sandstone up to 60 cm thick followed by a strong development of the Downton Castle Sandstone (King & Lewis, 1212, 1917; Whitehead & Pocock, 1947; Ball, 1951). Westwards between Much Wenlock and Ludlow, the Ludlow Bone Bed is poorly developed (Elles & Slater, 1906; Whitehead and others, 1928; Robertson, 1927a; Whitaker, 1962; Greig and others, 1968). It shows down-channelling into the Ludlow Shales only near Ludlow, but nevertheless denotes a significant change in sedimentary facies. Structural complications and poor exposure obscure relationships in the Silurian inliers south of Cleobury Mortimer and at Abberley, but it is certain that high Ludlovian beds are followed by the Ludlow Bone Bed at the base of a well-developed Downton Castle Sandstone (Mitchell and others, 1961).

The Silurian rocks of the Malvern Hills were recently described by Phipps & Reeve (1967). The highest beds are Whitcliffian, followed erosively by the Ludlow Bone Bed and a generally well-developed Downton Castle Sandstone, known locally as the Rushall Beds (Phillips, 1848; Stamp, 1923). The Ludlow Bone Bed is better developed westwards in the Woolhope inlier, where it is a locally conglomeratic and down-channelling sandstone up to 50 cm thick (Stamp, 1923; Gardiner, 1927; Squirrel & Tucker, 1960). At Gorsley and May Hill southeast of the Woolhope inlier, the local equivalent of the Ludlow Bone Bed is a thin phosphatized pebble bed which marks a pronounced change of facies (Lawson, 1954, 1955). The overlying Clifford's Mesne Beds, equivalent to the Downton Castle Sandstone of northerly areas, are locally as much as 25 metres thick. Lawson (1955) found these beds missing in the southern part of the May Hill inlier.

The base of the Lower Old Red Sandstone is locally exposed around the Mouth of the Severn. The Whitcliffian beds of Tites Point on the River Severn are overlain by 1·7 metres of Downton Castle Sandstone resting on a channelled surface (Cave and White, 1971). Southward at Wickwar, Wenlockian and Downtonian beds are in close contact, though the true nature of the relationship is unknown (Whittard and Smith, 1944). In the Usk inlier, the Speckled Grit Beds between 8 and 15 metres thick are the local equivalents of the Downton Castle Sandstone (Walmsley, 1959; Squirrell and Downing, 1969). Down-channelling and other evidence of an erosional break is seen at their contact with the underlying Whitcliffian siltstones and cal-

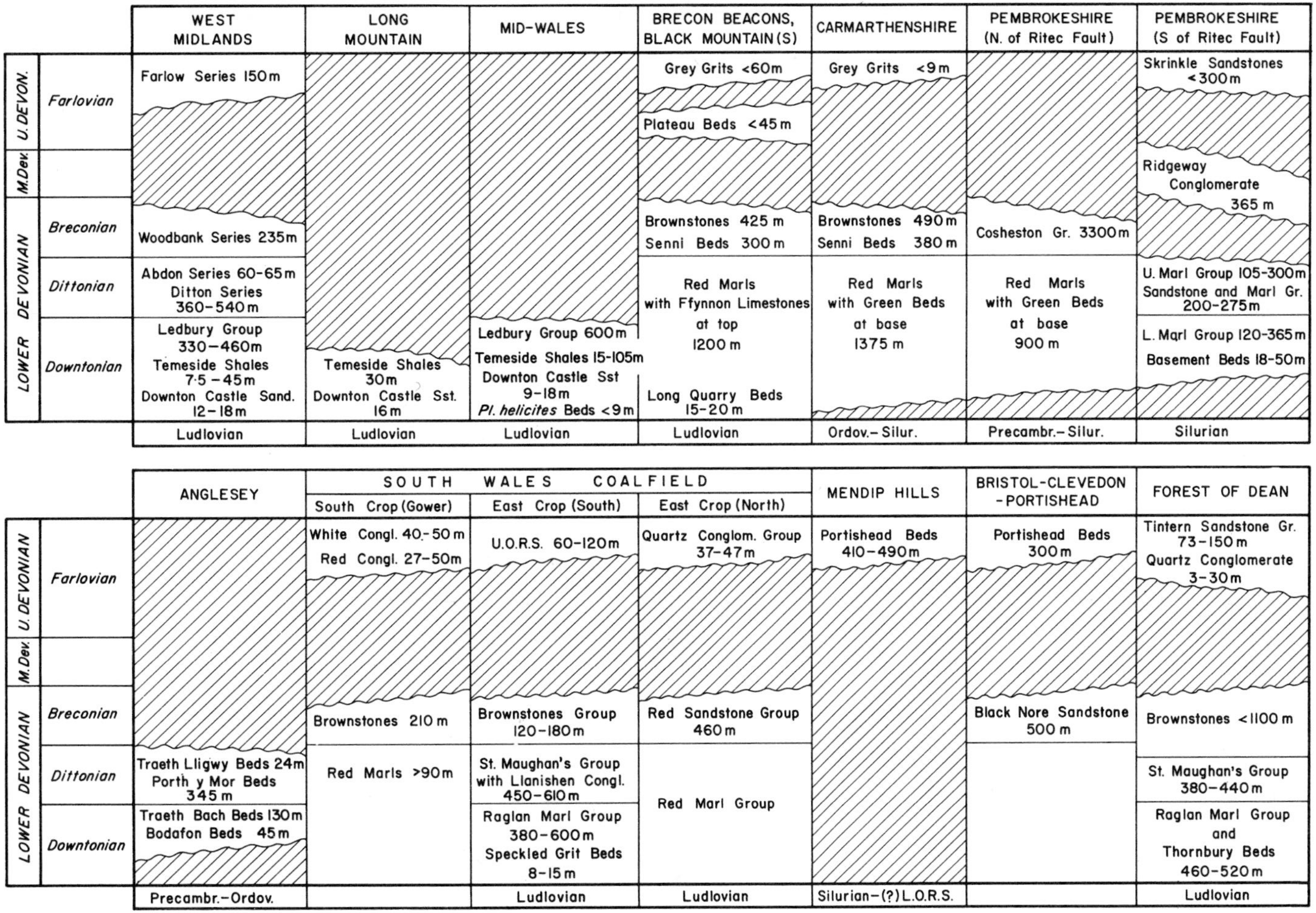

Fig. 5.   Correlation chart for Old Red Sandstone formations in Wales and the Welsh Borderland.

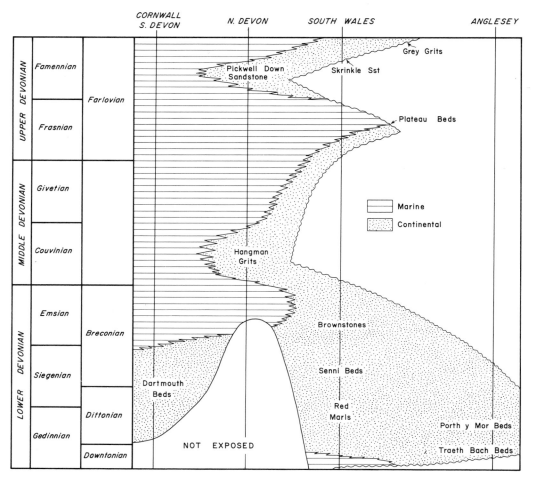

Fig. 6. Schematic relationship of the Old Red Sandstone of Wales to the marine Devonian rocks of Southwest England. The sections are not to scale as regards thickness (the Devonian succession in N. Devon is probably several times thicker than the maximum development of the Old Red Sandstone in South Wales).

careous shales. The Lower Old Red Sandstone is in contact with Silurian rocks near Cardiff (Sollas, 1879), but the details are obscure, and part of the Ludlovian could be absent (Walmsley, 1962; Holland and Lawson, 1963).

(b) Mid-Wales

The base of the Lower Old Red Sandstone here is again in contact with Whitcliffian calcareous shelly siltstones, though with faunal elements not found to the east (Holland and Lawson, 1963).

At Long Mountain west of Shrewsbury the Downton Castle Sandstone is well developed, with a metre or so of shales occurring above the Ludlow Bone Bed in contact with late Ludlovian rocks (Stamp, 1923; Austin, 1925; Das Gupta, 1932; Palmer, 1970).

In the hills around Knighton (Stamp, 1919; Earp, 1938, 1940; Holland, 1959), the base of the Lower Old Red Sandstone is strikingly different from that seen at Ludlow (Elles & Slater, 1906; Whitaker, 1962) or Kington (Banks, 1856; Stamp, 1923) to the east, on the other side of the Brecon Anticlinal. Although there is an horizon at which the faunal and lithological facies change markedly from those characteristic of the uppermost Ludlovian, the Ludlow Bone Bed is lacking and yellow sandstones similar to the Downton Castle Sandstone are not found for 6–12 metres above. Any depositional break that occurred was short-lived.

(c) South-Central Wales

In this area, roughly between Builth Wells and Carmarthen, the base of the Lower Old Red Sandstone steps westwards from Whitcliffian on to progressively older Ludlovian beds, and from them over Wenlockian and finally Ordovician rocks (Strahan and others, 1907, 1909; Stamp, 1923; Straw, 1930, 1953; Potter & Price 1965).

The overstep is at first gradual, and between Builth Wells and the neighbourhood of Llandeilo involves only Ludlovian rocks whose depositional thickness falls in the direction of the developing unconformity. The base of the Lower Old Red Sandstone lies beneath a series of yellow micaceous sandstones, formerly called Tilestones but now termed the Long Quarry Beds (Potter & Price, 1965) and measuring 15–2 metres thick. There is no detectable angular break between the Lower Old Red Sandstone and the older rocks.

Near Llandeilo the base steps on to Wenlockian beds and then, in the vicinity of Carmarthen, on to strongly folded and faulted Ordovician rocks. The Green Beds, at the base of the Lower Old Red Sandstone, are micaceous sandstones up to 150 metres thick.

(d) South-West Wales

The base of the Lower Old Red Sandstone is widely exposed west of the R. Taf, partly in the cores of Hercynian anticlines. Rocks of many different ages appear beneath the base in these outcrops, and the Lower Old Red Sandstone was clearly laid down on a surface of complex structure, an interpretation of which was offered by Sanzen-Baker (1972).

Between St. Clears and Narberth, in the most northerly outcrop, the base steps from Ordovician back on to Silurian (Llandoverian) rocks, the Green Beds reaching a maximum thickness of about 250 metres (Strahan and others, 1909, 1914). At the base of the Green Beds is a conglomerate nearly 2·5 metres in thickness. To the southwest at Llangwm, the Green Beds with a basal conglomerate unconformably overlie Precambrian igneous rocks of the Benton and Johnston Series (Strahan and others, 1914; Sanzen-Baker, 1972).

The base of the Lower Old Red Sandstone appears in two inliers, Winsle and Marloes, north of Milford Haven (Cantrill and others, 1916; Walmsley, 1962). All

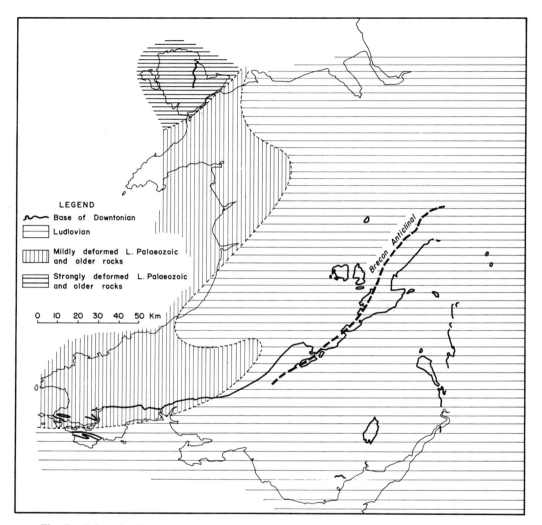

Fig. 7.   Inferred palaeogeology of the Downtonian floor. The distribution of Ludlovian rocks in West and North Wales is conjectural but not implausible.

stages of the Silurian are said to be represented at Marloes, though the contact with the overlying Lower Old Red Sandstone has never been satisfactorily fixed because of an apparent lithological transition between the Silurian and Devonian formations. A similar uncertainty attaches to the boundary between the Silurian and Devonian rocks at Lindsway Bay, St. Ishmael's, at the eastern end of the Marloes inlier. In the Winsle inlier, the Lower Old Red Sandstone can be seen in a road-cutting west of Sandyhaven Pill to lie unconformably on the Silurian. A breccia 2-3 metres thick, followed by unfossiliferous red siltstones and two more breccias, rests sharply and without signs of thrusting on richly fossiliferous Wenlockian mudstones with the topmost 0·5 metres stained pink beneath the Devonian. Elsewhere in the inlier, the breccias in the lower part of the Lower Old Red Sandstone can be seen within 2-3 metres stratigraphically of Wenlockian beds. The field evidence is hard to reconcile with Sanzen-Baker's (1972) stated view that the Old Red Sandstone "seems to continue from the Wenlock siltstones without a sedimentary break".

Dixon (1921, 1933) described the Silurian inliers of Freshwater East and Freshwater West, to the south of Milford Haven. Walmsley (1962) and Sanzen-Baker (1972) have more recently discussed them. Ludlovian beds appear to be present beneath the Lower Old Red Sandstone, which has thick conglomerates and quartzites at the base, and is almost certainly unconformable on the Silurian.

(e) Anglesey

It is finally necessary to refer to the base of the Old Red Sandstone in Anglesey (Greenly, 1919; Allen, 1965a), although the island is far distant from the other areas discussed. Here is a thick sequence of conglomerates followed by red beds, lying with strong unconformity upon Precambrian and Ordovician strata accompanied by infolded early Silurian rocks. The red beds are themselves folded and cleaved and overlain with flagrant unconformity by the Carboniferous Limestone. Thus far the red beds have proved unfossiliferous, but they are identical in facies and general sequence with the Downtonian and Dittonian rocks of South Wales and the Welsh Borderland. There is little resemblance between them and the Upper Old Red Sandstone of the surrounding region. On these grounds, combined with the evidence of pre-Carboniferous deformation, the beds are most probably of Lower Devonian age.

(d) Interpretation

The depositional break at the base of the Lower Old Red Sandstone marks, according to Rast & Crimes (1969), the final or Cymrian orogenic episode of the Caledonian orogeny in the British Isles. Jones (1956), however, interpreted the stratigraphical evidence to mean that in Wales the last major deformation occurred after the deposition of the Lower Old Red Sandstone.

The break recorded at the base of the Lower Old Red Sandstone clearly varies in magnitude from place to place (fig. 7). The Lower Palaeozoic beds missing on account of pre-Lower Old Red Sandstone erosion amount to perhaps 1000 metres in Southwest Wales and possibly as much as 2100 metres in Anglesey (Jones, 1956). East and south of the Brecon Anticlinal, the Lower Old Red Sandstone succeeds the highest known Ludlovian rocks, and the break is of unknown though probably small magnitude. The magnitude of the break may be smaller still in Mid-Wales to the west of the Brecon Anticlinal, where the Ludlow Bone Bed is lacking and the Lower Old Red Sandstone begins with shallow-marine shales. The area of minimum erosion could therefore have been a circumscribed one, bounded to the southeast and south by the former Lower Palaeozoic shelf, and to the northwest by a more strongly uplifted area of which Anglesey was a part.

IV. LOWER OLD RED SANDSTONE

(a) *Downton Castle Sandstone and Equivalents*

It seems possible that the oldest Downtonian beds are the Downton Castle Sandstone and its equivalents in Mid-Wales, South-central Wales, the Mouth of the Severn, and the West Midlands. The Green Beds and Basement Beds of Southwest Wales are probably slighter younger than the Downton Castle Sandstone, and the basal conglomerate in Anglesey is perhaps much younger. The Downton Castle Sandstone and its correlatives are broadly divisible into bone-bed, mudstone and sandstone facies (fig. 8).

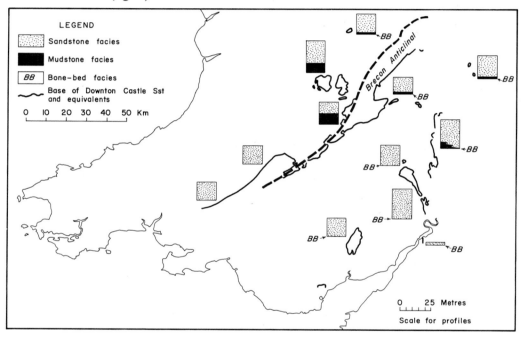

Fig. 8.   Lateral variation of facies developed in the Downton Castle Sandstone and its equivalents.

The bone-bed facies, represented mainly by the Ludlow Bone Bed, is restricted to the east of the Brecon Anticlinal. The facies is marked by a concentration of vertebrate debris, chiefly thelodont scales and various spines, together with phosphatic nodules, shell debris and carbonaceous materials. An indigenous bivalve-gastropod fauna, and sometimes derived and fragmentary Ludlovian brachiopods, are to be found. The Ludlow Bone Bed in some places is a cross-bedded sandstone, the vertebrate debris being mixed with plentiful quartz sand, and at others a pebbly deposit containing fragments of the underlying Ludlovian siltstones (Stamp, 1923; Lawson, 1954, 1955). Elsewhere the Ludlow Bone Bed consists of interbedded ripple marked bone sand and mudstone (Elles & Slater, 1906; Robertson, 1927a; Greig and others, 1968). Examples of the bone-bed facies, usually bands or stringers of bone sand, appear at levels above the Ludlow Bone Bed.

The facies clearly represents a time when terrigenous material was in such short supply that locally produced organic debris could become concentrated (Allen & Tarlo, 1963). The Ludlow Bone Bed, as the chief representative of the facies, may be a strand-line deposit formed during the transgression of a shelf area, as Allen & Tarlo (1963) argued, or a lag deposit formed in relatively shallow water.

The mudstone facies is thickest in Mid-Wales, where it is represented by the *Platyschisma helicites* Beds of Stamp (1919), Earp (1938, 1940) and Holland (1959). It is widely absent east and south of the Brecon Anticlinal (Lawson, 1954, 1955; Walmsley, 1959; Squirrell & Tucker, 1960; Potter & Price, 1965; Cave & White, 1971), and where found rarely exceeds 1 metre in thickness (Elles & Slater, 1906; King & Lewis, 1912; Stamp, 1923). At Long Mountain, however, the mudstone facies appears to be thin (Stamp, 1923).

The facies comprises green to gray mudstone bands associated with thin beds of very fine grained quartz sandstone, usually ripple-marked or parallel-laminated and on occasions graded, the thicker sandstones having erosional bases. The abundant fauna is dominated by the gastropod *Platyschisma helicites*, the bivalve *Modiolopsis complanata*, and the brachiopod *Lingula minima*. The latter two species occur both as drifted remains and in life position. There are also numerous ostracodes (Shaw, 1969) and, occasionally, stunted examples of the Ludlovian brachiopods *Salopina lunata* and *Protochonetes striatellus* (Holland, 1959). Bioturbation is common and widespread.

The mudstone facies in fauna and lithology compares well with the interbedded muds and sands found off prograding sandy shores at the present day (Allen and Tarlo, 1963). The fauna suggests if not full marine conditions, modified by abundant suspended sediment, then at least strong marine influences. To judge from the thickness distribution of the facies, these conditions persisted longest in Mid-Wales.

The overlying sandstone facies of the Downton Castle Sandstone is generally about 15 metres thick, and east of the Brecon Anticlinal often immediately follows the Ludlow Bone Bed (Lawson, 1954, 1955; Walmsley, 1959; Squirrell & Tucker, 1960; Potter & Price, 1965; Cave & White, 1971). The sandstones are usually yellow in colour and generally very fine grained and well sorted. Where the mudstone facies lies below, the transitional beds commonly are thick cross-bedded sandstones with erosional bases. The main sandstones, however, form parallel-laminated sets divided from each other by uneven scoured surfaces and associated with thin ripple-laminated beds. Oscillation, interference and current ripples can all be recognized. Only occasional sandstones are cross-bedded. Locally films of mudstone are present, at one locality showing suncracks (Allen, 1971), and concentrations of mica flakes and macerated plant debris ('coffee-grounds') are numerous. The sparse fauna is restricted to drifted lingulids. At the base of the Long Quarry Beds, however, a rich fauna resembling that of the mudstone facies is known (Potter & Price, 1965).

The sandstones are closely comparable with littoral sands being deposited today (Allen & Tarlo, 1963). The predominance of parallel-laminations indicates broad flat beaches subjected to vigorous waves, while the associated erosion surfaces suggest a periodic scour-and-fill process, perhaps seasonal. The ripple-lamination and the concentrations of mica and coffee-grounds indicate calmer but again wave-dominated conditions. The rocks could have originated as the littoral sand bodies of either a delta or coastal plain. How this complex spread over the region remains unknown, but it is tempting to think of the infilling of a marine-influenced gulf that was deepest in Mid-Wales.

(b) Lowermost beds in South-West Wales

The Green Beds of Strahan and others (1914) and Cantrill and others (1916), together with the Basement Beds of Dixon (1921), are little known stratigraphically and sedimentologically. Dixon (1921) postulated substantial overlaps between the various lowermost beds in Pembrokeshire, and King (1934) claimed that there were missing at Freshwater West beds with a thickness in the West Midlands of about 400 metres.

South of Milford Haven the basal beds begin with well-sorted pebble and cobble conglomerates of mainly local material accompanied by cross-bedded and parallel-laminated quartzites. Above lie interbedded red and green siltstones and green cross-bedded and parallel-laminated pebbly sandstones, together with thin pebble conglomerates. Suncracks are fairly common, but from these beds at Freshwater East, Dixon (1921) recorded frequent lingulids together with vertebrate and plant remains.

No firm interpretation can yet be placed on these beds, though two possibilities merit attention. The basal conglomerates may represent the littoral accumulations of an encroaching marine-influenced gulf, while the overlying siltstones and sandstones are regressive tidal-flat deposits. It is equally plausible that the basal conglomerates are fluviatile and the overlying finer grained beds with lingulids are the deposits of transgressive tidal flats. Sanzen-Baker (1972) writes of piedmont-fan and fluviatile accumulation.

(c) Anglesey

The Bodafon Beds, at the base of the Lower Old Red Sandstone, are conglomerates and conglomeratic sandstones with a maximum thickness of about 45 metres (Allen, 1965a). These were deposited on an irregular floor as a marginal facies to finer sediments represented in the Traeth Bach Beds and Porth y Mor Beds. The Bodafon Beds are possibly alluvial fan deposits (Allen, 1965a).

(d) Temeside Shales of the West Midlands and Mid-Wales

The Downton Castle Sandstone is followed by a higher group of non-red strata, the Temeside Shales, which appears to thin to south-west, south and east.

The Temeside Shales are thickest in the Knighton area, where they are named the Green Downtonian. Here Earp (1938, 1940) recorded a thickness of about 75 metres and Stamp (1919) mapped beds up to 105 metres thick. Holland (1959), working nearer the Brecon Anticlinal, found the Green Downtonian to be as thin as 15 metres. Eastwards near Ludlow and Much Wenlock, the Temeside Shales are between 12 and 45 metres thick (Elles & Slater, 1906; Robertson, 1927a; Whitehead and others, 1928; Greig and others, 1968). Near Dudley they are reduced in thickness to between 7·5 and 10 metres (King & Lewis, 1912; Whitehead & Pocock, 1947; Ball, 1951). The beds probably occur near Abberley (Mitchell and others, 1961), where they seem relatively thin, but are not known southwards at May Hill (Lawson, 1955), Tites Point (Cave & White, 1971), and Usk (Squirrell & Downing, 1969). The Downton Castle Sandstone appears to be followed directly by red rocks in South-Central Wales. There is, however, some lithological similarity between the Temeside Shales and the Green Beds of South-West Wales.

Lithologically the Temeside Shales are dominated by grey to olive-green siltstones, often coarse grained and with scattered calcareous concretions. These are associated with subordinate green to brown sandstones in beds rarely more than one metre thick. The sandstones are fine to very fine grained and often micaceous, with sharp lower contacts. Some sandstones rest on erosional bases strewn with siltstone pebbles, accompanied by drifted lingulids and valves of *Modiolopsis*, together with plant remains. Others have bases which, though sharp, preserve delicate organic and current marks and are probably little if at all erosional in character. The sandstones are mainly ripple-laminated and sometimes parallel-laminated. Cross-bedding is generally restricted to the lower parts of the sandstone beds. The sandstones usually grade up into the overlying siltstones. At some localities a small scale interleaving of the two lithologies occurs at the transition. Drifted shells, chiefly lingulids, *Modiolopsis* and other bivalves, and ostracodes, abound in the sandstones and in the lower levels of the overlying siltstones. The lingulids in the siltstones are sometimes found in life position. *Platyschisma helicites* occurs in the Temeside Shales to the west of the Brecon Anticlinal but apparently not to the east.

The depositional environment of the Temeside Shales is uncertain. Although the invertebrate fauna is marine-influenced, if restricted, the plant remains indicate proximity to land. However, the rocks have several features in common with modern tidal-flat sediments and, as Allen & Tarlo (1963) have argued, may have been deposited on broad mud flats associated with river mouths. If the numerous calcareous concretions should prove similar to the concretionary deposits known from higher levels in the Old Red Sandstone, we may have evidence for the repeated exposure of these mud flats to soil-forming processes, perhaps in response to changes of sea level.

Significantly, the Temeside Shales are thickest where the mudstone facies of the Downton Castle Sandstone is best developed. Again Mid-Wales seems to have be the focus of sedimentation and the area influenced most strongly by the sea.

(e) Higher Downtonian rocks

These crop out continuously from the West Midlands to the coastal cliffs of Pembrokeshire and the wooded hills of the Forest of Dean (Fig. 5). They surround the Usk and Cardiff Silurian inliers, and occur in places east of the River Severn. Being predominantly argillaceous, they are seldom well exposed, and consequently are generally little known stratigraphically and sedimentologically.

The higher Downtonian beds in the West Midlands are known stratigraphically from the work of King (1925, 1934), Whitehead and others (1928), Pocock and others (1938), Whitehead & Pocock (1947), Ball & Dinely (1952), Dinely & Gossage (1959), Ball and others (1961), Mitchell and others (1961), and Greig and others

(1968). The succession, between 330 and 460 metres in thickness, comprises red siltstones with subordinate sandstones and minor occurrences of other lithologies. The sandstones occur chiefly in the lower two-thirds of the sequence. Drifted lingulids and bivalves are common in sandstones from the lower part of the sequence, though only molluscs such as *Modiolopsis* can be found higher up.

The siltstones are chiefly red, though locally green or purple. They are coarse grained, often distinctly sandy, and poorly bedded. Careful examination reveals evidence of general bioturbation, and many outcrops show steeply inclined cylindrical tubes filled with calcite or silt, which may be roots or animal burrowings. Calcareous concretions are numerous in some beds.

The siltstone beds are typically many metres thick. They are associated with thin channel-filling intraformational conglomerates, usually unaccompanied by sandstone, and with thin very fine grained ripple-laminated sandstones. These sandstones commonly are burrowed and may infill suncracks, for example, near Ludlow (Ball and others, 1961).

The main sandstones lie between 1 and 5 metres thick and extend in outcrop for distances between a few hundred metres and several kilometres. Many such beds were, for example, mapped between Much Wenlock and Ludlow by Greig and others (1968). The rocks are medium to very fine grained and contain well rounded fragments up to cobble size of siltstone, usually arranged in seams or in beds of intraformational conglomerate, as well as occasional vein-quartz pebbles. A characteristic vertical sequence is shown by the sandstones. The base is sharp and erosional, in places chanelling deeply into the siltstones below. Above follow cross-bedded medium to fine grained sandstones associated with intraformational conglomerates and scattered siltstone debris. Although relatively coarse grained, the sandstones are often so rich in large flakes of white mica that they can be split into flexible laminae. The overlying beds are very fine grained sandstones with ripple-lamination, sometimes parallel-lamination, and commonly burrowings. These either grade up fairly rapidly into the overlying siltstones, or pass into them by interleaving. The drifted lingulids and molluscs generally occur in the conglomeratic parts of the sandstones.

The higher Downtonian rocks change in character to the southwest and south. On the eastern slopes of the Black Mountains there are sandstones similar to those found in the West Midlands, but they are less numerous (Clarke, 1955). From the Malvern Hills, Piper (1898) described a long section in the Downtonian above the Downton Castle Sandstone. There occur some 120 metres of interbedded red siltstones and cross-bedded with ripple-laminated sandstones which can be seen to rest on erosional bases. Drifted lingulids and vertebrate remains are numerous in the basal parts of the sandstones. The tops are gradational and the intervening siltstones are bioturbated. Further south still, in the Forest of Dean (Trotter, 1942; Welch and Trotter, 1961) and the Usk inlier (Squirrel & Downing, 1969), sandstones appear to be much fewer than in the West Midlands and restricted to the middle of the Raglan Marl Group, which broadly represents the Downtonian. Squirrell & Downing (1969) state that the sandstones occur less frequently as the Raglan Marl Group is reduced in thickness southward from 600 to about 300 metres. One sandstone is said to be 52 metres thick, but a group rather than a single bed may have been mapped.

Little is known of the higher Downtonian rocks north of the Brecon Beacons and the Black Mountain. On the coasts of Carmarthenshire and Pembrokeshire, however, there are magnificent exposures, described by Strahan and others (1909, 1914), Cantrill and others (1916), Dixon (1921), and Sanzen-Baker (1972). The faunal evidence suggests that in Pembrokeshire the Downtonian is represented broadly by Dixon's (1921) Basement Beds and Lower Marl Group (King, 1934; White, 1946). The Downtonian rocks are possibly several hundred metres thick around Carmarthen Bay, but seem to thin southwestwards. Dixon (1921) found the Lower Marl Group to be 275 metres thick at Freshwater East, but only 120 metres thick at Freshwater West.

The sequence is dominated by coarse red siltstones, often bioturbated and unbedded, but in places finely laminated, in beds from a few decimetres to many metres thick. Many beds have calcareous concretions, ranging from scattered irregular lumps to rod-like forms aligned perpendicular to the bedding. Toward the top of the Lower Marl Group, the concretions are abundant enough to form thick concretionary limestones which may be the equivalent of the *Psammosteus* Limestone of districts further east.

Sandstones and conglomerates, though few in number, are conspicuous on the cliffs by their generally green colour. The sandstones are usually a few metres thick and rest on sharp erosional bases, often followed by intraformational conglomerates. They are coarse to fine grained and cross-bedded in their lower parts, passing upward into ripple-laminated or parallel-laminated finer sandstones, often interleaved with thin suncracked siltstones. The conglomerates are generally thinner than the sandstones and often die out in the cliffs within a few tens of metres. They consist of pebbles of vein-quartz, igneous rocks and older sediments, in addition to intraformational debris. Cross-bedding is common, and some beds appear to be gravel bars preserved without modification of the original sedimentary form. Sanzen-Baker (1972), who measured the cross-bedding directions, found that the rocks in south Pembrokeshire had been transported from the south, in contrast with those in north Pembrokeshire, which had travelled from the north.

The higher Downtonian rocks in South-west Wales have not so far yielded either lingulids or molluscs, and in this respect differ from either equivalents to the east.

The Traeth Bach Beds in Anglesey to the north have a strong lithological resemblance to the later Downtonian rocks of Pembrokeshire, and could be of Downtonian age (Allen, 1965a). They consist of coarse red siltstones with calcareous concretions, subordinate concretionary limestones and dolomites, and occasional conglomerate or sandstone beds. Some of the conglomerates are intraformational in type, though others seem to have been derived from local Precambrian and Ordovician rocks. The concretionary limestones and dolomites are exceptionally thick, a maximum thickness of about 14 metres being recorded.

The deposition of the higher Downtonian rocks seems to have occurred at a time of decreasing marine influence, to judge from the upward faunal changes that have been described. No substantial alterations of lithology were introduced, however, and most sandstones are of the upward-fining type, with erosional bases and gradational tops. They are thought to have been formed in laterally wandering channels (Allen & Tarlo, 1963). The sandstones with drifted lingulids, in the lower part of the succession of the eastern outcrops, may represent the tidal lower reaches of rivers. The higher sandstones, yielding only molluscs or no invertebrates at all, resemble river channel deposits, and can be readily interpreted in the light of models presented by Allen (1965b). The siltstones that are so prevalent could therefore be either tidal mud-flat or river floodplain deposits. The prevalence of concretions, which will be discussed below, suggests that they were frequently exposed to atmospheric processes, a conclusion supported by the abundant suncracks

(f) Dittonian rocks

The Dittonian strata (Fig. 5) are better known stratigraphically than any other part of the Old Red Sandstone in this region, for they yield a rich variety of vertebrates, studied chiefly in the Clee Hills (White, 1950a; Ball and others, 1961). Even so, the upper limit of the division is not yet clearly defined faunally (Allen and others, 1968), and most of the outcrop remains undescribed from a modern standpoint.

In the Clee Hills and neighbourhood, the Dittonian is considered to include the Ditton Series and Abdon Series (Allen, 1961; Allen & Tarlo, 1963). The base of the Ditton Series is placed where red siltstones several to many tens of metres thick (Downtonian) are replaced by a facies in which sandstones, siltstones, concretionary limestones and intraformational conglomerates are interbedded. The top of the Dittonian is drawn arbitrarily at the top of a readily mapped limestone formation, in the absence of a faunal boundary. Other opinions regarding the stratigraphy at this general level are expressed by Ball and others (1961) and Greig and others (1968), who interpret the lower boundary as the base of the 'Psammosteus' Limestone (itself variously defined) and do not recognize the Abdon Series as such.

The thickness of the Dittonian appears to lie between 425 and 600 metres (King, 1921, 1934; Whitehead & Pocock, 1947; Dineley & Gossage, 1959; Ball and others, 1961; Mitchell and others, 1961; Greig and others, 1968). The lower part, perhaps 100–150 metres thick, consists of sandstones and siltstones in about equal proportions, together with subordinate concretionary limestones and intraformational conglomerates. Next follows a group of predominantly siltstones with few though individually thick sandstones. In the topmost 100–200 metres, sandstones grow in importance until, in the Abdon Series, they predominate.

These different lithologies are arranged according to a simple scheme, constituting a sedimentary cycle or cyclothem, examples of which can be found at most levels in the sequence (Fig. 9a). Generally the cycle consists of a group of sandstone beds, resting on a laterally extensive scoured surface, that grades up into siltstones (Allen & Tarlo, 1963). The sandstones, usually either parallel-laminated or ripple-laminated, are associated with intraformational conglomerates consisting of siltstone pebbles and limestone concretions in a sandy calcareous matrix. The conglomerates, frequently cross-bedded, often yield vertebrate debris and sometimes drifted plant remains. The siltstones are generally coarse grained, poorly bedded and often

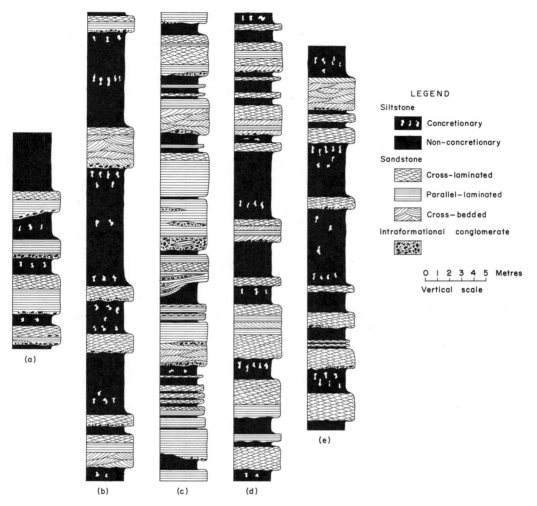

Fig. 9. Sections showing cyclothems in Dittonian formations of Wales and the Welsh Borderland. (a) Ditton Series at Lower Forge, southeast of Cleobury Mortimer, Clee Hills. (b) St. Maughan's Group exposed near Milepost 18 3/16 on Motorway 50 near Ross-on-Wye, Forest of Dean. (c) Lower Brownstone at Wilderness Quarry, Mitcheldean. (d) Sandstone and Marl Group, foreshore at Freshwater West, Pembrokeshire. (e) Porth y Mor Beds, cliffs between Dulas Bay and Lligwy Bay, Anglesey.

bioturbated. Thin ripple-laminated sandstones may appear near their bases, resting on suncracked or flute-marked soles. Most siltstones have one or more calcareous divisions, ranging from bands of scattered concretions, through rubbly limestones with a crude vertical rod-like structure, to massive concretionary limestones. The most calcareous beds are, however, restricted to the lower and upper parts of the Dittonian sequence, namely, the '*Psammosteus*' Limestone and Allen's (1961) Abdon Limestone at the top of the Abdon Series.

The concretionary limestones share many features with caliches, as described by numerous writers (Bretz and Horberg, 1949; Brown, 1956; Swineford and others, 1958; Moseley, 1965; Gile and others, 1965, 1966; Gile, 1967), and as compared with specimens from India, North Africa and North America. Each bed has a characteristic profile of scattered concretions passing up into close-packed concretions with patches of siltstone between and, at the highest levels, massive limestone with rare siltstone pods. In thin sections and polished slabs the rocks are very variable, but almost all yield ample evidence for the precipitation of calcite, the extensive replacement of detrital material and, in some cases, the *in situ* brecciation of the bed and the relative movement of the broken pieces. Since caliches originate in soils, the concretionary beds may record times of prolonged exposure to atmospheric processes, perhaps measuring hundreds or even thousands of years (Allen, 1965a).

Dittonian rocks crop out over large parts of Herefordshire, but are little known aside from the work of Clarke (1951, 1952), who concentrated on the '*Psammosteus*' Limestone and its associates.

Much more data are available for the Dittonian rocks of the Forest of Dean (Trotter, 1942; Welch & Trotter, 1961), where the lower Brownstones yields late Dittonian vertebrates (Allen and others, 1968), and the East Crop of the South Wales Coalfield (Strahan & Cantrill, 1912; Heard & Davies, 1924; Robertson, 1927b; Squirrell & Downing, 1969). The Dittonian vertebrate fauna is fairly well known from this region, largely through the work of White (1938, 1950a) and the Institute of Geological Sciences (Welch & Trotter, 1961; Squirrell & Downing, 1969). It occurs in the St. Maughan's Group and the lower Brownstones on the east side of the Forest of Dean, and in the St. Maughan's Group alone on the western side of the Forest and the East Crop. Near Llanishen and Risca on the East Crop, the Llanishen Conglomerate appears in the upper part of St. Maughan's Group. Within the Conglomerate is a laterally extensive limestone (Ruperra Limestone) which, together with limestones in the Monmouth area, may be correlative with the Abdon Limestone at the top of the Abdon Series to the north.

The St. Maughan's Group is broadly 400–600 metres thick and similar lithologically to the Ditton Series of the West Midlands. Concretionary limestones are well developed amongst the lower beds, where they are associated with thick siltstones and sandstones with intraformational conglomerates. Above come mainly siltstones, often concretionary, subordinate sandstones and conglomerates. Sandstones gradually replace siltstones in the higher part of the Group, and in the lower Brownstones, yielding Dittonian fish, are predominant. A cyclical arrangement of the strata like that in the Clee Hills is conspicuous, as can be seen along the M50 east of Ross-on-Wye and at Mitcheldean (Figure 9b, c). The cycles consist of very fine to fine grained sandstones resting on laterally extensive scoured surfaces and showing cross-bedding, parallel-lamination or ripple-lamination (Allen and others, 1968; Allen, 1964a, 1970). Often there is a basal intraformational conglomerate with vertebrate and sometimes plant remains. The sandstones grade up into siltstones with concretionary bands and, very commonly, thin sandstones lying on suncracked or fluted surfaces.

The Llanishen Conglomerate consists of red siltstones and sandstones associated with subordinate well-sorted pebble and cobble conglomerates between a few decimetres and 7 metres thick. The pebbles and cobbles are chiefly different types of sedimentary and igneous rocks. A few of the sandstone pebbles have yielded Silurian fossils, including Llandoverian elements. The formation, which has been compared with the Ridgeway Conglomerate of Pembrokeshire, is thought by Squirrell & Downing (1969) to have had a southerly provenance.

White (1938, 1946, 1950b) gave scattered faunal records showing the presence of Dittonian rocks in the Black Mountains, the Brecon Beacons and the Black Mountain. Detailed stratigraphical investigations, however, remain to be undertaken in this area.

The Dittonian rocks achieve their finest exposure around Carmarthen Bay and on the Pembrokeshire coast (Strahan and others, 1909, 1914; Cantrill and others, 1916; Dixon, 1921; King, 1934; Allen, 1963), where they are represented by the upper part of the Red Marls and by Dixon's Sandstone and Marl and Upper Marl Groups, with a combined thickness of 500–600 metres. The cliffs (figure 9d) reveal cyclically arranged red siltstones and sandstones with intraformational conglomerates (Allen, 1963, 1970). Many siltstones are concretionary and suncracks abound where they grade up from the sandstones. The sandstones are similar in sedimentary structures to the Dittonian sandstones elsewhere, and in many cases show clear evidence of their lateral deposition, in the form of epsilon cross-bedding.

A similar cyclical arrangement of red sandstones with intraformational conglomerates, siltstones, and concretionary deposits typifies the Porth y Mor Beds of Anglesey (figure 9e), which are 345 metres thick and thought to be of Dittonian age (Allen, 1965a). Many of the groups of sandstones were laterally deposited. These strata are followed without a break by the Traeth Lligwy Beds, measuring 24 metres in thickness, which consist of an alternation of red siltstones (without concretions) and massive-looking sandstones. This facies has no known parallels in the Lower Old Red Sandstone of the surrounding region, and its presence is one of the few points of differences between the beds in Anglesey and other parts of the Welsh region.

Although developed over a large area, the Dittonian rocks are remarkably constant in sedimentary facies. Nearly everywhere concretionary limestones are well developed near the base of the succession. If these rocks are correctly interpreted as caliches, then the Downtonian-Dittonian transition was associated with widespread and probably prolonged episodes when the sedimentary surface was exposed to atmospheric processes. The nature of that surface can be judged from the character of the sedimentary cycles so widely preserved in the Dittonian beds. Allen (1963, 1965a, 1970) showed that the sandstones and conglomerates in each cyclothem are very similar to the channel deposits of modern streams, and that the siltstones can be compared with floodplain sediments. The presence of epsilon cross-bedding is especially strong evidence suggesting deposition from streams, whose channels wander between laterally accreting sand bars. Thus during Dittonian times the area seems to have been part of a vast alluvial plain extending southward to the Devonian sea. Probably the concretionary limestones represent caliches formed in slightly elevated interfluvial areas, perhaps either in response to down-cutting by the rivers as sea level fell slightly, or due to climatic fluctuations.

(g) Breconian rocks

The Breconian formations are more restricted in distribution than the older parts of the Lower Old Red Sandstone (Fig. 5). They are confined to the higher slopes of the Clee Hills, and encircle Carboniferous outcrops in the Forest of Dean and the South Wales Coalfield. Isolated outcrops occur in Pembrokeshire and between Portishead and Clevedon on the River Severn. Considerable doubt attaches to the true age of these formations, since the Breconian division has never been properly defined and the beds yield few fossils.

Symonds (1872) found that in the Black Mountains the Red Marls beginning the Old Red Sandstone were followed by predominantly arenaceous deposits, which he divided between a largely green group below (later called Senni Beds) and the Brownstones above. Subsequently, these formations were recognised above the Red Marls in the Brecon Beacons and the Black Mountain to the west (Strahan and others, 1907, 1909; Robertson, 1932). The upper limit of the Senni Beds is marked by the replacement of green by predominantly red colours, though not by a change of facies. The lower limit, however, is far from clear. According to Ball and others (1961), W. N. Croft in the Black Mountains informally placed it a little below the lower of the two so-called Ffynnon Limestones (informal name). Nonetheless these limestones are associated with predominantly red and purple beds closely resembling the Abdon Series in the Clee Hills and the fish-bearing lower Brownstones in the Forest of Dean. A choice for the lower boundary of the Senni Beds in better accord with usage to the east might be above the Ffynnon Limestones, where the main colour change appears to lie.

The Senni Beds extend from the Black Mountains westward to the head of Carmarthen Bay, where they are overstepped by the Carboniferous (Strahan and others, 1907, 1909; Robertson, 1932). They are between 300 and 450 metres thick in the Brecon Beacons and Black Mountain, but thin to about 225 metres near Kidwelly, and in the Black Mountains are perhaps only 150–200 metres thick. The sequence is dominated by mainly green fine to very fine grained sandstones, either cross-bedded or parallel-laminated and only rarely ripple-laminated. The sandstones are accompanied by green intraformational pebble and cobble conglomerates, some beds reaching 2 metres in thickness. The conglomerates and sandstones are individually lenticular, with groups of beds associated together in broad channel-like structures. Siltstones are very subordinate and usually are green. They are generally found as concretionary beds a few decimetres thick grading up from very fine grained ripple-laminated sandstones, and their tops are usually defined by the scoured surface at the base of another group of conglomerates and sandstones. Many are suncracked. Thus the sequence gives indications of a cyclicity resembling that found in the Dittonian below.

In many places the Senni Beds yield large amounts of fragmentary land plants, partly pyritized and partly carbonized (Croft & Lang, 1942). The larger fragments occur in the sandstones and conglomerates, but the best preserved material is found in the finer sandstones and siltstones, as at Heard's (1927) famous quarry.

The Brownstones are more than 400 metres thick in the Brecon Beacons and Black Mountain, where they are magnificently exposed on north-facing escarpments; westwards they gradually thin due to Upper Old Red Sandstone and Carboniferous overstep (Strahan and others 1907, 1909; Robertson, 1932). Formerly the Brownstones were held to include at the top the formation known as the Plateau Beds. However, this unit is now known to be Upper Devonian and unconformable upon the Brownstones. Lithologically the Brownstones resemble the Senni Beds, except that fossils are fewer and red and brown colours prevail. The sandstones are fine to medium grained with scattered vein-quartz and other exotic pebbles, and the only evidence of plants is afforded by oxidized impressions. Although the siltstones are sandier than in the Senni Beds, and sandstones with intraformational conglomerates are often seen for many metres vertically, the different lithologies appear again to be cyclically arranged. Suncracks abound and rainprints can be found.

Between Llandybie and Kidwelly, Strahan and others (1907, 1909) mapped a series of pebbly sandstones (Pebbly Beds) as the topmost part of the Brownstones. The Pebbly Beds vary widely in position relative to the local top of the Senni Beds, whereas they always lie immediately beneath the Carboniferous. The Pebbly Beds may therefore be perhaps of Upper Old Red Sandstone or earliest Carboniferous age.

At the eastern end of Milford Haven, the Red Marls pass up without a break into the Cosheston Group (or Beds), correlative with the Senni Beds (Strahan and others, 1914; Cantrill and others, 1916). Most of the Cosheston Group consists of very fine to medium grained green sandstones, usually either cross-bedded or parallel-laminated, in alternation with much thinner green and sometimes red coarse siltstones. Intraformational conglomerates are associated with the sandstones, and both lithologies commonly yield fragmentary plants. Breccias and conglomerates appear in the upper part of the Cosheston Group. Strahan and others (1914) found these to contain 'subangular to angular fragments of a variety of rocks, igneous, sedimentary and metamorphic, besides white quartz.' The Cosheston Group is said to be more than 3000 metres thick.

Along the East and South Crops of the South Wales Coalfield the Breconian rocks are generally no more than 120–210 metres thick, on account of Upper Old Red Sandstone overstep (Strahan, 1907; Strahan & Cantrill, 1912; Robertston, 1927b; Squirrell & Downing, 1969). The Senni Beds are no longer recognizeable probably because of their lateral passage into a red facies, and a variety of usages have consequently been applied to the Breconian rocks, with confusing results (Croft, 1953). The succession, though not well exposed, consists of cross-bedded or parallel-laminated fine grained sandstones, with subordinate coarse grained red or sometimes green siltstones, in many place concretionary. Intraformational conglomerates are fairly common and exotic pebbles can be found. The sandstones can often be seen to rest erosively on the siltstones, reminding one of the fining-upward cycles widely prevalent in the Dittonian rocks. It was one of these erosional contacts that D. M. Williams (1926) erroneously described as the unconformable base of the Brownstones.

Breconian rocks are well developed in the Forest of Dean, where they are represented by all but the lowermost part of the Brownstones (Trotter, 1942; Welch & Trotter, 1961; Allen and others, 1968). The beds are at least 1100 metres thick in the vicinity of Mitcheldean and Ross-on-Wye, but thin westwards across the Forest, being overstepped by the Upper Old Red Sandstone near Monmouth.

In the Forest the Breconian rocks show two important vertical trends, a gradual upward coarsening of the sandstones and, in harmony, an upward diminution in the frequency and thickness of siltstones. The lower beds have strong affinities with the underlying Dittonian rocks. Clearly defined cyclothems are present, consisting of cross-bedded and parallel-laminated fine grained sandstones with intraformational conglomerates and subordinate coarse red siltstones. Suncracks abound and many siltstones are concretionary. These rocks give way upwards to a more confused association, of fine to coarse grained sandstones, generally cross-bedded and associated with intraformational pebble and cobble conglomerates, accompanied by lenticular siltstones. The sandstones channel down into each other and also any underlying siltstones, which may be suncracked. Some of the channels cut in siltstone have vertical walls 2–4 metres high and contain siltstone boulders one metre or more across. The intraformational conglomerates at the higher levels in the Brownstones include numerous exotic pebbles, of various igneous and sedimentary rock-types.

Across the R. Severn, at Portishead–Clevedon and Bristol, there are similar rocks to the Brownstones of the Forest of Dean (Wallis, 1928; Pick, 1964), called the Black Nore Sandstone. About 500 metres of beds are present, but their contact with rocks resembling the St. Maughan's Group is not seen.

In the Clee Hills to the north, King (1925, 1934) found a group of mainly arenaceous rocks to which he applied the name Brownstones. These rocks were later studied by Ball & Dineley (1952), Allen (1961, 1962), Ball and others (1961) and Greig and others (1968), all of whom reclassified and named the sequence in different ways. Allen gave to the rocks the name Woodbank Series, recognizing two subdivisions, distinguished on the basis of the grade and colour of the sandstones and the presence of siltstones. The sequence is at most 235 metres thick beneath unconformable Coal Measures.

The lower two-thirds of the Woodbank Series consists of cyclically arranged intraformational conglomerates, sandstones and siltstones (Allen, 1961, 1962). The sandstones are grey to green, calcareous, and usually fine or very fine grained. They are generally either cross-bedded or parallel-laminated. Many hold scattered siltstone pebbles and the sequence contains numerous intraformational pebble and cobble conglomerates. Scattered through these conglomerates are exotic pebbles representing a great variety of igneous and sedimentary rocks. The sandstones occur in groups of beds generally several metres thick and with the coarsest grained in the lower part of each group. These rest erosively upon either green, yellow or magenta siltstones grading up from sandstones, or directly upon the very fine grained sandstones at the tops of preceding cycles. Some of the siltstones are concretionary and others include thin beds of dolomitic limestone.

The upper one-third of the Woodbank Series consists of grey or pink fine to coarse grained pebbly sandstones, either cross-bedded or parallel-laminated. The pebbles are equally of intraformational and exotic origins, a great variety of rock types being represented.

Like the preceding Dittonian rocks, the Breconian sediments are fairly uniform in facies over a wide area. Allen (1962) interpreted the Woodbank Series as alluvial in origin, on the grounds that the sandstones resembled the channel deposits of modern rivers, while the intervening siltstones could be closely compared with floodplain muds. This general interpretation may be extended without difficulty to the Brownstones of the Forest of Dean and the Senni Beds and Brownstones of South-central Wales. Indeed, in these regions it is reinforced by the prevalence of suncracks and other proofs of exposure, features that are unknown from the Woodbank Series itself. Thus the Breconian sediments may also represent an extensive alluvial plain. However, in view of the greater abundance and coarser grade of the sandstones that are present, the plain may have been crossed by more steeply sloping and less strongly meandering rivers than existed in Dittonian times. Late in Breconian times the rivers could have been braided, to judge from the character of the sandstones and the paucity of siltstones (Allen, 1965b).

(h) Ridgeway Conglomerate

The Ritec Fault in Pembrokeshire is an important east-west disturbance ranging along Milford Haven and extending at least as far east as Tenby. It separates two contrasted developments of the Lower Old Red Sandstone. To the north the Red Marls grade up conformably into the Cosheston Group, with little doubt correlative with the Senni Beds to the east if not also with higher Breconian rocks. The Red Marls south of the Fault, however, are followed by the Ridgeway Conglomerate (Strahan and others, 1914; Dixon, 1921, 1933; B. P. J. Williams, 1971), a group little resembling any other formation in the Lower Old Red Sandstone. The Ridgeway Conglomerate is overlain unconformably by the Skrinkle Sandstones (Upper Old Red Sandstone), but has a maximum thickness of about 365 metres.

The Ridgeway Conglomerate consists of interbedded conglomerates, siltstones and sandstones, with occasional concretionary beds. The conglomerates are red to brown in colour and individually between a few decimetres and 9 metres thick. They consist of poorly to well sorted pebbles and cobbles of siltstone, quartzitic sandstone (with fragments of horny brachiopods and trilobites), vein-quartz and green phyllite, all set in a medium to very coarse grained sandstone matrix. Often the clasts are imbricated, and in some beds a large scale cross-bedding is seen and shows transport from the south. The bases of the beds are erosional and sometimes down-channelled, whereas thin medium to coarse grained sandstones may cap the tops. Between the conglomerate beds there are thick coarse red siltstones with scattered calcareous concretions and, rarely, thick concretionary limestones. Cross-bedded sandstones up to 2 metres thick are also found amongst the siltstones. Suncracks occur locally.

The age of the Ridgeway Conglomerate is in doubt, as indigenous fossils of stratigraphical value remain to be found. Dixon (1921, 1933) seems to have regarded the formation as conformable upon, and interleaving with, the Red Marls and as probably older than the Brownstones. It has been correlated with the conglomeratic upper beds of the Cosheston Group (Strahan and others, 1914) and with the Llanishen Conglomerate (Pringle and George, 1948), though these views are not supported by the radically differing compositions of the clasts in the three formations. The change of facies from the Red Marls to the Ridgeway Conglomerate is most striking, however, and on these grounds Allen (1965c) suggested that the base of the Conglomerate could record an important depositional break, extending perhaps into Middle or even Upper Devonian times. This suggestion is consistent with the locally erosion base of the Ridgeway Conglomerate and with the marked thickness variations shown by the Upper Marl Group beneath. B. P. J. Williams (1971) also thought the Conglomerate could be as young as Middle Devonian.

The Conglomerate beds in the Ridgeway Conglomerate are like modern braided-stream deposits. The facies as a whole has a general resemblance, however, to the playa-basin and alluvial-fan sediments of modern semi-arid regions, as described by Bull (1964a, 1964b), Ruhe (1964), Denny (1965), Lustig (1965) and G. E. Williams (1970). If this interpretation is correct, the debris in the conglomerates is probably of comparatively local origin.

V. PROVENANCE OF THE LOWER OLD RED SANDSTONE

Evidence concerning the provenance of the Lower Old Red Sandstone has come from the study of heavy minerals, the petrography of the sandstones, the composition of the exotic pebbles, and the orientation of directional sedimentary structures, chiefly cross-bedding.

The directions of cross-bedding in strata associated with the Downtonian-Dittonian boundary in the Clee Hills were found by Ball and others (1961) to be consistent with a northwesterly source area. Allen (1962) later established a similar direction of transport for the Abdon and Woodbank Series in this are and, in an unpublished study, for the Downtonian and the otherwise unsampled Dittonian levels. Similar transport directions have been established for the Lower Old Red Sandstone as far south as the Mouth of the Severn and as far west as the Brecon Beacons and north Pembrokeshire. In south Pembrokeshire, however, there is evidence for local southerly sources (Sanzen-Baker, 1972).

The heavy minerals are best known from the West Midlands (Fleet, 1925 1926; Walder, 1941), a little data being available from the East crop of the South Wales Coalfield (Heard & Davies, 1924) and the Mouth of the Severn (Wallis 1928). Garnet dominates the assemblages obtained by Fleet from the Downtonian and earliest Dittonian sediments, the other minerals, principally zircon, tourmaline and rutile, appearing in only very minor amounts. The later Dittonian and the Breconian sandstones are also rich in garnet, though much reduced from the levels found in the older beds. The Downtonian and early Dittonian garnets are but little worn and commonly idiomorphic. They seem to be of first-cycle origin, in contrast with the better rounded though otherwise similar garnets from the younger beds. The Downtonian and early Dittonian sediments are also richly micaceous, the mica being of metamorphic origin (Walder, 1941). To account for these features, Fleet (1926) suggested that the Lower Old Red Sandstone was freshly derived from garnetiferous rocks, possibly the Precambrian of Anglesey or another source, a view approached by Heard & Davies (1924). Wallis (1928) had little doubt that the source rocks resembled the Mona Complex of Anglesey, and Walder (1941) was even more definite in drawing this parallel.

The composition of sandstones from the West Midlands varies vertically in a similar manner to the heavy minerals. As noticed by Greig and others (1968), the Downtonian and early Dittonian sandstones abound in grains of phyllite and a variety of schists. The later sandstones are markedly poorer in these constituents, but contain much igneous and sedimentary material (Allen, 1962; Greig and others, 1968), some demonstrably at least second-cycle in origin.

Exotic pebbles brought to the depositional basin provide the most decisive evidence on the sources of the Lower Old Red Sandstone. Unfortunately, apart from the basal beds of the most westerly outcrops, they are confined to late Dittonian and Breconian horizons. Allen (1962) described from the Woodbank Series a varied suite of pebbles, including acid igneous rocks, Wenlockian limestones similar to those at outcrop in the West Midlands, graywackes resembling the Denbigh Grits (Cummins, 1957; Okada, 1967), and Ordovician tuffs. A related assemblage, rather richer in acid flow-textured lavas, was recently collected from the Brownstones of the Forest of Dean. The character of the pebbles, in combination with the palaeo-current evidence, suggests a provenance chiefly from the Lower Palaeozoic rocks of Wales and the Irish Sea area.

The late Dittonian Llanishen Conglomerate yields pebbles of many different rock types, although Heard and Davies (1924) and also Squirrell & Downing (1969) report only sandstones. In addition to several types of sandstone, one can find an abundance of flow-textured porphyritic igneous rocks of acidic to intermediate character. These somewhat resemble the Silurian igneous rocks of the Mendip and Bristol areas (Van de Kamp, 1969), and may be evidence for the southerly provenance of the Llanishen Conglomerate, as Squirrell & Downing (1969) have speculated.

The pebbles of the Ridgeway Conglomerate have not yet been successfully matched in older formations. According to palaeocurrent evidence collected by B. P. J. Williams (1971), they have come from the south.

There is little doubt that the Downtonian and early Dittonian sediments came from an extensive area dominantly of metamorphic rocks that lay somewhere to the northwest of the outcrops here considered. It is difficult to accept, however, that the source was some part of the Precambrian of the Irish Sea region. Insofar as we know them (Baker, 1969), these Precambrian rocks include insufficient garneti-ferous formations to have yielded the huge volume of sediment typified by garnet in the Lower Old Red Sandstone. We may prefer a source still further to the northwest even than the Irish Sea, perhaps a metamorphic terrain of Scottish or, continental drift being accepted, western Atlantic type.

There is equally no doubt that the later Dittonian and Breconian sediments came from a region mainly of sedimentary rocks, nearer at hand than the earlier source and including many familiar Lower Palaeozoic formations. However, the reason why garnet persists at these levels is not yet clear. It would seem that either the relative areal importance of the metamorphic source was diminished, or folding in the source region cut off that source entirely by reversing a part of the drainage, with the result that garnetiferous Downtonian deposits were recycled from the folded region. There is some evidence for the latter interpretation, since pebbles of what on mineralogical and faunal grounds appear to be Downtonian sandstones appear in small numbers in the Breconian conglomerates. Small local contributions may have come from southerly sources during early Downtonian, late Dittonian and Breconian times, in conformity with Stamp's (1923) suggestion of a ridge in the Bristol Channel.

VI. BASE OF THE UPPER OLD RED SANDSTONE

(a) West Midlands

The Farlow Series has a small outcrop at Titterstone Clee Hill, where the beds are unconformable on mid-Dittonian rocks and lie in faulted contact with rocks comparable with the lower part of the Woodbank Series (Whitehead and others, 1928; Ball and others, 1961; Greig and others, 1968). To the northwest, at Brown Clee Hill, Coal Measures now directly overlie the Woodbank Series. High Dittonian or early Breconian rocks reappear beneath Coal Measures at Cleobury Mortimer to the south of Titterstone Clee Hill (Mitchell and others, 1961). In this area, then, the Farlovian floor was composed of slightly deformed Lower Old Red Sandstone.

(b) Mouth of the Severn and the Mendip Hills

In the north, between Wickwar and Thornbury, the Quartz Conglomerate steps westwards from Wenlockian on to Downtonian rocks (Kellaway and Welch, 1955; Welch and Trotter, 1961; Curtis and Cave, 1964). The angular discordance between the formation, is small, however, and the overstep can be attributed in part to the action of one or more pre-Farlovian faults of north-south trend that downthrow to the west. A small and isolated outcrop of Downtonian rocks occurs at Wickwar (Whittard & Smith, 1944), but its position in relation to the associated Silurian beds suggests pre-Farlovian faulting rather than folding.

In the Avon Gorge at Bristol, and again on the coast between Portishead and Clevedon, the Portishead Beds lie unconformably upon the Black Nore Sandstone attributed to the Breconian (Wallis, 1928; Kellaway & Welch, 1955; Pick, 1964). Since these outcrops lie westward of the pre-Farlovian faults established in the Thornbury area, the overstep of the Upper Old Red Sandstone must be continued in a westerly direction, from Downtonian on to Dittonian and finally Breconian beds. The same inference can be drawn from the fact that, along the eastern side of the Forest of Dean, the Quartz Conglomerate rests on an unusually thick Breconian succession (Welch & Trotter, 1961). No significant pre-Farlovian faults have so far been found to affect the Dittonian and Breconian rocks. The Lower Old Red Sandstone of this area must in some measure have acquired a westward dip before Farlovian times.

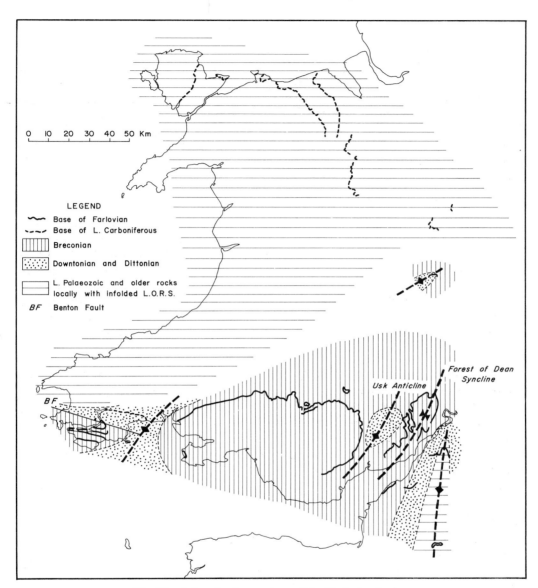

Fig. 10.   Inferred palaeogeology of the Farlovian floor.

At Beacon Hill near Shepton Mallet, almost due south of the Thornbury-Wickwar outcrops, the Portishead Beds unconformably succeed Silurian rocks, including volcanics, of Llandoverian or Wenlockian age (Reynolds, 1907, 1912; Curtis, 1955; Green & Welch, 1965; 1965; Van der Kamp, 1969). Whether the Lower Old Red Sandstone occurs at depth in this area is not known. The Silurian, at least, could have a considerable outcrop on the pre-Farlovian floor between Thornbury and Beacon Hill.

(c) Vale of Usk

The Vale of Usk is limited on the west by the hills of the East Crop of the South Wales Coalfield and on the east by the heights of the Forest of Dean. Near the axis of the Forest of Dean Syncline, the Quartz Conglomerate overlies upwards of 1100 metres of Breconian strata. Close to Monmouth, on the western limb, the Upper Old Red Sandstone oversteps from Breconian to Dittonian beds (Welch & Trotter, 1961). It is not seen again until the East Crop is reached, where the Quartz Conglomerate and its equivalents overlie a generally thin Breconian sequence (Strahan & Cantrill, 1912; Robertson, 1927b; Squirrell & Downing, 1969). Substantial pre-Farlovian movements are indicated by these relations.

(d) North and South Crops of the South Wales Coalfield

Between Kidwelly and Abergavenny on the North Crop of the Coalfield, and in Gower on the South Crop, the Upper Old Red is seen to overlie only Breconian rocks (Strahan, 1907; Strahan and others, 1907, 1909; Robertson, 1927b, 1932). The Breconian beds of the North Crop have a maximum thickness of about 900 metres but thin slowly westwards towards Carmarthen Bay, where they are eventually overstepped by the Carboniferous Limestone. There may be a southerly thinning of the beds beneath the South Wales Coalfield, since only 210 metres of Brownstones appear in the Gower.

(e) Pembrokeshire

In Pembrokeshire, as in South Wales generally, the Upper Old Red Sandstone grades up conformably into the Lower Carboniferous (Lower Avonian). The Skrinkle Sandstones in south Pembrokeshire thin northwards toward Milford Haven and eastwards toward Caldy Island (Cantrill and others, 1916; Dixon, 1921). They rest on the Ridgeway Conglomerate in most parts of the outcrop, but on Caldy Island overstep the Conglomerate to lie on the Red Marls. The Upper Old Red Sandstone is not known north of Milford Haven, where it is overstepped by the Lower Carboniferous.

Near Lawrenny the Lower Carboniferous oversteps the east-west Benton Fault, which down-throws the Cosheston Group on the south against Red Marls on the north (Strahan and others, 1914). The displacement of this fault may be 1500–2000 metres, according to Jones (1956). Further north, near Haverfordwest, the Lower Carboniferous unconformably succeeds the Llandoverian (Strahan and others, 1914).

(f) North Wales

The supposed Lower Old Red Sandstone of Anglesey (Allen, 1965a), in a cleaved and overfolded condition, is overlain together with Precambrian and Ordovician rocks by the Lower Carboniferous (Upper Avonian) with a thick basal conglomerate (Greenly, 1919). On the mainland of North Wales, from Colwyn Bay on the coast to the neighbourhood of Oswestry, the Lower Carboniferous (Upper Avonian) likewise overlies folded and cleaved Lower Palaeozoic rocks (Wedd & King, 1924; Wedd and others, 1927, 1929). Comparable relations exist in the West

Midlands where, near the Wrekin, the Lower Carboniferous (Upper Avonian) overlies beds of Lower Cambrian to Wenlockian age (Whitehead and others, 1928; Pocock and others, 1938).

(g) Interpretation

The preserved relationships of the Upper Old Red Sandstone and the succeeding Lower Carboniferous beds, suggest that the pre-Farlovian deformation of Wales and its surroundings was more intense than that preceding the deposition of the Lower Old Red Sandstone.

Several important structures can be glimpsed on the Farlovian floor in South Wales and the Welsh Borderland (Fig. 10). To the east of the R. Severn there is evidence for a faulted anticlinal structure of north-south trend which may indicate mid-Devonian movements on the Malvern Line. Mild folding took place in the West Midlands, though the caledonoid trend shown here is conjectural. The Usk Anticline, considered by George (1956) to be mainly of Armorican age, is now seen to date at least from mid-Devonian times. It follows that the Forest of Dean Syncline was also of mid-Devonian inception. Important fault movements, expressed by the Benton Fault, occurred in Pembrokeshire, to reinforce the impression of instability gained from the effects of the Ritec Fault and the relations of the Ridgeway Conglomerate.

The character of the supposed Lower Old Red Sandstone of Anglesey, and the basal relations of the Lower Carboniferous in North Wales, suggest that the Lower Palaeozoic rocks to the northwest of the main Old Red Sandstone outcrop could have experienced an orogenic deformation as late as the Middle Devonian (Allen, 1962, 1965c). Jones (1956), too, argued for a mid-Devonian age for the main movements, partly on these grounds and partly from the infolding of the Lower Old Red Sandstone and Lower Palaeozoic rocks east of the Towy Anticline. Rast and Crimes (1969), however, attribute the main deformation (Cymrian orogenic episode) to the end-Silurian, on the grounds that the Lower Old Red Sandstone is virtually undeformed except in Anglesey. It is precisely because of the pre-Carboniferous interpenetrative deformation of the Old Red Sandstone of Anglesey (granted its Lower Devonian age) that an end-Silurian date is perhaps unacceptable.

VII. UPPER OLD RED SANDSTONE: PLATEAU BEDS

The Plateau Beds (Fig. 5) are well exposed in the Black Mountains and Brecon Beacons, where they cap the somewhat softer Brownstones. The formation has a maximum thickness of about 45 metres in the Black Mountain, and thins to the west and east beneath the Grey Grits (Strahan and others, 1907; Robertson, 1927b, 1932; George, 1928; Allen, 1964b). The age and stratigraphical relations of the Plateau Beds were for long controversial, but we now know that the formation is Upper Devonian in age, unconformable upon the Brownstones, and in turn unconformably overlain by the Grey Grits which overstep it. Simpson (1951) reports that Dr. Gwyn Thomas found in the Plateau Beds C. verneuili, Leptodesma, Parallelodon, Bothriolepis and charaphyte algae. White (1951) and Croft (1953) noted also Holoptychius, Sauripterus and Coccosteus. Recently Taylor (1972) described some of these forms from further localities in the Plateau Beds. Hence the formation could be as old as Frasnian, though on balance a Frasnian-Famennian or earliest Famennian age seems more likely (Allen, 1964b; Webby, 1966). Thus the Plateau Beds belong to the second of the episodes of sedimentation distinguished in figure 6. That episode could date from early in the Upper Devonian.

On average the Plateau Beds become finer grained upwards (Allen, 1965c). The lower strata are well-sorted pebble conglomerates associated with mainly cross-bedded pebbly sandstones and sandstones. The beds are lenticular and in places fill channel-like structures. Cross-lamination and ripple-marked surfaces are found locally and bright red colours prevail. No fauna has so far been detected in these lower rocks, which are thought to be fluviatile (Allen, 1965c), having much in common with the channel deposits of modern low-sinuosity streams.

The upper Plateau Beds, yielding the fauna mentioned, consist of interbedded bright red quartzitic sandstones and coarse siltstones with very rare pebbly bands. The sandstones are very fine to fine grained and well sorted; most beds are parallel-laminated and show parting lineations. Other sandstones are cross-bedded, and the finer grained rocks may be cross-laminated, the tops of the beds revealing wave-current ripples. The siltstones are coarse grained and in places a few metres thick. Locally they are rapidly interbedded with thin layers and biscuits of ripple-marked sandstone, and may be crowded with trace fossils. The body fossils are represented mainly by drifted fragments in the sandstones. This association of rocks can be compared to the sediments accumulating in a coastal barrier complex, where the muddy to sandy deposits of barrier beaches, tidal flats and shallow bays and lagoons are closely associated (Allen, 1965c).

VIII. UPPER OLD RED SANDSTONE: DEPOSITS OF THE THIRD EPISODE

The Grey Grits which unconformably succeed the Plateau Beds pass up without a break into the Lower Carboniferous (Lower Avonian) and have correlatives in many parts of South Wales and the Welsh Borderland (Fig. 5). These beds belong to the third episode of sedimentation noted in Fig. 6, and episode with a prolonged extension into Carboniferous times.

The Upper Old Red Sandstone in Pembrokeshire is represented by the Skrinkle Sandstones, thinning northward and eastward from a maximum thickness of 300 metres at Freshwater West (Dixon, 1921). The Skrinkle Sandstones consist of red, green and grey pebble conglomerates, sandstones, shales and siltstones. The conglomerates comprise well rounded and sorted material and in many places fill channel-like structures. The sandstones are very fine to medium grained and usually quartzitic; cross-bedding, parallel-lamination and cross-lamination have all been found in them. Some sandstones contain bands of intraformational clasts and others seams or scattered examples of plant remains. The argillaceous rocks are very variable. A number of beds have calcareous concretions, others are suncracked, and a few show rootlets and so resemble fireclays.

The highest beds of the Skrinkle Sandstone are interbedded sandstones, shales and thin limestones, red colours often being shown. These rocks yield a marine fauna, principally of bivalves, gastropods, calcareous brachiopods and lingulids, details of which are given by Dixon (1921). The strata differ from the Lower Limestone Shales of the overlying Carboniferous mainly in their commonly red colouration and relatively high proportion of sandstones.

The Grey Grits of the Black Mountain, Brecon Beacons, and the outlier of Pen-Cerig-calch reach a maximum thickness of about 60 metres but are usually much thinner (Strahan and others, 1907; George, 1928; Robertson, 1932). The formation consists dominantly of grey to red quartzitic fine to medium grained cross-bedded sandstones. Towards the base are bands of pebble conglomerate consisting of well-sorted and rounded vein-quartz and small amounts of other debris. At higher levels the sandstones become olive-green or yellowish and inter-bedded with siltstones. Taylor (1972) records vertebrate remains and lingulids from the Grey Grits.

On the East and South Crops of the South Wales Coalfield the Upper Old Red Sandstone is represented by pebbly rocks, comparable in thickness with the Grey Grits, to which the local name Quartz Conglomerate Group may be given (Strahan, 1907; Strahan & Cantrill, 1912; Heard & Davies, 1924; Robertson, 1927b; Squirrell & Downing, 1969). The lower part of the formation on the East Crop, and apparently the whole of it in Gower, comprises conglomerates of well-sorted and rounded vein-quartz pebbles associated with cross-bedded sandstones. The higher beds are an alternation of sandstones with thin siltstones and, occasionally, shelly limestones. The sandstones are cross-bedded or parallell-laminated and may have plant impressions. Some beds yield a restricted marine fauna of brachiopods, bivalves, gastropods and crinoids, similar to that preserved in the limestones. There is a gradual passage up into the Lower Carboniferous.

The Forest of Dean and the Thornbury areas present a similar sequence, divided between the Quartz Conglomerate below and the Tintern Sandstone Group above (Trotter, 1942; Kellaway & Welch, 1955; Welch & Trotter, 1961). The Quartz Conglomerate is made up of thick beds of well-sorted and rounded vein-quartz, quartzite and other pebbles in a quartzitic matrix. The conglomerate beds are in places cross-bedded and commonly fill channel-like structures, in which drifted wood and other plant remains have been found. Associated with the conglomerates are cross-bedded or parallel-laminated fine to coarse sandstones. The Tintern Sandstone Group above is finer grained, though vein-quartz pebbles together with intraformational clasts occur in it sporadically. The formation appears to consist of an alternation of green, brown or red sandstones and siltstones on a scale of metres, similar to the cyclical alternations found in the Breconian rocks below. The sandstones are generally cross-bedded or parallel-laminated, and where they overlie the siltstones, an intraformational conglomerate or at least a seam of siltstone pebbles is usually to be found. The siltstones between are coarse grained and apparently massive; some contain scattered concretions. The highest beds of the Tintern Sandstone Group have a restricted marine fauna and include shelly limestones.

The Portishead Beds of the Bristol and Mendip area to the south are several times thicker than the Upper Old Red Sandstone of the Forest of Dean (Wallis, 1928; Kellaway & Welch, 1955; Pick, 1964; Green & Welch, 1965). At the type locality a rich Upper Devonian vertebrate fauna is known (Wallis, 1928). Unfortunately, the formation is rarely well exposed, though appearing to consist of thin siltstones alternating with much thicker sandstones and pebble and cobble conglomerates dominated by vein-quartz and quartzite. The sandstones are mainly cross-bedded and locally include seams or bands of intraformational clasts. The Lower Carboniferous is said to follow the Portishead Beds sharply (Green & Welch, 1965).

Although small in outcrop and a mere 150 metres thick, the Farlow Series of Titterstone Clee Hill in the West Midlands provide decisive evidence of the former northward extent of the Upper Old Red Sandstone (Ball and others, 1961; Greig and others, 1968). The lower part of the Series, with a conglomerate at the base, comprises pebble and cobble conglomerates in alternation with yellow to brown, fine to coarse grained, mainly cross-bedded sandstones. The conglomerates abound in fragments of vein-quartz and a number of different types of sandstone. Igneous pebbles, however, have not so far been found. The upper part of the Series consists of calcareous pebbly sandstones, and thin red or grey siltstones, some of which are concretionary. An extensive vertebrate fauna is known from the Farlow Series (Ball and others, 1961).

There is a clear parallel in the vertical pattern of lithologies between these Upper Old Red Sandstone deposits and the Plateau Beds representing the second sedimentary episode. As was demonstrated (Allen, 1965c), the lower, coarser grained part of the succession representing the third episode of sedimentation can be matched with the deposits of modern streams, perhaps braided in the case of the conglomeratic portions of the sequence, while the upper beds with their occasional marine faunas suggest the development of coastal barriers as the Carboniferous transgression advanced into the region.

IX.   PROVENANCE OF THE UPPER OLD RED SANDSTONE

To judge from the petrography of the rocks, the Upper Old Red Sandstone was derived from a region undergoing fairly intensive chemical weathering. Allen (1965c) found the sandstones to be more mature than those in the Lower Old Red Sandstone, though in the eastern parts of the outcrop the rocks are notably feldspathic and in Pembrokeshire rock fragments become common. Similar conditions are suggested by the conglomerates, which in most areas are dominated by highly resistant vein-quartz and quartzite pebbles.

The regional mapping of cross-bedding directions in the Upper Old Red Sandstone has shown that the beds were deposited by currents flowing southwards, with a slight fanning out to the southeast and southwest, off the Lower Palaeozoic massif to the north (Allen, 1965c). This evidence, together with the strong resemblance of the rocks to modern fluviatile sediments, suggests that South Wales and the Welsh Borderland was part of an extensive coastal plain during Upper Devonian times. That plain was transgressed twice from the south by the sea (Allen, 1964b;

Webby, 1966). The first transgression is represented today only in the Plateau Beds, and its complete extent cannot now be ascertained (Figure 6). The second transgression, which began to the south in the Famennian, extended into the Carboniferous and delimits in the region an entirely new palaeogeography, with little resemblance to that prevailing before Middle Devonian times.

X. REFERENCES

ALLEN, J. R. L. 1961. The highest Lower Old Red Sandstone of Brown Clee Hill, Shropshire. *Proc. Geol. Ass.*, **72**, 205–19.

——. 1962. Petrology, origin and deposition of the highest Lower Old Red Sandstone of Shropshire, England. *J. sedim. Petrol.*, **32**, 657–97.

——. 1963. Depositional features of Dittonian rocks: Pembrokeshire compared with the Welsh Borderland. *Geol. Mag.*, **100**, 385–400.

——. 1964a. Studies in fluviatile sedimentation: six cyclothems from the Lower Old Red Sandstone, Anglo-Welsh Basin. *Sedimentology*, **3**, 163–98.

——. 1964b. Pre-Pickwell Down age of the Plateau Beds (Upper Devonian) in South Wales. *Nature, Lond.*, **204**, 364–66.

——. 1965a. The sedimentation and palaeogeography of the Old Red Sandstone of Anglesey, North Wales. *Proc. Yorks geol. Polytech. Soc.*, **35**, 139–82.

——. 1965b. A review of the origin and characteristics of recent alluvial sediments. *Sedimentology*, **5**, 89–191.

——. 1965c. Upper Old Red Sandstone (Farlovian) palaeogeography in South Wales and the Welsh Borderland. *J. sedim. Petrol.*, **35**, 167–95.

——. 1970. Studies in fluviatile sedimentation: a comparison of fining-upwards cyclothems, with special reference to coarse-member composition and interpretation. *J. sedim. Petrol.*, **40**, 298–323.

——. 1971. The sedimentation of the Old Red Sandstone in the Forest of Dean. In *Geological Excursions in South Wales and the Forest of Dean*. Bassett, D. A. and M. G. (eds.) Cardiff. 9–19.

——, HALSTEAD (TARLO), L. B. and TURNER, S. 1968. Dittonian ostracoderm fauna from the Brownstones of Wilderness Quarry, Mitcheldean, Gloucestershire. *Proc. geol. Soc.*, No. 1649, 141–51.

—— and TARLO, L. B. 1963. The Downtonian and Dittonian facies of the Welsh Borderland. *Geol. Mag.*, **100**, 129–55.

AUSTIN, J. E. 1925. Notes on the highest Silurian rocks of the Long Mountain. *Proc. Geol. Ass.*, **36**, 381–82.

BAKER, J. W. 1969. Correlation problems of metamorphosed Pre-Cambrian rocks in Wales and S.E. Ireland. *Geol. Mag.*, **106**, 249–59.

BALL, H. W. 1951. The Silurian and Devonian rocks of Turner's Hill and Gornal, South Staffordshire. *Proc. Geol. Ass.*, **62**, 225–36.

—— and DINELEY, D. L. 1952. Notes on the Old Red Sandstone of the Clee Hills. *Proc. Geol. Ass.*, **63**, 207–14.

——, —— and WHITE, E. I. 1961. The Old Red Sandstone of Brown Clee Hill and adjacent area. *Bull. Br. Mus. nat. Hist.*, **A5**, 177–310.

BANKS, R. W. 1856. On the Tilestones, or Downton Sandstones, in the neighbourhood of Kington, and their constituents. *Quart. J. geol. Soc. Lond.*, **12**, 93–101.

BRETZ, J. H., and HORBERG, L. 1949. Caliche in southern New Mexico. *J. Geol.*, **57**, 491–511.

BROWN, C. N. 1956. The origin of caliche in the north-eastern Llano Estacado, Texas. *J. Geol.*, **64**, 1–15.

BULL, W. B. 1964a. Geomorphology of segmented alluvial fans in western Fresno County California. *Prof. Pap. U.S. geol. Surv.*, **352**-E, 89–129.

——. 1964b. Alluvial fans and near-surface subsidence in western Fresno County, California.

Prof. Pap. U.S. geol. Surv., **437**–A, 71 pp.

CANTRILL, T. C., DIXON, E. E. L., THOMAS, H. H. and JONES, O. T. 1916. The geology of the South Wales Coalfield. Part XII. The country around Milford. Mem. Geol. Surv. Gt. Britain, 185 pp.

CAVE, R., and WHITE, D. E. 1971. The exposures of Ludlow rocks and associated beds at Tites Point and near Newnham, Gloucestershire. *Geol. J.*, **7**, 239–54.

CHALONER, W. G., and STREEL, M. 1966. Lower Devonian spores from South Wales. *Argumenta Palaeobotanica*, **1**, 87–101.

CLARKE, B. B. 1951. The geology of Garnons Hill and some observations on the formation of the Downtonian rocks of Herefordshire. *Trans. Woolhope Nat. Fld Club*, **33**, 97–111.

——. 1952. The geology of Dinmore Hill, Herefordshire, with a description of a new myriapod from the Dittonian rocks there. *Trans. Woolhope Nat. Fld Club*, **33**, 222–36.

——. 1955. The Old Red Sandstone of the Merbach ridge, Herefordshire, with an account of the Middlewood Sandstone, a new fossiliferous horizon 500 feet below the Psammosteus Limestone. *Trans. Woolhope Nat. Fld Club*, **34**, 195–218.

CROFT, W. N. 1953. Breconian: a stage name of the Old Red Sandstone. *Geol. Mag.*, **90**, 429–32.

—— and LANG, W. D. 1942. The Lower Devonian flora of the Senni Beds of Monmouthshire and Breconshire. *Phil. Trans. Roy. Soc.*, **B231**, 131–63.

CUMMINS, W. A. 1957. The Denbigh Grits: Wenlock graywackes in Wales. *Geol. Mag.*, **94**, 433–51.

CURTIS, M. L. K. 1955. A review of past research on the Lower Palaeozoic rocks of the Tortworth and Eastern Mendip inliers. *Proc. Bristol Nat. Soc.*, **29**, 71–8.

—— and CAVE, R. 1964. The Silurian-Old Red Sandstone unconformity at Buckover, near Tortworth, Gloucestershire. *Proc. Bristol Nat. Soc.*, **30**, 427–42.

DAS GUPTA, T. 1932. The Salopian graptolite shales of the Long Mountain and similar rocks of Wenlock Edge. *Proc. Geol. Ass.*, **43**, 325–62.

DE LA BECHE, H. T. 1846. On the formation of the rocks of South Wales and southwestern England. *Mem. geol. Surv. Gt. Britain*. **1**, 1–296.

DENNY, C. S. 1965. Alluvial fans in the Death Valley region, California, Nevada. *Prof. Pap. U.S. geol. Surv.*, **466**, 62 pp.

DEWEY, J. F. 1969. Evolution of the Appalachian/Caledonian orogen. *Nature, Lond.*, **222**, 124–29.

DINELEY, D. L., and GOSSAGE, D. W. 1959. The Old Red Sandstone of the Cleobury Mortimer area, Shropshire. *Proc. Geol. Ass.*, **70**, 221–38.

DIXON, E. E. L. 1921. The geology of the South Wales Coalfield. Part XIII. The country around Pembroke and Tenby. *Mem. geol. Surv. Gt. Britain*, 220 pp.

——. 1933. Notes on the geological succession in South Pembrokeshire. *Proc. Geol. Ass.*, **44**, 402–11.

EARP, J. R. 1938. The higher Silurian rocks of the Kerry district, Montgomeryshire. *Quart. J. geol. Soc. Lond.*, **94**, 125–57.

——. 1940. The geology of the southwestern part of Clun Forest. *Quart. J. geol. Soc. Lond.*, **96**, 1–11.

ELLES, G. L., and SLATER, I. L. 1906. The highest Silurian rocks of the Ludlow district. *Quart J. geol. Soc. Lond.*, **62**, 195–221.

FLEET, W. F. 1925. The chief heavy detrital minerals in the rocks of the English Midlands. *Geol. Mag.*, **62**, 98–128.

——. 1926. Petrological notes on the Old Red Sandstone of the West Midlands. *Geol. Mag.*, **63**, 505–16.

GARDINER, C. I. 1927. The Silurian inlier of Woolhope, Herefordshire. *Quart. J. geol. Soc. Lond.*, **83**, 501–29.

GEORGE, T. N. 1928. The Carboniferous outlier at Pen-Cerig-calch. *Geol. Mag.*, **65**, 162–68.

——. 1956. The Namurian Usk Anticline. *Proc. Geol. Ass.*, **66**, 297–314.

GILE, L. H. 1967. Soil of an ancient basin floor near Las Cruces, New Mexico. *Soil Sci.*, **103**, 265–76.

——, PETERSON, F. F. and GROSSMAN, R. B. 1965. The K Horizon: a master soil horizon of carbonate accumulation. *Soil Sci.*, **99**, 74–82.

——, —— and GROSSMAN, R. B. 1966. Morphological and genetic sequences of carbonate accumulation in desert soils. *Soil Sci.*, **101**, 347–60.

GREEN, G. W., and WELCH, F. B. A. 1965. Geology of the country around Wells and Cheddar. *Mem. geol. Surv. Gt. Britain*, 225 pp.

GREENLY, E. 1919. Geology of Anglesey. *Mem. geol. Surv. Gt. Britain*, 980 pp.

GREIG, D. C., WRIGHT, J. E., HAINS, B. A. and MITCHELL, G. H. 1968. Geology of the country around Church Stretton, Craven Arms, Wenlock Edge and Brown Clee. *Mem. geol. Surv. Gt. Britain*, 379 pp.

HEARD, A. 1927. On Old Red Sandstone plants showing structure from Brecon, South Wales. *Quart. J. geol. Soc. Lond.*, **83**, 195–207.

—— and DAVIES, R. 1924. The Old Red Sandstone of the Cardiff district. *Quart. J. geol. Soc. Lond.*, **80**, 489–515.

HOLLAND, C. H. 1959. The Ludlovian and Downtonian rocks of the Knighton district, Radnorshire. *Quart. J. geol. Soc. Lond.*, **114**, 449–78.

—— and LAWSON, J. D. 1963. Facies patterns in the Ludlovian of Wales and the Welsh Borderland. *Lpool Manchr geol. J.*, **3**, 269–88.

JONES, O. T. 1956. The geological evolution of Wales and the adjacent regions. *Quart. J. geol. Soc. Lond.*, **111**, 323–51.

KELLAWAY, G. A., and WELCH, F. B. A. 1955. The Upper Old Red Sandstone and Lower Carboniferous rocks of Bristol and the Mendips compared with those of Chepstow and the Forest of Dean. *Bull. geol. Surv. Gt. Br.*, No. 9, 1–21.

KING, W. W. 1921. The geology of Trimpley. *Trans. Worcs. Nat. Club*, **7**, 319–22.

——. 1925. Notes on the "Old Red Sandstone" of Shropshire. *Proc. Geol. Ass.*, **36**, 383–89.

——. 1934. The Downtonian and Dittonian strata of Great Britain and Northwestern Europe. *Quart. J. geol. Soc. Lond.*, **90**, 526–70.

—— and LEWIS, W. J. 1912. The uppermost Silurian and Old Red Sandstone of South Staffordshire. *Geol. Mag.*, (5) **9**, 437–91.

—— and ——. 1917. The Downtonian of South Staffordshire. *Proc. Bgham nat. Hist. phil. Soc.*, **14**, 90–99.

LAWSON, J. D. 1954. The Silurian succession at Gorsley (Herefordshire). *Geol. Mag.*, **91**, 227–37.

——. 1955. The geology of the May Hill Inlier. *Quart. J. geol. Soc. Lond.*, **111**, 85–113.

LUSTIG, L. K. 1965. Clastic sedimentation in Deep Springs Valley, California. *Prof. Pap. U.S. geol. Surv.*, 352–F, 131–92.

MITCHELL, G. H., POCOCK, R. W. and TAYLOR, J. H. 1961. Geology of the country around Droitwich, Abberley and Kidderminster. *Mem. geol. Surv. Gt. Britain.* 137 pp.

MORTIMER, M. G. 1967. Some Lower Devonian microfloras from Southern Britain. *Rev. Palaeobotan. Palynol.*, **1**, 95–109.

MOSELEY, F. 1965. Plateau calcrete, calcreted gravels, cemented dunes and related deposits of the Maalegh-Bomba region of Libya. *Z. Geomorph.*, **9**, 166–85.

MURCHISON, R. I. 1839. *The Silurian System.* London, 523 pp.

OKADA, H. 1967. Composition and cementation of some Lower Palaeozoic grits in Wales. *Mem. Fac. Sci. Kyushu Univ.*, **D18**, 261–76.

PALMER, D. C. 1970. A stratigraphical synopsis of the Long Mountain, Montgomeryshire. *Proc. Geol. Soc.* No. 1660, 341–6.

PHILLIPS, J. 1848. The Malvern Hills, compared with the Palaeozoic districts of Abberley. Woolhope, May Hill, Tortworth and Usk. Mem. Geol. Surv. Gt. Britain, **2**, 1–330.

PHIPPS, C. B., and REEVE, F. A. E. 1967. Stratigraphy and geological history of the Malvern, Abberley and Ledbury Hills. *Geol. J.*, **5**, 339–68.

PICK, M. C. 1964. The stratigraphy and sedimentary features of the Old Red Sandstone, Portishead coastal section, north-east Somerset. *Proc. Geol. Ass.*, **75**, 199–221.

PIPER, G. H. 1898. The Passage Beds at Ledbury. *Trans. Woolhope Nat. Fld Club*, for 1895–97, pp. 310–13.

POCOCK, R. W., WHITEHEAD, T. H., WEDD, C. B. and ROBERTSON, T. 1938. Shrewsbury district, including the Hanwood Coalfield. *Mem geol. Surv. Gt. Britain*, 297 pp.

POTTER, J. F., and PRICE, J. H. 1965. Comparative sections through rocks of Ludlovian-Downtonian age in the Llandovery and Llandeilo districts. *Proc. Geol. Ass.*, **76**, 379–401.

PRINGLE, J., and GEORGE, T. N. 1948. *British Regional Geology: South Wales*. London, 2nd ed., 100 pp.

RAST, N., and CRIMES, T. P. 1969. Caledonian orogenic episodes in the British Isles and northwestern France and their tectonic and chronological interpretation. *Tectonophysics*, **7**, 277–307.

REYNOLDS, S. H. 1907. A Silurian inlier in the Eastern Mendips. *Quart. J. geol. Soc. Lond.*, **63**, 217–38.

——. 1912. Further work on the Silurian rocks of the Eastern Mendips. *Proc. Bristol Nat. Soc.*, (4) **3**, 76–82.

RICHARDSON, J. B. 1967. Some British Lower Devonian spore assemblages and their stratigraphic significance. *Rev. Palaeobotan. Palynol.*, **1**, 111–29.

—— and LISTER, T. R. 1969. Upper Silurian and Lower Devonian spore assemblages from the Welsh Borderland and South Wales. *Palaeontology*, **12**, 201–52.

ROBERTSON, T. 1927a. The highest Silurian rocks of the Wenlock district. *Mem. geol. Surv. Summ. Prog.*, for 1926. pp. 80–97.

——. 1927b. The geology of the South Wales Coalfield. Part II. Abergavenny. *Mem. geol. Surv. Gt. Britain*, 2nd ed., 145 pp.

——. 1932. The geology of the South Wales Coalfield. Part V. The country around Merthyr Tydfil. *Mem. geol. Surv. Gt. Britain*, 2nd ed., 283 pp.

RUHE, R. V. 1964. Landscape morphology and alluvial deposits in southern New Mexico. *Ann. Ass. Am. Geogr.*, **54**, 147–59.

SALTER, J. W. 1863. On the Upper Old Red Sandstone and Upper Devonian rocks. *Quart. J. geol. Soc. Lond.*, **19**, 474–96.

SANZEN-BAKER, I. 1972. Stratigraphical relationships and sedimentary environments of the Silurian–early Old Red Sandstone of Pembrokeshire. *Proc. Geol. Ass.*, **83**, 139–64, 479.

SCHMIDT, W. 1959. Grundlagen einer Pteraspiden-Stratigraphie im Unterdevon der Rheinischen Geosynklinale. *Fortschr. geol. Rheinld Westf.*, **5**, 1–82.

SHAW, R. W. L. 1969. Beyrichian ostracodes from the Downtonian of Shropshire. *Geol. För. Stockh. Förh.*, **91**, 52–72.

SIMPSON, S. 1951. Some solved and unsolved problems of the stratigraphy of the marine Devonian in Great Britain. *Abh. senckenberg. naturforsch. Ges.*, **485**, 53–66.

SOLLAS, W. J. 1879. On the Silurian district of Rhymney and Pen-y-lan, Cardiff. *Quart. J. geol. Soc. Lond.*, **35**, 475–507.

SQUIRRELL, H. C., and DOWNING, R. A. 1969. Geology of the South Wales Coalfield. Part I. The country around Newport (Mon.). *Mem. geol. Surv. Gt. Britain*, 3rd ed., 333 pp.

—— and TUCKER, E. V. 1960. The geology of the Woolhope inlier, Herefordshire. *Quart. J. geol. Soc. Lond.*, **116**, 139–81.

STAMP, L. D. 1919. The highest Silurian rocks of the Clun-Forest district (Shropshire). *Quart. J. geol. Soc. Lond.*, **74**, 221–44.

——. 1923. The base of the Devonian with special reference to the Welsh Borderland. *Geol. Mag.*, **60**, 276–82, 331–36, 367–72, 385–410.

STRAHAN, A. 1907. The geology of the South Wales Coalfield. Part IX. West Gower and the country around Pembrey. *Mem. geol. Surv. Gt. Britain*, 50 pp.

——, and CANTRILL, T. C. 1912. The geology of the South Wales Coalfield. Part III. The country around Cardiff. *Mem. geol. Surv. Gt. Britain*, 2nd ed., 157 pp.

——, CANTRILL, T. C., DIXON, E. E. L. and THOMAS, H. H. 1909. The geology of the South Wales Coalfield. Part X. The country around Carmarthen. *Mem. geol. Surv. Gt. Britain*, 177 pp.

——, CANTRILL, T. C., DIXON, E. E. L., THOMAS, H. H. and JONES, O. T. 1914. The geology of the South Wales Coalfield. Part XI. The country around Haverfordwest. *Mem. geol. Surv. Gt. Britain*, 262 pp.

——, CANTRILL, T. C. and THOMAS, H. H. 1907. The geology of the South Wales Coalfield. Part VII. The country around Ammanford. *Mem. geol. Surv. Gt. Britain*, 246 pp.

STRAW, S. H. 1930. The Siluro-Devonian boundary in south-central Wales. *J. Manchr. geol. Ass.*, **1**, 79–102.

——. 1953. The Silurian succession at Cwm Craig Ddu (Breconshire). *Lpool. Manchr. geol. J.*, **1**, 208–19.

SWINEFORD, A., LEONARD, A. B. and FRYE, J. C. 1958. Petrology of the Pliocene pisolitic limestone in the Great Plains. *Bull. Kans Univ. geol. Surv.*, **130**, 98–116.

SYMONDS, W. S. 1872. Records of the Rocks, London.

TAYLOR, K. 1972. New fossiliferous localities in the Upper Old Red Sandstone of the Ystradfellte —Cwm Taff district of Breconshire. *Bull. geol. Surv. Gt. Br.* No. 38, 11–14.

TROTTER, F. M. 1942. Geology of the Forest of Dean Coal and Iron Field. *Mem. geol. Surv. Gt. Britain*, 95 pp.

VAN DE KAMP, P. C. 1969. The Silurian volcanic rocks of the Mendip Hills, Somerset; and the Tortworth area, Gloucestershire, England. *Geol. Mag.*, **106**, 542–53.

WALMSLEY, V. G. 1959. The geology of the Usk Inlier (Monmouthshire). *Quart. J. geol. Soc. Lond.*, **114**, 483–521.

——. 1962. Upper Silurian-Devonian contacts in the Welsh Borderland and South Wales. *Symposium Silur-Devon-Grenze*, Stuttgart, pp. 288–95.

WEBBY, B. D. 1966. Middle-Upper Devonian palaeogeography of North Devon and West Somerset, England. *Palaeogeography, Palaeoclimatol. Palaeoecol.*, **2**, 27–46.

WALDER, P. S. 1941. The petrography, origin and conditions of deposition of a sandstone of Downtonian age. *Proc. Geol. Ass.*, **52**, 245–56.

WALLIS, F. S. 1928. On the Old Red Sandstone of the Bristol district. *Quart. J. geol. Soc. Lond.*, **83**, 760–87.

WEDD, C. B., and KING, W. B. R. 1924. The geology of the country around Flint, Hawarden and Caergwrle. *Mem geol. Surv. Gt. Britain*, 222 pp.

——, SMITH, B., KING, W. B. R. and WRAY, D. A. 1929. The country around Oswestry. *Mem. geol. Surv. Gt. Britain*, 234 pp.

——, SMITH, B. and WILLS, L. J. 1927. The geology of the country around Wrexham. *Mem. geol. Surv. Gt. Britain*, 179 pp.

WELCH, F. B. A., and TROTTER, F. M. 1961. Geology of the country around Monmouth and Chepstow. *Mem. geol. Surv. Gt. Britain*, 164 pp.

WHITAKER, J. H. McD. 1962. The geology of the area around Leintwardine, Herefordshire. *Quart. J. geol. Soc. Lond.*, **118**, 319–47.

WHITE, E. I. 1938. New pteraspids from South Wales. *Quart. J. geol. Soc. Lond.*, **94**, 85–114.

——. 1946. The genus *Phialaspis* and the "*Psammosteus*" Limestones. *Quart. J. geol. Soc. Lond.*, **101**, 207–42.

——. 1950a. The vertebrate faunas of the Lower Old Red Sandstone of the Welsh Borders. *Bull. Br. Mus. Nat. Hist.*, **A1**, 51–67.

——. 1950b. *Pteraspis leathensis* White: a Dittonian zone-fossil. *Bull. Br. Mus. nat. Hist.*, **A1**, 69–89.

——. 1951. The vertebrate faunas of the Old Red Sandstone of the Welsh Borders. *Rept. Intern. Geol. Congr.*, 18*th*, London, **11**, 21.

——. 1956. Preliminary note on the range of pteraspids in Western Europe. *Bull. Inst. Sci. nat· Belg.*, **32** (10), 1–10.

—— and TOOMBS, H. A. 1948. Guide to excursion C.16 Vertebrate Palaeontology. *Rept. Intern. Geol. Congr.*, 18*th*, London, pp. 4–14.

WHITEHEAD, T. H., and POCOCK, R. W. 1947. Dudley and Bridgenorth. *Mem. geol. Surv. Gt. Britain*, 226 pp.

——, ROBERTSON, T, POCOCK, R. W. and DIXON, E. E. L. 1928. The country between

Wolverhampton and Oakengates. *Mem. geol. Surv. Gt. Britain*, 224 pp.

WHITTARD, W. F., and SMITH, S. 1944. Unrecorded inliers of Silurian rocks, near Wickwar, Gloucestershire, with notes on the occurrence of a stromatolite. *Geol. Mag.*, **81**, 65–76.

WILLIAMS, B. P. J. 1971. Sedimentary features of the Old Red Sandstone and Lower Limestone Shales of South Pembrokeshire, south of the Ritec Fault. In *Geological Excursions in South Wales and the Forest of Dean*, Cardiff (Geologists Association, South Wales Group). pp. 222–39.

WILLIAMS, D. M. 1926. Note on the relation of the Upper and Lower Old Red Sandstone of Gower. *Geol. Mag.*, **63**, 219–33.

WILLIAMS, G. E. 1970. Piedmont sedimentation and Late Quaternary chronology in the Biskra region of the northern Sahara. *Z. Geomorph.*, **S10**, 40–63.

ZIEGLER, A. M. 1970. Geosynclinal development of the British Isles during the Silurian period. *J. Geol.*, **78**, 445–79.

(Professor Allen's manuscript was received in the autumn of 1971.)

LOWER CARBONIFEROUS ROCKS IN WALES

T. Neville George

I. ST. GEORGE'S LAND

(a) Devonian foundations

LOWER Carboniferous rocks, the 'Carboniferous Limestone series', have only a peripheral distribution in Wales. They follow the eastern fringe of North Wales from Llanymynech by Oswestry and Llangollen to the Eglwyseg escarpment and Minera and along the eastern flanks of the Clwydian range; they run into the Vale of Clwyd; they follow the coast from Abergele to the Great Orme's Head; and they occupy tracts in Anglesey and in neighbouring Arfon. They rim the South Wales coalfield, and form relatively wide outcrops in the Vale of Glamorgan, Gower, and southern Pembrokeshire. But they are not to be found in mid-Wales where, as their stratigraphical variations show, they formed at best a discontinuous cover, and were over a great part probably not deposited.

The empty region is St. George's Land, the 'massif' of palaeogeographical reconstruction. The origin of the massif as an upwarp above depositional level lay in the Caledonian movements, the more important of mid-Devonian age. The transition from Silurian to Devonian environments in Wales was spectacular in the replacement of grey richly fossiliferous marine shales by almost unfossiliferous red beds; but between Much Wenlock and Llandovery the Downtonian rocks of the Old Red Sandstone rest conformably on the Upper Ludlow Shales, and the transition was less the product of major deformation than of gentle emergence. In the most north-westerly outcrops along the Epynt front, as in Clun Forest, the Downtonian rocks are several hundred metres thick and at deposition continued some distance into what is now the heart of the massif. Thick Lower Old Red Sandstone in the Clee Hills also indicates continuity of deposition northwards far beyond the limits of present outcrop: it may well persist at depth beneath the thick sequence of Carboniferous rocks now preserved in the Lancashire basin.

In these relations there is not much sign of St. George's Land, except perhaps in the occurrence of conglomerates with pebbles of vein quartz and metamorphic rocks of northerly or north-westerly provenance. Only along the basal outcrop westwards from Llandeilo into Pembrokeshire is there strong overstep by the Lower Old Red Sandstone and clear sign of acute early-Devonian tectonic deformation by sharp folding and faulting of Silurian rocks, and an emergence of a core of St. George's Land—the deformation occurring along the margin of the Silurian geosyncline where earlier (pre-Llandovery) movement was also acute.

Mid-Devonian restlessness transformed the tectonic pattern, Middle Old Red Sandstone being unrepresented in Wales, and Upper Old Red Sandstone descending with strong overstep across all older rocks down to Precambrian in Anglesey. The amplitude of some of the contemporary Pembrokeshire folds is indirectly measurable in the order of thousands of metres, with evidence of complementary movement along the Ritec fault (beyond which Skrinkle Sandstones are not known to occur),

85

and the Benton fault. Northwards from the Brecon Beacons and the Black Mountains there is comparable discordance into the Clee Hills where the greater part of the Brecon Series is cut out by Farlovian overstep. How far north-westwards into the massif the Lower Old Red Sandstone survived transgression by the Upper, and the Farlovian rocks persisted to a terminal featheredge, is unknown; but along the north crop in South Wales the Upper formation is thin—locally less than 10 m— and in its lithology it is rich in quartz conglomerates whose pebbles may well have been derived from erosion of Lower Old Red Sandstone and Lower Palaeozoic rocks of the massif; and there is little doubt that the emerged St. George's Land was in its frame established during mid-Devonian times.

(b) Carboniferous transgression

Marine Carboniferous conditions were introduced into South Wales by minor subsidence and a northward advance of the Devonian sea in gentle and fluctuating overlap onto the wide, flat, almost horizontal spreads of Skrinkle and Farlow sandstones, the influx of terrigenes progressively diminishing and calcareous rocks, many of them biogenic, becoming the dominant kinds. 'Carboniferous Limestone' was the characteristic lithological type in the shelf seas flanking St. George's Land, a type in utter contrast to the terrigenes of the Old Red Sandstone, and formed in very different ways, but whose tectono-palaeogeographical context matched very closely the context in which the Upper Old Red Sandstone was laid down.

Within this cuvette of shelf-sea deposits the limestones differ very widely amongst themselves in their constituents and in the evidence they provide of contrasted factors controlling their formation; and in a variety of ways they display ecological and environmental influences that are to be interpreted in a stratigraphy of change in both vertical sequence and lateral facies. But however the rock-types in their differences may provide classifactory bases for the members and formations of a stratigraphical sequence, they all consistently demonstrate lateral passage northwards and north-westwards in a progressive thinning towards the shores of St. George's Land, from more than 1300 m in southernmost Pembrokeshire and not much less in Gower to a final disappearance in nil thickness near Haverfordwest and in the Clees.

Evidence of the continuity of advance of the Carboniferous seas in North Wales is imperfect. Old Red Sandstone being found only in a small outcrop in Anglesey, palaeogeographical inheritance from a Devonian environment is unknown. Moreover, the massif as land was long-sustained, Tournaisian rocks are not represented, and there is gross unconformity between the first members of the Carboniferous Limestone and the Lower Palaeozoic and Precambrian rocks (and the residual Old Red Sandstone): the northern flanks of St. George's Land were not submerged until relatively late in Dinantian (Viséan) times. Further, within the Viséan rocks preserved there is a southward and southwestward thinning of the formations from 700 m or more in the Clwydian range and in parts of Anglesey to little more than 200 m in the southernmost residual outcrops near Oswestry and Corwen; and the limestones display an accompanying internal overlap against a rising shore of St. George's Land, so that in places only the higher part of the *Dibunophyllum*

Zone (D_2) remains in unconformable contact with the floor of Silurian rocks.

The stratigraphical evidence is thus convincing basis on which the massif of St. George's Land may be reconstructed as a major barrier between the two relatively isolated cuvettes of sedimentation. It is strongly supported by the lithological and palaeontological evidence of contrasts in the varieties of limestone kinds and fossil associations to be found in the two cuvettes, South Wales closely linking with the Forest of Dean, the Bristol district, and the Mendips as part of the South-Western Province of Avonian sedimentation, North Wales linking closely with Derbyshire, the Clitheroe–Craven country, and the Dublin basin as part of the Central Province.

(c) The persistent barrier

The calcareous purity of the Carboniferous Limestone, relatively free of much terrigenous detritus, indicates shelf seas lapping against a hinterland of inconsiderable relief that was progressively diminished as marine transgression continued. The overlap displayed by the rocks particularly in north-east Wales is a stratigraphical sign of the drowning of the massif (see Wedd and others 1927, fig. 12, p. 118). Nevertheless, as the barrier between north and south persisted unbrokenly as a control on sedimentation, a core of St. George's Land remained unsubmerged probably throughout the Upper Palaeozoic era; and there was repeated renewal of the massif by pulsed upwarp to offset the regional subsidence beneath the Dinantian seas. In South Wales the 'hinge' between hinterland and cuvette remained surprisingly restricted to a narrowly fluctuating belt as Avonian times advanced, and it is probable that in combined non-deposition and intra-Avonian erosion the Avonian 'shore' was never far north of the present-day north crop (see George 1958, fig. 19; 1972, figs. 1, 2).

In both North Wales and South Wales St. George's Land neither grew nor was renewed by simple upwarp. The outcrops in Powys, from Llandegla to the head of the head of the Vale of Clwyd, show variations in thickness and in local stages of overlap pointing to minor swells and synclines in the Lower Palaeozoic floor that developed as sedimentation continued: it is perhaps significant that some of these variations are in subdued posthumous accordance with Caledonian structure in the underlying Silurian and Ordovician rocks (George 1961, fig. 25). In Anglesey also a thickening of the limestones in the Menaian outcrops implies a local sag between the anticlinal Bangor ridge of Precambrian rocks and the anticlinal Penmynydd zone to the north. (See Fig. 11.)

Structures supplementary to the main massif in South Wales are identified in renewed movement along the Ritec fault, and in minor tilts and sags athwart the main east-and-west 'shoreline' expressed in the variability of stages of thinning and internal non-sequence as the Avonian zones are traced from south to north (George 1969, fig. 5). The most considerable corrugation on the margin of the massif was the malvernoid Usk anticline, a structure that achieved its maximum expression in late- or post-Carboniferous times but that in continuing movement was instrumental in disrupting stratigraphical sequence probably from mid-Avonian times onwards, especially through Namurian overstep (George 1956).

The palaeogeographical profile of the cuvette in South Wales was controlled

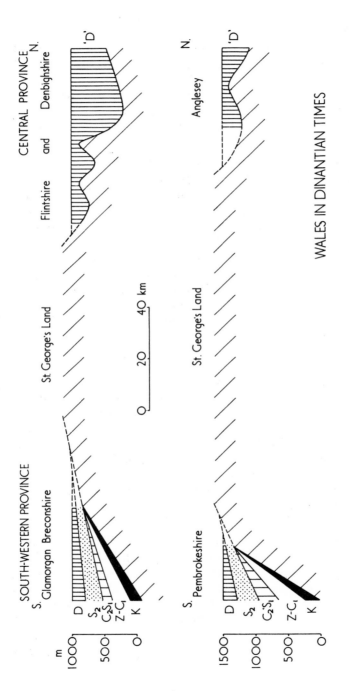

Fig. 11.   Dinantian contrasts on opposite flanks of St. George's Land. The Lower Carboniferous sequence in South Wales, deposits in full zonal range, contrasts with the restricted sequence, upper Visean only, of North Wales; but the thickness, despite the absence of most of the zones, is almost as great in North Wales as in South. Internal overstep is diagrammatically illustrated in South Wales.

by a northern 'shoreline' and a growing Usk axis to the east. The possibility is strong that a major promontory extending from the core of St. George's Land southwestwards into Dyfed was separated from a Leinster massif by a Dinantian basin in Cardigan Bay and the southern Irish Sea. Between the promontory and the Usk axis a broad embayment allowed deposition particularly of Seminulan rocks to extend for some distance north of the present north crop: the embayment, sharply contained by the Usk anticline to the east, was to continue as a prominent control on sedimentation into Upper Carboniferous times. (See George 1960, p. 361; 1972, fig. 1.)

Upwarping along the Usk axis began early, hints being seen in the Tournaisian rocks (see p. 94); but its effects became impressive only in later Viséan times, and the 'normal' north-crop sequence preserved in the K and Z sediments of the isolated outlier of Pen-cerig-calch is an indication of the north-easterly continuation of a depositional and facies trend approximately aligned towards the Titterstone outcrops and thence the Central Province. In mid-Dinantian movement an incipient Neath disturbance (later to become a major Hercynian zone of fracture) had greater effect as measured by the pulsed stages of overstep, against a rising St. George's Land, beneath Upper Caninian and Seminulan grits, and by variations in thickness of the *Dibunophyllum* Zone (George 1954, pp. 305, 311; Owen 1964, p. 303).

It is possible that along its eastern front St. George's Land was an incomplete barrier during early Tournaisian times, and the Lower Limestone Shales and perhaps the Oreton Limestone, like the Old Red Sandstone, may have continued northwards into the Central Province (George 1956, fig. 4). Similarly, the embayment of the southern Irish Sea may have separated the Leinster massif from St. George's Land and spilled through a strait into the Dublin basin (George 1960. p. 361). Otherwise types and fossils suggest that connection between South and North Wales was at best devious when the Carboniferous Limestone was being deposited, and only the *Dibunophyllum* Zone displays likenesses between the two cuvettes sufficiently close to suggest that links may have been relatively direct.

II. THE ROCKS IN SOUTH WALES

(a) Sequence and facies

Lower Carboniferous (Dinantian) sediments in South Wales begin in the Old Red Sandstone or the marine 'Devonian' Skrinkle Sandstones, lithostratigraphy not being accordant with chronostratigraphy (George 1972, p. 224): the base of the 'Carboniferous Limestone Series' is not the base of the Tournaisian Series, and the Lower Limestone Shales are a facies designedly coincident with the Avonian *Cleistopora* Zone. The upward transition from Old Red Sandstone is therefore to be interpreted as diachronous, the *Cyrtospirifer verneuli* band in the Plateau Beds of the Upper Old Red Sandstone at the head of the Tawe valley, some 10 m below the Lower Limestone Shales, being probably Famennian, the band with spirifers, *Syringothyris*, and camarotoechiids in the Upper Old Red Sandstone near Risca being probably Tournaisian (Squirrell & Downing 1969, p. 55). The Grey Grits, nominally Upper Old Red Sandstone, are, as Cantrill long ago surmised, almost certainly Carboniferous in age—but not because of their colour.

Unbroken transition, and thoroughly arbitrary distinction, between Skrinkle Sandstones or Plateau Beds and Lower Limestone Shales, are countered by abrupt change where the Upper Devonian and Upper Old Red Sandstones are missing (presumably through non-deposition and Carboniferous overlap); and an initiatory 'shoreline' of the Shales, the first sign of marine advance onto an eroded land surface, is to be recognised, locally with the development of a basal conglomerate, along the north crop for 50 km between Templeton (where the Shales rest on Dittonian Red Marls) and Llandebie (where they rest on Breconian Brownstones). Elsewhere, a passage to grey beds, and a first appearance or an increase in numbers of fossils, are usually the conventional determinants of formational names.

The Lower Limestone Shales in bulk commonly give an appearance of a monotonous alternation of calcareous mudstones, silty shales, impure muddy and sandy limestones, and clean oolitic limestones, to thicknesses of 150 m in southern outcrops. They are in detail, however, highly varied both in a cyclothemic lithology and in palaeoecological assemblage suites. The rhythms of banding especially in the lower part of the formation are probably to be attributed to pulses of terrigenous influx, the 'high-energy' intervals characterised by gritty and quartzitic layers, the quiescent intervals by purer limestones. Some of the beds, lubricated and unstable, collapsed to form slumped and brecciated rocks as they were deposited (see Kelling and Williams 1966).

Palaeoecologically a fossiliferous layer may be dominated by one or two species of brachiopod (bedding planes with abundant chonetids are common), or they may carry closely packed nests of camarotoechiids, or they may contain a drifted association of several species, or they may be virtually without brachiopods but swarm with a few species of bivalves; and it is evident that local biotopes were sharply different, under environmental controls not now evident, as sources of fossils preserved in the matrix of shales that otherwise are not readily distinguished. Conversely, coarse sandy layers may contain an association of common fossils scarcely different from an association in shales a metre or two above or below. Some of the beds are bioturbidites, the lamination almost completely destroyed and the texture disordered by a churned reworking of the constituent grains.

The great northward reduction in thickness of the Lower Limestone Shales, from some 150 m in the southern Pembrokeshire outcrops and in Gower to 30 m along the north-east crop and to 15 m at Pendine, appears to be due not to broken but to slow sedimentation, equated with proximity to the hinge 'shoreline' of St. George's Land. (See Fig. 12.)

The general uniformity of the Lower Limestone Shales is followed by the diversity of rock-types found in the *Zaphrentis* and Lower *Caninia* zones of the Lower Avonian (Main Limestone) sequence. In 'standard' facies the Main Limestone is dominated by crinoidal limestones in which an abundance of corals (the tabulates *Syringopora* and *Michelinia* with the rugosan zaphrentoids and caniniids) and brachiopods (notably chonetids, orthoids, spirifers, athyrids, and camarotoechiids) is compelling evidence of a well-aerated neritic sedimentary environment. Distinction from the Lower Limestone Shales is not absolute, many beds of calcareous shale

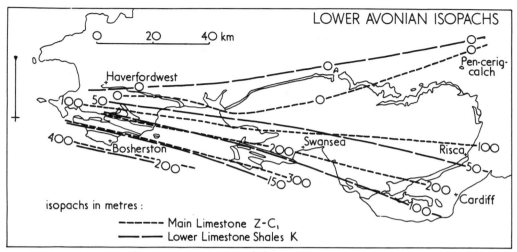

Fig. 12. Isopachs of the Lower Avonian rocks in South Wales, residual after Upper Avonian overstep. Thicknesses are not everywhere precisely determinable, and beyond the limits of outcrop, and at some places within them, the isopachs are conjectural; but the order of uncertainty in locating the isopachs lies within a few kilometres.

alternating with the more massive limestones, especially in the *Zaphrentis* Zone.

The 'standard' facies is found only in southern outcrops—southern Pembroke-shire, Gower, and the Vale of Glamorgan. Lateral passage into dolomites (especially in the Lower *Caninia* Zone, in which the *laminosa* Dolomite is prominent) is strong in the Tenby and Sageston synclines, in eastern Gower, and particularly along the south crop between Bridgend and Risca. The crinoid limestones, dolomitised or not, give place northwards to oolites formed in an evaporitic environment, the whole (residual) Tournaisian Main Limestone of the north-east crop being oolitic, and the *Caninia* Oolite being a major formation of the facies that extended southwards over the whole of the cuvette except southernmost Pembrokeshire. The lateral changes, spectacular in the contrasts in rock type as signs of shallows and calcite pans on the flat banks fronting St. George's Land, perhaps also with repeated cryptic nonsequences in the rocks of the north-east crop are reflected in a great reduction in thickness northwards from about 150 m to 20 m at Pendine (where it is residual beneath transgressive Upper Avonian rocks). (See Fig. 13.)

There was full transition from Lower to Upper Avonian only in southernmost Pembrokeshire (where, nevertheless, the only reefs known in South Wales mark the passage as having been in very shallow water): elsewhere a pulse of uplift caused a southward retreat of the sea, whose subsequent advance over an eroded surface of the Oolite Group is recorded in the peculiar facies of the Calcite-Mudstone Group (see p. 98), in the channelled floor of the Pendine Conglomerate, and in the 'mid-Avonian unconformity'. The Upper *Caninia* Zone has a more limited dis-tribution over most of South Wales than any other zone, in part because of post-depositional erosion to which renewed movement along lines of faulting may have contributed (Sullivan 1966, p. 232; Owen 1971, p. 1307); and although it is

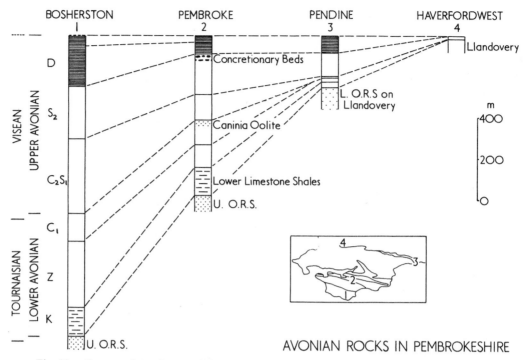

Fig. 13.   Comparative columns of the Avonian rocks in Pembrokeshire. The northward thinning is the product partly of differential subsidence, partly of the development of unconformity notably at the base of the *Seminula* (S₂) Zone. (Mainly after the memoirs of the Geological Survey.)

dominantly crinoidal with a rich mixed fauna and (like the Lower *Caninia* Zone in southern outcrop) is characteristically neritic, it is in residual development restricted to the more southerly part of the shelf. The underlying Calcite-Mudstone Group, despite its deposition at negligible depth, is in contrast found from Pembrokeshire and Gower into the extreme north-east crop, where it is at its thickest; and while the crinoidal limestones are absent along the north crop certainly in part because of overstep by the rocks above, they may also be absent in part by lateral change into the diachronous facies of the calcite mudstones.

A restlessness of the shelf floor was continued into Seminulan times, both in the lithological contrasts between the 'Seminula Oolite' and the associated pisolites, algal limestones, and calcite mudstones with the crinoid limestones beneath, and in major structural discordance with overstep by the *Seminula* Zone across all earlier Carboniferous rocks and Old Red Sandstone down to Silurian, along the whole of the north crop (George 1958, p. 261). The tectonic controls on sedimentation are very well illustrated in the thinning of the zone northwards from some 300 m in southernmost outcrops to 100 m in northern; in the intensity of overstep along the Haverfordwest outcrop (the zone resting on a Silurian floor), only 15 km (but with Hercynian foreshortening) from the unbroken and very thick Avonian sequence in the Bosherston syncline; in the intercalations of crinoidal and shelly limestones in the massive oolites of the zone in Gower; and in a replacement of the oolites of much of the zone by thick crinoid, coral, and shelly limestones in southernmost Pembrokeshire (see George 1970, fig. 21).

The Seminulan rocks display in closest correlation the kind of sedimentary facies and the rate of subsidence of the cuvette floor: mainly as oolites and calcite mudstones, deposited at very shallow depths—never more than a few metres— they show little change in the regional environment of 'banks' and 'lagoons' as the endogenic sediments, not linked in origin (as terrigenes would be) with pulsed revival of the hinterland, accumulated through hundreds of metres in a nice balance between evaporitic concentration, calcite-pan flats, bulk of accumulate, and subsiding floor.

The loss of that balance is indicated by the very rapid, almost abrupt, change from Seminulan rocks to the highly fossiliferous bioclastic rocks of the *Dibunophyllum* Zone, in which a great variety of corals, brachiopods, molluscs, polyzoans, and foraminifers is a sign of a reversion to open seas and refreshed waters. Even along the north crop the neritic character of much of the upper (D_2) part of the zone persists, the many kinds of fossils (including many species of corals) being perhaps surprising in such near-shore deposits (see George 1927, pp. 49, 77). Local variation in the neritic facies includes emphasis on productids in some beds, on zaphrentoid corals in others, and in a dominance of bivalves over both brachiopods and corals in part of the ground between Kidwelly and Llandebie.

The controls of St. George's Land are to be seen in a thinning of the zone northwards from over 200 m in the Bosherston syncline and in Gower to 85 m at Pendine and to 60 m at Kidwelly and Penwyllt (George 1958, fig. 19). The lower part of the zone (D_1), 160 m thick at Bosherston where it is dominantly bioclastic, shows the greatest change, being less than 30 m thick, in the east less than 20 m, along the north crop between Kidwelly and Penderyn, and becoming (as the Light Oolite) very strongly oolitic with specimens of *Linoproductus hemisphericus* some- times in profusion but with few other fossils. A more 'normal' crinoidal type of D_1 beds, with a 'standard' fauna, is a basal member to a thickness of 5 m at Kidwelly, but eastwards it is not recognised beyond Llandebie, its disappearance perhaps being due to lateral passage into the lower part of the Light Oolite or more probably to the development of non-sequence through gentle overstep. (See Fig. 14.)

The uppermost beds of the *Dibunophyllum* Zone (the Upper Limestone Shales, D_3) are everywhere present where they have not been cut out by transgressive Namurian, in singularly uniform facies from Gower and the Vale of Glamorgan to the north crop between Pendine and the Nedd valley. They thin towards St. George's Land to two or three metres, but in the rottenstones of the north crop they contain *Spirifer oystermouthensis*, *Martinia multicostata*, and *Amplexizaphrentis oystermouthensis*, if not quite so abundantly as in the Black Lias of Gower. In age, they are not proved not to continue into Namurian, the relations of the Plastic Clay Beds being ambiguous.

(b) *Arenites and conglomerates*

Calcareous sandstones, sandstones, and conglomerates are relatively common

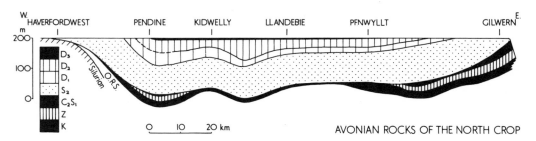

Fig. 14.   Section along the north crop from Pembrokeshire to Monmouthshire. The variation in thickness and sequence reflects repeated overstep, at the base of K, of C_2S_1, of S_2, and of the Namurian. It also reflects differential sagging, in the development of Z and C_2S_1 between Penwyllt and Gilwern. Regionally the broad synclinal form, particularly well shown by the *Seminula* and *Dibunophyllum* zones, marks an embayment into mid Wales that persisted as a palaeogeographical control on sedimentation into Namurian and Westphalian times.

in the sequence along the north crop, at horizons in all the zones present: some of them extend to the south crop. They particularly characterise the lower part of the Lower Limestone Shales, the lower part of the Calcite-Mudstone Group and of the *Seminula* Zone, and (as the Light Oolite) the lower part of the *Dibunophyllum* Zone. Occasionally they are very pure rocks, sometimes quartzitic, but mostly they are calcareous and contain not uncommon fossils in no exceptional facies association, and it is evident that most of them were not flushed-in terrigenes that as delta aprons locally replaced limestone deposits, but were reworked in shallow water with much bioclastic or evaporitic detritus—the Light Oolite, for instance, with abundant productids, the *Linoproductus* Oolite of the Calcite-Mudstone Group locally swarming with athyrids.

The uppermost beds of the Old Red Sandstone, lithologically distinguished from the Lower Limestone Shales, are the first of the sediments in South Wales to show the fluctuating proportions of quartz and detrital calcite that form the typical Lower Carboniferous arenites. In and about Risca they contain *Camarotoechia mithcheldeanensis* but lack the Famennian *Cyrtospirifer verneuli*, with anomalous diachronism in the lithological transition upwards from the Old Red Sandstone, the meaning of 'Carboniferous' not being restricted by lithology. Even as far south as Gower brown and buff basal sandstones contain common brachiopods; and the Skrinkle Sandstones of marine facies in Pembrokeshire cross the chronostratigraphical boundary between the two systems. A local definition of the base of the Avonian, as of the Dinantian, is correspondingly uncertain.

Sandstones and siltstones are diminishingly recurrent in the middle and upper part of the Lower Limestone Shales, and are almost completely absent from the Lower Avonian Main Limestone: even on the north crop the Oolite Group in its shallow-water proximity to the margins of the cuvette remains almost wholly free even of scattered quartz grains. A thin bed of quartz conglomerate in dolomitised members of the Oolite Group near Cwmbran (George 1956, p. 314; Squirrell & Downing 1969, p. 79) is exceptional, perhaps being an early pointer to an incipient Usk upwarp. (See Fig. 15.)

The Calcite-Mudstone Group is relatively clean in southern outcrops, although it carries a much greater abundance of scattered quartzes than the underlying limestones: the coarse conglomerates seen at and near its base particularly in Gower (George 1958, fig. 8) are very clear signs of strong contemporaneous erosion, but they are wholly calcareous and contain pebbles and boulders of the *Caninia* Oolite and associated algal beds immediately underlying, and do not mark the influx of terrigene material of distant provenance. The equivalent conglomerates and gritty beds on the north crop, on the other hand, are very mixed, with pebbles, some locally derived of underlying limestones, sone of vein-quartz, quartzite, and jasper presumably originating in eroded Old Red Sandstone not far to the north (and then implying overstep of all Lower Avonian rocks at no great distance) (George 1954, pp. 298 ff.: Sullivan 1965, pp. 289, 295). The *Linoproductus* Oolite, most of whose ooliths are quartz-cored, is a similar arenite with quartz pebbles: it may be followed along the greater part of the residual north-east crop, and its constituents indicate a similar proximate source.

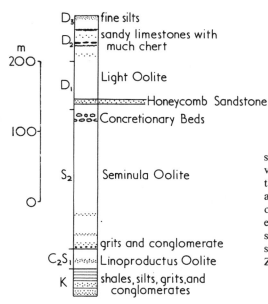

Fig. 15. Generalized column of the Avonian sequence on the eastern part of the north crop. The very mixed kinds of rocks in alternation are a sign of the fluctuating environments of deposition of shallow and near-shore sediments in which terrigenes and oolites are prominent. The Honeycomb Sandstone is extended to suggest the possibility of its continuing southwards into Glamorgan, where arenites are strongly developed at the base of the *Dibunophyllum* Zone locally on the south crop.

ROCK TYPES OF THE NORTH CROP

Seminulan beds are mainly clean limestones; but in their transgressive relations with all the rocks beneath their basal members are sandy and gritty along the greater part of the north crop, a quartz-pebble conglomerate at the northernmost outlier of Careg-yr-ogof (where the residual Lower Limestone Shales are only 5 m thick) confirming exposed Old Red Sandstone nearby at time of deposition.

Arenites in the *Dibunophyllum* Zone include the Honeycomb Sandstone at or near the base—a persistent facies of the Light Oolite along the greater part of the north crop that is perhaps continued southwards underneath the coalfield to appear on the south crop near Bridgend. It is the first of a suite of quartzose rocks that is especially to be seen in the upper part of the zone between Kidwelly and Penderyn, most of the limestones being sandy even when they are richly fossiliferous, and some of them being calcareous grits; but there are no massive sandstones comparable with the Drybrook Sandstone of the neighbouring Forest of Dean. A peculiarity of some of the pseudobreccias, suggesting contemporary fragmentation, is the concentration of quartz grains in the 'fragments'. Many of the quartzes are angular or subangular, transported only over short distances and not greatly reworked. (See George 1927, pp. 60 ff.)

The terrigenes extensively developed along the north crop eastwards from Pembrokeshire are clearly related to an inferred 'shore' of St. George's Land lying not far to the north, the depositional strike running more or less west-to-east. Terrigenous lenticles in the Upper Avonian rocks of the north-east crop appear to be more intimately linked with a growing Usk anticline and perhaps with contemporary movement along the Neath disturbance. The rocks in their matrix are 'normal' limestones, often fossiliferous, but there are included in variable proportion grains and pebbles of quartz sometimes in sufficient abundance to form a mildly calcareous sandstone or a quartz conglomerate not greatly different (except in being calcareous) from the basal grits of the Millstone Grit in superficial appearance. Such rocks are found away from the Usk axis in the *Seminula* Zone westwards beyond the Clydach valley, and in the *Dibunophyllum* Zone as far as the Vale of Neath (Owen & Jones 1961). They strongly imply the unroofing in Seminulan times of the Usk cover of pre-Seminulan rocks and the exposure in the Usk core of Old Red Sandstone—at least in the ground north of Usk itself, a few miles south-east of which a small outlier of Lower Limestone Shales still survives on the east flank of the fold (Welsh aad Trotter 1960, p. 138). If the pebbles of the quartz conglomerate in the dolomitised Oolite Group near Cwmbran (p. 95) are to be attributed to similar source, there was continuing growth of the Usk anticline through the greater part of Avonian times. (See fig. 16.)

(c) Oolite banks and algal flats

Some of the oolites form the 'purest' beds of the Carboniferous Limestone, not merely in consisting almost exclusively of carbonate (to 99 per cent) and in being strictly endogenic, but also in having as constituents only even-szied ooliths. This purity is particularly evident in the Pwll-y-cwm and Gilwern oolites of the north-east crop, and in the thin oolites found in the Lower Avonian beds of Z and lower C_1 age in Glamorgan, whose lithology implies very thorough washing and rewashing and a final resorting characteristic of oolite banks in which fine detritus is winnowed out, the influx of terrigenes is negligible, and a minimum of bioclastic debris suggests an inhospitable environment. Some of the sandy oolites, like the *Linoproductus* Oolite and the Light Oolite of the north crop, have a similar uniformity of grain size and provide evidence of sustained winnowing, the quartzes being of much the same size as the ooliths, and commonly being veneered by a thin oolitic film; but

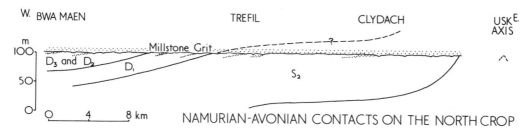

Fig. 16. Transgressive Millstone Grit on Carboniferous Limestone along the eastern part of the north crop. The major overstep, controlled by uplift along the Usk axis, is accompanied by anomalous occurrences of tongues of quartz conglomerate with a fossiliferous calcareous matrix, to be found at various horizons from D_3 down to S_2 at a number of localities. The relations suggest a repeated influx of terrigenes from the north-east related to the contemporaneous uplift of the Usk axis, and perhaps a feather-edge contact between successive Avonian members and the Namurian cover that diachronously migrated up the succession from east to west. (After Owen & Jones.)

the contaminant quartzes are complemented by the abundant indigenous productids, whose shell banks continuing for many kilometres show them to be virtually in place, as do their unabraded valves.

The thicker oolites—the *Caninia* Oolite and the *Seminula* Oolite—on the other hand, although they may be very pure carbonate rocks, are far more varied in their constituents, and include much bioclastic debris. In being well developed in southern outcrops they illustrate relations with laterally mergent biogenic facies that are not to be seen in the regional context of the oolites of the north crop, and they demonstrate that as the sediments of very extensive oolite banks they were not deposited in the isolation of shelves protected from the 'open' seas of the neritic zone by reefs or submarine dune fronts or other barriers to a fluctuating lateral passage. They are commonly richly crinoidal, the better-sorted members showing an inter-mixture of ooliths and crinoid plates of about the same size, and crinoid plates often forming oolith cores; and the 'purer' oolites alternate with beds in which neritic fossils— spirifers, athyrids, chonetids, orthotetids, tabulate and rugose corals, molluscs—lie in a mixed crinoidal-oolithic matrix or in a matrix as free of ooliths as a neritic 'standard' limestone of the *Zaphrentis* and *Caninia* zones. Locally it is possible to see an interfingering of oolitic with crinoidal-shelly beds; and regionally the *Caninia* and *Seminula* oolites lose their character in southern Pembrokeshire, where there is an unbroken thick sequence of bioclastic rocks almost throughout the Avonian series (George 1970, fig. 21). While within the classificatory rock-group of 'limestones' there are the greatest differences between biogenic and evaporitic kinds, both in their constituents and in the manner of their formation, comparatively slight differences in the environment of deposition, particularly of salinity, were sufficient to give rise to the contrasted rock-types, and to allow a palaeogeographical reconstruction that happily combines lithological product with tectonic control.

A third major rock-type is represented by the very mixed suite of algal, ostra-codal, pellet, and 'worm'-bored limestones and calcite-mudstones found at various horizons from the *Zaphrentis* Zone to the Concretionary Beds of the uppermost

Seminula Zone. In lacking 'standard' fossils—brachiopods, corals, crinoids—they provide clear signs of a special kind of environment characterised by very shallow water, probably high salinity, and at extremes the crystallisation of sulphates.

The most considerable of them and the most extensive, the Calcite-Mudstone Group at the base of the Upper *Caninia* Zone, is composed predominantly of algal limestones, and even in individual beds relatively free of algae, tufts and pin-heads of algal growth are usually scattered through the non-algal matrix. The kinds of algae present may show different proportions—mainly of codiaceans and stromatolites—in successive beds presumably in response to environmental controls; but algal colonies may be banded (perhaps seasonally) with different genera dominating different bands, and what ecological associations may occur, and what forms the algal growths may take, were perhaps trivially accidental. Some algal bands are composed of large nodules or biscuits that studded the contemporary sea floor: with contact growth they merged into continuous sheets that carpeted the floor for many kilometres and that, especially in stromatolitic growth, were anchored to the floor by penetrant threads of algal tissue descending into the intergranular voids of the underlying sediment. (See Fig. 17.)

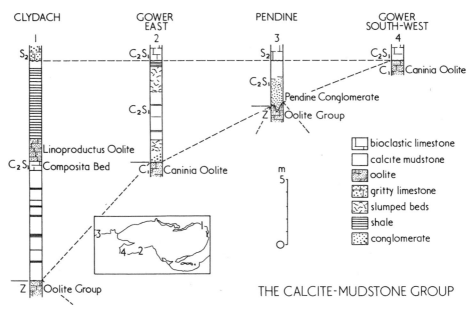

Fig. 17.   Variability in the basal beds of the Upper *Caninia* Zone. The facies contrasts are signs of a contemporary palaeogeography, the thickness contrasts of overstep mainly by the *Seminula* Zone (except in south-west Gower, where the formation is absent in a contact between crinoidal-shelly-coral limestones of the Upper *Caninia* Zone and the *Caninia* Oolite). The classifying of many of the beds as 'calcite mudstone' is generalised: the term covers rock types ranging from 'amorphous' lutites and algal 'dust' to calcisphere mudstones, pellet rocks, and ostracod limestones.

Many beds are uniformly 'amorphous', lutitic in grain size, compact and porcellanous without obvious internal texture. As typical calcite-mudstones they may be the winnowed washings derived from mixed sediments cleaned in a current-filtered environment; but some of them may be precipitates analogous to the whitings of present-day warm 'lagoonal' seas, precipitates inorganic-chemical in origin; and some, perhaps a majority, may be the disintegrated residue of stromatolitic algal tissue not cemented before death of the plant, and may then owe their origin to organic precipitation.

Beds composed in large part of organic debris, with fragmented brachiopod shells, or ostracods, or foraminifers, mixed in disturbed thanatocoenous association and in part presumably thrown into shallows by waves and currents, alternate with 'purer' calcite-mudstones: they commonly carry films or thicker wrappings of algal growth and contain the debris and the wrapped fragments in a lutitic matrix itself a product of disintegrated algal tissue.

The very great variety of sediments, many of them little more than a millimetre or two thick, showing very rapid alternations in physical composition and in grain size, in a formation that at maximum is less than 20 m thick, and in Gower is only 8 m thick, suggests that the Calcite-Mudstone Group was greatly affected by rapid changes as deposition continued in an environment perhaps best regarded as partly inter-tidal, partly 'lagoonal'. In rare circumstances of desiccation the flats, fed by mineralised capillary waters, encouraged the formation of sulphates, a thin bed packed with small gypsum crystals being known on the north crop above Crickhowell, where also there are strongly developed algal sheets. A depositional environment in some of its features not unlike the sabkha of the Persian Gulf is implied; and the Calcite-Mudstone Group may be looked on as a formation typifying the littoral margins of a subdued coast in a warm and perhaps semi-arid climate. (See George 1954, p. 306 and pl. xii; 1960, p. 354 and pls. xx, xxi; 1972, p. 236.)

The Concretionary Beds in the uppermost part of the *Seminula* Zone are also algal limestones: the algae, codiacean, spongiostrome, and stromatolitic, occur as pisoliths, biscuits, and nodules, some reaching the size of large pebbles, and neighbouring growths may fuse into small sheets usually less than a metre in diameter; but there is no extensive sheeted development like that occurring in the Calcite-Mudstone Group (at the base of which an algal carpet is known to extend unbrokenly for at least 20 km).

A peculiarity of the Concretionary Beds, very well seen in all the outcrops in South Wales except those of southernmost Pembrokeshire, is the strong rhythm that through a thickness of about 35 m shows an alternation of nodular algal beds and fine-grained calcite-mudstones, presumably in response to pulses of subsidence (and oscillating water-depth) each pulse perhaps of insignificant magnitude except in controlling algal growth. The rhythm includes the unexpected association of vast numbers of brachiopods—mainly athyrids (*Composita ficoides*) but also some linoproductids—with the algae, each shell or valve usually being an algal-wrapped core, and the majority of nodules having a shelly core: it is evident that the availability of the nuclei was an encouragement to algal growth. Many of the shells are well preserved and unabraded, and each bed being only a few centimetres thick and widely extensive—it is probable that individual beds may be identified in outcrops several kilometres apart—there appears to have been inconsiderable thanatocoenous dispersion of the shells, which are then virtually *in situ*, as the algal growths certainly are. Moreover, precisely the same lithology is seen in the Concretionary Beds at the same S_2 horizon in the Bristol district and in the Mendips; and although each individual bed has not yet been shown to be identifiable over the whole of the province, a regional environment of deposition in which a rich if restricted 'neritic' fauna was in nice ecological balance with algal gardens may be reconstructed in further confirmation of a recurrent uniformity of conditions along the coastal margins of the cuvette.

(d) Corals and coal seams

The rocks of the *Dibunophyllum* Zone are in sharp contrast to those of the *Seminula* Zone, not only in South Wales but almost everywhere in the South-Western Province (except in the sandy facies of the Forest of Dean and the Chepstow country). They contain massive oolites, not only along the north crop, and in many beds of the coral-shelly facies there are scattered ooliths sometimes in notable proportion; but the algal calcilutites are almost wholly absent, and there is little sign of 'lagoonal' flats. At the same time oolites are subordinate, and the rocks as a suite are very varied. Many of them are abundantly rich in a great variety of fossils especially in some of the D_2 crinoidal beds of the north crop: in the small thickness of 20 m of the rocks near Kidwelly, for instance, and in the 30 m near Penwyllt, many kinds of corals, simple and compound, include several species of *Lithostrotion*, *Orionastrea ensifer*, *Dibunophyllum* spp., *Corwenia rugosa*, *Lonsdaleia floriformis* and *duplicata*, *Nemistium edmondsi*, *Palaeosmilia murchisoni* and *regia*, koninckophyllids, *Amplexizaphrentis oystermouthensis* and *enniskilleni*, *Emmonsia parasitica*, *Michelinia tenuisepta*, and syringoporoids; and amongst the brachiopods are a dozen species of productoids and almost as many of spirifers (George 1927, pp. 48, 77). In a depositional environment at no great remove from the contemporary 'coast', such an abundant fauna is a pointer to clear open refreshed neritic waters.

Although not in such concentration, but still in abundance, similar faunas, especially of dibunophyllids and lithostrotiontids, are also to be found in both D_1 and D_2 beds in southern outcrops (Dixon 1921, pp. 74 ff.; George 1933, p. 248); and despite much diversity in detail, notably in the incoming of a molluscan fauna between Kidwelly and Llandebie, the regional development of the *Dibunophyllum* Zone is as widely persistent as the contrasted facies of the Concretionary Beds of the *Seminula* Zone beneath, from which it records a major palaeoecological change mainly (it is to be inferred) in salinity and aeration.

Nevertheless, an assumption of considerable depth of water in which the rocks accumulated is probably false; and 'neritic' is not to be interpreted as offshore.

The common occurrence of pseudobreccias in the zone—they are almost a lithological signpost to the zone—was interpreted by Dixon as the product of the recrystallisation of 'normal' limestones; but many of them have a peculiar association of characters that make them more complex in origin than his explanation allows for; and they give evidence of contemporaneous brecciation, even fragmentation, not by 'high-energy' erosion in turbulent water but by desiccation and some erosive solution. There are different degrees of pseudobrecciation, from massive limestones with a vague mottling to a fragmented bed in which bioclastic shards end abruptly at the margins of 'fragments', in complement to Dixon's evidence of grain-size and foraminiferal contrasts between 'fragments' and matrix; and many beds have a clay content that in abrupt lateral passage may become dominant.

The mode of origin of the pseudobreccias remains uncertain, but whatever it may have been the rock-type gives strong indication of exposure at sea surface and a negligible depth of sedimentation; and, as the pseudobrecciated limestones are widespread, from southern outcrops to the northernmost outpost in Careg-yr-ogof, they further imply shallows over a surprisingly uniform shelf—an environment in which the varied coral-brachiopod faunas thrived.

A further sign of the late-Avonian cuvette is seen in the coals of the *Dibunophyllum* Zone—coals seemingly quite out of place in being interbedded with 'normal' fossiliferous marine limestones (some brecciated) and not found as members of a yoredale-type cyclothem (Strahan 1907, p. 10; Dixon & Vaughan 1912, p. 490; Dixon 1921, p. 70). The coals are thin, sometimes no more than carbonaceous partings, and may rest on underclay; but although they are insignificant in thickness they prove contemporary waters so shallow as briefly to allow the growth of a peat *in situ*. The plants that contributed to the coals are unknown; but there is in the reconstructed environment obvious analogy with mangrove growths along the shoreward margins of sabkha and similar flats at the present day (where also in the multiplicity of lateral changes in facies there may be rich coral growths, and suites of other invertebrates, at depths scarcely greater than those in which the mangroves live).

The Calcite-Mudstone Group is notably different in lithology from the Concretionary Beds, and both formations offer still greater contrasts to the pseudobreccias; but repeatedly all the rocks provide evidence of radical facies changes not reflecting any great changes in depth of water, and 'neritic' and 'lagoonal' are terms more pertinent to lithological than to bathymetric factors.

III. THE ROCKS IN NORTH WALES

(a) The eastern outcrops

The Dinantian sequence in North Wales differs widely from the sequence in South Wales, despite the controlling influence on both of the hinterland of St. George's Land. Tournaisian rocks are unknown (although they may have been deposited in Powys and removed by early-Viséan erosion). Viséan rocks are restricted to late members, equivalents (on the evidence of the macrofossils) of the *Dibunophyllum* Zone of the South-Western Province. They are preserved in great thickness —far greater thickness than the beds in South Wales—and rests unconformably on a Lower Palaeozoic or Precambrian foundation. Earlier Viséan rocks are not known: a former attribution of a Seminulan age to some of the lowest beds, of the lower part of the Lower Brown Limestone at its thickest development, is suspect, athyrids referred to *Composita ficoides* being locally misidentified, and *Daviesiella llan-*

gollensis, sometimes given zonal significance, not being an index fossil in the Avonian sequence of the South-Western Province. It is possible that some of the poorly fossiliferous Basement Beds, and perhaps calcite-mudstones immediately above, may be older than the junction between D_1 and S_2 in South Wales, but there is as yet no proof that they are, and proof is not likely to be available until the microfossils are studied in detail[1].

The evidence that beds equivalent mainly or wholly to the *Dibunophyllum* Zone are the oldest preserved in North Wales leads to the inference that the emergence of St. George's Land was highly asymmetrical. In South Wales the *Seminula* Zone is in thickness the dominant zone: locally it is the sole survivor on the north crop, and in extrapolated isopachs it is inferred to have extended farther north than any other zone, including the *Dibunophyllum* Zone, onto the drowning flanks of St. George's Land (George 1972, fig. 2). The *Dibunophyllum* Zone on the north crop is in full expression (including D_1, D_2, and D_3) on the contrary very much thinner, no more and usually much less than 70 m. The equivalent beds in North Wales may be 850 m or more. A rocking about an axis in St. George's Land thus allowed alternations of thick sediments to accumulate—early in the south, later in the north (see Fig. 18).

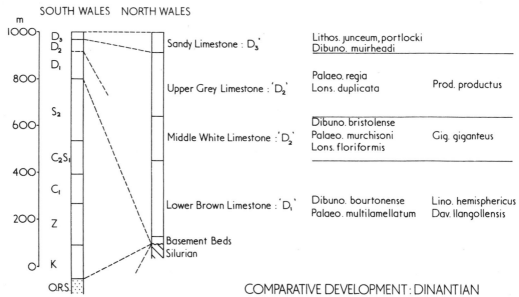

Fig. 18.   Correlation of Dinantian rocks. The similarity in thickness at maximum development contrasts with the disparity in sequence. Illustrative macrofossils justifying the zonal equivalence of the formations in North Wales are selectively listed.

---

[1] The Avonian zonal system is useful enough in its own terrain of the South-Western Province; but it partly rests on facies-controlled fossil assemblages that are not readily or convincingly transferable to other provinces, where Avonian terms have in fact caused some confusion in correlation. While it is highly desirable to determine (by whatever means) the equivalence of beds in different provinces, the allocation of rocks in North Wales to Avonian zones is best avoided (see George 1972, p.223). A revised chronostratigraphical system applicable to the Lower Carboniferous rocks of the whole of the British Isles is urgently needed, and should be forthcoming shortly.

The sequence in North Wales not as yet being readily divisible into faunal zones despite its great thickness, it continues to be described lithostratigraphically in terms of members and formations, although there is not much doubt, even if it cannot always be proved, that diachronism of rock units is inevitable when correlation rests on lithological facies; and the classification first established nearly a hundred years ago by Morton (1878; 1886; 1897; 1898) is still applied in the successive formations of the Basement Beds, the Lower Brown Limestone, the Middle White Limestone, the Upper Grey Limestone, the Sandy Limestone, and the Black Limestone[1]. The divisions are best seen in the eastern outcrops for 65 km between Llanymynech and Prestatyn, along which they show marked variations in thickness and lithology.

The Basement Beds are a mixed suite of sandstones and shales with some conglomerates, sometimes over 30 m thick, sometimes thinning to disappearance. In being red, the washings of an old land surface, they strongly hint that they were not superimposed on the Lower Palaeozoic floor by the erosive stripping of earlier Viséan limestones but were the first sediments to encroach onto a subsiding coast. The general absence of fossils in them, and the occasional occurrence of cornstones, suggest not fully marine, or only intermittently marine, conditions along the margins of the cuvette, in a facies having some analogy with, but very different in age from, the uppermost Farlovian or Skrinkle sediments of the Upper Old Red Sandstone in South Wales. The variable development of the group—relatively thick in the contemporary troughs, thin to disappearing on the intervening swells—also suggests topographically controlled fluvial or estuarine or embayed containment of the terrigenes.

The classification of the overlying limestones mainly on colour is crude, and rests partly on subjective judgement. Random analyses of samples of the Lower Brown Limestone, with nearly 98 per cent of calcite, show them to be little different in composition and in 'purity' from samples of the Middle White, with nearly 99 per cent of calcite (Morton 1897, p. 196). In places there is only a vague boundary between the formations; and locally an apparent thinning or absence of one or other formation may reflect no more than a changing shade of colour. Nevertheless, along much of the eastern outcrop there is a general consistency in the stratigraphical relations that allows a convincing palaeogeography to be established, and in broad terms the fossils suggest that the Lower Brown Limestone equates approximately with the lower part (D_1) of the *Dibunophyllum* Zone of South Wales, and the Middle White Limestone and the formations above with the upper part (D_2 and D_3).

Most of the rocks are bioclastic (see Jones 1921), many of them with abundant crinoid debris, many kinds of brachiopods, and many kinds of simple and compound corals including lithostrotiontid and lonsdaleoid masses (Wedd, Smith, and Wills 1927, pp. 121 ff; Wedd and Others 1929, pp. 95 ff.). They are characteristically neritic; oolites are few, and calcite-mudstones and algal pisolites are mainly restricted to

---

[1] There is some variability in the terminology: the Lower Brown Limestone is sometimes called the Lower Grey and Brown; the Middle White sometimes the White; the Upper Grey sometimes the Intermediate; the Sandy Limestone the Arenaceous Limestone; the Black Limestone the Aberdo Limestone.

the lower part of the Lower Brown Limestone. On the other hand, as in the *Dibuno-phyllum* Zone of South Wales, pseudobreccias are recurrent, and there are thin carbonaceous seams, both rock-types indicating shallows and near-shore conditions. Not many beds are sandy except in the uppermost formation (the Sandy Limestone); and for the greater part of the time the hinterland was subdued.

A general impression is given by the sweep of the present-day outcrops that they conform broadly with the depositional trends along the north-eastern margins of the hinterland massif; but the great variations in thickness displayed by the formations as they are traced from south to north show the impression to be misleading (Neaverson, 1946; George 1958, fig. 18). At the southern limits, terminated by the transgressive Trias, the full thickness of the Carboniferous Limestone is probably not much greater than 220 m—the combined Lower Brown and Middle White Limestones being no more than 40 m near Llanymynech and Pant. There is fluctuating increase northwards to the neighbourhood of Llawnt, where a thickness of about 460 m is found; but at Fron Cysyllte, a few miles still farther north, all the Lower Brown Limestone and almost all of the Middle White are missing by overlap. There is expansion to 425 m in Eglwyseg Rocks, but again a great thinning, also a product in part of overlap, to Minera, where the basal members of the local sequence belong to the uppermost 50 m of the Middle White Limestone. Beyond the Bryn-eglwys (Bala) fault zone, in the Llandegla outcrops on the eastern flank of the Clwydian Hills, the thickest developments of the Carboniferous Limestone in North Wales are seen: they reach nearly 900 m, and continue northwards to the coast with only relatively minor reduction to about 730 m at Mold and 720 m at Halkyn. (See Wedd and Others 1923, pp. 14 ff.; Wedd, Smith, and Wills 1927, pp. 108 ff.; Wedd and Others 1929, pp. 88 ff.) (See Fig. 19.)

The variations in thickness are in part due to the drowning of a hinterland of some relief, residual hills no doubt for a time standing up as islands above the advancing sea; but some measure of the variation is attributable also to contemporary deformation, thinning being seen not only in successive lowest beds but also in relative subsidence, the thicker accumulations occupying downwarps in the Lower Palaeozoic floor, the thinner resting on swells. In a generalised relationship it is perhaps possible to recognise posthumous Caledonian movement as contributing to the variations, notably in the coincidence of the Minera sequence with the Cyrn-y-brain anticline (see George 1961, fig. 25); but the thinning at Fron Cysyllte, in the heart of the Llangollen syncline, lacks such a link with older structures.

The Lower Brown and Middle White limestones retain their individuality (even if they are not always readily distinguished) over the greater part of the eastern outcrops; but the Sandy Limestone changes significantly as it is followed into north Flintshire. In the Oswestry country the alternation of arenaceous beds with 'normal' limestones, the typical Sandy Limestone, allows a ready separation from the Upper Grey, although the more calcareous beds show little difference between the one formation and the other. The sandstones, some of which are red, indicate a revival of relief in the hinterland, and some of them are coarsely gritty and contain strings of quartz pebbles reflecting pulses of intensified erosion. The suite, variable in thickness, continues without much change to Minera, a sandy oolite being well

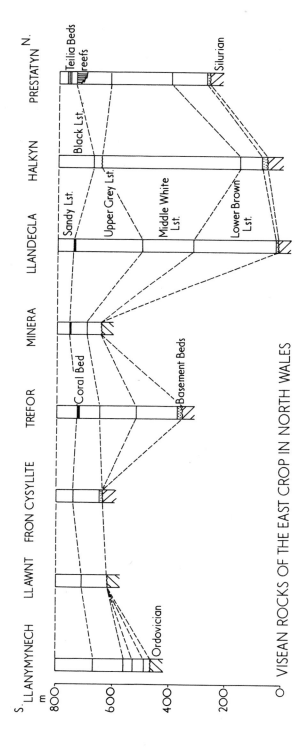

Fig. 19. Comparative columns summarising the development of the Carboniferous Limestone in part of Flintshire and Denbighshire. The uncertainties of formational correlation based on crude lithology are well illustrated between Llandegla and Halkyn, where the gross thickness shows little change but the Middle White Limestone is greatly expanded northwards, the Lower Brown correspondingly contracted.

developed at its base. Beyond the Bryneglwys fault zone the oolite is not seen, and although the sediments remain in more or less degree sandy toward Mold and Halkyn, the sandstones alternate with limestones that become increasingly dark, fine-grained, cherty, and muddy and there is lateral passage into the Black Limestone, a rock formation having some analogy with the Upper Limestone Shales of South Wales, and other analogy with the lower part of the Bowland Shales of Lancashire (Schnellman 1939, p. 2; Khosrovani 1940, p. 475; Banerjee 1959; Oldershaw 1969). The Black Limestone includes the highly fossiliferous Teilia Beds, in which near Prestatyn abundant well-preserved plants are in unusual occurrence in Carboniferous Limestone. (See Hind & Stobbs 1906, pp. 387, 396 ff., 455; Walton 1926; Sargent 1923.)

Uncertain in horizon because of the incidence of much minor faulting in ground where contacts are poorly exposed, reefs apparently high in the Middle White Limestone or the Upper Grey Limestone, or perhaps as high as the Black Limestone, are strongly developed in the northernmost outcrops near Prestatyn. They are closely similar in lithology to some of the Derbyshire and Clitheroe knolls, fine 'amorphous' calcite-mudstone being the dominant matrix, and very well preserved brachiopods being common fossils. They also contain goniatites including beyrichoceratids, *Goniatites hudsoni*, and *Merocanites henslowi*, referable to the B_2 zone of the Cracoean: local rocks of Black Limestone facies, apparently overlying the knolls, yield *Goniatittes* cf. *falcatus* and *sphaericostriatus*, which, with *Posidonia becheri* and *Pterinopecten persimilis*, probably indicate P_1 horizons. It is possible, therefore, that the Upper Grey and Middle White limestones are locally older than has hitheto been supposed. (See Neaverson; 1930; 1943; 1946.)

The Carboniferous Limestone of Powys thus combines in rock types and in lateral passage characteristics offering close comparison with some of the rocks of Derbyshire and of the Bowland trough; and in its associations as well as in its southwestward backing against St. George's Land it has few links with the Carboniferous Limestone of South Wales.

(b) The Vale of Clwyd

The rift of the Vale of Clwyd, arcuately downfaulted on its eastern flank, step-faulted on its western, contains Carboniferous rocks in its floor partly covered by thick Trias. Slivers narrowly elongate at the foot of the Clwydian Hills nowhere show a full succession of Carboniferous Limestone. Near Dyserth at the head of the Vale, where the nose of the Clwydian anticline brings unfaulted basal beds into the Vale, a thin development of red beds, sandy and conglomeratic, rests with the usual sharp unconformity on Silurian, in a manner uniform along the eastern outcrops. They are followed by Morton's Lower Brown Limestone, divided by Neaverson (1930, p. 187) into subsidiary groups distinguished on fossil content: the lowest members, which contain sandy beds with plant remains, appear to equate with the lower *Dibunophyllum* Zone of South Wales, abundant brachiopods in them being inaptly compared with 'Composita ficoides' and not being a sign of equivalence with the *Seminula* Zone (Neaverson 1929, p. 116). Farther south the slivers appear to bring mainly Lower Brown Limestone to outcrop between Tremeirchion, Bodfari, and Llanfair-dyffryn-Clwyd (Morton 1897, pp. 3–6; Neaverson 1929, p.119); but at the head of the Vale, where the north-plunging nose of the faulted syncline brings a full sequence of the rocks to the surface on the north flank of the Llanelidan fault, the main formations may still be recognised, in a thickness of about 390 m (Morton 1897, p. 55; Wedd, Smith, and Wills 1927, pp. 126 ff.; Neaverson 1929, p. 114). At the well-known Faenol quarries and in neighbouring exposures the great abundance of corals in the bioclastic rocks of the Middle White Limestone includes masses of lonsdaleoid and other compound forms including *Corwenia;* and a notable record is of *Saccamminopsis*, with allusion to marker beds in northern England (Morton 1897, pp. 53; Neaverson 1929, p. 125). Beyond the head of the Vale, 10 km. to the south-west, the isolated outlier of Hafod-y-calch above Corwen also appears to carry a full formational sequence, the Lower Brown Limestone not being over-lapped despite its location far into the massif; and the abundance of fossils there matches in facies the development at Faenol: but the thickness is reduced to not much more than 220 m (Morton 1878, p. 313).

The post-Carboniferous faulting at the head of the Vale, much of the movement having a strike-slip component, makes difficult a reconstruction of the palaeo-geographical relationships of the broken outcrops. In particular, the great thickness at Llandegla of 900 m is in anomalous contrast to the 530 m of Faenol–Nantclwyd only 5 km away to the west, and to the 175 m (or thereabouts) at Minera 6 km to

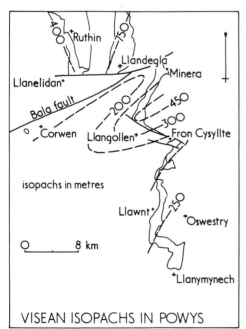

VISEAN ISOPACHS IN POWYS

Fig. 20. Structural relations of the Lower Carboniferous rocks between Llanymynech and the Vale of Clwyd. The isopachs are generalised, and (along outcrops mainly linear) their closures are conjectural; but the order of range of thickness is to be accepted, and sufficiently illustrates the effects of contemporary movement on sedimentation along the north-east front of St. George's Land. The influence of the Cyrn-y-brain axis on the sequence at Minera is particularly noticeable, and the development of a trough in the Eglwyseg country between Minera and Fron Cysyllte equally so. The great reduction in thickness between Llandegla and Minera may suggest a front steeply falling northwards, perhaps along a growing Llanelidan fault; or, more probably, may imply a faulted contiguity of present outcrops relatively displaced for perhaps a score of kilometres along the Bala and Llanelidan faults (in the angle between which Carboniferous rocks are not now preserved).

the east. When the inferred (but not measurable) effects during sedimentation of the Cyrn-y-brain and Mynydd Cricor anticlines are also taken into account, the disparities appear as the combined product of contemporaneous differential subsidence and oblique subsequent faulting. There is also the possibility that, the outcrops of the thin developments at Minera and Fron Cysyllte swinging far to the east, north-and-south isopachs may imply a major swell beneath them as a salient of St. George's Land. (See Fig. 20.)

On the western flank of the Vale of Clwyd the highest formation is mostly overstepped by Trias; but it is to be seen at the head of the Vale where it is typical Sandy Limestone composed of alternating limestones, calcareous sandstones, shales, and quartz conglomerates, in a facies comparable with that of Llandegla. To the north, about Ruthin and Denbigh, the Lower Brown Limestone, with relatively thick red sandy basement Beds beneath, maintains its essential characters of bioclastic sediments, but some beds are arenaceous and a few conglomeratic, drifted plants are common, and intercalated oolites and calcite-mudstones (with spirorbids) suggest deltaic or near-shore environments rather more strongly than are to be seen elsewhere. The Middle White Limestone in its lithology matches very closely its development in the eastern outcrops. Thickness (of the incomplete sequence) is of the order of 520 m towards Abergele, rather less at Denbigh).

Morton's formational divisions begin to break down from Abergele westwards, both in their range and in their zonal allocation (Neaverson 1935, p. 223; 1936, p. 119), although the Middle White Limestone near Llandulas includes reefs like those at Prestatyn. The Basement Beds (which at Ffernant are notable for including pebbles of Upper Ludlow rocks not otherwise known at outcrop in North Wales) reach a thickness of about 20 m; the Lower Brown Limestone becomes massively dolomitic in the Gloddaeth syncline and at Great Orme, and may continue upwards (in a thickness exceeding 275 m) to include a development of alternating bioclastic limestones and calcite mudstones; the White Limestone is a mixed group of nodular and pseudobrecciated limestones, products in part of shelf desiccation, that include some oolitic and sandy lenticles and pebble beds; and the Black Limestone may be represented in dark fine-grained 'Aberdo' mudstones with cherts. At Great Orme the thickness (base and summit not seen) is more than 460 m. (See Morton 1898, p. 385 ff.; Smyth 1925.)

(c) Anglesey and Arfon

Morton (1901) continued his work into Anglesey, but there his formational system became strained in its application, and it was left to Greenly (1919, pp. 600 ff.) to bring order to a complex stratigraphy that gave little support to regional preconceptions. Despite the very wide variations in thickness and sequence to be seen on the mainland in the long outcrops between Llanymynech and the Great Orme, Morton's unifying system proved to be generally applicable partly because of a relatively constant (if accidental) place of the outcrops in relation to the cuvette of sedimentation and the hinterland massif. Anglesey, however, lay precisely and persistently at the margin of the cuvette against the massif: its present-day outcrops reveal the rapidly variable kinds of sediments and sedimentary relations that characterise the immediate vicinity of a coast, and that are not accommodated to a simplified fourfold formational sequence inherited from the mainland.

Greenly (1919, p. 626), although repeatedly puzzled by structure and rock-type, showed that the greater part of Anglesey was land during latest Dinantian times, and that the land surface was highly irregular in physique during its drowning by the Dinantian seas. He also showed that, as elsewhere in North Wales, the drowning was very late, all the rocks being correlated with the *Dibunophyllum* Zone of the South-Western Province; and that the earlier beds (correlated with D_1) were restricted to very small pockets on the east of the island and were in lithology very different from 'normal' Lower Brown Limestone. Even in the Penmon ground, there is little similarity in detail between the sequence in the upper beds (equated with D_2) and that recognised by Smith and Neaverson in the Great Orme and the Little Orme, only some 15 km away. Incidentally, the stratigraphical contrasts throw a revealing light on the subjective manner in which a stratigraphy grows, and on a desire for order in the influence of persistent stratigraphical terms insinuated into an imposed system of correlation.

The oldest rocks, found in Lligwy Bay, are a basement conglomerate of cryptic origin, in which large boulders approaching a metre in diameter include material mainly derived from the Mona Complex, but in which there are also subrounded to angular boulders of limestone not all of local provenance. The immediately overlying Lligwy Bay Conglomerate on the other hand, also with large boulders, includes much material not derived from the Mona Complex but from Lower Palaeozoic sources; and anomalous directions of powerful current-flow are indicated to form very mixed delta or longshore spreads whose kind it not recognised elsewhere in Anglesey.

Rocks of about the same age but very different in lithology form the basal members of the sequence across the Menai Strait in Arfon: they constitute the Loam-Breccia Formation thought by Greenly (1928, p. 386) to be a terrestrial deposit brockram-like in its nature, with angular boulders of Penmynydd-Zone metamorphics and with ferric-oxide ooliths: they are an indication of a margin to the local cuvette. They are followed by water-borne conglomerates, analogous to many found in Anglesey, containing quartz pebbles a majority of which can be identified as having their source in the Gwna Beds of the Mona Complex. Like the Lligwy conglomerates, the basal rudites in Arfon are limited in outcrop and presumably were restricted in their area of sedimentation.

Inland, the main development of Carboniferous Limestone occupies a narrowing basin running from the Lligwy–Red Wharf coast on the north-east to the Malldraeth depression on the south-west. It progressively thins in doing so, from about 400 m at Lligwy, about 150 m near Llangefni, to nil near Bodorgan. The thinning is due to overlap against a rising floor, and in a general sense the form of the present outcrop reflects an original basin or gulf of deposition that opened towards the north-east. In Greenly's analysis, a sequence equated with the *Dibunophyllum* Zone, and divided by him into D_1, D_{2a}, D_{2b}, and D_3 partly on lithofacies partly on a distribution of macrofossils, shows a limitation of D_1 to the ground north of Llangefni, and of D_{2a} north of Llangristiolus. A diachronous rise in the horizon of the basal conglomerates accompanies the overlap. The rocks show great lithological variation. Many of them are 'normal' calcarenites with abundant bioclasts, and with many kinds of fossils including a variety of corals and a greater variety of brachiopods indicating a refreshed neritic environment. Some beds are calcite mudstones, comparatively few are oolites: there is little indication of sustained bank environments. Many limestones are pseudobreccias, and (as on the mainland in the crop running east from the Great Orme) there are repeated signs of their fragmentation into nodular limestones in some of which the nodules may be isolated and rolled to give the appearance of conglomerates—a sign of desiccation and reworking. The uppermost beds contain common cherts, which sometimes fuse into mammillated and tabular sheets: they then compare with the black limestones and cherts of north Flintshire.

The sequence is one of alternating limestones and sandstones. Many of the bioclastic limestones carry scattered quartz grains, and there is a gradation between very 'pure' beds with over 99 per cent of calcite through sandy limestones to calcareous sandstones, and terrigenes were constantly brought into the cuvette by streams off the nearby hinterland. At several horizons the terrigenes become dominant, to the exclusion of calcareous debris; and in a number of sandstones, notably the Helaeth Sandstone in 'D$_1$' and the Benllech Sandstone in 'D$_2$', the strings of pebbles, the cross-bedding, the repeated signs of washouts, and the absence of fossils (except drifted plants) point to delta spreads thick and massive enough to exclude limestones altogether. A peculiar feature of many of the sandstones is the downward extension from their base of long tongues descending as pipes into the eroded beds of limestones beneath, the walls of the pipes being sharply defined and often irregularly overhanging: an origin of the pipes by fresh-water solution is indicated, and a consequent presumption that the surface of the sands at time of deposition was above sea-level—a sign of the generally shallow water in which all the sediments accumulated, including the neritic limestones. The sandstones are lenticular, and although they are recurrent in the sequence any one is not traceable for long distances, partly because of south-western overlap, partly because of changing routes of sediment transport.

The small easternmost outcrop of Carboniferous Limestone in Anglesey, from Careg-onen to Penmon and the outlier of Puffin Island, is without an expected basal conglomerate of any considerable thickness, but in general lithology the succession of limestone and sandstones otherwise compares with that of the main outcrop. In the western exposures about Careg-onen the lowest beds, to a thickness of about 90 m, are ascribed by Greenly to 'D$_1$', and are comparable to, and of much the same thickness as, the beds of similar age south of Lligwy; and higher beds, for 135 m, are placed in 'D$_{2a}$'. In the exposures about Penmon, massive limestones and chert beds, about 45 m thick, fall into 'D$_{2b}$' and 'D$_3$'. The cumulative thickness, about 180 m, of the beds above 'D$_1$' is much reduced from the 270 m of the Lligwy ground, apparently because of overlap against a swell rising eastwards: in the easternmost outcrops about Penmon and Puffin Island, there is no representative of either 'D$_1$' or 'D$_{2a}$', the residual thickness being reduced to 45 m. There are also rapid variations in lithology, notably in the incoming of the Parc Sandstone about Penmon and the Fedw Sandstone farther west, both at high horizons perhaps replacing, not merely being interbedded with, 'D$_3$' limestones. Many of the limestones are arenaceous, and transgressive stratal breaks are recurrent. Dolomitised beds on Puffin Island compare with those (part of the Middle White Limestone) seen on

Great Orme, but there is no proof of an equivalence of age. (See Fig. 21.)

The Straitside Carboniferous outcrop in Anglesey, with its outpost in Arfon, shows at its base a diachronism of overlap analogous to the relations in the main and Penmon basins (Greenly 1928, p. 412). 'D₁' is recognised only on the mainland

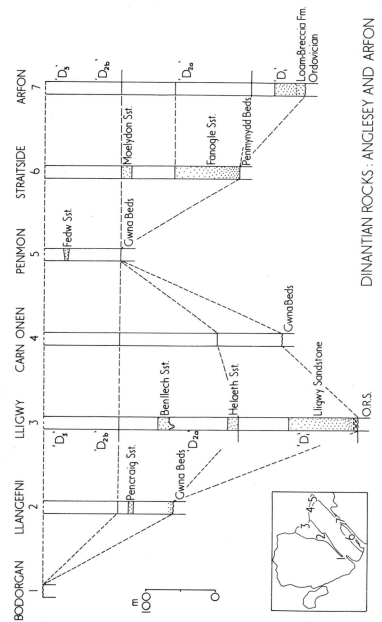

Fig. 21. Comparative columns in the Carboniferous Limestone of north-west Wales. The great variations in thickness, and the occurrence of internal overlap, prove the drowning of a corrugated terrain of a salient of St. George's Land. Pulses of uplift of the hinterland during sedimentation are illustrated by the repeated development of local sandstones. (After Greenly.)

where, above the basal conglomerates (30 m thick), about 30 m of fossiliferous limestones, some bioclastic and pseudobrecciated, some lutitic, include sandy and conglomeratic lenticles and are themselves often strongly arenaceous. It thins rapidly by overlap westwards, the basal conglomerate rising in the sequence, and finally disappears where it runs into the Menai Strait: it is not found on the Anglesey shore, where the basal beds were referred by Greenly to 'D_{2a}'. South-westwards by Bryn Siencyn to Dwyran the rocks reach a thickness of about 270 m, ranging from 'D_{2a}' through 'D_{2b}' to 'D_3'. At their base the Fanogle Sandstone, a pebbly and conglomeratic rock 90 m thick, is unusual in containing interbedded soft red marls. Its upper layers alternate with sandy limestones, and it is overlain by fossiliferous bioclastic limestones including the Edwen Oolite in which strings of scattered pebbles are not uncommon. At higher levels the red Edwen Sandstone, rippled and conglomeratic, and the Moelydon and Carnedd sandstones are arenaceous tongues, forming with cherty limestones the uppermost beds. The recurrence of impressively thick sandstones, as in the main outcrop, marks the temporary flooding of the neritic seas by terrigenes deposited in an environment of pulsed 'high-energy' transport.

(d) Anglesey in its regional context

The mainland outcrops from the Great Orme to Llanymynech provide a generalised frame of Viséan sedimentation, but except in isolated fragments—at Fron Cysyllte, at Minera—they say little that is precise of the relations of cuvette and hinterland. The Anglesey outcrops, on the other hand, show much of the island to have been above depositional level until the close of Viséan times while the basins subsided as elongate gulfs that penetrated deeply into a hinterland of rough relief, to define very precisely the relations of structure to sediments.

The incidental evidence is locally sharply explicit, not merely in terms of mapped formational relations but more directly where rock contacts are seen in exposure. Thus successive beds of the Fanogle Sandstone along the Menai coast pick out in close definition stages of subsidence where in a few metres one bed overlaps another and is itself overlapped by a third as the formation encroaches onto the floor of Penmynydd metamorphics. In the terrain about Llangristiolus, cliffs of Gwna rocks 12 or 15 m high formed recognisable limits to some of the limestones of 'D_{2a}' age; and similar cliffs between Llangefni and Lligwy appear to have been incipient fault scarps, contemporaneous with the sediments, that were to be intensified by later Hercynian movement. In the immediate neighbourhood of Llangefni a knob of Precambrian Gwna beds rose as an island, scores of metres in height, that was gradually lapped and drowned by the accumulating limestones.

These local hints of a diversified landscape promote a reconstruction of a margin to the massif characteristic of a coast of subsidence subject to contemporary deformation. The regional evidence of the place of Anglesey in a wider context is even more impressive in providing larger structural links between island and mainland. Thus the featheredge of uppermost limestones near Bodorgan, where they pass under Upper Carboniferous rocks that come to rest on Precambrian, lies only 6 km from the relatively thick Viséan sediments (reaching several hundred metres) of the Straitside basin; and an intervening barrier of Penmynydd rocks, a peninsula or an island, not passively residual but arching as a posthumous anticline, must be supposed. Its southern margin may well have been determined in part by an incipient (or revived) Berw fault. The Straitside rocks thicken north-eastwards, with 'D₁' appearing near Bangor as the deposits of a contemporary sea that presumably opened into the region of the Great Orme and the Gloddaeth syncline; and conversely, in thinning south-westwards, they show internal overlap that, in analogy with the Bodorgan overlap, may have resulted in their final disappearance against a rising coast at some point under or beyond Newborough Warren. (See Fig. 21.)

The limestones of the Penmon outcrop are not readily accommodated. The 'D₁' beds about Careg-onen may originally have been directly continuous with the Lligwy rocks; but as there are appreciable differences in sequence, the link may have been indirect, the intervening Gwna beds of the present outcrop being a reflection of a partial barrier in Viséan times. The eastward disappearance of 'D₁' and 'D₂' between Careg-onen and Penmon against a rising Precambrian floor is, however, in an opposite sense to the emergence of 'D₁' near Bangor, and the Straitside gulf, surmised to have opened north-eastwards, had at least for part of its life a barrier separating it from the Penmon terrain. It remains unknown how far the 'D₁' beds of Careg-onen continued southwards without being overlapped, and how effective the barrier was between Penmon and Straitside in 'D₂ᵦ' times. (Fig. 22.)

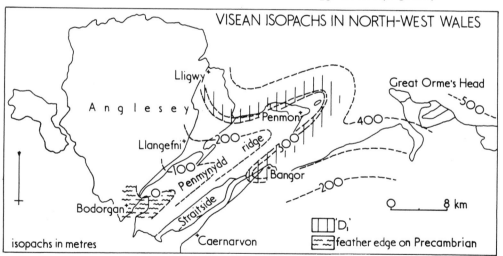

Fig. 22.   The corrugated terrain of Lower Carboniferous deposition in Anglesey and Arfon. Variations in thickness and in 'zonal' sequence allow a convincing palaeogeographical restoration in terms of isopachs, which, although conjectural in many places, leave no doubt of the 'anticlinal' effect of the Penmynydd ridge in separating the Straitside gulf from the Lligwy embayment; and the zero isopach near Bodorgan defines a promontory locally persistent above depositional level until the close of Lower Carboniferous times. (Mainly after Greenly.)

North-western Anglesey, and the submarine platform of Precambrian rocks to the north, may always have been without Lower Carboniferous cover; but the abundance of cherts in the drifts of the area suggested to Greenly (1919, p. 776) that a synclinal basin lay no great distance away; and recent geophysical survey and exploratory drilling have shown that Carboniferous rocks floor the Irish Sea to the south of the Isle of Man as part of the Central Province (Eden, Wright, & Bullerwell 1971, p. 123).

The evidence is partial and indicative rather than comprehensive; but the corrugated foundation on which the Lower Carboniferous sediments lie in Anglesey is sufficiently revealed in the variations in rock sequence; and a caledonoid pattern emerges as a structural control on deposition and delineates the high irregularities of the massif where it met the Carboniferous sea. The mainland outcrops, being mostly farther offshore, lack such detailed evidence of the effects of the foundation; but analogously the swells at Fron Cysellte and Minera may be regarded as the cores of caledonoid anticlinal tongues like the Penmynydd tongue in Anglesey.

IV REFERENCES

BANERJEE, A. 1959. Petrography and facies of some upper Viséan (Mississippian) sediments of North Wales. *J. sed. Petrol.*, **29**, 377–90.

——. 1969. Carboniferous lithofacies, Rhydymwyn to Erryrys, North Wales, *Geol. Jl*, **6**, 181–4.

DIXON, E. E. L. 1921. The country around Pembroke and Tenby. *Mem. Geol.* Surv. Gt. Britain.

—— and VAUGHAN, A. 1912. The Carboniferous succession in Gower' *Quart. J. geol. Soc. Lond.*, **67**, 477–571.

EDEN, R. A., WRIGHT, J. E., and BULLERWELL, W. 1971. The solid geology of the eastern Atlantic continental margin adjacent to the British Isles. *Rep. Inst. geol. Sci.*, **70/14**, 111–28.

GEORGE, T. N. 1927. The Carboniferous Limestone (Avonian) succession of a portion of the north crop of the South Wales coalfield. *Quart. J. geol. Soc. Lond.*, **83**, 38–95.

——. 1954. Pre-Seminulan Main Limestone of the Avonian Series in Breconshire. *Quart. J. geol. Soc. Lond.*, **110**, 283–322.

——. 1956a. Carboniferous Main Limestone of the east crop in South Wales. *Quart. J. geol. Soc. Lond.*, **111**, 309–22.

——. 1956b. The Namurian Usk anticline. *Proc. Geol. Assoc.*, **66**, 297–316.

——. 1958. Lower Carboniferous palaeogeography of the British Isles. *Proc. Yorks. geol. Soc.*, **31**, 227–318.

——. 1960. Lower Carboniferous rocks in County Wexford. *Quart. J. geol. Soc. Lond.*, **116**, 349–64.

——. 1961. North Wales. *Geol. Surv. Mus.*

——. 1969. British Dinantian stratigraphy. *C. R. 6me Congr. intern. Strat. Géol. Carb. Sheffield 1967*, **1**, 193–218.

——. 1970. South Wales. *Inst. geol. Sci.*

——. 1972. The classification of Avonian limestones. *J. geol. Soc.*, **128**, 221–56.

GREENLY, E. 1919. Geology of Anglesey, *Mem. geol. Surv. Gt. Britain*, 2 vols.

——. 1928. The Lower Carboniferous rocks of the Menaian region of Caernarvonshire. *Quart. J. geol. Soc. Lond.*, **84**, 382–439.

HIND, W., and STOBBS, J. T. 1906. The Carboniferous succession below the Coal Measures in north Staffordshire, Denbighshire, and Flintshire. *Geol. Mag.*, **5** (3), 387–400, 445-59, 496–507.

JONES, T. A. 1921. A contribution to the microscopic study of the Carboniferous Limestone of North Wales. *Proc. Lpool geol. Soc.*, **13**, 78–99.

KELLING, G., and WILLIAMS, B. P. J. 1968. Deformation structures of sedimentary origin in the Lower Limestone Shales (basal Carboniferous) of south Pembrokeshire, Wales. *J. sed. Petrol.*, **36**, 927–39.

KHOSROVANI, K. 1940. The correlation of strata at Halkyn mines, North Wales, by the study of insoluble residues. *Trans. Inst. Min. Met.*, **49**, 473–511.

MORTON, G. H. 1878. The Carboniferous Limestone and Millstone Grit in the country around Llangollen. *Proc. Lpool geol. Soc.*, **3**, 152–205, 299–325, 371–428.

——. 1886. The Carboniferous Limestone and Cefn-y-fedw Sandstone of Flintshire. *Proc. Lpool geol. Soc.*, **4**, 297–320, 381–403.

——. 1897. The Carboniferous Limestone of the Vale of Clwyd. *Proc. Lpool geol. Soc.*, **8**, 32–65, 181–204.

——. 1898. The Carboniferous Limestone of the country around Llandudno. *Quart. J. geol. Soc. Lond.*, **54**, 382–400.

——. 1901. The Carboniferous Limestone of Anglesey. *Proc. Lpool geol. Soc.*, **9**, 25–48.

NEAVERSON, E. 1929. Faunal horizons in the Carboniferous Limestone of the Vale of Clwyd. *Proc. Lpool geol. Soc.*, **15**, 111–33.

——. 1930. The Carboniferous rocks around Prestatyn, Dyserth, and Newmarket. *Proc. Lpool geol. Soc.*, **15**, 179–212.

——. 1937. The Carboniferous rocks between Llandudno and Colwyn Bay. *Proc. Lpool geol. Soc.*, **17**, 115–35.

——. 1943. Goniatites from the Carboniferous Limestons of Prestatyn and Newmarket (Flintshire). *Proc. Lpool geol. Soc.*, **18**, 135–43.

——. 1945. The Carboniferous rocks between Abergele and Denbigh. *Proc. Lpool geol. Soc.*, **19**, 52–68.

——. 1946. The Carboniferous Limestone Series of North Wales: condition of its deposition and interpretation of its history. *Proc. Lpool geol. Soc.*, **19**, 113–44.

OLDERSHAW, A. E. 1969. Carboniferous lithofacies, Halkyn Mountain, North Wales. *Geol. Jl*, **6**, 185–92.

OWEN, T. R. 1964. The tectonic framework of Carboniferous sedimentation in South Wales. *In* van Straaten, L. M. J. U. (editor): *Deltaic and shallow marine sediments*, 301–7. Amsterdam.

——. 1971. The relationship of Carboniferous sedimentation to structure in South Wales. *C. R. 6me Congr. intern. Strat. Géol. Carb. Sheffield 1967*, III, 1305–16.

——, and JONES, D. G. 1961. The nature of the Millstone Grit—Carboniferous Limestone junction of a part of the north crop of the South Wales coalfield. *Proc. Geol. Ass.*, **72**, 239–49.

SARGENT, H. C. 1923. The Massive Chert Formation of north Flintshire. *Geol. Mag.*, **60**, 168–83.

——. 1927. The stratigraphical horizon and field relations of the Holywell Shales and 'Black Limestone' of north Flintshire. *Geol. Mag.*, **64**, 252–63.

SCHNELLMAN, G. A. 1939. Applied geology at Halkyn District Applied Mines Limited. *Bull. Inst. Min. Met.*, 415.

SMYTH, L. B. 1925. A contribution to the geology of Great Orme's Head. *Proc. roy. Dublin Soc.*, **18**, 141–64.

SQUIRRELL, H. C., and DOWNING R. H. 1969. The country around Newport (Mon.). *Mem. geol. Surv. Gt. Britain.*

STRAHAN, A. 1885. Geology of Rhyl, Abergele, and Colwyn. *Mem. geol. Surv. Gt. Britain.*

——. 1890. Geology of Flint, Mold, and Ruthin. *Mem. geol. Surv. Gt. Britain.*

——. 1907. The country around Swansea. *Mem. geol. Surv. Gt. Britain.*

SULLIVAN, R. 1965. The mid-Dinantian stratigraphy of a portion of central Pembrokeshire. *Proc. Geol. Ass.*, **76**, 283–300.

——. 1966. The stratigraphical effects of the mid-Dinantian movements in south-west Wales. *Palaeogeogr., Palaeoclimatol., Palaeoecol.*, **2**, 213–44.

WALTON, J. 1926. Contributions to the knowledge of Lower Carboniferous plants. *Phil. Trans. roy. Soc.*, **B211**, 210–24.

WEDD, C. B. and others. 1923. The geology of Liverpool. *Mem. geol. Surv. Gt. Britain.*

—— and ——. 1929. The country around Oswestry. *Mem. geol. Surv. Gt. Britain.*

——, SMITH B. and WILLS L. J. 1927. The geology of the country around Wrexham. *Mem. geol. Surv. Gt. Britain.*

WELCH, F. B. A. and TROTTER, F. M. 1960. Geology of the country around Monmouth and Chepstow. *Mem. geol. Surv. Gt. Britain.*

THE NAMURIAN SERIES IN SOUTH WALES

D. G. Jones

I. GENERAL STRATIGRAPHY

THE Millstone Grit as now recognised in Britain has the same biostratigraphical limits as the Namurian Series and recently in South Wales there has been considerable progress in the detailed stratigraphical study of the outcrop encircling the coalfield—see Ramsbottom (1954), Woodland & Evans (1964), Archer (1965 and 1968), Squirrel & Downing (1969), Jones (1958, 1969 and 1971), Jones & Owen (1956 and 1966), Owen & Jones (1961), George (1956 and 1970). This work has extended and refined the lithological, faunal and, to a lesser extent, the floral sequences proposed by Evans & Jones (1929), Robertson (1929 and 1933) and Dix (1931 and 1933). Especial mention must be made of the classic work of Ware (1939) in Carmarthenshire which can be regarded as the starting point for the more recent studies. While a great deal of stratigraphical information is now available, absence of data concerning the character and age of sub-Westphalian rocks beneath the coalfield greatly handicaps stratigraphical correlation and interpretation of facies change, between the North and South Crops.

The initial lithological subdivision of the Millstone Grit was made for convenience of mapping the outcrops north of the Newport district by the second major geological survey in the eighteen nineties. Later, by convention and usage the type lithological designations for what was then roughly regarded as a formation, were derived from this area. The original sub-division was:—

FAREWELL ROCK ...	...	(A sandstone which between Risca and Brynmawr apparently lay below the lowest workable coal)
SHALE GROUP ...	...	(Sometimes called Middle Shales)
BASAL GRITS ...	...	(Now called Basal Grit)

Subsequently these lithological groupings were regarded as representative of the Millstone Grit succession throughout South Wales. The problems of recognising them lithologically and of defining them in relation to the faunal and floral succession in Carmarthenshire was fully discussed by Ware (1939). He encountered particular difficulty in identifying and defining the Farewell Rock subdivision. It is now known that in the past this name was given to various sandstones ranging from low Namurian to lower Westphalian A in age. Furthermore, as Leitch *et al.* (1958, p. 462), Archer (1968, p. 5) and Squirrel & Downing (1969, p. 2) have shown that the term lacks stratigraphical significence, it is plain it should be discarded. Ware added a fourth member to the lithological succession—the Plastic Clay Beds, lying between the Upper Limestone Shales (then regarded as Lower Carboniferous) and the Basal Grit. The Basal Grit and Shale Groups as general lithological subdivisions and thus as mappable units are recognisable in some areas of the North Crop but when related to biostratigraphical units, they show much lithological deviation from the original lithological designation. On the South Crop the Basal Grit and Shale Group were shown to be inapplicable as formational terms in the mainly shale and mudstone succession between Risca and Gower, by Woodland & Evans

(1964, p. 10) and Squirrell & Downing (1969, p. 91). In Pembrokeshire, on both North and South Crops, the same difficulty of applying the original tripartite sub-division is encountered (Archer 1965, Jones 1969).

The confusion arising out of the use of the older lithostratigraphical terms and subsequent miscorrelations has led to an increasing reliance on correlation by biostratigraphical units, based on goniatites. The zonal refinement of the northwest European standard goniatite scheme in some stages now equals that attained in the ammonite succession of the Jurassic (Ramsbottom 1969, p. 2240). The stages and zones recognised to date in South Wales are given in fig. 23, in a comparison with the accepted standard scheme. As yet, there is no designated local "stratotype" for

STAGE	ZONE	SUBZONE
YEADONIAN (G_1)	($\underline{G}$. $\underline{cumbriense}$(G_{1b}))* ($\underline{G}$. $\underline{cancellatum}$(G_{1d}))*	($\underline{G}$.$\underline{crencellatum}$ * ($\underline{G}$.$\underline{cancellatum}$ * ($\underline{G}$.$\underline{branneroides}$
MARSDENIAN (R_2)	($\underline{R}$. $\underline{superbilingue}$($R_{2b}$))* ($\underline{R}$. $\underline{bilingue}$(R_{2b}))* ($\underline{R}$. $\underline{gracile}$(R_{2a}))	($\underline{Donetzoceras}$ $\underline{sigma}$ * ($\underline{R}$. $\underline{superbilingue}$ * ($\underline{R}$. $\underline{metabilingue}$ * ($\underline{R}$. $\underline{bilingue}$ *
KINDERSCOUTIAN (R_1)	($\underline{R}$. $\underline{reticulatum}$(R_{1c}))* ($\underline{R}$. $\underline{nodosum}$(R_{1b})) ($\underline{R}$. $\underline{circumplicatile}$($R_{1a}$))*	($\underline{R}$. $\underline{coreticulatum}$ ($\underline{R}$. $\underline{recticulatum}$ * ($\underline{R}$. $\underline{nodosum}$ ($\underline{R}$. $\underline{dubium}$ ($\underline{R}$. $\underline{todmordenense}$ ($\underline{R}$. $\underline{circumplicatile}$ *
ALPORTIAN (H_2)	($\underline{Ht}$. $\underline{prereticulatus}$($H_{2c}$))* ($\underline{H}$. $\underline{undulatum}$(H_{2b}))* ($\underline{Hd}$. $\underline{proteus}$(H_{2a}))	($\underline{Ht}$. $\underline{prereticulatus}$ * ($\underline{H}$. $\underline{eostriolatum}$
CHOKIERIAN (H_1)	($\underline{H}$. $\underline{beyrichianum}$.H_{1b})* ($\underline{H}$. $\underline{subglobosum}$(H_{1a}))*	
ARNSBERGIAN (E_2)	($\underline{N}$. $\underline{nuculum}$(E_{2c}))* ($\underline{Ct}$. $\underline{nitidus}$(E_{2b})) ($\underline{E}$. $\underline{bisulcatum}$(E_{2a}))*	($\underline{N}$. $\underline{nuculum}$ * ($\underline{N}$. $\underline{stellarum}$ * ($\underline{Ct}$. $\underline{nititoides}$ ($\underline{Ct}$. $\underline{nitidus}$ ($\underline{Ct}$. $\underline{edalensis}$ ($\underline{E}$. $\underline{bisulcatum}$ * ($\underline{C}$. $\underline{cowlingense}$ *
PENDLEIAN (E_1)	($\underline{C}$. $\underline{malhamense}$(E_{1c})) ($\underline{E}$. $\underline{pseudobilingue}$($E_{1b}$))* ($\underline{C}$. $\underline{leion}$(E_{1a}))	($\underline{E}$. $\underline{hudsoni}$ ($\underline{E}$.$\underline{stubblefieldi}$ ($\underline{E}$.$\underline{tornquisti}$

C.	= $\underline{Cravenoceras}$	H.	= $\underline{Homoceras}$
Ct.	= $\underline{Cravenoceratoides}$	Hd.	= $\underline{Hudsonoceras}$
E.	= $\underline{Eumorphoceras}$	Ht.	= $\underline{Homoceratoides}$
G.	= $\underline{Gastrioceras}$	R.	= $\underline{Reticuloceras}$
*	= Recognized in S. Wales	N.	= $\underline{Nuculoceras}$

Fig. 23.   The Namurian stages and goniatite zones and subzones recognised to date in South Wales, in comparison with the internationally agreed standard scheme. Standard scheme after Ramsbottom.

the Namurian in South Wales and pending a reclassification, the stage divisions should be preferred to the older rock stratigraphical names. Where goniatites occur they are pre-eminent as time markers but in their absence plants can locally be of great value. In South Wales their remains are common throughout the succession either as driftwood or comminuted plant debris but beds with well-preserved material and diversity of genera and species are confined to the eastern half of the North Crop. The succession of fossil plants at the head of the Swansea Valley was long ago described by Robertson (1933) and Dix (1933), the latter erecting a sequence of floral zones which were regarded as standard for the British Isles. Two of the zones were regarded as characteristic of the Millstone Grit, while a third ranged into the Lowest Coal Measures; the zones were:—

Floral C   Zone of *Lyginopteris hoeninghausi* (Range into Westphatian)
            and *Neuropteris schlehani*
Flora B   Zone of *Pecopteris aspera*
Flora A   Zone of *Lyginopteris stangeri* and *Alethopteris parva*

Subsequently, Ware (1939) showed that the plant assemblage with *Lyginopteris stangeri* as the principal constituent occurred in association with *Nuculoceras nuculum* in Carmarthenshire and were thus Arnsbergian in age. Jones (1958) recognised them in the lower Namurian succession at the head of the Neath Valley. Plant assemblages of the Chokierian, Alportian and Kinderscoutian stages are poorly represented in South Wales—in common with the general dearth of assemblages throughout Britain at these levels. The assemblages of the Marsdenian and Yeadonian stages are fairly diversified, and reasonably well related to goniatite-bearing marine bands. In spite of later additions to the Namurian floras in South Wales, it is difficult to support Dix's recognition of three zonal divisions—a major drawback being the paucity of Chokierian, Alportian and Kinderscoutian records. However, with the present state of knowledge there are sufficient records to show that the early Namurian floras in South Wales (mainly Arnsbergian) were characterised by species of *Lyginopteris*, *Rhodea* and by *Diplotmema adiantoides* and *Sphenophyllum tenerrimum* followed by later assemblages with neuralethopterids and species of *Neuropteris*. At present, little microfaunal and no palynological data is available in published form to add further stratigraphical refinement, or to supplement, the goniatite-based biostratigraphy.

(a) The Pendleian Stage and the base of the Namurian

Paucity of exposure and fossils, folding and faulting, developing nonsequence and disconformity all result in the Pendleian being poorly defined in South Wales and on the North Crop its presence is inferred. Although there are records of *Eumorphoceras* sp. from the Plastic Clay Beds, the first definitely recognised zone is that of *Eumorphoceras bisulcatum* in shales within the overlying local Basal Grit indicating an Arnsbergian (E2$_a$) age near Capel-hir-bach in Carmarthenshire (Ware 1939, p. 200). The apparent conformity of the Plastic Clay Beds to a thin sequence of Upper Limestone Shales (Upper *Dibunophyllum* subzone—'D$_3$' of the Avonian scheme) and the occurrence of bivalvia such as *Leiopteria longirostris*, which was recorded in Gower in association with *Eumorphoceras* cf. *pseudobilinque* by Dix (1931, p. 533), suggests that the Pendleian may be in part represented in the Plastic Clay Beds. However, before definite conclusions can be drawn, there is need for a re-examination of the stratigraphical position of the Upper Limestone Shales and also limestones allocated in some areas to 'D$_2$' (Middle *Dibunophyllum* subzone) so that an upper limit may be placed to the Viséan in South Wales. The Plastic Clay Beds are confined in outcrop to the area between Drefach and Llygad Llŵchŵr. East and west of these localities the basal Namurian beds are apparently everywhere disconformable to underlying strata on the North Crop. Restriction of the stage to the Gower area of the South Crop was suggested by Woodland and Evans (1964, p. 9) in a comprehensive discussion of the stratigraphy and fauna around the Namurian-Visean junction. Although goniatites e.g. *Sudeticeras* sp. ranging from Upper *Posidonia* (P$_2$) to Arnsbergian (E$_2$) age were recorded from a borehole near Aber Kenfig they concluded that non-sequence resulted in the absence of the E$_1$ stage (Pendleian) between Gower and Bridgend. Eastwards along the South Crop, although the details are often obscured by Mesozoic unconformity, the sub-Namurian disconformity increases in magnitude until near Risca the Namurian basal beds are of Yeadonian age (Hall & Squirrell 1972) and rest on Tournaisian limestones.

Poor exposures and often severe folding and faulting prevent definite conclusions as to the exact age of the lowest Namurian on the South Crop in Pembrokeshire. At Tenby the lowest goniatites indicate an upper Chokierian (H$_{1b}$) age while the uppermost Viséan has been allocated to the Middle *Dibunophyllum* subzone (D$_2$) by Sullivan (1964). The presence of *Nuculoceras nuculum* at Lydstep in beds towards the base of the local succession overlying a thin development of the Lower *Dibunophyllum* subzone indicates an Arnsbergian age for the lowest Namurian and supports the conclusion of Dixon & Pringle (1935, p. 17) that disconformity is present on the South Crop in Pembrokeshire. Archer (1965, p. 148) suggested that disconformity prevails everywhere on the Pembrokeshire North Crop but Kelling & G. T. George (1971, p. 242) record the occurrence of conodonts (names unlisted) indicative of an E$_1$ age and thus suggestive of non-sequence rather than disconformity at the Namurian base of the Pembrokeshire eastern North Crop at Ragwen Point. If this is sustained slight modifications of the isopachytes and basin expansion are needed in figs. 25 and 26.

(b) The upper limit of the Namurian

By international agreement the upper limit of the Namurian in western Europe is placed at the base of the *Gastrioceras subcrenatum* Marine Band and as it is widespread in South Wales recognition of the upper limit usually presents no difficulty. However, it should be noted that a marine band, stratigraphically higher than the *Gastrioceras subcrenatum* Marine Band, yields *G. subcrenatum* in an area of overthrusting at Waterwynch in Pembrokeshire.

II.   CONTROLS ON PALAEOGEOGRAPHY AND FORM OF THE BASIN

(a) General

The major controls on early Namurian palaeogeography were mostly inherited

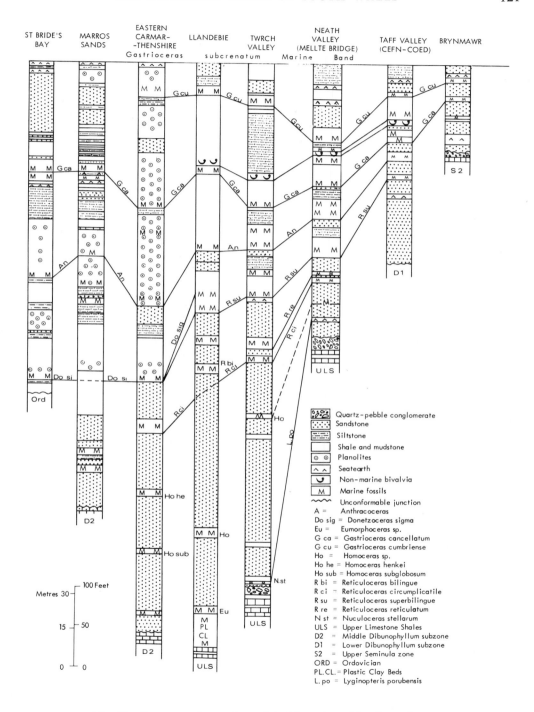

Fig. 24.   Representative sections of the Namurian of the North Crop.

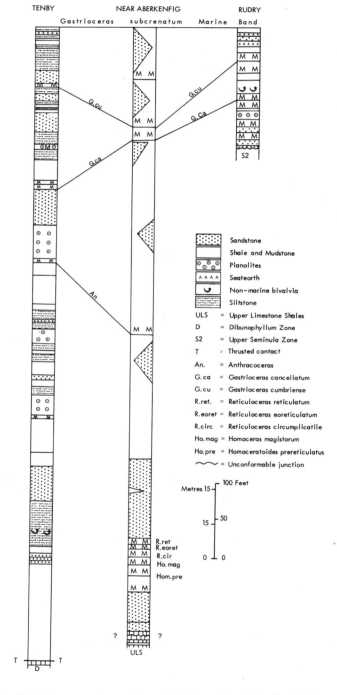

Fig. 25.   Representative sections of the Namurian of the South Crop.

from Upper Viséan times and were movements in the hinterland of the St. George's Land massif (Robertson 1933, George 1956, 1958, 1970 and Owen 1971). These led to shifts of its southerly margin and consequently of its depositional edge. In early Visean times the massif had a broad east-west trend which was later modified by movement along lines such as the malvernoid Usk axis and caledonoid Neath Disturbance. The southerly fringe thus became indented, with shallow gulfs occupying sags between local elevated or shoal-type promontories. Although the growth and varying influence of the external controls can be traced outside of the Namurian basin, the details and causes of subsidence on the actual site of sediment accumulation remain vague because of the lack of borehole and geophysical information of the sub-Namurian foundation of the South Wales coalfield.

(b) The late Viséan-early Namurian basin

Appreciable amounts of quartz sand in the limestones of the Upper *Seminula* Zone (S_2) and in those of the Lower and Middle *Dibunophyllum* subzones (D_1 and D_2) north of the coalfield testify to contemporary uplifts of St. George's Land. On the line of the Neath Disturbance these become pebbly and together with seatearths and thin coals chow early southward directed encroachment of coastal flat and deltaic conditions into the carbonate basin of the central North Crop. The eastern margin of the basin was conditioned by a southerly projection of St. George's Land, identified with the Usk axis. First hints of movement along this line are suggested by by the presence of quartzose sands in the Upper *Seminula* Zone (S_2) near Brynmawr and in deflection of the zonal isopachytes (George 1958, fig. 9) but main effects of uplift occurred in later Visean and early Namurian time. The parallelism of easterly directed Namurian overstep overlap and attenuation between the North and South Crops of the eastern half of the basin indicate regional uplifts on a much greater scale than local upwarpings along the line of the present Usk anticline (George 1956).

In east Carmarthenshire, although a southwesterly trend of the margin of St. George's Land can be deduced from the trend of the late Viséan and early

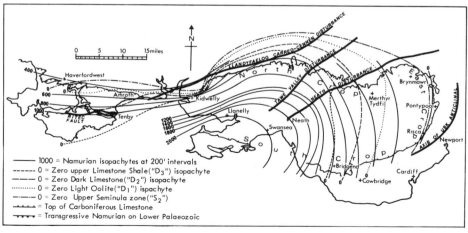

Fig. 26.   Combined late Viséan and Namurian isopachytes for South Wales.

Namurian isopachytes (fig. 26), recognisable local structural control along structural lines such as the Llandyfaelog-Carreg-Cennen Disturbance or Pontesford Hill Fault is less in evidence. The Upper Limestone Shales-Plastic Clay Beds lithology here suggests a contemporary hinterland of low mature relief.

The major influence on the form and margin of the early Namurian basin in Pembrokeshire was late Viséan uplift of St. George's Land, probably locally emphasised by hinge-type movements along structures such as a proto-Ritec Fault system. These led to an extension of the southern edge of the massif to a position certainly to the south of the present main Namurian outcrop in south Pembrokeshire, in earliest Namurian time.

The older interpretations of the disposition of the upper Viséan zonal units in South Wales regarded the form of their sub-Millstone Grit outcrop entirely as a reflection of early Namurian tectonic movements and resultant sub-Millstone Grit unconformity (Dixey & Sibly 1919, Robertson & George 1929, Pringle & George 1937). However, there is increasing evidence that late Viséan marine regression led to an off-lap situation whereby carbonate sediments were deposited in successively shrinking areas. On the North Crop the regression was accompanied by a facies change whereby coarse quartzitic gravels replaced the carbonate sediments in areas not suffering actual emergence. These gravels are now preserved as irregular patches of quartz-pebble conglomerate where they are of late-Visean age (Owen & Jones 1961). They often contain marine fossils such as foraminifera, productoid and spiriferoid brachiopods, crinoidal ossicles, fragments of corals, polyzoan and algal tissue and fish teeth. These serve to distinguish them from overlying transgressive Namurian sandstones, which carry only occasional crinoidal ossicles and more abundantly lycopod and calamatid stems. The quartz-pebble conglomerate is a markedly diachronous phase in the sense that influxes of pebbly sands began as early as mid-Upper *Seminula* (S_2) times on the north-easterly fringes of the Viséan basin e.g. at Garn Caws and subsequently spread intermittently south-westwards (cf. George 1954, p. 302). Thus in the Taff Valley the uppermost oolitic and crinoidal layers of the Light Oolite subdivision (of Lower *Dibunophyllum* subzone D_1 age) are abruptly replaced by coarse pebbly sands. These are overlain, in turn, by at least 21 m. of sandstones in the Dowlais area. Evidence of age of these sandstones is only indirectly provided by the occurrence of *Gastrioceras cancellatum* in Nant Morlais 28 m. above the Viséan limestones (Jones 1970, p. 1026). It is not improbable that apart from the lowermost pebbly beds that some of the overlying sandstones may be of Viséan rather than of Namurian age in this area. The Dark Limestone subdivision (Middle *Dibunophyllum*—D_1 subzone) likewise has coarse pebbly sands in the uppermost beds, Owen & Jones (1961) and the thinning of the limestones noted near Penderyn (Owen 1954, p. 551) is probably due to contemporaneous replacement of limestones by quartzitic sediments as well as pre-Namurian arching along the Neath Disturbance.

West of the Neath Valley to Llygad Llŵchŵr pebbly sands overlie a thin development of Upper Limestone Shales and the presence of *Lyginopteris porubensis* indicative of an Arnsbergian age at many localities in beds everlying the pebbly sands shows no further marked diachronism of the basal beds. The pebbly sands die out above the Upper Limestone Shales west of Llygad Llŵchŵr, their place being taken by the Plastic Clay Beds which conceivably may be of the same age.

The disposition of late Visean sediments in early Namurian times was thus a consequence of a rise of St. George's Land and also, more in evidence in the eastern half of the basin, a complementary rise of the Usk uplift. A further consequence was that the early Namurian basin was a considerably shrunken remnant of the earlier Visean one and with a facies of quartzose gravels in the north-east and a marine but non-calcereous facies of the Plastic Clay Beds type in east Carmarthenshire and chert-shale facies in the Gower area.

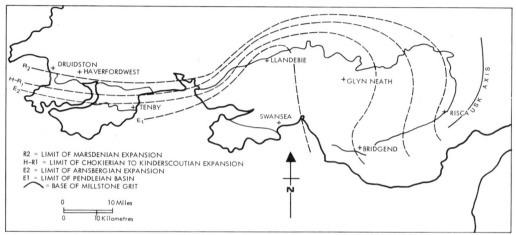

Fig. 27. Development of the Namurian basin in South Wales. (Should the Quartz-pebble Conglomerate prove to be of Pendleian age between the Neath Valley and Llygad Llŵchŵr, the E_1 limit would have to be taken eastwards to Glynneath.)

(c) Later development of the Namurian basin

In post-Pendleian times there was a marked eastward expansion of the area of deposition and to a lesser extent northwards (Fig. 27). The thin and mainly arenaceous succession of the central North Crop suggests a close northerly margin within miles of the present outcrop. Furthermore, no Namurian is known in the Welsh borderland but a possible link with the Clee Hills may be suggested by the occurrence of Namurian in the lower levels of the Cornbrook Sandstone of Titterstone Clee (inferred from palynological evidence—B. Owens personal communication).

Post-Pendleian expansion was not extensive in east Carmarthenshire and may have been closely controlled by movement along the line of the Llandyfaelog-Carreg-Cennen Disturbance as indicated by Arnsbergian and Chokierian overstep of Visean limestones in the area between Mynydd-y-garreg and Mynydd Llangyndeyrn. A further local but marked overstep near this area may be indicated by problematical patches of quartzite of presumed Millstone Grit age resting on the Lower Old Red Sandstone at Penlan (Dixon & Pringle 1927, Archer 1968, George 1927 and 1970).

In Pembrokeshire northerly and northwesterly expansion of the basin was accompanied by increasing unconformity which reaches an apparent maximum time magnitude for the Namurian of South Wales near Druidston. Here late Marsdenian (R_2) shales overlie Ordovician sediments and may even rest on Precambrian. Although this is spectacular overstep the sub-Namurian surface beneath the unconformity was a result of earlier Devonian and Dinantian movements as well as Namurian transgression. The presence of basal Namurian shales suggests gentle drowning of a low-lying mature topography rather than comparatively rapid burial of an unstable area that the overstep first suggests (Archer 1965, p. 49).

While the northern and eastern limits of the Namurian depositional area can be fairly well defined, the southern and south-eastern limits are problematical. The thin Namurian succession in the Bristol district and its absence from the Forest of Dean is regarded by Ramsbottom as indicating deposition in a basin separated from South Wales. As the Namurian of Bristol thins and is overstepped by Westphalian A–B, which in turn is overlapped by Westphalian C, all generally northwards toward the Lower Severn Axis, while near Risca thin Namurian is apparently overstepped by attenuated Westphalian A (Jones & Squirrel 1969) these could all be complementary. If this interpretation can be sustained, it indicates a merged Usk-Lower Severn uplift veering west-southwest and providing a southerly margin to the Namurian basin as far west as Cowbridge. At present there is no evidence of a major or any other type of barrier between South Wales and North Devon in Namurian times. The western extension of the basin is even more conjectural. It may have been open ended as in Dinanthian times (George 1969) and linked with a contemporary Atlantic (Ramsbottom 1970, Lavrov 1967, and Webb 1968).

Although there was marked post-Pendleian expansion it did not result in migration of the axis of maximum sedimentation. This remained generally aligned along the trend of the early Namurian trough in the main basin of the coalfield (fig. 26).

While it is certain that the Namurian basin subsided on the stabilised continental crust of the Midland Platform (Dewey 1969, fig. 1 and Dunning 1966), much conjecture obtains as to the nature and causes of actual subsidence of the site. The increasing evidence of close control on the form and development of the basin and the cyclic nature of the sedimentary infilling by renewals of movement along old established structural lines, makes it tempting to adopt the model whereby the subsidence was a consequence of ductile flow in the upper mantle of isostatic origin, as has been proposed for the development of other British Carboniferous intracratonic basins by Bott (1964) and Bott & Johnston (1967). This model has the added attraction of explaining why post-Pendleian expansion was greater on the western edge of the Midland Platform and much less on the fringes of the Caledonian orogenic zone of south and mid Wales. It could provide further clues as to the location and proximity of rising source areas for the sediments. However, while a tectonic control over the general form of the basin is probable and possibly also on its sedimentary fill, the major marine invasions of Yeadonian time show that regional rise and fall in sea-level must be considered a major factor in type of deposition.

III. LITHOFACIES AND BIOFACIES

(a) The Plastic Clay Beds of the North Crop

These mainly consist of fine grained cherty sandstones and clays with sub-ordinate shale bands. The origin of the clays has been ascribed to weathering and disintegration of chert (Strahan 1907, Ware 1939). Bedded vitreous chert, however, is uncommon and most of the siliceous beds are very fine grained sandstones with over 65 per cent detrital quartz grains cemented by colloidal silica and are thus cherty sandstones. The thicker clay members are often up to 3 m. thick and X-ray examination (A. Hall personal communication) shows them to be mainly composed of illite with some quartz. The illite is the 2M polytype characteristic of derivation from igneous and metamorphic rocks and not the 1Md authigenic polytype. Furthermore, kaolin is rare in the clays. A normally preserved benthonic fauna of fenestrate polyzoa, spiriferoid and productoid brachiopods, crinoidal ossicles and zaphrentoid corals is present as also are primary bedding planes. There seems therefore little doubt that the thicker bedded clays are primary marine argillaceous sediments and not decomposed weathered residuum. The Plastic Clay Beds represent a northerly and more littoral, benthonic extension of the chert-shale facies of Gower but still laid down in quiet water as is suggested by the occurrence of phosphatic nodules in the fine grained greasy clays. In Gower, the chert-shale lithofacies lies in the goniatite-pectinoid biofacies belt of Ramsbottom (1970, p. 151).

(b) Arenaceous beds

These can be conveniently placed in the four basic categories (Moore 1959) of pebbly sandstone, flaggy sandstones, sheet and channel sandstones together with finer material of silt grade. Pebbly sandstones are common in the Arnsbergian to Kinderscoutian stages of the North Crop but absence of characteristic or fossiliferous pebbles give no hint of provenance although their close resemblance to Upper Old Red Sandstone sediments has led to the suggestion that they may have been derived from the Old Red Sandstone by erosion or further by erosion of Old Red Sandstone and Lower Palaeozoic and Pre-Cambrian areas to the north and north-east (George 1970, p. 75). The pebbles, which are well sorted, are overwhelmingly composed of vein quartz with rare orthoquartzite, jasper and felsite. Pebbles of Carboniferous Limestone are unknown. Many of the finer sandstones, particularly on the North Crop, are remarkably mature orthoquartzites with up to $98 \cdot 5$ per cent detrital quartz and formerly much worked for refractory material. They show rhythms of sedimentation first described by Robertson (1933, p. 54) and conceivably may be interpreted as fining-up sequences (cf. Reading 1971, p. 1402) as a result of a delta-constructive phases which are also seen in the Marsdenian and Yeadonian stages.

While the sedimentary structures and cross-stratification in the Namurian are typical of the north-west European paralic setting much work is needed to determine the detailed palaeoenvironments for the whole of South Wales on the lines of Reading (1970) and Kelling & G. T. George (1971) in Pembrokeshire. These authors have recorded in great detail the lithology of cliff sections at Marros sands and Telpyn Point and their environmental interpretation shows an overall simplicity in the advance and retreat of a delta and sub-deltas, i.e. in delta-constructive and delta-destructive phases. However, their bed by bed interpretation shows considerable complexity of palaeoenvironmental development. Thus sediments show features indicating accumulation as on-delta distributory, flood plain, advancing coastal bars, delta front, lagoonal and brackish marine deposits.

(c) *Argillaceous sediments and faunal bands*

Mudstones and shales of varying grade and colour are the most persistent lateral members of the succession and in bulk show evidence of accumulation in a marine environment. The use of the term 'Marine Band' in Millstone Grit stratigraphy in South Wales has been discussed by Archer (1968, p. 7), Squirrel & Downing (1969, p.82) and in general by Ramsbottom (1969, p. 84) who concluded, for convenience of description, that it should refer to the richest and most marine part of a marine sequence. They further considered that with a few exceptions such as seatearths and non-marine bivalvia bands all other argillaceous sediments were deposited in an essentially marine environment. The marine phases are interpreted as episodes occurring within a broader pattern of cyclic sedimentation. Ramsbottom (1969, p. 227 and 1970, p. 151) has demonstrated that three broad faunal belts *viz* the *Lingula* biofacies belt, the shelly benthonic boifacies belt and the goniatite-pectonoid facies belt could be distinguished in the marine episodes and arranged according to distance from the contemporary shoreline (fig. 28). Cyclic changes in deposition caused migration of these belts during marine advance and retreat which are now discernible in vertical lithological and faunal profiles of the argillaceous beds.

In the mainly shale and mudstone succession of the South Crop between Gower and Pembrokeshire, the main marine phases remained almost entirely within the goniatite-pectinoid belt from Pendleian to late Marsdenian time and even in Yeadonian time, establishment of benthonic faunas was sporadic. This poor representation of benthonic faunas has been regarded as indicating deoxygenating conditions over most of the southern portion of the basin. Linked controlling factors probably also included rate and type of mud sedimentation, particle size and bottom current velocity, as thick sequence of characteristically grey and 'soapy' mudstones are barren in the sense of not even having visible trace fossils.

Although the present outcrops trend broadly parallel to the depositional strike, far more complex litho- and biofacies patterns can be discerned on the North Crop throughout the Namurian period in the argillaceous beds. The complexity in development of faunal phases is well illustrated on the North Crop in Carmarthenshire and the upper Swansea valley to Taff Valley areas. Archer (1968, p. 8) distinguishes eight phases, the order of occurrence of which shows increasing evidence of "open-sea" conditions and thus broadly of palaeosalinity. These phases are:

1. Barren phase
2. *Planolites* phase
3. *Lingula* phase

4. Spat phase
5. Lamellibranch phase
6. *Anthracoceras* or *Dimorphoceras* phase
7. 'Goniatite' phase
8. Brachiopod phase

The recognition of regional biofacies belts and of their detailed phasal development enables previously puzzling variations in both lateral and vertical lithological and

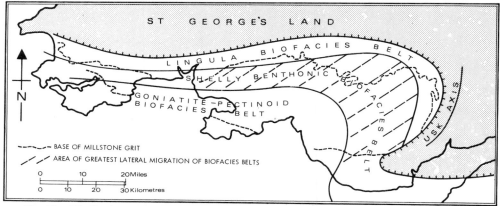

Fig. 28.   Idealised arrangement of biofacies belts in relation to contemporary shorelines—adapted from Ramsbottom.

faunal profiles to be convincingly interpreted in environmental terms. Thus the former problems of 'inconstancy of the marine bands' (Robertson 1933, p. 53 and Ware 1939, p. 198) are now resolved. Again marked contrasts in the richness of the benthonic faunas of the *Gastrioceras cancellatum* and *G. cumbriense* Marine Bands particularly between the Swansea and Taff Valleys can be shown to be due to changing geographical position of the biofacies belts. While differential development of the combined litho- and biofacies explains local differences in the thickness of the general marine profile, complete absence of marine argillaceous sediment is often the result of removal by channelling during deposition of later sandstones as a result of a delta constructive phase.

Quasi-marine conditions in mudstones are represented in the Marsdenian and Yeadonian stages by phases with *Planolites opthalmoides* and in the South Crop show that fluctuations of palaeosalinity possibly of cyclic origin can be detected even in the predominantly shale and mudstone sequences. Non-marine bivalvia occur at two levels, in the Kinderscoutian of Pembrokeshire, and in a more widespread phase in the Yeadonian of the central North Crop. In this latter phase sections in the Neath Valley at right angles to the depositional strike show size, shape and shell thickness variations in *Carbonicola* and *Anthraconaia* to be linked with lithological changes. These are similar to those described by Eagar (1960) and in South Wales may be related to distance from the contemporary shoreline and suggest that a facies belt distribution may apply to these genera also.

Apart from the almost general absence of turbidites and feldspathic sandstone, the Namurian lithofacies in South Wales differs little from that north of St. George's Land in being a record of delta advances and retreats. It is in the distribution and thickness pattern of the main sedimentary types that major differences are seen. In South Wales there is a simple, almost classical relationship between coarsening and thinning northwards and eastwards toward the contemporary shorelines, in marked contrast to the Central Province where sandstones often cause a considerable thickening of the succession at the expense of mudstones, both near and away from the margins of the basin (Ramsbottom 1969, p. 222 and Rayner 1967, p. 208).

IV. SUMMARY

The Namurian lithofacies in South Wales resulted from the south and southwesterly spread of coastal flat and deltaic conditions which had already begun in mid-Visean times on the north-eastern fringes of the basin. The influx of terrigenous sediment led to a progressively westward and thus diachronous cessation of shallow marine carbonate sedimentation in late Visean and early Pendleian time. Isopachyte data, facies change, non-sequences and disconformities suggest that the initial influx of terrigenous material was mainly controlled by related hinterland uplift and intra-cratonic subsidence during this period. The relatively thin though widespread development of Marsdenian and Yeadonian sediments suggests that uplifts in the hinterland area were not so severe or had even ceased. Support for this conclusion may be found in the greater lateral extent of argillaceous sediments and in greater migration of biofacies belts so that eustatic rise and fall of sea level was a major control on both litho- and biofacies development in Marsdenian and Yeadonian times.

V. REFERENCES

ARCHER, A. A. 1965. Notes on the Millstone Grit of the North Crop of the Pembrokeshire Coalfield. *Proc. Geol. Ass.*, **76**, 137–59.

——. 1968. The Upper Carboniferous and later formations of the Gwendraeth valley and adjoining areas. Mem. Geol. Surv. Gt. Britain.

BOTT, M. H. P. 1964. Formation of sedimentary basins by ductile flow of isostatic origin in the upper mantle. *Nature* **201**, 1082–84.

—— and JOHNSON, G. A. L. 1967. The controlling mechanism of Carboniferous cyclic sedimentation. *Quart. J. geol. Soc. Lond.* **122**, 421–41.

DEWEY, J. F. 1969. Evolution of the Appalachian/Caledonian orogen. *Nature* **222**, 124–29.

DIX, E. 1931. The Millstone Grit of Gower. *Geol. Mag.* **68**, 529–43.

——. 1933. The succession of fossil plants in the Millstone Grit and the lower portion of the Coal Measures of the South Wales Coalfield (near Swansea) and a comparison with that of adjacent areas. *Palaeontographica* **78**, 158–202.

DIXEY, F., and SIBLY, T. F. 1919. The Carboniferous Limestone series of the south-eastern margin of the South Wales Coalfield. *Quart J. geol. Soc. Lond.* **58**, 111–64

DIXON, E. E. L., and PRINGLE, J. 1927. The Penlan Quartzite. *Sum. Prog. Geol. Surv. G.B.* for 1926, 123–26.

——. 1935. Guide to geological excursion to South Wales and Bristol. H.M.S.O.

EAGAR, R. M. C. 1960. A summary of the results of recent work on the palaeoecology of Carboniferous non-marine lamellibranchs. *Compt Rend. Quat. Cong. pour l'avanc. des etud.*

strat. et du Geol. du Carb. **1**, 137–99.

EVANS, D. G., and JONES, R. O. 1929. Notes on the Millstone Grit of the North Crop of the South Wales Coalfield. *Geol. Mag.* **68**, 164–77.

GEORGE, T. N. 1927. The Carboniferous Limestone (Avonian) succession of a portion of the North Crop of the South Wales Coalfield. *Quart. J. Geol. Soc. Lond.* **83**, 38–95.

——. 1954. Pre-Seminulan Main Limestone of the Avonian Series in Breconshire. *Quart. J. Geol. Soc. Lond.* **110**, 283–321.

——. 1956. The Namurian Usk anticline. *Proc. Geol. Ass.*, **66**, 297–316.

——. 1958. Lower Carboniferous palaeogeography of the British Isles. *Proc. Yorks. Geol. Soc.* **31**, 227–318.

——. 1969. British Dinantian Stratigraphy. *Compt Rend. 6me Congr. Strat. Carb.*, **1**, 193–218.

——. 1970. South Wales. *Brit. Reg. Geol.* 3rd ed.

HALL, I. H. S., and SQUIRRELL, H. C. 1972. New sections in the Basal Westphalian and uppermost Namurian strata at Risca and Abersychan, Monmouthshire. *Bull. Geol. Surv. G.B.* **38**, 15–41.

JONES, D. G. 1958. A note on new Namurian plant localities at the head of the Neath Valley, South Wales. *Geol. Mag.* **95**, 77–81.

——. 1969. The Namurian succession between Tenby and Waterwynch, Pembrokeshire. *Geol. J.* **6**, 267–72.

——. 1971. The base of the Namurian in South Wales. *Compt Rend. 6me Congr. Strat. Carb.* **111**, 1023–29.

—— and OWEN, T. R. 1957. The rock succession and geological structure of the Pyrddin, Sychryd and Upper Cynon Valleys, South Wales. *Proc. Geol. Ass.*, **67**, 232–50.

—— and OWEN, T. R. 1967. The Millstone Grit succession between Brynmawr and Blorenge. *Proc. Geol. Ass.*, **77**, 187–98.

—— and SQUIRRELL, H. C. 1969. The discovery of the *Gastrioceras subcrenatum* Marine Band at Risca, Monmouthshire. *Bull. Geol. Surv. G.B.* **30**, 65–70.

KELLING, G., and GEORGE, G. T. 1971. Upper Carboniferous sedimentation in the Pembrokeshire coalfield. In *Geological Excursions in South Wales and the Forest of Dean.* Bassett, D. A. and M. G. (eds.) Cardiff.

LAVROV, V. M. 1967. North Atlantic latitudinal, trough. *Dok. Acad. Sci. U.S.S.R.* **174**, 99–101.

LEITCH, D., OWEN T. R. and JONES, D. G. 1958. The basal Coal Measures of the South Wales Coalfield from Llandybie to Brynmawr. *Quart. J. Geol. Soc. Lond.* **133**, 461–86.

MOORE, D. 1959. Role of deltas in the formation of some British Lower Carboniferous cyclothems. *Journ. Geol.* **67**, 522–39.

OWEN, T. R. 1954. The structure of the Neath Disturbance between Bryniau Gleision and Glynneath. *Quart. J. Geol. Soc. Lond.* **109**, 333–65.

——. 1971. The relationship of Carboniferous sedimentation to structure in South Wales. *Compt. Rend. 6me Congr. Strat. Carb.* **111**, 1305–15.

—— and JONES, D. G. 1961. The nature of the Millstone Grit-Carboniferous Limestone junction of a part of the North Crop of the South Wales Coalfield. *Proc. Geol. Ass.*, **72**, 239–49.

PRINGLE, J., and GEORGE, T. N. 1937. South Wales. *Brit. Reg. Geol.* 1st ed.

RAMSBOTTOM, W. H. C. 1954. In *Sum. Prog. Geol. Surv. G.B.* for 1953, 54.

——. 1969. The Namurian of Britain. *Compt Rend. 6me Congr. Carb. Strat.* **1**, 219–32.

——. 1970. Carboniferous faunas and palaeogeography of the South West of England region. *Proc. Ussher Soc.* **2**, 144–57.

RAYNER, D. 1967. The stratigraphy of the British Isles. *Camb. Univ. Press.*

READING, H. G. 1971. Sedimentation sequences in the Upper Carboniferous of Western Europe. *Compt Rend. 6me Congr. Carb. Strat.* 1967, **4**, 1401–12.

ROBERTSON, T. R. 1927. The geology of the South Wales Coalfield. Part II. The country around Abergavenny. 2nd ed. *Mem. geol. Surv. Gt. Britain.*

——. 1933. The country around Merthyr Tyffil. 2nd ed. *Mem. geol. Surv. Gt. Britain.*

—— and GEORGE, T. N. 1929. The Carboniferous Limestone of the North crop of the South Wales Coalfield. *Proc. Geol. Assoc. Lond.* **49**, 18–40.

SQUIRRELL, H. C., and DOWNING, R. 1968. The geology of the South Wales Coalfield. Part I. The country around Newport (Mon.). 3rd ed. *Mem. geol. Surv. Gt. Britain.*

STRAHAN, A., and CANTRILL, T. C. 1907. Part VII. The country around Ammanford. 1st ed. *Mem. geol. Surv. Gt. Britain.*

SULLIVAN, R. 1965. The mid-Dinantian stratigraphy of a portion of central Pembrokeshire. *Proc. Geol. Ass.*, **31**, 76–92.

—— 1966. The stratigraphical effects of the mid-Dinantian movements in south-west Wales. *Palaeogeog., Palaeoclimat., Palaeoecol.* **2**, 213–44.

WARE, W. D. 1939. The Millstone Grit of Carmarthenshire. *Proc. Geol. Ass.*, **50**, 168–204.

WEBB, G. W. 1968. Palinspastic restoration suggesting late Palaeozoic North Atlantic rifting. *Science N.M.* **159**, 875–78.

WOODLAND, A. W., and EVANS, W. B. 1964. The geology of the South Wales Coalfield. Part IV. The country around Pontypridd and Maesteg. 3rd ed. *Mem. geol. Surv. Gt. Britain.*

THE WESTPHALIAN (COAL MEASURES) IN SOUTH WALES

L. P. Thomas[1]

I. INTRODUCTION

THE Westphalian rocks of South Wales are of an entirely sedimentary character, being composed of argillaceous and arenaceous coal-bearing sequences. Strata of Westphalian age occur in three distinct areas (see Fig. 29): the Pembrokeshire coalfield in the west, the South Wales Coalfield main basin (the largest area) in the middle; and the small Forest of Dean Coalfield in the east.

These coalfields are the eroded remnants of an area of Carboniferous sediments affected by post-Westphalian earth movements, which resulted in a downfold with an east-west trend across the middle part of South Wales, and led to the preservation of a large basin of Westphalian strata, containing many minor folds and a well-developed fault pattern. Smaller areas of similar strata are found in Pembrokeshire, here however the tectonic influence has been more severe and major thrusting, as well as faulting and folding, has complicated the Westphalian sequence. The remaining area, the Forest of Dean Coalfield, contains a less complete Westphalian sequence which has undergone gentler synclinal deformation and simple faulting. Subsequent erosion has produced a topography which, especially in the case of the central basin area, is reflective of its geological structure. The geology of each area is best considered separately but correlation between the areas has been attempted where possible.

(a) Cyclic Sedimentation

The Westphalian sequence in South Wales consists of a series of sediments arranged in successions of repetitive units. These repetitive units are termed rhythms, cycles or cyclothems and have been described by Trueman (1947), Woodland & Evans (1964), Squirrell & Downing (1969) and George (1970). George (1970 p. 83) describes a 'typical' cycle as beginning with a coal, representing a dense vegetation area formed at or near water level, succeeded by a marine shale or impure limestone containing goniatites, brachiopods, foraminifera, gastropods and bivalves, which represents a subsidence phase. This is followed by shales becoming increasingly shallow-water and non-marine in character, having a non-marine mussel fauna, and passing upwards in turn into sandy shales, sandstones and occasionally coarser arenites. The cycle is completed by underclays and rootlet beds, with their overlying coal, this last phase representing shallow water mud deposition and the re-establishment of vegetation cover. Woodland and Evans (1964, p. 20) show that a 'typical', 'complete' or 'ideal' cyclothem rarely occurs, local variations in conditions of subsidence and deposition inevitably causing modifications to the pattern, and producing a cyclothem which differs widely in many cases from the "ideal" described above. Woodland and Evans (1964, p. 21) have described four main variations in the Coal Measures of the Maesteg-Pontypridd

---

[1] Published by permission of the Director, Institute of Geological Sciences.

district. These are illustrated in Fig. 30, the 'ideal', cyclothem being best represented by that of the Upper Middle Coal Measures. The typical 'pennant' cyclothems are

Fig. 29. Sketch map showing the outcrops of the Coal Measures in South Wales. The main basin is divided into six areas in the text.

markedly different, indicating that they were formed under different conditions of deposition (Woodland & Evans 1964, p. 21, Owen 1964, p. 306), and the incoming of this type of cyclothem is a reflection of major geographical changes in the morphology of the sedimentary basin in South Wales. Such changes were also inferred by Squirrell & Downing (1969, p. 94). They considered the 'complete' cyclothem to represent 'a period of subsidence allowing transgression of the strandline, followed by a regression as the delta front gradually advanced. The sequence of environments caused by these fluctuations was, shallow sea, subaqueous delta with brackish water, subaerial delta and coal swamp. A final stage was that of an inland flood plain which was probably attained at the close of some cycle in the upper part of the Middle Coal Measures and was largely a feature during the formation of the Upper Coal Measures'.

(b) Stratigraphy

The Coal Measures of South Wales have been examined by numerous workers and the sequence was classified on the basis of non-marine mussels (Davies and Trueman, 1927) and flora (Dix 1934). Assemblages of non-marine mussels are found at many horizons throughout the Lower and Middle Coal Measures, and less abundantly in the Upper. The mussel species show changes in upward sequence that enabled Davies and Trueman to divide the sequence into zones (see Fig. 31). Dix (1934) proposed seven plant assemblage zones and related them to continental stages of the Westphalian (see Fig. 31). Later classification has been based on the discovery and correlation of marine bands, the advantage of these horizons being that they are thin and widely developed, and are believed to represent single

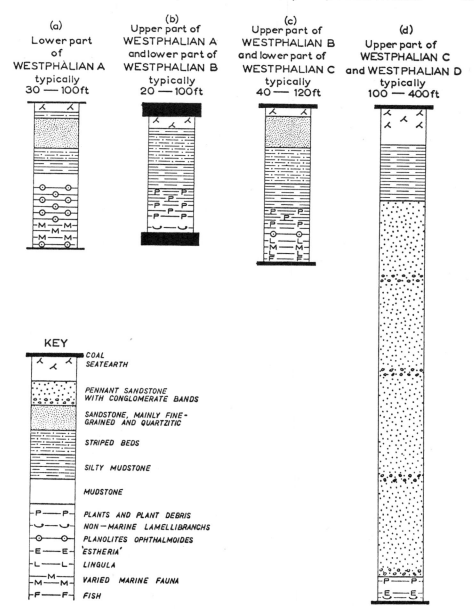

Fig. 30.   Ideal cyclothems in the central part of the South Wales Coalfield (reproduced from Woodland and Evans, 1964, fig. 5).

depositional episodes over the whole coalfield. The more important marine bands are mainly characterised by an association of particular species of fossils. Woodland and Evans (1964), Archer (1968), Squirrell and Downing (1969) and George (1970) have published descriptions of the known marine bands and have listed their typical faunas. A summary of the fauna of the more important marine bands is given below:

Principal Marine Bands of South Wales Coal Measures

Marine Band

Upper Cwmgorse
Anthracoceras cambriense, Politoceras (Homoceratoides) kitchini, Peripetoceras (Cyclonautilus) dubium, Huanghoceras postcostatum, Dunbarella macgregori, Myalina compressa, nuculids, productids, spirifers, ostracods and foraminifera.
(It is equivalent to the Top Marine Band of the Pennine Coalfield and *Anthracoceras cambriense* Marine Band of the Continent and is the highest marine band known in the British Coal Measures.)

Lower Cwmgorse
Coelogasteroceras, Lingula mytilloides, L. cf. *squamiformis, Orbiculoidea, Euphemites anthracinus, Dunbarella macgregori, Posidonia, Hollinella, Naiadites* cf. *daviesi, Geisina subarcuata*, gastropods, palaeoniscid scales.

Five Roads
Geisina subarcuata, Anthraconaia pruvosti, Myalina compressa, Edmondia, Euphemites.

Foraminifera
Foraminifera including *Agathamminoides* sp. *Glomospira* sp., *Hyperammina* sp., *Tolypammina* sp., *Planolites ophthalmoides, Lingula mytilloides, Nuculopsis* cf. *gibbosa, Schizodus axiniformis*, palaeoniscid scales, *Rhabdoderma* sp.

Cefn Coed
Anthracoceras aegiranum, Gastrioceras aff. *globulosum, Politoceras politum, (Homoceratoides jacksoni)*, brachiopods including *Productus, Rugusochonetes skipseyi, Tornquistia diminuta*, bivalves including nuculids, end pectinids, the corals *Zaphrentis postumis*, crinoids, trilobites, ostracods, conodonts, and the fish *Edesia pringlei, Rhabdoderma (Coelacanthus)* and *Platysomus.*
(It is equivalent to the Mansfield, Gin Mine and Dukinfield marine bands of England and to the Aegir marine band of the Continent.)

Britannic or Trimsaran
Foraminifera *Planolites ophthalmoides, Lingula, Orbiculoidea.*

Hafod Heulog or Mole
Foraminifera *P. ophthalmoides, Sphenothallus, Lingula, Orbiculoidea.*

Amman
Foraminifera, *Lingula, Orbiculoidea*, calcareous brachiopods, such as *Spirifer pennystonensis*, bellerophontid and turreted gastropods, crinoids, sponge spicules, ostracods.
(It correlates with the Clay Cross and Pennystone marine bands and equates with the Katharina (*Anthracoceras vanderbecki*) Marine Band of the Continent.)

Margam
Foraminifera, *Lingula mytilloides, Orbiculoidea* sp. bellerophontid gastropods, *Dunbarella papyracea, Posidonia* sp., *Myalina*, conodonts, fish debris.

Cefn Cribbwr
Gastrioceras listeri, G. circumnodosum, nautiloids, *Lingula mytilloides, Dunbarella, Edmondia*, ostracods, fish denris.
(It is correlated with the Halifax Hard Bed and the Bullion Mine marine bands of Yorkshire and Lancashire.)

Gastrioceras subcrenatum
Gastrioceras subcrenatum, Anthracoceras sp., *Homoceratoides divaricatus, Dunbarella* cf. *papyracea, Myalina* cf. *sublamellosa, Posidonia, Lingula mytilloides*, palaeoniscid scales.
(It is correlated with the Pot Clay Marine Band of the Pennines and Sarnsbank Marine Band of the Continent, and is the horizon chosen in Britain and the Continent as demarcation between the Namurian and the Westphalian.)

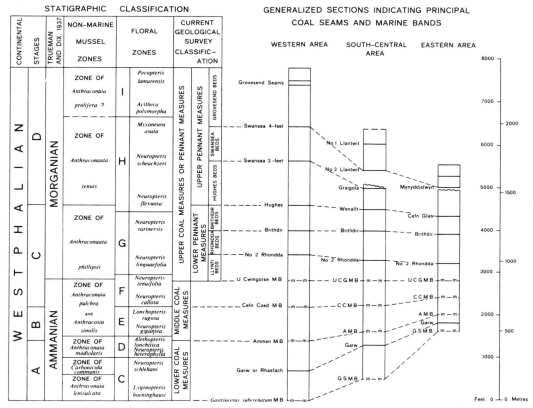

Fig. 31. Classification of the Westphalian (Coal Measures) in South Wales, with outline vertical sections of the Measures in different parts of the main basin. Adapted from George (1970) and Woodland and Evans (1964).

The marine bands exhibit many variations in thickness and faunal content (Calver 1969) throughout the coalfields, and on occasion may be absent altogether e.g. the Britannic Marine Band which has not been proved in the South-East Crop area (Squirrell & Downing 1969, p. 111). Classification, by the use of the distribution of plant spores and pollen in the measures, has been outlined by Smith and Butterworth (1967) and has shown that palynological zones may be used to supplement the macrofloral zones.

These zonal schemes are not all precisely integrated one with another as the boundaries between each mussel zone and each plant zone tend to have been placed at coal seams rather than at marine horizons. The various classifications, together with sections across the main Coalfield basin of South Wales showing the principal Upper Coal Measures coal horizons can be seen in Fig. 31. Trotter and Stubblefield (1957), and Woodland and others (1957) defined the broad divisions of the Coal Measures for use on Geological Survey maps as follows (see Fig. 31):—

3.   Upper Coal Measures or Pennant Measures—all of the sequence above the Upper Cwmgorse Marine Band.

2. Middle Coal Measures—the sequence between the top of the Upper Cwmgorse Marine Band and the base of the Amman Marine Band.

1. Lower Coal Measures—the sequence between the base of the Amman Marine Band and the base of the *Gastrioceras subcrenatum* Marine Band.

This classification has been applied throughout the main coalfield area, and is probably applicable to the Pembrokeshire Coalfield, but in the case of the Forest of Dean Coalfield the absence of most of the Lower and Middle Coal Measures (above which marine bands do not occur) means that an older lithological classification is still applicable and therefore the accepted correlation is tentative. The relationship of the South Wales area to other British coalfields, and to those of the Continent has been described by Calver (1969).

II. THE MAIN BASIN OF THE SOUTH WALES COALFIELD

The main part of the South Wales Coalfield is a structurally elongate basin approximately 1080 square miles (2690 km²) in area. It is a broad east-west synclinal structure closing at the eastern end, with the western edge hidden under Carmarthen Bay. Within this are several smaller anticlines and synclines, all having a broad ENE-WSW trend. The main fault pattern consists of a set of NW-SE trending faults, with several larger faults trending WNW-ESE. There are two lines of major disturbance, roughly following the Tawe and Neath Valleys, these having a NE-SW trend (see Fig. 32). The Westphalian or Coal Measures strata lie on Namurian (Millstone Grit Series) rocks in all areas of the basin, and the latter crop out on all exposed sides of the basin with the exception of the area south of Llantrisant, where the Coal Measures are overlain unconformably by Mesozoic strata (see Fig. 32). The Lower and Middle part of the sequence is of a predominantly argillaceous character, and the Upper part of the sequence is a mainly arenaceous one. The main coalfield sequence has been studied in greater detail than that in the smaller fields and for this reason emphasis has been placed on this area in the account which follows.

(a) History of Research

The Coal Measures of the Main Basin were first mapped by H. T. de la Beche, W. E. Logan and D. H. Williams, and their work was published on the Old Series one-inch-to-one-mile scale in the 1830's and 1840's. The Coalfield was then resurveyed on the six-inch-to-one-mile scale by the Geological Survey under the supervision of Dr. A. Strahan, and published in the 1890's and 1900's. With these maps, accompanying memoirs were published, describing the stratigraphy, palaeontology and correlation of the Coal Measures. Two of these were subsequently revised; the Abergavenny memoir (sheet 232) in 1927 and the Merthyr Tydfil memoir (sheet 231) in 1932, both by T. Robertson. A systematic resurvey of the Main Basin, begun in 1946, has led to the publication of new editions of the Pontypridd and Maesteg memoir (sheet 248) by A. W. Woodland and W. B. Evans in 1964, the Newport memoir (sheet 249) by H. C. Squirrell and R. A. Downing in 1969, and a special memoir by A. A. Archer on the Gwendraeth Valley (parts of sheets 229, 230, 246) in 1968. Maps of the Swansea (sheet 247) and Ammanford (sheet 230) areas are also being produced. Numerous other workers have con-

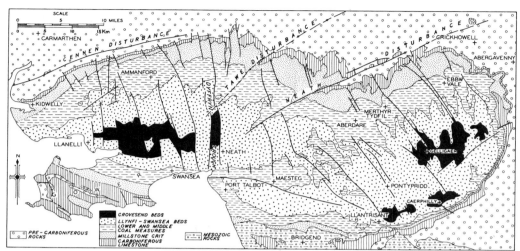

Fig. 32. Map of the main basin of the South Wales Coalfield (Reproduced from George, 1970 fig. 28.)

tributed to the understanding of the geology of the Main Basin area, these include Davies & Trueman (1927), Dix & Trueman (1928), Dix (1933, 1934, 1937), Jones (1935), Richards (1936), Moore & Cox (1943), Moore (1945, 1947), Trotter (1947), Blundell (1952), Stubblefield & Trotter (1957), Woodland and others (1957), Leitch *et al.* (1958), Squirrell & Downing (1964).

The Coal Measures of the Main Basin are best described in ascending order and, for descriptive purposes, the coalfield is divided into five areas as shown on Fig. 29.

(b) Lower Coal Measures (Westphalian A, Gastrioceras subcrenatum Marine Band— Amman Marine Band)

The lowest Coal Measures are in some ways transitional between the mainly marine deposits of the Namurian and the mainly non-marine beds of the greater part of the Coal Measures (Archer 1968, p. 33, Woodland & Evans 1964, p. 33). This transition can be considered to be complete at the horizon of the Garw Coal (also known as Cnapiog or Rhasfach Vein).

The cyclothems below the Garw Coal are characterised by frequent marine strata, and by few coals, whilst those above the Garw Coal have few marine horizons and thick coals. The sequence begins with the marine incursion of the *Gastrioceras subcrenatum* Marine Band, this being followed by an arenaceous phase of varying development known as the 'Farewell Rock'. Then a series of marine and non-marine intercalations with coal occurs, the marine horizons being usually five in number. Leitch *et al.* (1957) have designated them Marine Bands M_1 to M_5 in ascending order, and the highest is followed by a thick cyclothem capped by the Garw Coal. The Garw Coal is followed by a series of shales, seatearths and coals with ironstones extensively developed in the shales. Surface exposures are confined to the perimeter of the Coalfield basin, and the following description is based on these surface section supplemented by deep borehole information.

The Lower Coal Measures of the Ammanford Area are poorly exposed, being much obscured by drift deposits. The most complete sequence is recorded from an underground borehole near Cynheidre Colliery Shaft 3. (Woodland and others 1957 p. 40, Archer 1968 p. 36—for general sequence see Fig. 33). The *Gastrioceras subcrenatum* Marine Band is well developed in this area, being over 80 ft (24·4 m) thick in the Cynheidre borehole. The marine band is overlain by siltstone followed by the 'Farewell Rock', which is composed of sandstone and striped beds inter

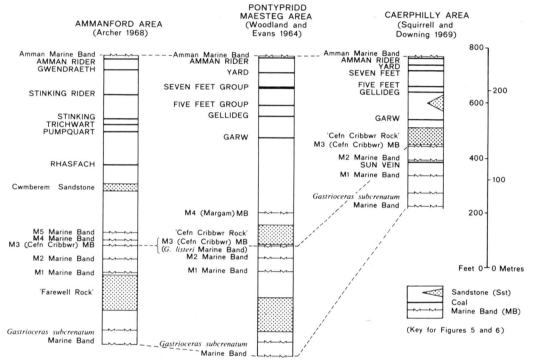

Fig. 33. Generalised sections of the Lower Coal Measures in three areas of the main basin of the South Wales Coalfield showing principal horizons.

bedded with thin bands of quartzitic sandstone. The following 200 ft (60·9 m) contain six marine horizons in grey mudstones and siltstones separated by thin sandstones and mudstones and all the bands contain *Lingula mytilloides* and *Planolites opthalmoides*. Some have fish scales and plant fragments. These horizons have been designated Marine Bands M_1 to M_5 as described above, and the top horizon is also known as the Cwm Berem Marine Band. Above these up to the horizon of the Garw Coal or Rhasfach Vein, sandstone and striped beds predominate. The total thickness of measures below the Rhasfach is about 675 ft (206 m) (Archer 1968, p. 36). Boreholes in the Cynheidre area prove the sequence from the Rhasfach up to the base of the Amman Marine Band to consist of 400 ft

(122 m) of mudstones containing mussels of the *Carbonicola pseudorobusta/martini/communis* groups and *Naiadites*, thin sandstones, seatearths and coals. Two important coals are the Pumpquart and Gwendraeth seams, which have been extensively mined. Moving eastwards along the northern edge of the Coalfield, Leitch *et al.*(1958) have described exposures of the above sequence in stream sections between Llandybie and the Swansea (Tawe) Valley.

The sequence below the Garw Coal has been calculated to be about 350 ft (106·6 m) thick in the River Llech (Leitch *et al.* 1958, p. 469). The Farewell Rock, which consists of a series of olive and grey grits, flags and shales, is 120 ft (36·6 m) in thickness. The marine horizons (M_1-M_5) have been found and above these occur two mussel bands containing *Carbonicola*, and designated C_1 and C_2. The overlying measures comprise 400 ft (122 m) of mudstones, seatearths and coals, (Dr. B. Kelk, personal communication).

In the Merthyr Area, the Lower Coal Measures are exposed in numerous stream sections. In the Upper Neath Valley, good exposures in the River Gwrelech show that the *Gastrioceras subcrenatum* Marine Band to Garw sequence is about 300 ft (91·4 m) thick and contains all the horizons mentioned above, the Farewell Rock being about 140 ft (42·7 m) thick (Leitch *et al.* 1958, p. 465). Above this, the sequence to the Amman Marine Band is approximately 300 ft (91·4 m) thick (K. Taylor, personal communication). Exposures of the Lower Coal Measures are poor in the vicinity of Aberdare and Merthyr. The *Gastrioceras subcrenatum* Marine Band is seen in Nant Melyn near Aberdare where it is over 30 ft (9·1 m) in thickness, and is also seen in Nant Morlais near Merthyr where it is overlain by a series of sandstones and pebbly grits comprising the Farewell Rock. An estimated thickness of 200 ft (60·9 m) for the measures up to the Garw Coal has been given by Leitch *et al.* (1958, p. 474) for this area. The chief development of ironstone occurs above the Garw in this area and was once extensively mined. Sections showing the principal ironstones in this sequence have been described by Robertson (1932, p. 101).

When traced eastwards into the Brynmawr Area, the sequence noticeably thins, the measures below the Garw being about 150 ft (45·7 m) in the Beaufort-Brynmawr vicinity (Leitch *et al.* 1958, p. 475). The Farewell Rock here becomes the predominant stratum at the expense of the marine horizons. Above the Garw the measures again consist of shales, ironstones and coals of about 120 ft (36·6 m) in thickness (Robertson 1927, p. 72). At the north-eastern edge of the basin, the Lower Coal Measures are exposed in the Clydach Valley, the sequence here being a predominantly arenaceous one with thin argillaceous intercalations. The marine horizons seen to the west are not well developed here, a dark shale containing *Lingula* (M_1) is overlain by mudstones which contain a few fish scales at two horizons (M_3 and possibly M_4, Leitch *et al.* 1958, p. 476). This sequence becomes much thinner to the south-east but the ironstone sequence is still in evidence.

In the Caerphilly area, the only sequence of which there is a complete record is from the Rudry Borehole (Woodland *et al.* 1957, p. 50, Squirrell & Downing 1969, p. 97, for generalised sequence see Fig. 33). The *Gastrioceras subcrenatum* Marine Band is nearly 50 ft (15·25 m) thick in the Borehole and contains a rich fauna of horny brachiopods, mollusca and fish (Woodland and others 1957, p. 54).

It is followed by 60 ft (18·3 m) of mudstones and silty mudstones containing *Planolites ophthalmoides*, this being overlain by a thin marine band containing *Edmondia* and *Nuculana*. The Farewell Rock, developed in other parts of the Coal-field at this horizon, is absent, and the thin marine band is thought to be equivalent to the M_1 horizon (Leitch *et al.* 1958) of the North crop. Sandstone and striped beds 50 ft (15·25 m) in thickness overlie the marine horizon and in turn are overlain by a coal, known as the Sun Vein, the roof mudstones of which have yielded *Lingula*, *Orbiculoidea* and Productids. This is correlated with the M_2 horizon of Leitch and others (1958) by Squirrell & Downing (1964, p. 127). The Ms horizon is absent in the Rudry Borehole but it occurs in a nearby exposure in a position approximately 30 ft (9·1 m) above M_2. M_3 is succeeded by a thick arenaceous phase, namely the Cefn Cribbwr Rock, which is 65 ft (19·8 m) thick at Rudry. A thin band some 35 ft (10·6 m) above the Cefn Cribbwr Rock containing *Planolites ophthalmoides* may be the equivalent of the M_4 horizon of Leitch *et al.* (1958) and the Margam Marine Band of Woodland & Evans (1964, p. 37). The remainder of the sequence up to the Garw Coal is of a mainly arenaceous character. The succession from the Garw to the Amman Marine Band is known from mining information, as it contains several workable coal seams. In the Caerphilly Area this part of the sequence thins markedly to the east and south, the above mentioned coals being separated by mudstones in some areas and almost coalescing in others.

In the Pontypridd Area, surface outcrop is limited to a narrow strip along the southern margin and a small faulted area near Maesteg. Details of the Lower Coal Measures (see Fig. 33) are almost entirely derived from the Margam Park No. 1 Borehole (Woodland and others 1957, p. 44). The following details are from Woodland & Evans (1964, p. 33–36).

The sequence below the Garw is approximately 1100 ft (335 m) thick. The *Gastrioceras subcrenatum* Marine Band is separated from the Farewell Rock by 80 ft (24·4 m) of mudstones and ironstones. The Farewell Rock itself is a flaggy sandstone 110 ft (33·5 m) thick and is followed by a series of thin sandstones and mudstones containing four marine bands. These may be correlated with those described by Leitch (1958) as follows: 120 ft (36·6 m) above the Farewell Rock (M_1), 200 ft (61 m) above the Farewell Rock, immediately overlying a coal (M_2), 76 ft (23·2 m) above M_2 in a 20 ft (6·1 m) band of dark mudstone called the Cefn Cribbwr Marine Band and notable for the presence of *Gastrioceras listeri* (M_3). The M_3 Marine Band is followed by sandstone 120 ft (36·6 m) thick, the Cefn Cribbwr Rock, overlain by the Margam Marine Band (M_4). Some 300 ft (91·4 m) of argillaceous beds separate this from the Garw Coal, and from the Garw Coal to the Amman Marine Band there are 200-600 ft (60·9-183 m) of mudstone, thin sandstones and thick coals.

In the Swansea Area, the Lower Coal Measures extend in a belt between the Loughor estuary in the west and Swansea Bay in the east. Stevens (Geological Survey six-inch maps SS 59 SW, SE 69 SW) estimates that the Lower Coal Measures have a thickness of 1500 ft (457 m). Above the *Gastrioceras subcrenatum* Marine Band there is a series of mudstones, sandstones and coals. The arenaceous phase of the Farewell Rock is not strongly developed. Two marine bands and two

Lingula bands occur in this sequence, the upper marine band being the *Gastrioceras listeri* Marine Band (the Wernffrwyd Marine Band of Jones 1935, p. 320) which Woodland & Evans (1964, p. 35) correlate with the Cefn Cribbwr Marine Band. Between this horizon and the Amman Band the strata include several good coals which have been worked in the past in the North Gower area.

(c) *Middle Coal Measures (Westphalian B—C, Amman Marine Band to Upper Cwmgorse Marine Band)*

The Amman Marine Band, at the base of the Middle Coal Measures, is followed by a succession of coal-bearing cyclothems (which contain the majority of the thickest and most-worked coals in the basin). Local arenaceous phases occur at intervals, as do mussel-bearing mudstones, until the onset of the next marine incursion, represented by the Hafod Heulog or Mole Marine Band. Conditions in the basin appear to change temporarily at this point, as two more marine phases (the Britannic or Trimsaran Marine Band and the Cefn Coed Marine Band) are found not far above, coals are thin and sandstones impersistent. However, there is locally a thick sandstone termed the Cockshot Rock. Above the Cefn Coed Marine Band, the sequence is one of coal-bearing cyclothems associated with regional sandstone formation and four marine phases. Of these the uppermost two, the Lower and Upper Cwmgorse Marine Bands are important.

In the Ammanford Area, colliery workings have provided most of the available information, since surface exposures are poor and discontinuous. The Amman Marine Band is well developed, consisting of over 10 ft (3 m) of dark grey mudstone containing *Lingula, Orbiculoidea, Dunbarella, Edmondia,* foraminifera and gastropods. This is followed by light grey mudstones having a good mussel fauna of the *Anthracosia* group. The sequence then passes up into 620 ft (189 m) of coal-bearing cycles (Archer 1968, p. 64, 80, 98), many coals having good mussel faunas in their roofs. The principal seams are the Gras Uchaf, Ddugaled, Big, Pennypieces, Soap and Graigog. Sandstones are locally developed, the most important of these occurring above the Big. Above the Graigog are silty and micaceous mudstones containing *Lingula* and palaeoniscid scales, these being known as the Graigog Marine Band. This horizon is thought to equate with the Two-Foot Marine Band of the Pennines, but is unknown elsewhere in South Wales (Archer 1968, p. 98). The succeeding coal, the Graigog rider, has a similar roof fauna which Archer (1968, p. 98) suggests might compare with the Clown Marine Band of the Pennines. The Mole Marine Band (known as the Hafod Heulog Marine Band in the rest of the Coalfield) occurs some 70 ft (21·3 m) above this.

The sequence above the Mole Marine Band consists of about 900 ft (274 m) of measures which contain several marine horizons, the most important ones being shown in Fig. 34. Notable are the Cefn Coed Marine Band which is well developed and has a varied fauna including horny brachiopods, goniatites (including *Anthracoceras aegiranum*), crinoids, corals, echinoids and calcareous brachiopods, and the Upper Cwmgorse Marine Band, which also has a well established marine fauna, including the goniatites, *Anthracoceras cambriense* and *Politoceras kitchini.* A number of coals occur between these marine horizons, in particular the Carway

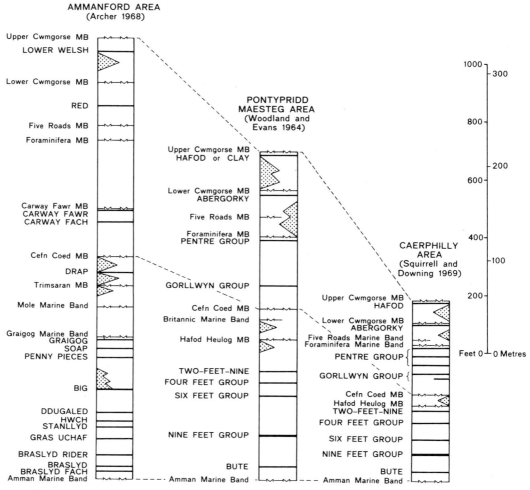

Fig. 34.   Generalised sections of Middle Coal Measures in three areas of the main basin of the South Wales Coalfield showing principal horizons.

Fawr and the Red Vein Group, these having been extensively mined in the Ammanford area.

In the Swansea Area (Geological Survey six-inch maps SS 59, 69 and 79), the Middle Coal Measures have a thickness of about 1800 ft (549 m) and contain most of the major elements of the sequence already mentioned for Ammanford to the north.

In the Pontypridd and Merthyr areas, the thickness of the Middle Coal Measures varies between 700 and 1600 ft (213-487 m), the lower part of the Middle Coal Measures containing many important workable coals. The importance of the Amman Marine Band lies in the fact that it is the only known marine stratum associated with this main group of coals (Woodland & Evans 1964, p. 45). The major coals, listed here in ascending order, are the Bute, Nine Feet Group, Six Feet Group, Four Feet Group and the Two Feet Nine, the 'Groups' being made up of major seams which are split into several workable coals. These coals are widespread throughout the area, absent only when local washouts occur, and they have been the main mining resource in this part of the Coalfield. They are usually accompanied by roof shales with good mussel faunas. Above these coals conditions of deposition changed and a period of coal formation gave way to one of marine transgression and regression. The following marine bands are present, the Hafod Heulog, Britannic and Cefn Coed, which correlate with the Haughton, Sutton and Mansfield marine bands respectively of the Pennines (Woodland & Evans 1964, p. 59). The Cefn Coed Marine Band is the best developed of these, and at Aberbaiden Farm, in the southern part of the area, it has yielded the richest fauna in any Coal Measure marine band in Britain (Ramsbottom 1952, p. 8). This includes corals, crinoids calcareous brachiopods, goniatites, gastropods, marine mussels, ostracods and conodonts.

Between the Cefn Coed and Upper Cwmgorse marine bands there are some 3–800 ft (91·4–244 m) of measures that include the Foraminifera and Lower Cwmgorse marine bands, the Gorllwyn, Pentre, Abergorky and Hafod workable coals and several variable sandstones.

In the Caerphilly Area, the Middle Coal Measures become thinner overall, varying between 400 and 800 ft (122–244 m) but still retain the main sequential elements as in other parts of the Coalfield described above. The Amman Marine Band persists over the greater part of the area, as does the Cefn Coed Marine Band, but the marine horizons which occur between them are more restricted locally, (Squirrell & Downing 1969, p. 103). A major development of important coals persists above the Amman Marine Band, these being the Bute, the Nine Feet, Six Feet and Four Feet Groups and the Two Feet Nine, similar to those in the Pontypridd Area. Above the Cefn Coed Marine Band, the Gorllwyn, Pentre and Hafod coals persist, and the Foraminifera and Lower Cwmgorse marine bands are also present. Sandstones are frequent, usually of local development, and many become conglomeratic in places. The sequence is completed by the Upper Cwmgorse Marine Band, which overlies the Hafod Coal, and varies in thickness from 2 to 28 ft (0·61–8·5 m) (Squirrell & Downing 1969, p. 115). It contains a fauna consisting of horny and calcareous brachiopods, sponge spicules, gastropods, goniatites and fish.

The sequence in the Brynmawr Area is basically similar to that in the Caerphilly Area, becoming thinner likewise as the eastern margin of the Coalfield is approached.

(d) *Upper Coal Measures* (*Westphalian C—D, Upper Cwmgorse Marine Band and Above*)

Above the Middle Coal Measures, a marked alteration in the depositional history of the Coalfield takes place. Marine transgression phases are nowhere in evidence; coal deposition, although present, is nowhere as prominent as before. Instead, an influx of clastic material which has gone to form very thick sandstones and conglomerates dominates the sequences of the Upper Coal Measures. Woodland, Evans and Stephens (1957) reclassified the Upper Coal Measures of South

Wales, into the primary subdivisions, Lower and Upper Pennant Measures, which were subdivided as shown in Table 1.

TABLE 1
(from Woodland and others 1957, p. 11)

		Principal Coal Horizons
Uppper Pennant Measures 3100 ft	Grovesend Beds	Base at Swansea Four Feet, Wernffraith, Llantwit No. 3 or Mynyddislwyn Seams, Strata include Penyscallen, Gelli and Grovesend seams.
	Swansea Beds	Base at Swansea Three Feet or Graigola seam.
	Hughes Beds	Base at Hughes, Wenallt, Cefn Glas, Bettws Four Feet etc. Strata include Mountain, Daren-ddu, Maesmelyn, Swansea Two Feet etc.
Lower Pennant Measures 1350–2300 ft	Brithdir Beds	Base at Brithdir. Strata include Brithdir Rider, Glyngwilym.
	Rhondda Beds	Base at No. 2 Rhondda, Wernddu, Malthouse, Penlan, Castell-y-weiver. Strata include Daren Rhestyn, No. 1 Rhondda, No. 1 Rhondda Rider.
	Llynfi Beds	Base at top of the Upper Cwmgorse Marine Band. Strata include White, Lower Pinchin, Tormynydd, No. 3 Rhondda, Wernpistyll seams.

The boundary between the Lower and Upper Pennant Measures is taken at the base of the Hughes Vein (or its equivalent). This horizon coincides with the junction between the Westphalian C and D floras and is taken as the arbitrary boundary between the mussel zones of *Anthraconauta phillipsii* and *Anthraconauta tenuis*. The further subdivision was based on the selection of two type areas, the Pontypridd area for the Lower Pennant Measures and the Swansea area for the Upper Pennant Measures (Woodland *et al.* 1957, p. 11).

The typical lithology of Pennant cyclothems is shown in Fig. 30. Most are extremely thick and made up for the greater part of sandstones, this lying immediately on or close to a coal and frequently having a conglomeratic base. In certain cases mudstones may be present between the coal and the sandstone, and may contain non-marine mussels and ostracods. Varied and abundant plant debris is a feature of these cycles (Woodland & Evans 1964, p. 21–22).

In the Ammanford Area, the Llynfi Beds are approximately 700 ft (213 m) thick (Archer 1968, p. 123) and are made up of alternating sandstones and thin coals. A persistent band of ironstone occurs about 160 ft (48·7 m) above the Upper Cwmgorse Marine Band. There is a sparse fauna, *Anthraconauta phillipsii* being most common. The Brondini veins of this area are thought to approximate to the No. 2 Rhondda seam to the east. The Rhondda Beds are about 850–1100 ft (259–335 m) thick (Archer 1968, p. 129) and are similar in character to the Llynfi Beds, but with thicker sandstones. The Brithdir beds commence at the Goodig or Clay seam, thought to equate with the Brithdir seam to the east, and end with the Pwll Big Vein (the Hughes Vein). They are about 900 ft (274 m) thick (Archer 1968, pp. 132–133) and are essentially a sequence of thick sandstones with some shale and thin coals. The Hughes Beds are about 8–900 ft (244–274 m) thick in the Amman-ford Area (Archer 1968, p. 134). The Hughes or Pwll Big Vein together with the Pwll Little Vein, 100 ft (30·5 m) above, have been mined in the south western part of this area. In addition the sandstones lying above have been quarried near Llanelli. The Swansea beds are present only in the core of the Llanelli Syncline. They are thought to be about 3–400 ft (91·4–122 m) thick (Archer 1968, p. 136) though there are no exposures.

In the Swansea Area, the Llynfi Beds are around 1300 ft (386 m) thick, the Rhondda and Brithdir Beds being around 1400 ft (417 m) thick. They have a similar character to those described for the Ammanford Area, the Brithdir seam not being readily identified. The Swansea Area is the type area for the Upper Pennant Measures (Woodland and others 1957, p. 11). The Hughes Beds are about 1000 ft (304 m) thick and contain local coals and shales replaced laterally by thick sand-stone. The Swansea Beds are around 750 ft (228 m) thick and are lithologically comparable with the underlying Hughes Beds. 1200 ft (356 m) of Grovesend Beds complete the sequence, these containing numerous coals and thick shale sequences. Sandstones are less prominent here than in the sequence below. (Thicknesses obtained from Geological Survey published six-inch sheets SS 59, 69 and 79.)

In the southern part of the Merthyr Area, only the Llynfi and Rhondda beds are present, these showing an overall similarity in their basic character to those of Pontypridd just to the south, that were selected by Woodland and others (1957, p. 11) for their type area.

In the Pontypridd Area the Llynfi Beds thin towards the north-east from over 700 to 250 ft (213–76 m) (Woodland & Evans 1964, p. 65). The lowest part is an argillaceous sequence around Maesteg containing several coals which have been worked locally, e.g. the Tormynydd. This is overlain by massive pennant-type sandstone known as the Llynfi Rock, which reaches a thickness of 180 ft (54·9 m) in some localities and in others is absent altogether. Close above the Llynfi Rock lies the No. 3 Rhondda Seam. This is best developed in the southern part of the area where it has been worked extensively by many colleries (Woodland & Evans 1964, p. 66). The sequence continues with sandstone and local coals, and the highest Llynfi Beds are essentially argillaceous, and contain the Wernpistyll coals in the south-west of the area. These yield a roof fauna of *Anthraconauta phillipsii* and *Carbonita* sp. (Woodland & Evans 1964, p. 67). The Rhondda Beds have their most extensive outcrop in the Pontypridd Area, forming much of the high ground in the central part, and vary in thickness between 1100 ft (335 m) in the south-west and 600 ft (183 m) in the north-east (Woodland & Evans 1964, p. 67). At the base is the No. 2 Rhondda Coal which is the most extensively worked of all the Pennant Coals in the area, (Woodland & Evans 1964, p. 67). It has many local names (see Table 1) but becomes less significant in the Merthyr Area to the north. The No. 2 Rhondda is succeeded by massive pennant-type sandstones, with argillaceous measures partly replacing them in several areas. These contain local coals such as the Daren Rhestyn. The next identifiable coal horizon, although it is not developed everywhere, is the No. 1 Rhondda, which is overlain by ferruginous mudstones containing *Anthraconauta phillipsii* and ostracods (Woodland & Evans 1964, p. 70). In most of the area a thin coal, the No. 1 Rhondda Rider, lies above these mudstones, this having a roof fauna of *Anthraconauta phillipsii* and *Anthraconauta* cf. *tenuis* (the lowest occurrence of *Anthraconauta* cf. *tenuis* in the area, Woodland & Evans 1964, p. 70). Above this, the sequence is composed of between 200 and 500 ft 60·9–152 m) of massive sandstone followed by argillaceous sediments which complete the Rhondda Beds succession. The Brithdir Beds vary in thickness from 500 ft (152·3 m) in the north to 800 ft (244 m) in the south-west. Unlike the Llynfi and Rhondda Beds, the measures are predominantly arenaceous, and coals, where developed, are thin and unimportant (Woodland & Evans 1964, p. 71). The Brithdir seam is less well developed in this area than in that to the east, as is the Brithdir Rider, although the latter, where present, has a mudstone roof containing *Anthraconauta phillipsii* and *Anthraconauta tenuis* (Woodland & Evans 1964, p. 71). The succeeding sandstones are upwards of 400 ft (122 m) thick with thin impersistent mudstones. The Hughes Beds are preserved in three synclinal areas, the highest beds being absent through erosion in most places. A complete sequence is seen on Cefnmawr where it is about 800 ft (244 m) thick (Woodland & Evans 1964, p. 71). The base of the group is characterised by a complex coal horizon which has many local names (see Table 2). It has been worked at several localities, and from its roof shales, a fauna of large *Anthraconauta tenuis* associated with *Anthraconauta phillipsii* and ostracods is widely known. Above this, sandstones and mudstone intercalations, with the Daren-ddu or Mountain coal, are in turn followed by thick sandstones, the highest beds recorded being argillaceous.

The Swansea Beds commence with the Graigola Vein which lies conformably on the Hughes Beds, and is followed by 80 ft (24·4 m) of sandstone (Woodland & Evans 1964, p. 72). The Grovesend Beds are restricted to the Llantwit Syncline. They are argillaceous in character, containing several coals, the No. 3, No. 2 and No. 1 Llantwit seams, and appear to rest unconformably on the underlying Swansea Beds.

The Pennant Measures of the Caerphilly Area are about 2800 ft (853 m) thick (Squirrell & Downing 1969, p. 115). Past workers, Moore (1945, 1948) and Blundell (1952) considered that intra-Coal Measures unconformities have reduced or removed much of the Lower Pennant Measures in this area. However, evidence to show that attenuated representatives of these beds are present has been put forward by Squirrell & Downing (1964, p. 121–125). The Swansea Beds may be absent because of an unconformity beneath the Mynyddislwyn (Grovesend) Seam.

In the Caerphilly Area, red and green mudstones occur at several horizons in the Lower Pannant Measures. Downing & Squirrell (1965) suggest that these represent flood plain deposits derived from red soils weathered on an upland source area in an oxidising environment and deposited on the floodplain under oxidising (red) or reducing (green) conditions. They also suggest that grey sediment lying on a subaerial delta or plain could have been wholly or partially oxidised by tropical weathering conditions, if slight uplift occurred.

The Llynfi Beds are exposed in a narrow belt around the rim of the Coalfield and reach a maximum thickness of 340 ft (104 m) in the south-west; their minimum thickness of less than 30 ft (9·1 m) is in the south-east (Squirrell & Downing 1969, p. 117). They comprise a variable sequence of mudstone, some sandstones and mainly local coals. The No. 3 Rhondda coal occurs almost everywhere.

The Rhondda Beds are 670 ft to 150 ft (204–45·7 m) thick (Squirrell & Downing 1969, p. 118) and commence with the No. 2 Rhondda seam, which has been extensively worked. Arenaceous and argillaceous rocks occur in approximately equal proportions, except in the east where arenaceous strata predominate. The argillaceous strata include red and green mudstones, as mentioned above, which are present throughout the succession. The arenaceous members comprise conglomerates and pennant-type sandstones, which may be replaced laterally by mudstones. These pass upwards into a sequence of mudstones containing several coals, the No. 1 Rhondda and the No. 1 Rhondda Rider being the most important. *Arthraconauta phillipsii* and *Anthraconauta tenuis* have been recorded in the roof mudstones of these seams.

The Brithdir Beds have a thickness of 800 ft (244 m) in the south-west and less than 200 ft (60·9 m) in the east (Squirrell & Downing 1969), p. 120). The sequence is made up largely of pennant-type sandstones with subordinate mudstones, seatearths and a few coals. In the northern part of the area, the Brithdir coal at the base has been worked extensively. To the south, it deteriorates and may be locally absent. Argillaceous strata topped by the Brithdir Rider pass upwards into pennant-type sandstones containing some mudstones with thin coals, these being followed by more argillaceous beds underlying the Cefn Glas.

The Hughes Beds commence with the Cefn Glas and their upper limit is the Mynyddislwyn Seam. They outcrop over a large part of the Caerphilly Area, their thicknesses ranging from 400 to 650 ft (122·198 m) (Squirrell & Downing 1969, p. 122). The Cefn Glas or Hughes Vein has a mudstone roof containing large *Anthraconauta tenuis* associated with *Anthraconauta phillipsii*, *Spirorbis*, *Carbonita* spp. and fish (Squirrell & Downing 1969, p. 122). Mudstones and massive pennant-type sandstones containing some coals, notably the Daren-ddu, comprise the remainder of the Hughes Beds in this area.

The Grovesend Beds extend from the Mynyddislwyn seam to the highest Coal Measures known in the area. They occur in several synclinal areas, the thickest development being 500 ft (152 m) in the south. The Mynyddislwyn (equivalent to the No. 3 Llantwit of the Pontypridd area) has been worked extensively. Mudstones and sandstones including a number of thin coals make up the rest of the Grovesend Beds in the Caerphilly Area.

In the Brynmawr Area, Upper Coal Measures are present towards the south, and the sequence bears a close relationship to that in the northern part of the Caerphilly Area.

(e) Thickness Variations

The Coal Measures exhibit considerable changes in thickness within the Main Basin, the sequence as a whole as well as its component parts thickening and thinning in definite directions. In areas of relative attenuation, the principal

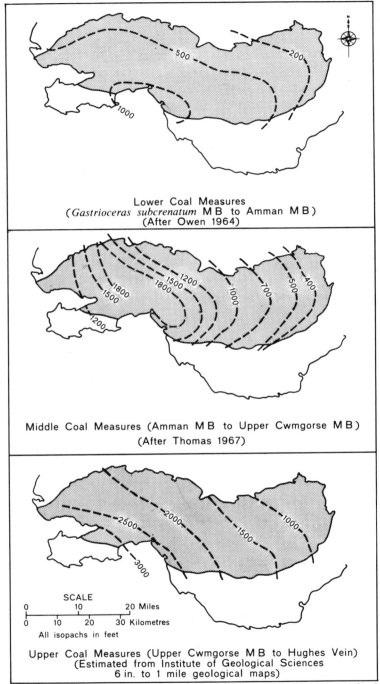

Lower Coal Measures
(*Gastrioceras subcrenatum* M B to Amman M B)
(After Owen 1964)

Middle Coal Measures (Amman M B to Upper Cwmgorse M B)
(After Thomas 1967)

SCALE

| 0 | 10 | 20 Miles |

| 0 | 10 | 20 | 30 Kilometres |

All isopachs in feet

Upper Coal Measures (Upper Cwmgorse M B to Hughes Vein)
(Estimated from Institute of Geological Sciences
6 in. to 1 mile geological maps)

Fig. 35.   Isopachyte diagrams of the three main divisions of the Coal Measures in the main basin
of the South Wales Coalfield. (Only the Lower Pennant Measures are included in the third diagram because
the sequence above is not preserved over much of the Coalfield.)

stratigraphic horizons are usually persistent showing that thinning is not mainly due to erosion or non-deposition, but to such factors as the nature of the sediment, compaction and the rate of basin subsidence. Numerous workers have examined the variations in thickness of the Coal Measures sequence, but rarely for the coalfield as a whole.

Thicknesses within the Lower Coal Measures have been studied by Leitch and others (1958) and Owen (1964) has evaluated them for the entire Coalfield area. The general pattern is that thicknesses for the Basal Coal Measures below the Garw Coal increase towards the south-west (see Fig. 35a), and a sequence less than 200 ft (60·9 m) thick in the eastern and north-eastern edges of the basin changes to one of over 1000 ft (304 m) on the south-western rim of the coalfield (Owen 1964, p. 305). Variations in the overall thickness of the Middle Coal Measures have been examined by Owen (1964) and Thomas (1967), and the same general pattern seems to prevail. These measures are less than 400 ft (122 m) thick at the eastern margin of the basin increasing to over 1800 ft (549 m) in the western area between Maesteg and Ammanford (see Fig. 35b), the area of maximum thickness lying slightly westward of that for the Lower Coal Measures. Thomas (1967) also noted that some gradual reduction in thickness appeared to occur on the very western edge of the Coalfield although the sequence was still 1200 ft (356 m) thick, (see Fig. 35b).

No single study has been made of thicknesses of the Upper Coal Measures for the coalfield as a whole. Studies have either dealt with individual groups of beds (Kelling 1964, 1968) or with variations throughout the sequence in a limited area (Woodland & Evans 1964, Squirrell & Downing 1965, 1969). An attempt to show overall thickness changes for the Lower Pennant Measures is made in Fig. 35c, the thicknesses being compiled from Geological Survey six-inch maps. The Upper Pennant Measures cannot be illustrated in this way because of their limited local preservation and the probable presence of an unconformity at the base of the Grovesend Beds. However, the Lower Pennant Measures do show a similar trend to that of the lower parts of the sequence in that the area of maximum thickness is in the west of the coalfield, but is probably south of the areas of Lower and Middle Coal Measures maxima, in the region of the Gower. Owen (1964) considered that for the Upper Coal Measures the maximum thickness was likely to be in the north-west of the basin.

The above indicates that the same major tectonic control of sedimentation persisted throughout the Lower and Middle Coal Measure times, and the pattern of sedimentation underwent little change. A change occurred early in the sedimentary history of the Upper Coal Measures, and although the resultant pattern of sediment accumulation remained similar, the character and quantity of sediment coming into the basin altered. The predominance of sandstones over argillaceous and coal deposits, together with the absence of marine horizons, suggests that uplift around the basin margins led to more vigorous erosion and to the filling-in of the basin area. These inferred earth movements were operative (as suggested by Owen, 1964) before the Coal Measure sedimentation began, and reached their climax in the Armorican Orogeny.

III. THE PEMBROKESHIRE COALFIELD

The Pembrokeshire Coalfield is separated from the Main Basin of the South Wales Coalfield by Carmarthen Bay. The Coal Measures occupy an area of approximately 50 square miles (125 km^2) and occur in two separate coalfields, namely the Little Haven-Amroth and Nolton-Newgale coalfields (see Fig. 36).

The Little Haven-Amroth Colafield is the more southerly and constitutes a narrow belt of measures of east-south-east to west-north-west orientation. The southern boundary is for the most part a faulted one, bringing Pre-Cambrian intrusives, Silurian and Upper Carboniferous rocks in contact with the Coal Measures. The northern boundary is entirely in contact with Namurian rocks, and is less severely distributed. Within the coalfield, as a result of deformation, folds have been displaced by thrusts, many of which have an east-south-east to west-north-west trend.

The Nolton-Newgale Coalfield is an elongate belt of measures running roughly north to south along the north-east side of St. Brides Bay (see Fig. 36). Its landward boundaries are defined by faults in the north, bringing the Coal Measures in contact with Ordovician and Cambrian strata, and, in the south, in contact with Namurian rocks. A narrow belt of Ordovician and Namurian rocks separates it from the Little Haven-Amroth Coalfield to the south. Within the coalfield area, the measures are made up of several faulted sequences.

(a) Stratigraphy

In these structurally complex coalfields, stratigraphic work has been undertaken by Goode (1913), Strahan and others (1914), Cantrill and others (1916), George & Trueman (1925), Dixon (1933), Dix (1933, 1934), Trueman (1934) and Jenkins (1960, 1962). A sedimentological study was carried out by Williams (1966, 1968).

George and Trueman (1925) studied the non-marine mussel faunas from Coal Measures in the Little Haven-Amroth Coalfield, and Trueman (1934) and Jenkins (1960, 1962) extended this work in both Pembrokeshire coalfields. Jenkins (1960, 1962) has produced the most recent comprehensive faunal study, in which he describes assemblages from twelve horizons in the Ammanian of the Little Haven-Amroth Coalfield and four in the Morganian of the Nolton-Newgale Coalfield. The Little Haven-Amroth Coalfield contains measures representative of the lower four zones of Davies and Trueman (1927, see Fig. 31), although the highest of these namely the *C. similis-A. pulchra* zone is only in part recognised, due to faulting. The Nolton-Newgale Coalfield contains strata representative of the next two zones i.e those of *Anthraconauta phillipsii-A. tenius* (see Fig. 31). Jenkins (1962) has recognised the following major marine bands, the *Gastrioceras subcrenatum* Marine Band, the horizons M_1–M_4 which are correlated with four of the horizons M_1–M_5 of the Main Basin (Leitch and others, 1958, Williams 1968), the Amman Marine Band and the Picton Point (Cefn Coed) Marine Band.

Floral studies of the Pembrokeshire Coal Measures were carried out by Goode (1913) and Dix (1933, 1934) and the plant assemblages of the Little Haven-Amroth Coalfield were assigned to Dix's floral zones C and D and those of the Nolton-Newgale Coalfield to zones G and H (see Fig. 31).

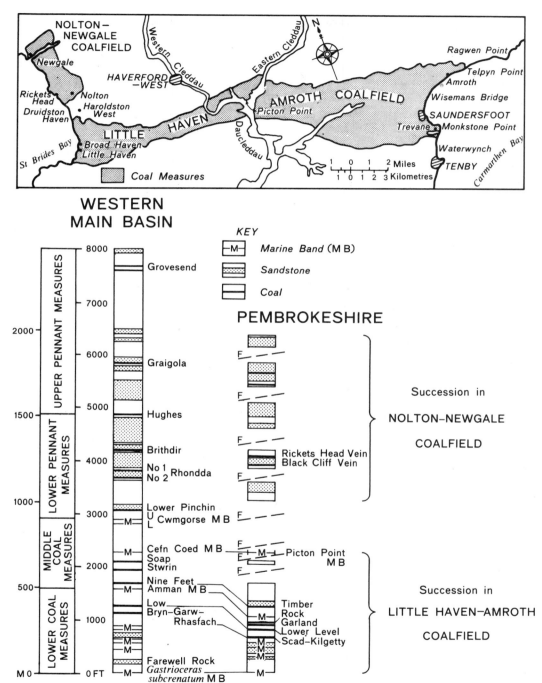

Fig. 36.   Sketch map of the Pembrokeshire Coalfield and generalised sections showing the correlation of the Pembrokeshire Coal Measures with those at the western end of the main basin (adapted from Jenkins 1962).

The two coalfields, one of which consists of Lower and Middle Coal Measures and the other of Upper Coal Measures, also display significant differences in lithology comparable to the Main Basin area (see above). The Little Haven-Amroth Coalfield comprises sediments of predominantly argillaceous character with coals and subsidiary sandstone, while the Nolton-Newgale Coalfield has an essentially arenaceous aspect. The sedimentary interrelationships and depositional environments of the Pembrokeshire Coal Measures have been described in detail by Williams (1966, 1968).

(b) Details

The succession in the Little Haven-Amroth Coalfield is seen in a series of faulted and folded coastal sections, and inland in the Daucleddau region (see Fig. 36). The base of the Westphalian is marked by the *Gastrioceras subcrenatum* Marine Band, consisting of 50 ft (15·25 m) of dark grey shale at Telphyn Point (Jenkins 1962, p. 71) which contain the index goniatite, *Anthracoceras* sp., *Lingula mytilloides*, nuculid bivalves and bellerophontid gastropods. This is overlain by grey, barren shales followed by a layer of 'slump' material described by Kuenen (1949, p. 373). This layer approximates in horizon to part of the Farewell Rock of the Main Basin. At Trevane, about 40 ft (12·2 m) of shaly measures with thin coals are seen above this 'slumped' horizon, containing the marine horizons M_1 to M_4 together with mussel bands yielding *Anthraconaia lenisulcata* and *Carbonicola*, and plant beds with *Mariopteris acuta*. The marine horizons are dark shales containing *Lingula mytilloides*, *Orbiculoidea*, fish fragments and foraminifera (Jenkins 1962, p. 79). Sandstones occurring within the sequence are variable in thickness and lateral extent. The sequence between the M_4 horizon and the Amman Marine Band is characterised by shales bearing non-marine lamellibranch horizons which have yielded *Carbonicola* cf. *communis*, *C. pseudorobusta*, *C. oslancis*, abundant seat-earths and coals, and plant-bearing horizons which contain *Neuropteris schlehani*, *Lepidodendron obovatum*, *Sphenopteris* sp. and *Mariopteris acuta*. The coals (Scad Vein to Rock Vein—see Fig. 36) have yielded the bulk of the coal mined in this coalfield. The Amman Marine Band occurs as a dark shale roof to a coal and contains *Lingula mytilloides*, *Dunbarella* sp. and *Conularia*. The position of these measures in the mussel and floral zonations is seen in Fig. 36.

Above the Amman Marine Band the measures are severely faulted and a complete sequence is not seen, but representative measures occur in the west and east coastal sections and also in the central part of the Coalfield at Picton Point. At the latter locality about eight inches of marl and grey clay overlying a thin coal, contain *Productus*, *Chonetes laguessiana*, *Crurithyris carbonaria*, *Aviculopecten* sp., comparable with the Cefn Coed Marine Band fauna of the Main Basin.

Fossil plants in shales south-east of the above locality have yielded *Neuropteris tenuifolia* and *Sphenophyllum saxifragaefolium*. Non-marine mussels are also present and are representative of the Lower *similis-pulchra* Zone (Jenkins 1962, p. 82). Williams 1968 (p. 356) estimated a minimum thickness figure of 40 ft (122 m) for the sequence from the Amman Marine Band to the Picton Point (Cefn Coed) Marine Band.

The Upper Coal Measures of Pembrokeshire are confined to the Nolton-

Newgale Coalfield. Jenkins (1960, 1962) allocated these beds to the zones of *Anthraconauta phillipsii* and *Anthraconauta tenuis* and this is supported by plant evidence which indicates the floral zone G (Dix 1933, 1934) for the coalfield south of Maidenhall Point, and the floral zone H (Jenkins 1962, p. 90) for the measures north of Maidenhall Point, the latter measures presumably representing the highest coal measures exposed in Pembrokeshire. The southern sections between Druidston Haven and Nolton Haven expose sandstones with thin shales, coals and seatearths which have been folded and faulted. Jenkins (1962, p. 88) has identified four faunal bands in this sequence, the lowest containing *Anthraconauta phillipsii* with rare *Anthraconauta* cf. *tenuis* and *Anthraconaia* cf. *pringlei*. The next two bands contain *A.* aff. *phillipsii* and *A.* cf. *tenuis*, while in the uppermost band *A. phillipsii* and *A. tenuis* are equally common associated with *Spirorbis* sp. and *Euestheria* sp. This sequence is 400 ft (122 m) thick (Jenkins 1962, p. 88), and is faulted against 285 ft (86·9 m) of sandstones containing *A. phillipsii*. North of Rickets Head (see Fig. 36) 400 ft (122 m) of sandy strata containing a fauna indicative of the *A. phillipsii* zone and floral zone G (Dix 1934), are exposed. North of Maidenhall Point, about 300 ft (91·4 m) of sandstones, thin shales and coals (Jenkins 1962, p. 90) yield a suite of plants which include *Mixoneura* (*Neuropteris*) *ovata*, *Odontopteris lindleyana*, and *Annularia sphenophylloides*, which Jenkins considers to be a similar flora to that above the Hughes Vein of the Swansea District.

(c) Thickness Variations

The Little Haven-Amroth Coalfield.

Strahan and others (1914), without the aid of faunal zones, estimated the thickness of the Lower Coal Measures to be approximately 6300 ft (1920 m) (without dip correction) from colliery records. Jenkins (1962) on the basis of his faunal classification estimated the thickness of the *lenisulcata* Zone to be about 1400 ft (417 m) in the east, thinning to 360 ft (110 m) in the west, the *communis* Zone to be consistently 340–400 ft. (104–122 m) in the east and similarly in the west, and the *modiolaris* Zone to be 180 ft (55 m) in the east and west. The Lower *similis-pulchra* Zone has been assigned a thickness of at least 400 ft (122 m) by Williams (1968, p. 356), who also suggests that the thickness of the *lenisulcata* Zone in the east has been overestimated by Jenkins. In his appraisal of the Lower Coal Measures, Williams (1968) gives thicknesses for the sequences *Gastrioceras subcrenatum* Marine Band to M_1, 164–330 ft (50–100 m); M_1 to M_4, 104–200 ft (32–61 m) and M_4 to the Amman Marine Band more than 600 ft (183 m), in which variations take place from north to south.

The Nolton-Newgale Coalfield.

Because of its structure and lack of readily identifiable horizons within the sequence, it is only possible to indicate a probable order of superposition of the faulted sequence. Approximately 2000 ft (609·5 m) of measures have been recorded by Jenkins (1962) but it is not known how much repetition or loss of sequence has been caused by faulting.

IV. THE FOREST OF DEAN COALFIELD

The Forest of Dean Coalfield, roughly triangular in outline, extends for a distance of some 10 miles (16·1 km) from Ruardean in the north to Lydney in the south, and for about 6½ miles (10·5 km) in an east-west direction. The strata are Upper Coal Measures and have a total thickness of about 2000 ft (609·5 m). They are similar to the Upper Coal Measures in the South Wales Coalfield, showing a predominance of arenaceous sediments, with shales and coals; red beds also occur in the sequence. They lie unconformably on Lower Carboniferous and Devonian sediments and were described in detail by Trotter (1942) who divided them into three main groups (see Fig. 37).

Sullivan (1964), through the use of Miospores, has found that the Upper part of the Drybrook Sandstone in the Wigpool Syncline, included by Trotter in the Carboniferous Limestone, is of Westphalian A age. This part of the sandstone

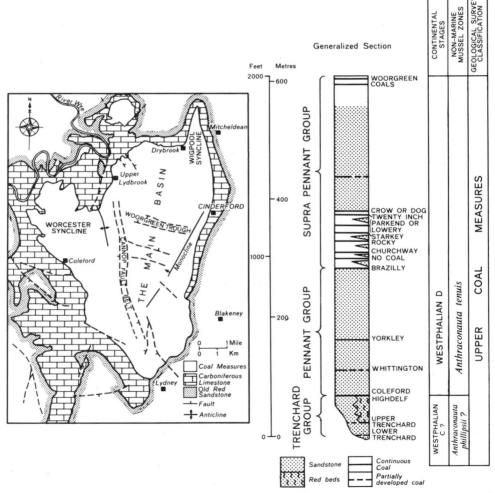

Fig. 37. Geological sketch map of the Forest of Dean Coalfield and generalised vertical section of the Coal Measures.

presumably rests unconformably on Viséan Drybrook Sandstone, and is overlain by the lowest strata mapped as Coal Measures, the Trenchard Group. In other parts of the Coalfield the Trenchard Group lies directly on much older rocks of Lower Carboniferous and Devonian age.

The Trenchard Group, mainly variable sandstones, is of doubtful age, since no fossil evidence as to age has yet been obtained. However the group is tentatively correlated with the lower part of the Upper Coal Measures (*Anthraconauta phillipsii* Zone). The Trenchard Group is followed by the Coleford High Delf seam overlain by massive sandstones of the Pennant Group, similar in appearance to much of the Pennant Measures of the South Wales Coalfield. The uppermost or Supra-Pennant Group, the measures from the Brazilly seam upwards, may equate with the Grovesend Beds of the South Wales Coalfield.

(a) Trenchard Group

The Group oversteps and overlaps the Carboniferous limestone onto older formations. This relationship accounts in part for the great variations in thickness within the Group, between 50 and 400 ft (15·25 to 122 m) in a roughly south-north direction. There is also a marked change in facies from north to south. In the north, the Group is composed largely of coloured beds, whereas, from Coleford south-wards, the coloured beds are absent and the measures consist mainly of sandstone. The Trenchard coal seam is present near the base of the Group, this having an average thickness of 4 ft 6 in (1·37 m) in the south. To the north it splits into two seams, each about 2 ft (0·61 m) thick. The top of the group is the base of the Coleford High Delf seam (see Fig. 37).

In the northern part of the Coalfield, the Trenchard measures have a conglomeratic base succeeded by reddish-purple and greyish-yellow sandstones and grits. These are overlain by the Upper Trenchard Seam and then by a series of variegated clays and shales with subordinate gritty sandstones. The Trenchard seams have not been proved in the central part of the Coalfield, though 250 ft (76·15 m) of measures consisting mostly of red beds, but without any coal, occur in colliery boreholes. The Trenchard Group at Coleford is the sandy southern facies. Red beds are absent, and the two Trenchard coals lie near the base of the Group. A 60 ft (18·3 m) sandstone overlies the Upper Trenchard seam, followed by blue and grey-blue shales and mudstones. The proved thickness of the group is 200 ft (60·9 m).

In the south, the Trenchard Group outcrops around the margin of the Coalfield. The measures, about 50 ft (15·25 m) thick, represent only the Upper part of the Trenchard Group of the north. Between the Coleford High Delf Seam and the Trenchard Seam there is a series of grey argillaceous shales with alternating grey sandstones; similar strata also occur below the Trenchard Seam.

(b) Pennant Group

The Pennant Group embraces the strata from the floor of the Coleford High Delf to the floor of the Brazilly Seam. These strata consist of massive sandstones with subordinate shales, and include three seams of workable thickness. The Group increases in thickness from 600 ft (183 m) in the north to 800 ft (244 m) in the south.

The Coleford High Delf Seam is a bright bituminous coal with an average thickness of 3 ft 6 in to 5 ft (1·07–1·52 m). Over most of the area this seam has a sandstone roof, but occasionally shales intervene between the two. These shales have yielded a flora indicating a Westphalian D age, *Asterotheca* cf. *arborescens*, *Neuropteris ovata*, *Sphenopteris neuropteroides*, and *Annularia stellata* (Welch and Trotter 1961, p. 90), and the following fauna, *Anthraconauta tenuis*, *Anthraconaia* aff. *pruvosti*, *Carbonita* cf. *agnes*, *Cypridina* cf. *radiata*, *Hemicycloleaia boltoni*, *Leaia bristolensis* (Welch & Trotter 1961, p. 108). Calver considers that this horizon should be referred to the *Anthraconauta tenuis* Zone. The measures between the Coleford High Delf and the Yorkley seam consist of 240 ft (73·2 m) of sandstone. In the south these are divided by 70 ft (21·35 m) of shales in which occurs the Whittington Seam, 3 ft (0·91 m) thick, beneath which the shales have yielded *Anthraconauta phillipsii*. The Yorkley Seam is a bituminous coal reaching its maximum development in the south, where it averages 2 ft 9 in (0·83 m) in thickness, and contains a good flora (Trotter 1942, p. 36). Between the Yorkley and Brazilly Seams are 330 ft to 450 ft (100–137 m) of strata consisting of massive sandstones in the north, but including two belts of shale in the south, these being up to 80 ft (24·4 m) in thickness.

(c) The Supra Pennant Group

The Supra Pennant Group can be divided into two parts, the lower part consisting of relatively argillaceous measures in which eight workable coals occur, and the upper which contains thick sandstones and subordinate coal seams. The group is confined to the central area of the Coalfield.

The lower division includes the measures from the Brazilly to the Crow Delf Seam and is about 300 ft (91·4 m) thick. The Brazilly Seam is 2 ft to 2 ft 6 in thick (0·61–1·07 m) and is overlain by 50 ft to 70 ft (15·25–21·35 m) of flaggy sandstones locally passing into sandy shales. A succeeding seam called the No Coal is variable in development having a thickness ranging between 8 in and 2 ft 3 in (0·02–0·68). Above the No coal is the Churchway High Delf, which splits into several seams in the south. A sandstone separates this seam from the Rocky Seam above. This latter seam is overlain by a massive sandstone 30 ft to 40 ft (9·1–12·2 m) thick, which does not appear to be present along the northern outcrop of the Coalfield. The overlying Starkey and Lowery Seams are both splitting seams and about 50 ft (15·25 m) above these, the Crow Delf occurs with a thickness of 1 ft to 1 ft 4 in (0·30–0·31 m). A good flora has been obtained from the measures between the Brazilly and the Crow Delf Seams (Trotter 1942).

The upper division of the Supra Pennant Group is estimated to be 800 ft (244 m) thick. Arenaceous measures, named the Serridge Sandstones, characterise the lower half of the division. North-west of a line from Ruspidge to Edge End, they are some 350 ft (106·6 m) thick and split by a thin coal. Developments of shale are present within this sequence in several parts of the coalfield. Above the Serridge Sandstones, the measures consist largely of shales with thin beds of coal and bands of ironstone. At the top are the Crabtreehill Sandstones, these being two developments of sandstone each 25 ft (7·6 m) thick, separated by 30 ft (9·1 m) of shale. The Lower Woorgreen Coal lies 30 ft (9·1 m) above these, and occurs around the central part of the basin. The Upper Woorgreen Coal however, being 30 ft (9·1 m) above the Lower, occurs only within the Woorgreen Trough. A good flora has been obtained from these two seams (Trotter 1942).

REFERENCES

ARCHER, A. A. 1968. The Geology of the South Wales Coalfield. Special Mem. The Upper Carboniferous and later formations of the Gwendraeth Valley and adjoining areas. *Mem. geol. Surv. Gt. Britain.*

BLUNDELL, C. R. K. 1952. The Succession and Structure of the North-eastern area of the South Wales Coalfield. *Quart. J. geol. Soc. Lond.* **107**, 307–33.

CALVER, M. A. 1969. Westphalian of Britain. In *Compte Rendu 6me Congres Intern. Strat. Géol. Carbonif. Sheffield* 1967, **1**, 233–54.

CANTRILL, T. C., DIXON, E. E. L., THOMAS, H. H. and JONES, O. T. 1916. The Geology of the South Wales Coalfield. Part XII. The Country around Milford, *Mem. geol. Surv. Gt. Britain.*

DAVIES, J. H., and TRUEMAN, A. E. 1927. A revision of the non-marine lamellibranchs of the Coal Measures and a discussion of their zonal sequence. *Quart. J. geol. Soc. Lond.*, **83**, 210–59.

DIX, E. 1933. The Succession of Plants in the Millstone Grit and the Lower portion of the Coal Measures of the South Wales Coalfield (near Swansea) and a comparison with that of other areas. *Palaeontographica*, **78B**, 158–202.

——. 1934. The sequence of floras in the Upper Carboniferous with special reference to South Wales. *Trans. Roy. Soc. Edin.*, **57**, 789–821.

——. 1937. The Value of Non-marine lamellibranchs for the correlation of the Upper Carboniferous. *Compte Rendu 2me Congr. Aranc. Et. Stratigr. Carbonif. Heerlen* 1935., **1**, 185–201.

DIX, E., and TRUEMAN, A. E. 1928. Marine Horizons in the Coal Measures of South Wales, *Geol. Mag.*, **65**, 356–63.

DIXON, E. E. L. 1933. Some recent stratigraphical work and its bearing on South Pembrokeshire problems: *Geol. Soc. Lond. Proc.*, **44**, 217–25.

DOWNING, R. A., and SQUIRRELL, H. C. 1965. On the Red and Green Beds in the Upper Coal Measures of the Eastern part of the South Wales Coalfield. *Bull. Geol. Surv. G.B.*, **23**, 45–56.

GEORGE, T. N. 1970. *Brit. Reg. Geol. South Wales. H.M.S.O.*

GEORGE, T. N., and TRUEMAN, A. E. 1925. The Correlation of the Coal Measures in the western portion of the South Wales Coalfield IV: notes on the Coal Measures of east Pembrokeshire. *Proc. S. Wales Inst. Eng.*, **41**, 409–15.

GOODE, R. H. 1913. On the fossil flora of the Pembrokeshire portion of the South Wales Coalfield. *Quart. J. geol. Soc. Lond.*, **69**, 252.

JENKINS, T. B. H. 1960. Non-marine lamellibranch assemblages from the Coal Measures (Upper Carboniferous) of Pembrokeshire, West Wales. *Palaeonotology*, 3, 104–23.

——. 1962. The sequence and correlation of the Coal Measures of Pembrokeshire. *Quart. J. geol. Soc. Lond.*, **118**, 65–101.

JONES, S. H. 1935. The Lower Coal Series of north-western Gower. *Proc. S. Wales Inst. Eng.*, **50**, 317–81.

KELLING, G. 1964. Sediment Transport in part of the Lower Pennant Measures of South Wales. *Developments in Sedimentology* 1, *L.M.J.U. van Straaten (Ed.), Deltaic and shallow Marine Deposits*, 177–84.

——. 1968. Patterns of Sedimentation in Rhondda Beds of South Wales. *Am. Ass. Petrol. Geol. Bull.*, **52**, 2369–86.

KUENEN, P. H. 1948. Slumping in the Carboniferous rocks of Pembrokeshire. *Quart. J. Geol. Soc. Lond.*, **104**, 365–80.

LEITCH, D., OWEN, T. R. and JONES, D. G. 1958. The Basal Coal Measures of the South Wales Coalfield from Llandebie to Brynmawr. *Quart. J. Geol. Soc. Lond.*, **113**, 461–86.

MOORE, L. R. 1945. The Geological Sequence of the South Wales Coalfield, the South Crop and Caerphilly basin, and its correlation with the Taff Valley sequence. *Proc. S. Wales Inst. Eng.*, **60**, 141–252.

——. 1947. The sequence and structure of the southern portion of the East Crop of the South Wales Coalfield. *Quart. J. Geol. Soc. Lond.*, **103**, 261–300.

—— and COX, A. H. 1943. The Coal Measures sequence in the Taff Valley and its correlation with the Rhondda Valley sequence. *Proc. S. Wales Inst. Eng.*, **59**, 189–265.

OWEN, T. R. 1964. The tectonic framework of Carboniferous sedimentation in South Wales. *In L.M.J.U. van Straaten (Ed.), 'Deltaic and shallow marine deposits'. Elsevier Amsterdam*, 301–07.

RAMSBOTTOM, W. H. C. 1952. The Fauna of the Cefn Coed Marine Band in the Coal Measures at Aberbaiden near Tondu, Glamorgan. *Bull, Geol. Surv. G.B.*, **4**, 8–30.

RICHARDS, R. 1936. An investigation of the Measures exposed in mine drivages at Bedwas Colliery with special reference to marine beds. *M.Sc. Thesis Univ. of Wales.*

ROBERTSON, T. 1927. Geology of the South Wales Coalfield Part II. Abergavenny. *Mem. geol. Surv. Gt. Britain.*

——. 1932. Geology of the South Wales Coalfield Part V. Merthyr Tydfil. *Mem. geol. surv. Gt. Britain.*

SMITH, A. H. V., and BUTTERWORTH, M. A. 1967. Miospores in the coal seams of the Carboniferous of Great Britain. *Spec. Pap. Palaeont.* 1.

SQUIRRELL, H. C., and DOWNING, R. A. 1964. The attenuation of the Coal Measures in the S.E. part of the South Wales Coalfield. *Bull. Geol. Surv. G.B.*, **21**, 119–132.

—— and DOWNING, R. A. 1969. Geology of the S. Wales Coalfield Part 1. The Country around Newport (Mon.). *Mem. geol. Surv. Gt. Britain.*

STRAHAN, A., CANTRILL, T. C., DIXON, E. E. L., THOMAS, H. H. and JONES, O. T. 1914. The Geology of the South Wales Coalfield Part II: the Country around Haverfordwest. *Mem. geol. Surv. Gt. Britain.*

SULLIVAN, H. J. 1964. Miospores from the Dryborok Sandstone and associated measures in the Forest of Dean Basin, Gloucestershire. *Palaeontology* 7, 351–92.

THOMAS, L. P. 1967. A sedimentary study of the sandstones between the horizons of the 4 ft 0 in Coal and the Gorllwyn Coal of the Middle Coal Measures of the South Wales Coalfield. *Ph.D. Thesis Univ. of Wales* (unpub.).

TROTTER, F. M. 1942. Geology of the Forest of Dean Coal and Iron-ore Field. *Mem. geol. Surv. Gt. Britain.*

——. 1947. The structure of the Coal Measures in the Pontardawe-Ammanford area, South Wales. *Quart. J. Geol. Soc.*, **103**, 89–133.

TROTTER, F. M. and STUBBLEFIELD, C. J. 1957. Divisions of the Coal Measures on Geological Survey Maps of England and Wales. *Bull. Geol. Surv. Gt. Brit.*, **13**, 1–5.

TRUEMAN, A. E. 1934. The age of the highest Coal Measures in W. Pembrokeshire. *Geol. Mag.*, **71**, 116–118.

——. 1947. Stratigraphical problems in the coalfields of Great Britain. *Quart. J. Geol. Soc. Lond.*, **103**, 65–104.

WELCH, F. B. A., and TROTTER, F. M. 1960. Geology of the Country around Monmouth and Chepstow. *Mem. geol. Surv. Gt. Britain.*

WILLIAMS, P. F. 1966. The Sedimentation of the Pembrokeshire Coal Measures. *Ph.D. Thesis Univ. of Wales.*

——. 1968. The Sedimentation of Westphalian (Ammanian) Measures in the Little Haven-Amroth Coalfield, Pembrokeshire. *J. Sed. Pet.*, **38**, 332–62.

WOODLAND, A. W., EVANS, W. B. and STEPHENS, J. V. 1957. Classification of the Coal Measures of South Wales with special reference to the Upper Coal Measures. *Bull. Geol. Surv. Gt. Brit.*, **13**, 6–13.

——, ARCHER, A. A., and EVANS, W. B. 1957. Recent Boreholes into the Lower Coal Measures below the Gellideg-Lower Pumpquart Coal horizon in South Wales. *Bull. Geol. Surv. Gt. Brit.*, **13**, 39–60.

—— and EVANS, W. B. 1964. The Geology of the South Wales Coalfield Part IV. The Country around Pontypridd and Maesteg. *Mem. geol. Surv. Gt. Britain.*

THE NAMURIAN OF NORTH WALES

W. H. C. Ramsbottom[1]

I. INTRODUCTION

IN North Wales Namurian rocks crop out only in the area to the east of the Carboniferous Limestone from near Prestatyn in the north to Ruabon and Oswestry (actually in Shropshire) in the south (fig. 38). Within this distance of some 56 km there is a change of facies from the thicker predominantly shaly succession of the

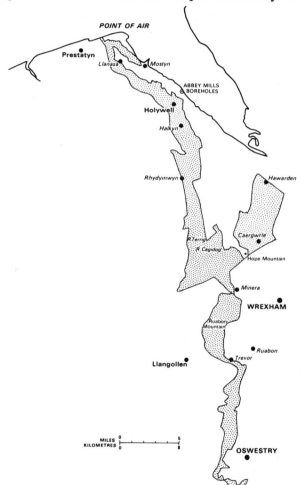

Fig. 38. Map showing the extent of the Namurian outcrops in North Wales and the locations of places mentioned in the text.

[1] Published by permission of the Director, Institute of Geological Sciences.

northern area (the Holywell Shales) to the thinner predominantly sandy succession of the south, where the sandstones form a part of the Cefn-y-Fedw Sandstone Group.

Geological Survey maps have also shown areas mapped as Millstone Grit in Anglesey and the Vale of Clwyd, but none of these is currently considered to be of Namurian age (see p. 182).

II. HISTORY OF RESEARCH

The classification of the Namurian rocks of the area has had a somewhat chequered history which has been given by Wood (1936) in greater detail than is here necessary. Early on Conybeare & Phillips (1822, p. 419) had recognised that the shaly beds overlying the Carboniferous Limestone and now forming part of the Holywell Shales, corresponded in character and stratigraphical position to the similar shaly and sandy beds of the Millstone Grit of Derbyshire. But Green (1867) allocated these same shales to the Yoredale Series, of Lower Carboniferous age, by analogy with similar beds of Staffordshire also placed by him in the Yoredales.

Then, in 1878, Walker on the basis of (as we now know misidentified) fossils placed the Holywell Shales in the Lower Coal Measures and, in spite of the lack of coal seams, this interpretation was accepted by G. H. Morton in 1888 and by the Geological Survey (Strahan, 1890). In various Geological Survey publications until 1924 the Holywell Shales remained in the Lower Coal Measures, even though in the southern part of their outcrop their correlation with the Millstone Grit had been established as early as 1913 (King 1914). When the Holywell Shales were regarded as of Lower Coal Measures age the Millstone Grit of the northern area was supposed to be represented by the underlying Chert Beds which are now believed to be of Viséan age.

In the southern area G. H. Morton (1876) proposed the term Cefn-y-Fedw Sandstone, for all the beds between the top of the Carboniferous Limestone (which he took at the base of the Sandy Limestone) and the base of the Coal Measures. On Cefn-y-Fedw (Ruabon Mountain) he divided the Sandstone into Lower (comprising the Lower Sandstone and Cherty Shale), Middle (Middle Sandstone) and Upper (Lower Shale to top of Aqueduct Grit). The fossiliferous nature of some of these sandstones in this southern area early attracted the attention of Davies (1865; 1870), Prosser (1865), and Aitken (1870), but most such sandstones are now regarded as Viséan in age. In this area the Geological Survey regarded the Cefn-y-Fedw Sandstone as coincident with the Millstone Grit and it is so shown on their current maps.

Only when goniatites came to be used in Carboniferous stratigraphy was real progress made in recognising the true limits and stratigraphy of the Namurian. King's work of 1913 has already been mentioned. Jackson (1925) recorded the Namurian marker-goniatite *Hudsonoceras proteus* (Brown) from the Holywell Shales, and Sargent (1927) gave an almost complete zonal succession of goniatites in these shales. Jones & Lloyd (1930, but not published in full until 1942) recorded many sections and goniatite localities from the Flintshire area, and attempted to place the various faulted areas in correct succession. From Flintshire, too, but also from farther south in Denbighshire, Wood (1936) recorded a number of goniatite horizons from within the Cefn-y-Fedw Sandstone, and these finally proved the contemporaneity of the Sandstone with the Holywell Shales.

The present brief account attempts to being together the available published information but has also involved a re-interpretation and collation of information derived from boreholes especially in the northern part of the district. Since it has now been shown that the cherty beds above the Carboniferous Limestone in Flint-shire are of Viséan age, and since there is no reason to doubt the traditional correlation of these cherts with the Cherty Shale of Denbighshire, it seems reasonable to suppose that the Cherty Shale is also of Viséan age. For this reason that part of the Cefn-y-Fedw Sandstone below the top of the Cherty Shale is excluded from the Namurian in this account.

III. PALAEOGEOGRAPHY

North Wales is situated on the south-western edge of the Central Province Namurian basin. The northern coast of St. George's Land passes through the region and the sandy nature of the Namurian rocks of Denbighshire reflects its proximity to the coastline. In the north, basinal shaly deposits predominate though it is likely that these were a number of islands (as near Rhydymwyn) or, at least, an uneven coast-line.

Two types of sandstone occur in the British Namurian—quartzitic and felds-pathic. It is known that sandstones derived from St. George's Land are of the quartzitic type—as are the Cefn-y-Fedw Sandstones (except for the Aqueduct Grit at the top). It is probable that the greatest influx of sandstones from this southern source—the Middle Cefn-y-Fedw Sandstone—came into the basin in the early Namurian and that, as in areas to the south and east, denudation of St. George's Land was reduced in the later Namurian (Ramsbottom 1969, pp. 226, 228; 1970, p. 152).

Feldspathic sandstones came from the north-east into the Central Province basin, but did not reach North Wales until R_2 and G_1 times and they include the Lower Gwespyr Sandstone and Aqueduct Grit.

Deposition was possibly continuous with the underlying Viséan only in the basinal areas, that is so far as North Wales is concerned, around Holywell and Flint, though even here the lowest Namurian horizons have not been recognised by means of goniatites. In other circumbasin areas the Namurian sea was transgressive, and so it is in North Wales. Around Hope Mountain E_2 beds apparently rest on Carboniferous Limestone, and at Rhydymwyn the *Gastrioceras cancellatum* Zone of G_1 age likewise rests directly on the limestone. It is likely too, though it has not yet been adequately proved, that the lower Namurian horizons are missing around Oswestry. In fact E_1 beds are unknown except in the north.

In the Namurian of Britain the near coastal shallow water faunas are charac-terised by the presence of a benthonic fauna—usually of productoid, spiriferoid and other brachiopods and of benthonic mollusca. This is in contrast to the basinal faunas in which the only macrofossils are the non-benthonic goniatites and bivalves. As one would expect, shelly benthonic faunas occur in the southern part of the district within the Cefn-y-Fedw Sandstone, but are also found at some horizons within the Holywell Shales. Occasionally these shelly bands become limestones, especially at the *G. cancellatum* horizon—a feature noted at this level farther east along the northern shore of St. George's Land in Lincolnshire (Ramsbottom 1969, p. 225).

The most north-westerly outcrop of Namurian beds in North Wales is at Prestatyn, though no significant fossils have been found there. But at Bodelwyddan, only 7 km to the south-west, Upper Coal Measures appear to rest on D_1 Zone Carboniferous Limestone, and this provides a local southern limit to the area of deposition. Further west the supposed 'Millstone Grit' of Anglesey is apparently of Westphalian age, and presumably the late Viséan regression took the coast line to the north of the present Carboniferous outcrops on Anglesey.

IV. STRATIGRAPHY

The age of the Sweeney Mountain Sandstone, a quartzitic, pinkish-white sandstone 30 m (100 ft.) thick near Oswestry, is unknown, and it may represent only

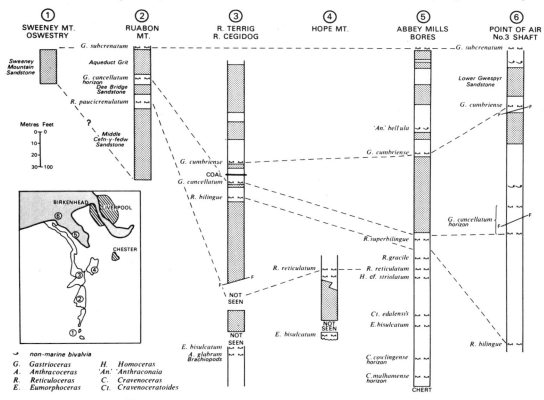

Fig. 39.   Vertical sections of Namurian rocks of North Wales.

---

  * It should be noted that Wood (1936) gives details of the No. 4 borehole at Abbey Mills, whereas Jones and Lloyd (1942) give details of the No. 1 borehole, and this explains the discrepancies between the two accounts. In Fig. 39 both boreholes are combined to give a single section.

the uppermost beds of those present farther north. At Trevor to the north of Ruabon Mountain, between Ruabon and Llangollen, beds above the Middle Cefn-y-Fedw Sandstone (itself a quartzitic sandstone) are exposed in the banks of the River Dee (Wedd and others 1927, pp. 165-7). There, Morton's (1876, p. 195) 'Lower Shale', between the Dee Bridge Sandstone and the Middle Cefn-y-Fedw Sandstone, has yielded a fauna of R_{1a} age with brachiopods and mollusca including *Reticuloceras paucicrenulatum* Bisat and Hudson. It may be noted that Wood (1936, p. 18) gave the horizon of this fauna in error as being above rather than below the Dee Bridge Sandstone. The *G. cancellatum* band (G_{1a}) occurs in Morton's 'Upper Shale' at a nearby locality, and the *G. subcrenatum* band was collected by H. H. Simpson above the Aqueduct Grit (a feldspathic sandstone about 21 m thick) in the brick pit at Trevor. The *G. cancellatum* Marine Band is also known from Minera Mill whence Bisat (1940) described *G. branneroides*, the goniatite characteristic of the lowest fauna of this band.

Some 8 km to the north, in faulted ground near Nant-y-Frith, Wood records E_2 beds in a shaly band within the Cefn-y-Fedw Sandstone, and the fauna evidently represents the main ($E_{2a}{}^2$) *bisulcatum* horizon as the fauna includes *E. bisulcatum* Girty and *Kazakhoceras hawkinsi* (Moore). Nearby, in Glascoed Quarry, Jackson (1946) recorded what is probably an E_{2b} fauna comprising *Eumorphoceras sp.*, *Tylonautilius nodiferus* (Armstrong), '*Pseudamussium*' *jacksoni* (Demanet), *Posidonia corrugata* (Etheridge jun.), *Solemya* cf. *primaeva* (Portlock), *Sanguinolites sp.*, *Schizodus sp.*, *Paraconularia* aff. *quadrisulcata* (J. Sowerby), *Stenoscisma* cf. *papyracea* (Roemer), *Lingula mytilloides* (J. Sowerby) and *Orbiculoidea sp.* It is probable that these E_2 faunas lie within the lower part of the Middle Cefn-y-Fedw Sandstone.

Still farther north Namurian rocks are well displayed in sections in the Rivers Terrig and Cegidog (Fig. 39), but there are faults and gaps in the sections. E_2 beds are known near the base (Wood 1936, p. 16; Jones & Lloyd 1942, pp. 254, 258) and benthonic fossils occur in this area too at this level. *E. bisulcatum* of probable E_{2b} age has also been collected on the east side of Hope Mountain, and E_2 fossils are known from several places in the vicinity (Jones & Lloyd 1942, p. 259). In all these cases quartzitic sandstones of the Cefn-y-Fedw Sandstone Group overly the E_2 beds. In the R. Terrig these are probably several hundred feet thick and are overlain by a marine band containing *Reticuloceras bilingue* (Salter) of R_{2b} age. On Hope Mountain the sandstones are overlain by high R_1 shales containing *R. reticulatum* (Phillips) of R_{1c} age. This thick sandstone must be, at least in part, the lateral correlative of the Dee Bridge Sandstone farther south.

In higher Namurian beds of this area marine bands are seen in the R. Terrig' 2 km West of Tryddyn Church, with *G. cancellatum* Bisat (occurring in a limestone band) and *G. cumbriense* Bisat (associated with benthonic mollusca), both indicating a near-shore facies. They are overlain by sandstones which are the correlatives of the Aqueduct Grit of the Trevor section, and the Lower Gwespyr Sandstone of the north of the district.

Around Hawarden, about 8 km north-east of the sections in the R. Terrig there are no thick sandstones in the lower part of the succession, which is nearly all shaly with only a few thin quartzitic sandstone beds. These latter evidently represent the most north-easterly extensions of tongues of Cefn-y-Fedw Sandstone reaching into the Namurian basin. In Warren Dingle, 4 km south of Hawarden, goniatites collected by W. B. R. King in 1913 prove beds of R_1 and R_2 age, all in goniatite/bivalve facies, though the thickness of the local succession is difficult to estimate because of faulting.

In the northern part of Flintshire, where no Cefn-y-Fedw Sandstone occurs, and all the lower part of the sequence is of Holywell Shale facies, the succession is best known from the boreholes at Abbey Mills, Greenfield, and from No. 3 Shaft at Point of Air Colliery. There are, however, more or less isolated exposures in the Shales near Holywell itself and between Holywell and Halkyn, detailed by Jones and Lloyd (1942, locs. 2–14), which provide most of the goniatite horizons from upper E_2 to the *G. cancellatum* horizon in goniatite/bivalve facies.

The lowest beds are known only in Abbey Mills No. 4 borehole* and these have recently been re-examined. The *Cravenoceras* cf. *stellarum* of Wood at 585 ft (178·3 m) is reinterpreted as *Cravenoceratoides edalensis* (Bisat) of basal Ess age. The next goniatite horizon down, at 615 ft, (187 m) is the main *E. bisulcatum* band (Esss) and contains *E. bisulcatum, Dimorphoceras* sp. s.l. *Cravenoceras sp.,* crinoid columnals, and *Posidonia corrugata.* Below this, at 707 ft (215 m), the *C. cowlingense* horizon (Esss) contains *Eumorphoceras sp., Cravenoceras* cf. *cowlingense* Bisat and *P. corrugata.* At 760 ft (232 m) *P. membranacea M'Coy* occurs and this is interpreted as the *C. malhamense* horizon (Ess), at which this fossil is usually abundant. Below this only fish remains and much collophane were collected above the cherts which occur at a depth of 789 ft (240 m). Continuity with the underlying Viséan cannot be proved and the lowest part of the E_1 succession is apparently condensed.

In the Abbey Mills boreholes the succession consists entirely of shaly material up to the horizon of *R. superbilingue* (R_{2c}), above which there is a non-quartzitic sandstone nearly 60 m thick; the *G. cancellatum* Marine Band was not found though its horizon is probably below this sandstone. In Point of Air Colliery No. 3 Shaft the R_2 shales are much thicker (Fig. 39) though the sandstone below the *G. cumbriense* horizon is much thinner. The *G. cumbriense* marine band occurs in both bores at Abbey Mills ('goniatites' were recorded by Strahan 1919, p. 60) and the fauna is of goniatite/bivalve facies. In Point of Air No. 3 Shaft, however, both the *G. cumbriense* band and the *G. cancellatum* band (which is repeated by faulting) contain a rich shelly benthonic fauna including calcareous brachiopods, pleurotomarians and nuculoid bivalves. The feldspathic sandstone beds above the *G. cumbriense* band, called the Lower Gwespyr Sandstone by Jones and Lloyd (1942, p. 252), has *Gastrioceras subcrenatum*, the basal Westphalian goniatite, immediately above it at its exposure at Nant Felin-blwm, west of Mostyn Station. In the Point of Air Shaft, however, non-marine bivalves occur a short distance below the *G. subcrenatum* band, and '*Anthraconaia' bellula* (Bolton) is also known from what is probably a slightly lower horizon in Abbey Mills No. 4 borehole (see Fig. 39). Both the *G. cumbriense* and *G. cancellatum* marine bands are known in the boring at Heswall (Strahan 1914) on the other side of the Dee Estuary at depths of 3172 ft (1067 m) and 3264 ft (1095 m) respectively, and the *G. cancellatum* band, with the *G. cren-*

cellatum fauna is known in a boring near Gyrn Castle, Llanassa, where non-marine bivalves occur at several horizons above it. It seems clear that the beds above the *G. cumbriense* band are highly variable in character and thickness.

REFERENCES V

AITKEN, J. 1870. The grit-rocks of the eastern border of North Wales. *Geol. Mag.*, **7**, 263–5.

BISAT, W. S. 1940. An early *Gastrioceras* (*G. branneroides* sp. nov.) from North Wales. *Trans. Leeds geol. Ass.*, **5**, 330–5.

CONYBEARE, W. D., and PHILLIPS, W. 1822. *Outlines of the geology of England and Wales.* London.

DAVIES, D. C. 1865. On the discovery of fossils in the Millstone Grit near Oswestry. *Rep. Oswestry and Welshpool Nat. Field Club*, 41–2.

——. 1870. On the Millstone Grit of the North Wales border. *Geol. Mag.*, **7**, 68–73, 122–7.

GREEN, A. H. 1867. On the Lower Carboniferous rocks of North Wales. *Geol. Mag.*, **4**, 11–14.

JACKSON, J. W. 1925. Sabden Shale fossils near Holywell, Flintshire. *Naturalist*, No. 821, 183.

——. 1946. *Tylonautilus nodiferus* (Armstrong) from the Cefn-y-Fedw series at Nant-y-Ffrith (new to North Wales). *Proc. Lpool geol. Soc.*, **19**, 161–4.

JONES, R. C. B., and LLOYD, W. 1942. The stratigraphy of the Millstone Grit of Flintshire. *J. Manchr geol. Ass.*, **1**, 247–62.

KING, W. B. R. 1914. *Summ. Prog. geol. Surv. Gt. Br.* for 1913, 12–14.

LLOYD, W., and JONES, R. C. B. 1930. The Upper Carboniferous of Flintshire (abstract). *Geol. Mag.*, **67**, 45.

MORTON, G. H. 1876–8. The Carboniferous Limestone and Millstone Grit of North Wales. *Proc. Lpool geol. Soc.*, **3**, 152–205, 299–325, 371–428.

——. 1879. *The Carboniferous Limestone and Cefn-y-Fedw Sandstone of the country between Llanymynech and Minera, North Wales.* London.

——. 1883. On the strata between the Carboniferous Limestone and the Coal Measures in Denbighshire and Flintshire. *Trans. Manchr geol. Soc.*, **17**, 74–86.

PROSSER, W. 1865. On the fossiliferous character of the Millstone Grit at Sweeney, near Oswestry, Shropshire. *Geol. Mag.*, **2**, 107–10.

RAMSBOTTOM, W. H. C. 1969. The Namurian of Britain. *C. R. Congr. Stratigr. Geol. carbonif., Sheffield 1967*, **1**, 219–32.

——. 1970. Carboniferous faunas and palaeogeography of the south-west England region. *Proc. Ussher Soc.*, **2**, 144–57.

SARGENT, H. C. 1927. The stratigraphical horizon and field relations of the Holywell Shales and 'Black Limestone' of North Flintshire. *Geol. Nag.*, **63**, 252–63.

STRAHAN, A. 1890. The geology of the neighbourhoods of Flint, Mold, and Ruthin. *Mem. geol. Surv. Gt. Britain.*

——. 1914. The Heswall boring. *Summ. Prog. geol. Surv. Gt. Br.* for 1913, 95–7.

——. 1919. On a boring at Abbey Mills, Holywell, 1917–18. *Summ. Prog. geol. Surv. Gt. Br.* for 1918, 58–62.

WALKER, A. O. 1878. Notes on the Lower Coal Measures between Bagillt and Holywell. *Proc. Chester Soc. Nat. Sci.*, No. 2, 9.

WEDD, C. B., SMITH, B. and WILLS, L. J. 1927. The geology of the country around Wrexham, Part 1. Lower Palaeozoic and Lower Carboniferous rocks. *Mem. geol. Surv. Gt. Britain.*

WOOD, A. 1936. Goniatite zones of the Millstone Grit Series in North Wales. *Proc. Lpool geol. Soc.*, **17**, 10–28.

THE WESTPHALIAN OF NORTH WALES

M. A. Calver and E. G. Smith[1]

I. INTRODUCTION

THE Westphalian rocks of North Wales consist of a lower sequence of productive grey measures and an upper sequence of barren, largely red measures. Over much of the region the red measures correspond approximately to the Upper Coal Measures of the standard classification used in England and Wales, but their base is not constant in horizon and the Top Marine Band, marking the base of the Upper Coal Measures, has not been found in North Wales. The relationship of the two classifications to the standard Westphalian stages is shown below:

TERMINOLOGY ADOPTED IN THIS ACCOUNT	WESTPHALIAN STAGES	STANDARD CLASSIFICATION ENGLAND AND WALES	PRINCIPAL MARINE BANDS
RED MEASURES	D	UPPER COAL MEASURES	
	C		*A. cambriense* M.B.
		MIDDLE	*A. hindi* M.B.
Warras M.B.	B	COAL MEASURES	
GREY MEASURES Llay M.B.			*A. vanderbeckei* M.B.
	A	LOWER COAL MEASURES	
G. subcrenatum M.B.			*G. subcrenatum* M.B.

Both grey and red measures occur in the Flintshire and Denbighshire coalfields and in Anglesey, and the latter are also found in the Vale of Clwyd and along the Menai Strait in Caernarvonshire (Fig. 40). All the outcrops are extensively obscure by drift deposits, and knowledge of the Westphalian rocks is therefore largely dependent on boreholes, mine-shafts and underground workings, most of which pre-date modern studies in stratigraphical palaeontology.

The grey measures consist largely of a cyclic sequence of coals, mudstones with siltstones, sandstones and seatearths, occurring ideally in that upward order, and considered to have been deposited in an intermittently subsiding basin, for the most part in fresh or brackish, shallow water. Bands of dark mudstone with marine fossils, usually occurring in the roofs of coals, bear witness to occasional incursions of the sea and enable correlations to be made with other British and European sequences. Non-marine bivalves (mussels) are also valuable for correlation purposes and are the basis of a zonal classification (Fig. 42).

The red measures consist essentially of red and purple mudstones and sandstones with rare thin limestones. There are minor intercalations of grey beds and a few

[1] Published by permission of the Director, Institute of Geological Sciences.

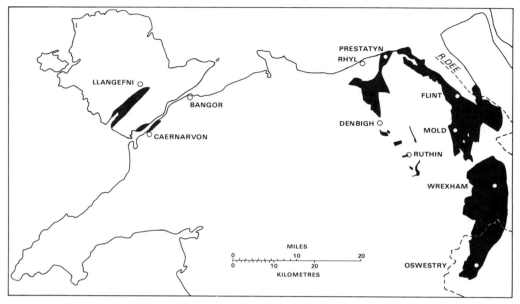

Fig. 40.   Distribution of Westphalian outcrops in North Wales.

thin coals, and a cyclic pattern is discernible in parts, but animal fossils are scarce and few of these are of stratigraphical value. Initially the red measures are believed to have formed around the margins of the Westphalian basin, but eventually they extended across and filled it, and the suggestion is made below that the lower parts pass laterally into grey measures. The red measures were evidently laid down under oxidising conditions, probably in water even shallower than that in which the bulk of the grey measures was deposited.

Included with both the grey and the red measures are strata of intermediate character, the Buckley Fireclay Group at the top of the former, and the Coed-yr-allt Beds within the latter. They are interpreted as transitional deposits between the grey facies and the red facies.

II. FLINTSHIRE AND DENBIGHSHIRE COALFIELDS

(a) Introductory

The stratigraphy and structure of the Westphalian rocks in these coalfields have been described by the Geological Survey (Wedd et al. 1923, Wedd & King 1924, Wedd et al. 1928 and Wedd et al. 1929), and more recent brief accounts have been provided by Wood (1954) and Smith & George (1961).

The Flintshire Coalfield, extending from Point of Air in the north to the neighbourhood of Caergwrle in the south, is separated from the Denbighshire Coalfield by a narrow outcrop of Namurian and Viséan rocks brought to the surface by the Llanelidan (Bala) Fault and associated anticline (Fig. 41). The Denbighshire Coalfield is continuous southwards with the Oswestry Coalfield of north Shropshire, which is included in this account. The total area of the coalfields is about 380 square

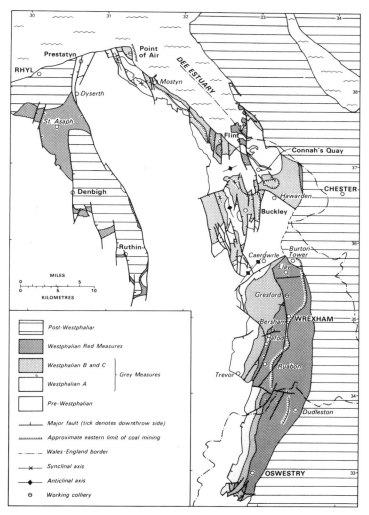

Fig. 41. Sketch-map of the Flintshire and Denbighshire coalfields, including the Vale of Clwyd.

kilometres, rather more than a third of which is covered by red measures.

The Westphalian rocks of North Wales and Lancashire are probably continuous at depth below the Permo-Trias of the Cheshire Plain, though the only published evidence of this is the occurrence of an upfaulted inlier of Erbistock Beds near Aldford, south of Chester (Poole and Whiteman 1966, pp. 9–10). There may also be continuity between North Wales and north Staffordshire, but possibly only of the red measures, for Newcastle Beds rest on pre-Westphalian rocks south-west of Newcastle-under-Lyne (Wills 1956, pp. 56, 171).

The Westphalian rocks have a general easterly dip, with a northerly component in the north and a southerly component in the south. Superimposed on this regional picture (Fig. 41) are folding in the north and faulting with associated block-tilting. Folding occurs on two sets of axes, north-easterly or east-north-easterly and north-westerly. The Horseshoe Anticline of Wedd *et al.* (1924, pp. 10–11, pl. II), swinging round from parallelism with the north-easterly-trending Llandelidan Fault in south Flintshire to follow the course of the Dee Estuary in the north, incorporates both these trends, but is of doubtful validity. A number of small structures are probably involved, and the inlier of Namurian rocks between Hawarden and Caergwrle may be the result of block-faulting rather than upfolding. Moreover, recent boreholes in the Dee Estuary show little evidence of a major anticline there. Much of the faulting in both coalfields is parallel or subparallel to the strike of the rocks, but there are a number of cross-faults with very large throws.

The grey measures, containing the coal-bearing strata and forming the lower part of the Westphalian sequence, are upwards of 500 m thick at Point of Air, less than 300 m between Mostyn and Flint and beneath the Dee Estuary, up to nearly 600 m in Denbighshire, and about 250 m around Oswestry. The overlying red measures are everywhere incomplete, but the thickest preserved sections, over 1300 m, occur in Denbighshire south of Wrexham (Fig. 43).

In the following description of the stratigraphy the sequence is dealt with in ascending order. There is little up-to-date information from south Flintshire, and the account therefore concentrates on north Flintshire and the Denbighshire Coalfield (Fig. 42).

(b) Grey Measures

Westphalian A

It follows from the international acceptance of the *Gastrioceras subcrenatum* Marine Band as the base of this division that the Lower Coal Measures of the Geological Survey's Flint (108) Sheet are almost entirely of Namurian age (p.182). and the bulk of the actual Lower Coal Measures are included in the 'Middle Coal Measures' of both the Flint and Wrexham (121) sheets.

The *G. subcrenatum* Marine Band consists of 1 m of dark mudstone occurring 20 m above the Lower Gwespyr Sandstone in north Flintshire (Shanklin 1956, pp. 539–41) and resting on the Aqueduct Grit in Denbighshire (Wood 1937, p. 14); the horizon is also recorded from a borehole at Abbey Mills, near Holywell (Jones & Lloyd 1942, fig. 1) and a shaft sinking at Point of Air, and was proved at a depth of about 870 m in the Heswall Borehole, Wirrall (Strahan 1914). The fauna is of goniatite-pectinoid facies; *G. subcrenatum* (Frech) is the typical goniatite, and *Homoceratoides* aff. *divaricatus* (Hind) and rare *Reticuloceras superbilingue* Bisat also occur. The characteristic bivalves are *Dunbarella papyracea* (J. Sowerby), *Posidonia gibsoni* Salter and *Caneyella multirugata* (Jackson). Calcareous and horny brachiopods as well as conodonts are recorded in addition from a stream exposure and shallow boreholes at Nant Felin-blwm, near Mostyn, (Shanklin 1956, p. 540).

In north Flintshire the 30 m of measures separating this band from the overlying *Gastrioceras listeri* Marine Band include the Upper Gwespyr Sandstone, which is flaggy in parts and up to 13 m thick. Poorly preserved *Carbonicola* spp. were recorded by Wood (1937, pp. 3–4) in silty partings near the top of the sandstone.

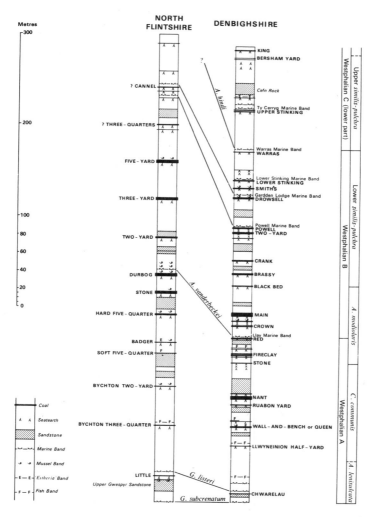

Fig. 42. Comparative sections of the Westphalian grey measures of north Flintshire and Denbighshire.

The *G. listeri* Marine Band overlies the Little Coal, a short distance above the top of the Upper Gwespyr Sandstone. The marine fossils range through nearly 2 m of dark mudstones recorded from shallow boreholes and trial pits at Nant Felinblwm and Connah's Quay. The characteristic goniatite, *G. listeri* (J. Sowerby), is associated with *Anthracoceratites* sp., pectinoid bivalves and conodonts (Shanklin 1956, p. 538). This northern facies is an extension of that known in the south Pennine area (Calver 1968, pp. 25–7); in contrast, in Denbighshire the band is largely of *Lingula* facies, for example in the Australia clay-pit, near Trevor, where it lies 2·5 m above the Chwarelau Coal (Wood 1937, p. 15).

The measures between the *G. listeri* Marine Band and the Llwyneinion Half-Yard Coal are not known in detail. They are of the order of 50 m in thickness and consist largely of mudstones with thin coals and sandstones. The uppermost of three *Lingula* bands near the bottom of the Dudlestone A5/6 Bore, south-east of Ruabon, probably belongs to the lower part of this sequence and could be the equivalent of Tonge's Marine Band of the Lancashire succession.

In the absence of diagnostic mussels the position of the *A. lenisulcata/C. communis* zonal boundary cannot be established, but it must lie below the Llwyneinion Half-Yard Coal of Denbighshire, a seam possibly equivalent to the Bychton Three-Quarter of Flintshire. Both these coals have abundant fish remains in their roof measures. This feature and the stratigraphical position of the coals suggest possible correlation with the Arley Mine of Lancashire.

Wood (1937, p. 5) records mussels above the Wall-and-Bench (Queen) Coal at several localities around Wrexham. The fauna includes *Carbonicola communis* Davies and Trueman and *C. pseudorobusta* Trueman, as well as small forms of *Carbonicola* referred to the *C. bipennis* (Brown)—*C. antiqua* (Brown) group. Wood also noted the presence of *C. communis* preserved in ironstone occurring as nodules in black shale above the Bychton Two-Yard Coal of Point of Air Colliery. Further collecting from this colliery has shown that between the coal and the mussel-band there are grey shales with rare *Planolites ophthalmoides* Jessen and sporadic fish remains. The overlying measures contain well-developed sandstones, and these are succeeded by the Soft Five-Quarter Coal with the Badger Coal supposedly a short distance above. However, there is some doubt as to whether two distinct coals are involved, and these may be different names for the same seam. In Point of Air Colliery workings the Badger (here 0·75 m thick) was seen to be overlain by grey mudstones containing several examples of a large estheriid associated with mussel fragments. The estheriid recalls the species identified as '*Estheria*' sp. nov. from the Low '*Estheria*' Band of the Yorkshire and East Midlands Coalfield. These respective occurrences may represent the same horizon, for there is indirect evidence that both lie in the upper part of the *C. communis* Zone. Wood (1937, pp. 6–7) records several bands with mussels in Nant-y-ffynon Lwyd, Mostyn; these probably represent *C. communis* Zone faunas but they cannot be assigned to named coals.

The top of the *C. communis* Zone is placed tentatively at the Hard Five-Quarter Coal, a decision based on the faunal assemblage occurring in ferruginous mudstones above this coal in Point of Air Colliery. The fauna includes *Anthraconaia* sp., *Carbonicola* cf. *robusta* (J. de C. Sowerby), *Naiadites flexuosus* Dix & Trueman, *Geisina arcuata* (Bean) and fish remains. The roof measures of the overlying Stone Coal contain abundant plants, but in the abandoned Neston Colliery on the Wirral the supposed equivalent seam, known as the Six-Foot, is overlain by ferruginous mudstones containing a mussel-band (Wood 1937, p. 7). The shells are well preserved and represent a *Carbonicola oslancis* Wright fauna, which is consistent with a position near the base of the *A. modiolaris* Zone.

The coal occurring some 20 m higher in the sequence is named the Durbog in north Flintshire and the Red in Denbighshire. The mussel-band found in the roof of the Durbog Coal of Point of Air Colliery, although within Westphalian A, is included for convenience of description in the following section.

Westphalian B

The Llay Marine Band at the base of this division was named after Llay Main Colliery, where marine fossils, here comprising *Lingula* and gastropods, were first recorded in the roof measures of the Red Coal (Magraw & Calver 1960, p. 336). In Point of Air Colliery this same marine band occurs in the measures above the Durbog Coal, immediately overlying the mussel-band referred to above, and yields a more varied assemblage. The full sequence is as follows: a 10-cm canneloid mudstone with fish remains forms the roof of the coal and passes up into dark grey mudstones, nearly 1 m thick, containing abundant and well-preserved *Anthracosia* spp. belonging to the *A. regularis* (Trueman) group. Rare examples of *Anthraconaia salteri* Leitch are also present, associated with the ostracod *Geisina arcuata*. The marine fossils occur in up to 4 m of overlying dark to grey, slightly silty mudstones; the fauna includes *Glomospira* sp., sponge spicules, *Lingula mytilloides* J. Sowerby, *Spirifer pennystonensis* George, *Posidonia* cf. *sulcata* Hind, *Hollinella* (*Praehollinella*) *claycrossensis* (Bless & Calver) and *Hindeodella*. There are also some ghost-like impressions which may represent *Anthracoceratites*.

Nearby boreholes at Talacre show that a thick mussel-band overlies the mudstones with marine fossils; *Anthracosia ovum* Trueman & Weir and *A.* aff. *phrygiana* (Wright) are the dominant forms.

The succeeding measures are largely mudstones, but there are several coals, the most important of which is the Main. A distinctive mussel-band occurs above this coal, the fauna representing an *Anthracosia phrygiana* assemblage (Wood 1937, p. 9 and fig. 6; Magraw & Calver 1960, p. 337). Typical species include *Anthracosia beaniana* King, *A.* aff. *retrotracta* (Wright), *Anthracosphaerium turgidum* (Brown) and *Naiadites* aff. *quadratus* (J. de C. Sowerby); in addition *Carbonita humilis* (Jones & Kirkby) and fish remains are recorded. The faunal evidence places this horizon in the upper part of the *A. modiolaris* Zone.

The base of the Lower *similis-pulchra* Zone is taken in this account at the Black Bed Coal horizon. Mussels from the roof measures of this coal include *Anthraconaia pulchella* Broadhurst, *A.* sp. nov. cf. *williamsoni* (Brown), *Anthracosia aquilinoides* (Chernyshev) and *Naiadites productus* (Brown) together with *Carbonita humilis*.

The Powell Marine Band in the roof of the Powell Coal is known from several boreholes and drivages in the Bersham–Gresford district. The fauna is usually confined to *Lingula*, but locally *Hollinella* and *Myalina* are recorded. A feature of this horizon is the close association of the marine fossils with elements from the non-marine environment such as stunted *Anthracosia*, *Curvirimula* and *Geisina*. At Tan Llan Opencast site a mixed marine/non-marine fauna of this nature ranged through nearly 5 m of mudstone (Magraw & Calver 1960, pp. 338, 346). The marine band is locally succeeded by mudstones containing a varied *Anthracosia* assemblage including *A. acutella* (Wright), *A. concinna* (Wright) and *A. lateralis* (Brown) (Wood 1937, pp. 11–12). The faunal sequence shows that this horizon is the correlative of the Two-Foot Marine Band of the main Pennine coalfields.

In the Wrexham area the measures between the Powell Coal and the overlying Drowsell Coal thin to the south, so that the Drowsell forms the upper leaf of a group of seams known as the Powell Group (Magraw & Calver 1960, p. 340).In this area the Powell Marine Band is absent, but a higher marine band occurs above the Drowsell and overlies a mussel-band. This marine band is named after Gardden Lodge Opencast Site, south-west of Wrexham, where it was first recognized (Magraw and Calver 1960, pp. 339, 347). At this locality the non-marine fauna immediately above the coal included *Anthracosia atra* (Trueman), *A*. aff. *aquilinoides*, *Carbonita humilis* and scales of *Rhabdoderma*, *Rhizodopsis* and *Megalichthys*. The overlying marine fossils ranged from 0·6 m to 2·0 m above the coal, and the fauna was restricted to *Lingula* and rare foraminifera. This horizon is the correlative of the Clown Marine Band of the mid-Pennines.

Both the Powell and Gardden Lodge marine bands were recorded in the Talacre No. 2 Borehole, located on the shore north-west of Point of Air Colliery, but their relationship to the proved seams in the colliery has not been established. The boring showed that the seam underlying the higher of the two marine bands was an inferior cannel coal nearly 1 m thick, and this could possibly be the Cannel or Hollin Coal of the Mostyn area.

In the Buckley district a thick sandstone known as the Hollin Rock succeeds the Hollin Coal and locally replaces all other beds almost up to the base of the Buckley Fireclay Group, but the sandstone thins both westwards and eastwards (Wedd & King 1924, p. 68).

In Denbighshire the measures equivalent to the Hollin Rock contain coals and several important faunal bands. The lowest of these is an *Anthracosia atra* fauna in the roof measures of the Smith Coal at several localities in the Wrexham area (Magraw & Calver 1960, pp. 339–42). The succeeding horizon is the Lower Stinking Marine Band, best known from the well-preserved fossils found on the tips at Brynmally (Wood 1937, p. 14), and also proved at several other localities in the Wrexham area—Gresford, Ifton and Hafod collieries, Black Lane Drift and Gardden Lodge Opencast Site (Magraw & Calver 1960, pp. 339, 342, 348). The marine fossils occur above the Lower Stinking (or Four-Foot) Coal and include *Glomospira* sp., sponge spicules, *Lingula mytilloides*, *Orbiculoidea* cf. *nitida* (Phillips), *Levipustula* cf. *rimberti* (Waterlot), cf. *Pernopecten carboniferus* (Hind), crinoid remains, *Serpuloides stubblefieldi* (Schmidt and Teichmüller), conodonts and faecal pellets. The fauna shows close agreement with that recorded from the Haughton/Bradford Marine Band of the East Pennines/Lancashire coalfields.

Westphalian C (lower part)

The Warras Marine Band is the correlative of the Mansfield Marine Band of the Pennine coalfields, and is thus the arbitrary base of Westphalian C and of the Upper *similis-pulchra* Zone. It occurs in the roof of the Warras Coal, 20m above the Lower Stinking Marine Band, at Hafod Colliery, where confirmation of the occurrence of marine fossils at this horizon was first obtained. It was subsequently proved in the workings of Gresford and Ifton collieries and was also identified in boreholes at Dudleston, south-east of Ruabon. The band is about 1 m thick and notable for the rich fauna of calcareous brachiopods included in the following composite list: *Crurithyris carbonaria* (Hind), *Levipustula* cf. *rimberti*, *Lingula mytilloides*, *Orbi-*

culoidea cf. *nitida*, *Productus carbonarius* de Koninck, *Reticulatia craigmarkensis* (Muir-Wood), *Rugosochonetes*? *skipseyi* (Muir-Wood), *Spirifer* sp., *Tornquista diminuta* (Demanet), *Aviculopecten delepinei* Demanet, *Pernopecten carboniferus*, cf. *Septimyalina* sp., an orthocone nautiloid, coiled nautiloids, '*Anthracoceras*' sp., *Hollinella* sp., crinoid columnals [circular], *Serpuloides stubblefieldi* and conodonts.

About 100 m of measures containing thin coals separate the Warras Marine Band from the Upper Stinking Coal and overlying Ty Cerryg Marine Band. To date, this marine band has not been recorded outside Ty Cerryg Opencast Site, north-west of Wrexham (Magraw and Calver 1960, pp. 349–50). Boreholes at that site showed that the marine fossils ranged over some 5 metres and were distributed in two main bands. The lower band contained abundant foraminifera, rare *Lingula*, *Euphemites* and small, pyritized *Nuculopsis* and *Edmondia*. The higher band was characterized by foraminifera, *Myalina compressa* Hind, *Hollinella claycrossensis*, and fish remains; closely associated or interbedded with this fauna were the non-marine forms *Curvirimula*, *Naiadites* and *Anthraconaia persulcata* Weir. The ostracod *Geisina subarcuata* (Jones) was also present in both leaves of the band. The combined assemblage is closely comparable with that from the composite Five Roads Marine Band and Foraminifera Marine Band of South Wales (Archer 1968, p. 111) and from the *Edmondia* Band of the Pennines (Calver 1968, pp. 51–3).

In this same area a carbonaceous mudstone overlying a thin coal 15 m above the Upper Stinking Coal contains *Lioestheria vinti* (Kirkby), macrospores and abundant fish remains. The succeeding measures include the Cefn Rock, a massive quartzose sandstone 25 m thick (Wedd *et al*. 1928, p. 45). Between this sandstone and the Ruabon Marl in Denbighshire are a series of mudstones with variable coals and seatearths. In the Bersham area a lower composite coal is known as the Bersham Yard and a higher coal is called the King. No faunas are known from this group of beds.

In Flintshire, the highest beds assigned to the grey measures are known as the Buckley Fireclay Group, apparently about 100 m thick. They consist of pale and pinkish grey quartzose sandstones and grey, red and purple mudstones and fireclays. The only recorded fossils are rare *Anthraconauta phillipsii* (Williamson) and *Carbonita salteriana* (Jones & Kirkby) (Wood 1937, p. 12), indicative of the upper part of Westphalian C. It is possible that the horizon of the Top Marine Band, marking the base of the Upper Coal Measures, lies towards the base of this group.

(c) *Red Measures, including the Coed-yr-allt Beds*

These measures, designated Upper Coal Measures on the Geological Survey maps, comprise in upward succession up to 330 m of Ruabon Marl, up to 160 m of Coed-yr-Allt Beds and up to 900 m of Erbistock Beds (Fig. 43). These formations are broadly comparable to the Etruria Marl, Newcastle Beds and Keele Beds respectively of north Staffordshire.

The Ruabon Marl, not recognized north of the Wrexham area, consists largely of red or purple mudstones with green or yellow banding and mottling. Thin beds of grey to black mudstone and coal smuts are occasionally found, thin espleys

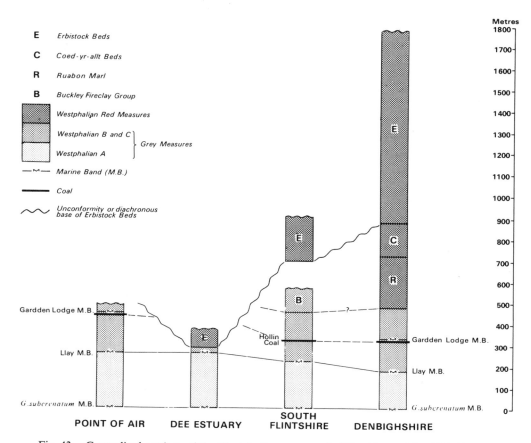

Fig. 43. Generalised sections of the Westphalian sequence in Flintshire and Denbighshire.

(sandstones or breccias) occur, chiefly in the lower part and most abundantly in south Denbighshire, and there are thin *Spirorbis* limestones in the upper part. The only fossils recorded are plants, *Spirorbis*, ostracods, fish and rare mussels. *Anthraconauta phillipsii* has been collected 25 m above the base of the Ruabon Marl in Hafod Colliery workings, and collections from the uppermost beds of the formation at Pont-y-Cyfflogyn, near Ruabon, (Simpson 1937, pp. 193–4) consist largely of the same species though they include some variants approaching *A. tenuis* (Davies & Trueman). Associated ostracods at the latter locality include *Carbonita pungens* (Jones & Kirkby) and *C. salteriana*. The Ruabon Marl, 330 m thick in the south, shows a northward thinning to about 200 m around Wrexham.

The Coed-yr-Allt Beds consist essentially of grey mudstones, sandy mudstones and sandstones, but parts of the succession show red and purple bands and mottling. They have the same cyclic sequence as the productive grey measures, but coals are generally thin and there are limestone bands, one of which forms the basal member of the formation. Mussels are recorded in the beds above the basal limestone at Ponty-y-Cyfflogyn (Wedd *et al.* 1928, p. 115; Simpson 1937, pp. 193-5). They comprise

A. phillipsii and *A. aff. tenuis* as well as numerous small shells previously referred to *A. calcifera* (Hind) (Simpson *idem*), but probably representing immature or dwarfed *A. phillipsii*. Simpson (1936, p. 280) records *A. tenuis* from a tip associated with a working in the Morlas Coal towards the top of the formation near Chirk. Other recorded fossils from the Coed-yr-allt Beds are plants, *Spirorbis* (in the limestones), ostracods and fish debris. The ostracods are common at several horizons, particularly in the limestone bands, and include *C. evelinae* (Jones), *C. salteriana*, *C.* sp. nov. [elongate] and *C.* sp. [coarsely punctate]. The faunal evidence suggests that the base of the Coed-yr-allt Beds approximates to the Westphalian C-D boundary. The Coed-yr-allt Beds have not been found north of Caergwrle, except possibly in the Vale of Clwyd (see below), nor in the south near Oswestry; they thin northwards from 160 to about 90 m.

The Erbistock Beds have a wide outcrop from Llay, north of Wrexham, to Oswestry, and red measures occurring along and beneath the Dee Estuary are also assigned to the formation. They consist of red and purple mudstones, with subordinate green banding and mottling, and a considerable thickness of coloured and grey, generally feldspathic, sandstones. Characteristic features are marl-breccias and beds containing abundant haematite grains or 'shot'. There are intercalated grey measures, locally with a few centimetres of coal, and limestone bands like those of the Ruabon Marl and Coed-yr-allt beds (see above) are also found. The only fossils recorded are plants, *Spirorbis* and tetrapod footprints. Because the Erbistock Beds are unconformably overlain by Permo-Triassic rocks and are therefore incomplete, no thickness trends have been established.

To explain the relationships of the red measures to the productive grey measures, and of the Erbistock Beds to underlying formations which they apparently overstep, unconformities have been invoked at the base of the Ruabon Marl and at the base of the Erbistock Beds (Davies 1877, pp. 10, 14; Wedd & King 1924, pp. 132–6; Wedd *et al.* 1928, pp. 91–4; Wedd *et al.* 1929, pp. 123–4, 144–5). The lower unconformity at least is now in doubt. The similarity of the North Wales Westphalian sequence to those of south Lancashire and north Staffordshire suggests that similar explanations apply. In the latter coalfields recent work by Mr. W. B. Evans (personal communication) indicates that the red beds are a facies, the onset of which occurs at different times in different places. Thus in north Staffordshire the Black Band Group passes laterally to the south and west into the Etruria Marl (*Summ. Prog. geol. Surv. Gt. Brit.* for 1964, 1965, p. 56), and the Newcastle Beds are the lateral equivalents of the lower part of the Keele Beds (Evans 1966, p. 54). The same relationships are present in south Lancashire, where the Bradford Coal Group passes westwards into the Barren Red Measures below the Ardwick Limestones, and the Ardwick Limestones pass into the red Upper Group of Trotter (1952, p. 263). It is therefore not unreasonable to interpret the Ruabon Marl of Denbighshire, which is lithologically similar to the Etruria Marl, as the equivalent of grey measures in south Flintshire (Fig. 42). In the latter area the Buckley Fireclay Group (see above), the highest division of the productive grey measures proved in North Wales, has much in common with the Black Band Group of north Staffordshire (Wedd & King 1924, p. 70), although these are also similarities to the underlying Great

Row Measures of Earp (1961, p. 158).

Whether the Erbistock Beds are also diachronous, and more markedly so than the Ruabon Marl, or whether, as has long been supposed, they rest with strong unconformity on the rest of the Westphalian, is not clear. Underground sections between Mostyn and Flint and recent boreholes in the Dee Estuary show that rocks of Erbistock Beds facies overlie productive grey measures only slightly younger than the Llay Marine Band (Fig. 43), suggesting post-Westphalian A uplift in this area. The shafts and boreholes provided no direct evidence of a major unconformity, which must be present if the uplift took place after the deposition of the productive grey measures and was followed by extensive erosion. The area could, however, have been a positive one throughout post-Westphalian A times, resulting in the deposition of red measures there while sediments of grey measures facies were accumulating to the north and south. If the latter explanation is correct, the post-Westphalian A grey measures at Point of Air (Figs. 42 and 43) were probably deposited in a basin effectively separate from that of south Flintshire and Denbighshire. It is tempting to attribute difficulties of correlation to such a separation.

(d) Exploitation

Coal has been mined over a long period and from numerous collieries, only three of which, Point of Air, Gresford and Bersham (Fig. 41) are now active. The principal seams worked or working at Point of Air are those from the Hard Five-Quarter to the Five-Yard (Fig. 42). The most important seams formerly mined farther south in Flintshire are the Nine-Foot, Wall-and-Bench, Brassy and Main in the same part of the succession, but they also include the much higher Hollin Coal. In Denbighshire the most extensively mined seams are the Wall-and-Bench (a lower seam than the coal of this name in Flintshire), Nant, Red, Main, Brassy, Crank, Two-Yard and Powell (Fig. 42). Since 1943 coal has also been extracted by opencast methods in thirty sites scattered across the coalfields and covering much of the grey measures succession. Two and a half million tons have been won by these means, but there is no production at present.

The exposed coalfields have thus been extensively exploited, and workings continue eastwards below the cover of barren red measures, underlying the Triassic outcrop in the Burton Tower area (Fig. 41). Remaining reserves of workable coal, in seams 0·6 m (2 ft) or over to a depth of 1100 m, are estimated at about 760 million tons. Reserves, classified and unclassified, accessible from the three remaining collieries, however, are only 60 million tons.

Other economic products are bricks and tiles, with the Ruabon Marl as the most notable source, and refractory materials, derived especially from the Buckley Fireclay Group, whose clays have also been used for pottery making. Building stone, particularly from the Cefn and Hollin Rocks, and ironstone were once widely worked, and, in the upper part of Westphalian A in Flintshire, oil has been extracted from the Cannel Coal and its immediate roof.

III. VALE OF CLWYD

Westphalian rocks are believed to occur throughout the Vale north of Ruthin, although little is known about them, for, even where they are not concealed beneath Permo-Triassic rocks, there is an extensive drift cover. There is some evidence from old lead-mine drivages to suggest that grey measures occur between Dyserth and Prestatyn (Fig. 41), where they are presumed to succeed conformably the Namurian rocks proved there (Geological Survey 1-in Sheet 95, 1970), but there are no exposures, boreholes or workings. Small exposures of red mudstone and sandstone occur in the banks of the River Elwy south of St. Asaph, and at a number of places along the western side of the Vale between Denbigh and Ruthin, where the outcrop is faulted against older strata. Mudstone, purple sandstone and black shale with a thin coal crop out on the eastern side of the Vale north of Ruthin (Wedd & King 1924, p. 60), and there are small outcrops of purple sandstone and red and blue shale at the southern end of the Vale (Wedd *et al.* 1928, pp. 18–21) which also are probably of Westphalian age. A number of boreholes have reached Westphalian rocks, proving up to nearly 200 m of red, or red and grey banded, mudstones, marls and sandstones with, in some instances, seatearths, thin coals or traces of pre-existing coals, and thin limestones. The picture is complicated by probable staining beneath the Permo-Triassic unconformity, but some sections of wholly red measures are reminiscent of the Erbistock Beds, and the red and grey measures may be of Coed-yr-allt facies. The St. Asaph Borehole (*Ann. Rep. Inst. geol. Sci. for 1968*, 1969, p. 29) proved red measures overlying red and grey measures, the latter containing *Anthraconaia* sp., *Leaia* sp. and *Anomalonema reumauxi* (Pruvost), a fauna which is indicative of Upper Coal Measures and suggests an horizon near the Westphalian C–D boundary.

The total thickness of the Westphalian rocks in the Vale is unknown. Powell (1956) calculated from gravity anomalies that about 760 m of Upper Carboniferous rocks are present in the north, but he assumed their absence in the south. Wilson (1959), who carried out a seismic refraction survey, estimated that over 300 m of Upper Carboniferous rocks occur on the eastern side of the Vale as far south as the Denbigh area. Namurian rocks, which occur near Prestatyn, may extend at depth below the Rhyl area, but to the south the Upper Carboniferous rocks appear to be entirely of Westphalian age and to rest unconformably on the Viséan. If productive grey measures exist in the Vale outside this northern area, and there is no evidence that they do, they are likely to be thin. Red measures apparently form the whole or the bulk of the succession over most of the Vale and suggest, as in the Mostyn-Flint and Dee Estuary region (see above), that the area was marginal to the basin of deposition in Westphalian times.

IV. ANGLESEY AND ARVON

Westphalian rocks crop out over about 25 square kilometres to the south-west of Llangefni in southern Anglesey (Fig. 40). The outcrop, roughly corresponding to the low-lying area of Malldraeth Marsh, is however almost entirely drift-covered, and information about the rocks is largely derived from shafts and boreholes, all old and some unsited. The Westphalian rocks rest for the most part on Viséan limestones, but in the south-west on the Pre-Cambrian. They are preserved on the north-westerly, upthrow side of the large north-easterly-trending dislocation known as the

Berw Fault. Their overall dip apparently decreases towards the fault, and the Westphalian rocks can be looked upon as occupying a syncline with a south-westerly plunge and an axial plane inclined to the north-west.

The succession consists of over 200 m of red measures on about 450 m of grey measures (Greenly 1919, p. 666). The latter include, at the base, 120 m of so-called Millstone Grit, which consists largely of false-bedded grey or brown sandstone with a basal conglomerate, but also contains shale and coal. About half way up the 'Millstone Grit' dark marine shales occur (Greenly 1919, p. 662), and fossils have been collected from them at several localities near Glan traeth. A Westphalian age is attributed to the 'Millstone Grit' because of this marine fauna. It includes goniatites, which are definitely *Gastrioceras*, but are poorly preserved; they are thought most likely to be *G. listeri*. The associated fossils include *Caneyella multirugata*, *Dunbarella papyracea*, *Posidonia gibsoni* and a distinctive *P.* sp. nov. which has a prominent anterior ear on the right valve. This assemblage is typical of the *G. listeri* Marine Band of Flintshire (p. 173) and the nearest coalfields of the Pennine province. The rest of the grey measures appear to be a normal cyclic Westphalian sequence; they contain at least 13 coals (Greenly 1919, p. 663), of which 4 are more than a metre thick. Coal was mined intermittently from the fifteenth century until 1875. The only fossils of stratigraphical significance collected from these measures are mussels found in spoil-banks at Morfa-mawr and, according to Greenly (1919, p. 665), probably originating from beds about 245 m above the top of the 'Millstone Grit'. The fauna includes *Carbonicola bipennis*, *C. browni* Trueman & Weir, *C. martini* Trueman & Weir, *C. polmontensis* (Brown) and *Curvirimula* cf. *trapeziforma* (Dewar), suggesting that the horizon lies in the lower part of the *Carbonicola communis* Zone. The red measures consist of red-mottled and grey mudstones and chiefly red, sometimes conglomeratic, sandstones. Greenly (1919, pp. 672, 821) produces some evidence that they rest unconformably on the grey measures, but it is not considered conclusive.

Red measures, apparently overlying Viséan limestone and cropping out (Fig. 40) over nearly 2 square kilometres along the Menai Strait opposite Caernarvon (Greenly 1919, pp. 668–74), are assigned on lithological evidence to the Westphalian. They are known as the Ferry Beds. A few metres of red mudstones and marls with some pale grey-green mottling and thin bands of pebbly sand are exposed.

Red measures on the Arvon side of the Menai Strait are also assigned to the Westphalian, partly on lithological grounds and partly because they are reputed to contain thin coals locally (Ramsay 1881, p. 261; Greenly 1938, p. 340). Their outcrop covers nearly 1½ square kilometres along the Strait to the north-east of Caernarvon (Fig. 40) and is bounded on the south-east by the Dinorwic Fault. In the north the beds are apparently involved in thrusting (Greenly 1938, pp. 340–4). They consist of red mudstones or marls with green bands, conglomerates and a boulder bed. The last consists of sub-angular blocks of volcanic and local sedimentary rocks in an argillaceous matrix, with bedding varying from horizontal to chaotic. Its relationship to the argillaceous measures is obscure, but it is assumed to occur locally at their base. There are few good exposures of the red measures, and the best section is provided by Plas Brereton Borehole which was drilled in 1903 and proved 177 m resting on probable Viséan beds.

V.   REFERENCES

ARCHER, A. A. 1968. Geology of the Gwendraeth Valley and adjoining areas. *Mem. geol. Surv. Gt. Britain.*

CALVER, M. A. 1968. Distribution of Westphalian marine faunas in northern England and adjoining areas. *Proc. Yorks. geol. Soc.*, **37**, 1–72.

DAVIES, D. C. 1877. On the relation of the Upper Carboniferous strata of Shropshire and Denbighshire to beds usually described as Permian. *Quart. J. geol. Soc. Lond.*, **33**, 10–28.

EARP, J. R. 1961. Exploratory boreholes in the North Staffordshire Coalfield. *Bull. geol. Surv. Gt. Br.*, No. 17, 153–90.

EVANS, W. B. 1966. In *Summ. Prog. geol. Surv. Gt. Br.* for 1965, 54.

GREENLY, E. 1919. The geology of Anglesey. *Mem. geol. Surv. Gt. Britain.*

——. 1938. The red measures of the Menaian region of Caernarvonshire. *Quart. J. geol. Soc. Lond.*, **94**, 331–45.

JONES, R. C. B., and LLOYD, W. 1942. The stratigraphy of the Millstone Grit of Flintshire. *J. Manchr geol. Ass.*, **1**, 247–62.

MAGRAW, D., and CALVER, M. A. 1960. Faunal marker horizons in the Middle Coal Measures of the North Wales Coalfield. *Proc. Yorks. geol. Soc.*, **32**, 333–52.

POOLE, E. G., and WHITEMAN, A. J. 1966. Geology of the country around Nantwich and Whitchurch. *Mem. geol. Surv. Gt. Britain.*

POWELL, D. W. 1956. Gravity and magnetic anomalies in North Wales. *Quart. J. geol. Soc. Lond.*, **111**, 375–97.

RAMSAY, A. C. 1881. The geology of North Wales. 2nd edit. *Mem. geol. Surv. Gt. Britain.*

SHANKLIN, J. K. 1956. New record of the *Gastrioceras listeri* Marine Band in Flintshire. *Lpool. Manchr geol. J.*, **1**, 536–42.

SIMPSON, H. H. 1936. Note on the Zone of *Anthraconauta phillipsi* in the Denbighshire Coalfield. *Geol. Mag.*, **73**, 278–80.

——. 1937. Note on *Anthraconauta calcifera* in the Upper Coal Measures of Denbighshire. *Geol. Mag.*, **74**, 193–5.

SMITH, B., and GEORGE, T. N. 1961. North Wales, 3rd edit. revised by T. N. George. *Br. reg. Geol.*

STRAHAN, A. 1914. The Heswall Boring. In *Summ. Prog. geol. Surv. Gt. Br.* for 1913. 95.

TROTTER, F. M. 1952. Exploratory boreholes in south-west Lancashire. *Trans. Instn. Min. Engrs.* **112**, 261–81.

WEDD, C. B., SMITH, B., SIMMONS, W. C. and WRAY, D. A. 1923. The geology of Liverpool with Wirral and part of the Flintshire Coalfield. *Mem. geol. Surv. Gt. Britain.*

—— and KING, W. B. R. 1924. The geology of the country around Flint, Hawarden and Caergwrle. *Mem. geol. Surv. Gt. Britain.*

——, SMITH, B. and WILLS, L. J. 1928. The geology of the country around Wrexham. Part II. *Mem. geol. Surv. Gt. Britain.*

——, ——, KING, W. B. R. and WRAY, D. A. 1929. The geology of the country around Oswestry. *Mem. geol. Surv. Gt. Britain.*

WILLS, L. J. 1956. *Concealed coalfields.* Blackie and Son Ltd., London and Glasgow.

WILSON, C. D. V. 1959. Geophysical investigations in the Vale of Clwyd. *Lpool Manchr geol. J.*, **2**, 253–70.

WOOD, A. 1937. The non-marine lamellibranchs of the North Wales coalfield. *Quart. J. geol. Soc. London.*, **93**, 1–22.

—— 1954. The coalfields of North Wales. *In* Trueman, Sir Arthur, *The coalfields of Great Britain.* Arnold, London.

UPPER CARBONIFEROUS SEDIMENTATION IN SOUTH WALES

Gilbert Kelling

I. INTRODUCTION

(a) Stratigraphical Considerations

THE succession of rocks considered here comprises the Namurian and West-phalian divisions of the Upper Carboniferous. In South Wales this Upper Carboniferous sequence is entirely composed of sedimentary rocks, dominantly clastics but including important coal seams. The stratigraphical framework of Upper Carboniferous sedimentation is dealt with elsewhere in this volume (see Thomas, Chapter 7) and it will be sufficient for the present purpose to summarise the principal facts which relate to the present discussion.

In the traditional view, the Upper Carboniferous rocks of South Wales were deposited in a region dominated by three major structural elements:

(i) To the north, an emergent positive area or barrier ("St. George's Land").
(ii) To the east, a north-south trending axis of uplift (the Usk Axis), which at times appears to have continued in a southwestward arc extending through the Cardiff area and south of the present South Crop of the coalfield.
(iii) To the south, a rapidly subsiding geosynclinal trough in which flysch-type sequences were deposited.

In addition to these bounding lineaments, thickness considerations and strati-graphical evidence point to the existence of a number of structural elements *within* the South Wales depositional area, which were active during Upper Carboniferous time (Owen 1964, 1970). Some of these features may have resulted from posthumous movement on basement fractures (Owen, 1970).

In broad terms, the South Wales Namurian sequence reflects the rapid south-ward advance of a paralic facies-complex, presumably fringing the uplifted St. George's Land massif. In this facies, marine faunas are confined to bands of dark mudstones, calcareous in places, which are intercalated with terrigenous and subordinate chemical/organic sediments representing a variety of littoral and coastal plain environments.

In the southern outcrops of South Wales this style of sedimentation is developed, with little or no break or disconformity, above the shelf-carbonates of the Lower Carboniferous but the northern outcrops reveal evidence of a sub-Namurian unconformity of variable magnitude (George, 1927; Ware, 1939). The diachronous appearance of sands and gravels within the late Visean carbonates (Owen & Jones, 1961) has undoubtedly complicated the effects of this unconformity but the Namurian development of the Usk Axis is attested by the eastward overstep and overlap of successively higher Namurian units (George, 1956; Jones, 1970). The absence of Namurian strata from the Forest of Dean area provides further evidence of contemporary instability in the upper Severn region.

The Namurian sequence of South Wales achieves a maximum thickness of 1800 feet (550 m) in Gower, where it consists mainly of argillaceous sediments.

However, elsewhere in South Wales, and especially in the thinner Namurian successions of the northern and eastern margins of the basin, lenticular bodies of orthoquartzite, often pebbly, form a significant part of the sequence.

Westphalian sedimentation in southern Britain was heralded by the widespread *Gastrioceras subcrenatum* marine transgression, and the basal Westphalian rocks of South Wales were deposited within a structural framework broadly similar to that which existed during late Namurian time. Thus the relatively narrow zone of lower Westphalian paralic facies which lies to the north of the geosynclinal tract of Devon and Cornwall and south of the Leinster-Wales-Brabant massif is interrupted, like its Namurian counterpart, by several important transverse upwarps or positive axes. One important line of this type was the convoluted Usk-Malvern Swell which simultaneously separated and defined the margins of the contiguous South Wales and Bristol-Somerset basins.

The Lower and Middle Coal Measures (Westphalian A, B, and lower part of C) of South Wales contain the principal development of economic coals and the thickest and most widespread marine horizons in the Upper Carboniferous sequence. The sediments of this stratigraphic interval thus represent a persistently paralic or deltaic complex of environments, marked by a generally low relief in which even minor adjustments of the relative sea level (due to eustatic, tectonic or sedimentary causes) resulted in substantial lateral migrations of the contemporary strand-line and associated facies belts. Similarly, the commonly observed cyclic arrangement of the Lower Westphalian strata in this area (Robertson, 1948) may be attributed to the relative sensitivity of this environment to changes in sediment supply, or in eustatic or tectonic behaviour. A striking feature of the Lower Westphalian rocks of South Wales is the marked and uniform decrease in thickness of the component formations both northward and eastwards (T. N. George, 1970; Leitch *et al.*, 1958). A cumulative maximum thickness for the Lower and Middle Coal Measures of some 3000 ft (915 m) near Swansea is thus diminished to 1400 ft (427 m) around Merthyr Tydfil and north of Cardiff, to 800 ft (244 m) on the eastern margin near Pontypool, and is finally reduced to less than 200 ft (61 m) on the southeastern flanks of the coalfield near Risca (fig. 44). Much of this eastwards attenuation previously has been ascribed to a Malvernian phase of deformation, uplift and erosion which preceded overstep by the late Westphalian Pennant Measures (Moore, 1948; Blundell, 1952). However, Squirrell & Downing (1964) have demonstrated the presence in the attenuated sequence of most of the key horizons of the lower Westphalian succession and have attributed most of the decrease in thickness to condensation of the sequence resulting from proximity to the Usk-Cardiff positive axis.

The Upper Coal Measures (mid- and upper Westphalian C, Westphalian D) of the paralic belt of southern Britain were deposited under conditions unique in the British Carboniferous. In the first place, this period witnessed a great extension of the area of deposition, principally to the north and east of existing basins. Thus Upper Coal Measures sediments are found, resting unconformably upon Dinantian or Devonian rocks, in the Forest of Dean and below the Mesozoic cover of the Newent area of Gloucestershire and in north Oxfordshire (Poole, 1969; Worssam,

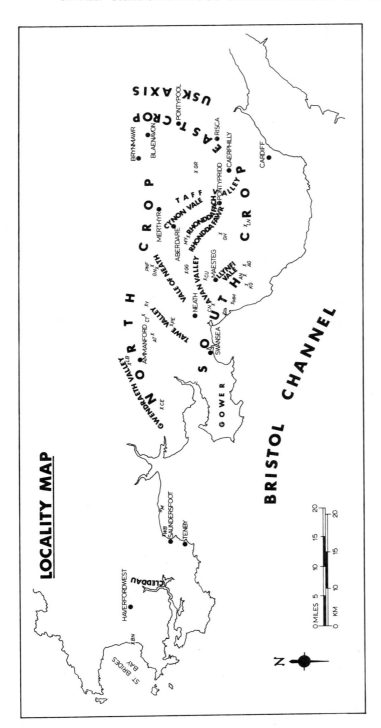

Fig. 44. Generalised map of South Wales, showing localities named in the text. Key: AG—Aberkenfig; AN—Aberbaiden; AT—Abernant; BN—Broadhaven; CE—Cynheidre; CN—Cwmavan; CT—Cwm Twrch; CU—Caerau; GG—Glyncorrwg; GH—Gilfach Goch; GN—Glynneath; GR—Gelligaer; KG—Kenfig Hill; LB—Llandebie; LN—Llanharan; M—Marros; MM—Margam; MY—Maerdy; PE—Pontardawe— PNF—Pontneddfechan; WB—Wisemans Bridge; Y—Ystradgynlais.

1963). The location and relative thickness of these sequences and the correlative successions of the south Midlands provide strong evidence for the progressive extinction of the Wales-Brabant massif as a barrier to sedimentation in this region. Secondly, faunal evidence of marine deposition is entirely lacking in the preserved outcrops of Upper Coal Measures in southern Britain: the animal or plant fossils present are exclusively of fresh water or brackish aspect. Finally, the late Westphalian sequences of this region are predominantly arenacous. The sandstones are petrographically less mature than the underlying sediments and most are referable to lithic arenite or subgreywacke categories. Such rocks are traditionally referred to as 'pennant' sandstones in the coalfields of South Wales, the Forest of Dean and Bristol-Somerset.

In South Wales the Upper Coal Measures (Pennant Measures) attain a maximum thickness of some 6000 ft (1830 m) and the basal formations (Llynfi Beds and Rhondda Beds) again display the marked eastwards attenuation observed in the earlier Westphalian. However, the overlying units are marked instead by a general, less conspicuous thinning northwards. Moreover, the characteristic thick subgreywacke sands behave diachronously, appearing earliest in the south-west around Swansea and Margam (low mid-Westphalian C) and latest on the eastern and northeastern margins of the basin (upper Westphalian C). Thus the early Pennant Measures formations exhibit two petrographic facies: a lithic arenite assemblage to the south and west and a pebbly quartz-arenite lithology to the east and north. However, later Westphalian sequences reveal a uniform development of the lithic arenite sands, across the South Wales region and the disappearance of the mature orthoquartzites.

The Upper Carboniferous sequence in South Wales is completed by the Grovesend Beds, which occupy the cores of the Gowerton, Gelligaer and Llantwit-Caerphilly synclines. This formation is largely argillaceous, and is accompanied by red sands and marls which become more common upwards. Coals are well developed in the west but fail eastwards.

(b) Sedimentological Considerations

Speculations and inferences concerning the mode and environment of sedimentation of the Upper Carboniferous rocks of South Wales have a venerable history, dating back to De la Beche (1846). In more recent times, petrological studies by Heard (1922) and careful faunal and lithological analyses made by several workers and summarised by Robertson (1948) and Trueman (1941; 1946; 1954), have contributed substantially to our knowledge of the Upper Carboniferous history of this area. However, sedimentological techniques which have been developed in recent years have now furnished the means for recognising and interpreting specific sedimentary environments and for delineating both local and regional patterns of sediment transport and dispersal.

During the past decade several of these sedimentological tools have been applied to the Upper Carboniferous sequences of South Wales. A particularly valuable technique is that of palaeocurrent analysis, which utilises the orientation of current-formed features in a sedimentary unit to deduce the direction (and, in some cases, the relative energy) of the fluid flow which was responsible for transporting and depositing the sediment. Used on a regional scale such vector-analyses may delineate ancient patterns of fluvial drainage, or marine and atmospheric (wind) circulation systems (Potter & Pettijohn, 1963). However, it is now recognised that the value of this technique may be reduced or entirely vitiated by indiscriminate and uncritical measurement of every vector property in a given sequence. Different types of directional feature exhibit differing degrees of reliability as estimators of the regional transport path (cf. Allen, 1967) and may result from contrasted styles of current activity (e.g. erosion versus deposition). Moreover, the nature of the vector-distribution and the significance which may be attached to the mean or modal values differs substantially in different environments or sub-environments (Klein, 1967; Selley, 1968). Thus accurate assessment of regional sediment transport patterns can only be achieved by a comparison of vectors derived from similar structures formed within the same general environment. The recognition and distinction of the sedimentary environments represented in a particular group of rocks is thus a pre-requisite for any paleocurrent analysis of that group. Conversely, the vector-pattern may assist in the diagnosis of specific environments (Selley, 1967).

Interpretation of the depositional milieu of a sequence of sediments ultimately depends on assessment of all the physical, chemical and biological attributes displayed by that sequence. Environmentally significant physical parameters include the lateral geometry of particular lithotypes, such as sand-bodies, and the vertical arrangement of lithological units. In recent years it has become recognised that cyclic behaviour in sedimentary sequences is frequently related to the nature of the depositional environment; thus the character of the vertical lithological transitions within such sequences may be used as environmental criteria. For example, repetitive fining-upwards sequences of clastic sediments which rest on prominent erosion surfaces generally indicate a fluvial environment (Allen, 1965a), whereas successions which commence with a marine mudstone and gradually coarsen upwards into cross-stratified sandstones may result from the advance of a distributary across a delta platform (cf. De Raaf and others, 1965) or the growth of a coastal barrier-bar (Davies and others, 1971). In assessing such sequences the succession of sedimentary structures encountered is at least as important as the changes in grain-size and texture and careful evaluation must take full account of the rapidly increasing body of information available on modern environments and the associated sediments.

A final point concerns the petrographic composition of the Upper Carboniferous sediments, particularly the sands. The size and character of clasts and minerals in the sands reflect not only the nature of the source-rocks but also the nature of the processes which have modified the original detritus during its dispersal history. Thus elucidation of the provenance characteristics of the sediment must take due account of textural and mineralogical changes superimposed during transport and deposition, and here again the correct identification of the environmental associations present within the sequence studied becomes a paramount consideration.

The account which follows is predicated on these principles and is essentially a summary of the main results achieved from a series of studies designed to provide an integrated synthesis of the sedimentary history of this part of Britain during the Upper Carboniferous. Because of space limitations this discussion is concerned almost entirely with conclusions and interpretations; the detailed evidence on which the vector patterns and environmental interpretations are based generally must be sought in the publications and theses listed. However, additional information is provided in a few cases where new interpretations, not previously published, are presented.

II. NAMURIAN FACIES AND ENVIRONMENTS

Along the northern margin of the South Wales basin the Namurian sequence commences with an orthoquartzite facies (the Basal Grits formation). This rock-group is here regarded as being essentially Lower Namurian in age but the diachronous behaviour of this facies must be acknowledged. Only in the northwest portion of the main basin (and possibly in west Carmarthenshire) are the E and H stages present in the Basal Grits sequence. Further to the west (Haverfordwest) only the R_1 and R_2 stages are represented in the attenuated Basal Grits while at Brynmawr, on the northeast corner of the coalfield, the base of this facies appears to lie within the R_2 stage of the upper Namurian.

Although the sedimentology of the Basal Grits has not yet been studied in detail within the main basin, the probably littoral character of this formation is attested by the lenticular to tabular geometry of the sand-bodies, which are separated by goniatite-bearing dark mudstones, occasional underclays, and carbonaceous silts or rare coal-smuts. The sand-bodies are characterised by low-angle trough cross-stratification and parallel lamination and the mineralogically mature sands are generally coarse to medium in grade and are moderately to well-sorted ($\sigma I = 0 \cdot 3$ to $0 \cdot 6$). Conglomeratic and pebbly sands are common, especially near the base of the Basal Grits sequence. The pebbles are rounded to well-rounded and consist almost entirely of vein-quartz with a few metaquartzite clasts. These pebbles are rarely present in sufficient proportions to constitute a framework-conglomerate, but tend to occur in a sand matrix. In many sections there is a clear tendency for pebbles to be concentrated near the top of the sandstone units in a form of inverse grading (see also Woodland and Evans, 1964, p. 10). Such a structure is observed in many modern beach or barrier sands as they advance regressively seaward.

The poorly exposed Basal Grits sequences of the southern margin of the main basin are proportionately thicker and contain higher proportions of mudstone and shale than the equivalent formation to the north. An interesting feature of the Gower sequence is the development of a group of radiolarian cherts in the basal portion of the local Namurian. These observations suggest that the littoral environment represented in the northen exposures gives way southwards to quieter, sublittoral marine conditions. An important feature of Basal Grits sedimentation is the marked lateral variation in thickness, particularly along the North Crop. The thickest development of the Lower Namurian occurs in the Llandebie-Cwm Twrch area and is associated with a prominent axis of maximum thickness which extends southwards to Gower (T. N. George, 1970, fig. 24). The marked decrease in thickness of the Basal Grits both to east and west of this 'Llandebie embayment' is attributed to successive overlap of the E and H stages by the R stages, associated with the pronounced unconformity between the Namurian and the Dinantian, mentioned earlier (p. 185). The implication of these observations is that during early Namurian time continued uplift of the Usk axis prevented significant preservation of sediments to the east while active subsidence in the Llandebie-Gower area permitted substantial accumulation of these near-shore sediments.

In Pembrokeshire and west Carmarthenshire the basal Namurian of the northern outcrops again rests with apparent unconformity upon the Visean shelf-limestones, although the recent discovery of a conodont fauna of E_1 age in the Basal Grits of the Marros region (G. T. George, 1970, p. 27) indicates that the magnitude of the stratigraphical break at this locality is relatively small. However, further to the west the Basal Grits lithofacies oversteps successively older formations, finally resting on Upper Ordovician rocks in the Rosehill area, south of Haverfordwest (George, 1962). These relationships indicate a significant extension of the area of sedimentation during early Namurian time, but probably subsequent to late Dinantian warping.

The Marros section provides the most complete exposure of Namurian sediments in southwest Wales (G. T. George, 1970; Kelling & George, 1971). At Ragwen Point the uppermost Carboniferous Limestone ($D_2 - D_3$ zones) is erosively succeeded by a thin (1 m) group of trough cross-stratified orthoquartzites with intervening mudstone laminae which in turn are overlain erosively by a thicker (4 m) channel-filling sequence of trough cross-stratified orthoquartzites. The scoured base of this channel-fill is marked by a lag deposit composed of rounded vein quartz pebbles, angular mudstone pellets and rolled sideratic ironstone nodules. Carbonaceous mudstones associated with abundant rootlet development in the topmost arenite unit, cap the fill-sequence. Five such sequences can be recognised in the basal 40 m of the Marros section, and marine faunas of goniatites and conodonts have been discovered in the intensely burrowed mudstones which intervene between several of the channel-fills (see Kelling & George 1971, p. 242).

G. T. George (1970) has interpreted this basal Namurian succession as a regressive delta-floodplain sequence, dissected by major distributary channels which became emergent and colonised by plants from time to time. Unimodal current-vectors obtained from the channel-fill sands support a distributary origin and indicate general flow towards the south (G. T. George 1970, fig. 43).

Poor exposures of attenuated lower Namurian sequences which occur further west, in the Cleddaus and on the shores of St. Bride's Bay, display a similar type of facies development and comparable, south-directed fluvial flow-vectors. However, the correlative succession in the Cresswell-Tenby area, to the south, is appreciably thicker and more argillaceous and the intercalated thin, orthoquartzitic sands include representatives of distributary channel and coarsening-upward barrier-bar facies (G. T. George, 1970, fig. 40).

In a regional context, the early Namurian sediments of west Wales appear to have been deposited in a northwest-trending embayment, flanked to the north and the west by deltaic lobes (fig. 45). Within this embayment, lagoonal and prodelta muds and silts accumulated to considerable thickness, occasionally coarsening

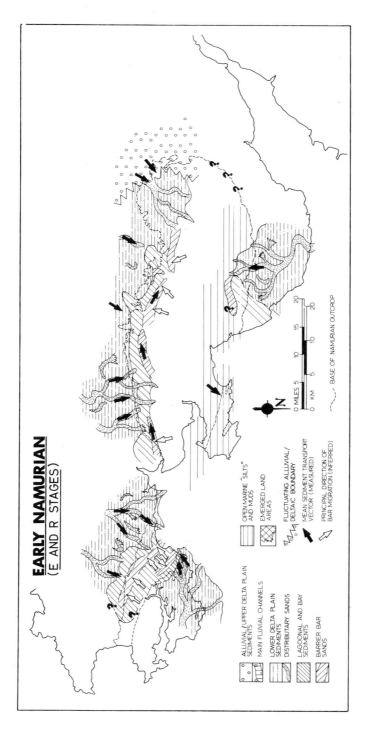

Fig. 45. Environmental reconstruction of South Wales in the early Namurian (E and R stages), based largely on the work of G. T. George (1970) and R. O. Oguike (1969). Note that in this and the following reconstructions the position of secondary environmental features such as distributary and stream channels is indicated only schematically.

upwards into impersistent barrier-bar sands which constantly migrated landward (G. T. George, 1970, pp. 178–180).

In the main coalfield basin, the sedimentology of the later Namurian rocks (Reticuloceras and Gastrioceras stages) has been elucidated by Oguike (1969). Sequences assigned to this stratigraphic interval display evidence of deposition in a variety of environments, including fluvial, deltaic, littoral and subittoral (open marine) milieux. Three major factors play a significant role in the sedimentation of these later Namurian rocks.

(i) Axes of differential subsidence: two major areas of enhanced accumulation are recognised. One, in the Llandebie area, trends north-south and is accompanied by a complementary trough of similar trend in the Twrch region. The other, trending east-west, occurs near the southern margin of the coalfield syncline, in a zone which includes Pontypridd, Margam and south Gower. Conversely, in addition to the general attenuation which occurs near the East Crop, there is a marked thinning of the later Namurian in the region between the Tawe and Neath valleys (Oguike, 1969, fig. 12), which may be ascribed to contemporaneous activity along the line of the Neath Disturbance.

(ii) Marine incursions: widespread units of mudstone or shale, sometimes calcareous, carrying a marine fauna of goniatites and brachiopods, recur several times in the later Namurian sequences. These marine bands provide a convenient means of dividing the Upper Namurian succession into four major 'cyclothem' phases, each of which forms a laterally variable but distinguishable lithological complex, marking the stratigraphic response of a contiguous group of environments to changes in tectonic, eustatic and possibly climatic factors.

Some marine bands are localised in their occurrence and occupy specific niches in the lithological sequence, for example, immediately above a coal. Such units presumably result from an excess of subsidence over the local sediment supply. However, the widespread marine horizons tend to be less rigorously constrained in their sequential position, and succeed a variety of facies types in different parts of the basin, suggesting a general eustatic flooding of the marginal areas with relatively rapid marine transgression across a variety of environments. Faunal studies (Ramsbottom, 1969) reveal that in the south and southwest these major marine bands are characterised by a 'basinal' or open-marine assemblage of goniatites and pectinoids whereas the correlative units in the North and East Crops contain benthonic faunas of brachiopods, crinoids and bivalves, indicative of a more shallow, near-shore environment. Such evidence, taken in conjunction with the more sandy and fluvial aspect of the North and East Crop sequences, suggests that during the Namurian open sea still lay to the south and southwest of the present South Wales area.

(iii) Location and nature of sediment sources: the principal source of most of the sediment supplied to the South Wales area during the Namurian lay to the north. This conclusion is sustained by the thin and coarse aspect of the North Crop sequence and the faunal evidence, already cited, of more marine conditions in the south, and is substantiated by the current vectors within the North Crop fluvial facies, which consistently indicate flow towards the south (Oguike, 1969, fig. 2/17).

The attenuated Namurian sequences on the eastern and southeastern margins of the basin furnish comparable evidence of transport from the east and southeast, while intermittent and localised supply from a southerly source is indicated for the fluvio-deltaic sediments in the Aberkenfig-Margam sector of the South Crop (Oguike, 1969, pp. 135–136). Apart from variations in texture and composition related to environmental factors, the petrology of the sands derived from these varied sources is relatively uniform and the aggregate suite of minerals and clasts is consistent with derivation from sedimentary source-rocks (see below, pp. 212–216).

The interaction of the factors outlined above has resulted in the following sequence of events, as interpreted by Oguike (1969). During the early part of the R_2-stage, littoral conditions prevailed briefly in the western part of the North Crop but fluvial and deltaic sediments were being deposited at this time in the northeastern part of the basin and these conditions quickly spread both west and south, but did not fully impinge on the Gower area where 'basinal' muds continued to accumulate in prodelta and interdistributary bay environments (fig. 46). Following the widespread marine transgression which gave rise to the *Reticuloceras superbilingue* marine band, a thick deltaic lobe prograded southwards from the Llandebie area. This delta was flanked to the east by areas of coastal zone sedimentation in which deposits formed in lagoons, interdistributary bays and sandy beaches can be recognised. Longshore currents, generally directed towards the east or east northeast, affected these littoral sediments. However, in the Brynmawr area of the northeastern region, a southwest-flowing alluvial system was initiated at this time. Along the East Crop, erosion and scouring were prevalent, dispersing sediment westwards, particularly into the 'Pontypridd-Margam Trough'. In the vicinity of Aberkenfig, on the South Crop, a localised influx of sand and silt marks the northward progradation of a small delta (fig. 46), whereas, 'basinal marine' muds continued to accumulate in the Gower region (Oguike, 1969, pp. 244–245).

The middle part of the R_2 stage was marked by a widespread regression which led to the development of an extensive littoral zone, including accretionary beaches, bars and spits, well represented in the upwards coarsening sand unit known as the 'Twelve Feet Sandstone'. Subsequently pene-marine conditions returned, with widespread deposition of muds, ultimately succeeded by a thin (3–4 ft) littoral sheet sand, found over most of the North Crop.

The marine transgression associated with the *Gastrioceras cancellatum* band ushered in the G_1-stage and may have been preceded by a general phase of subsidence. This apparently led to deposition of muds on an almost basin-wide scale, mostly in interdistributary bays which were frequently isolated from the open sea, forming brackish lagoons and freshwater lakes in which mussels flourished. West of Llandebie, frequent emergence of the delta-top region is recorded by numerous underclays, rootlet-beds and thin coals. Input of sand and silt was confined to distributary channels, mainly preserved in the Llandebie, Twrch and Neath Valley areas. These record a general southward and southwestward advance of the delta lobes structurally located in these areas throughout the remainder of upper Namurian time. This regressive advance was interrupted only briefly by the *G. cumbriense* marine incursion.

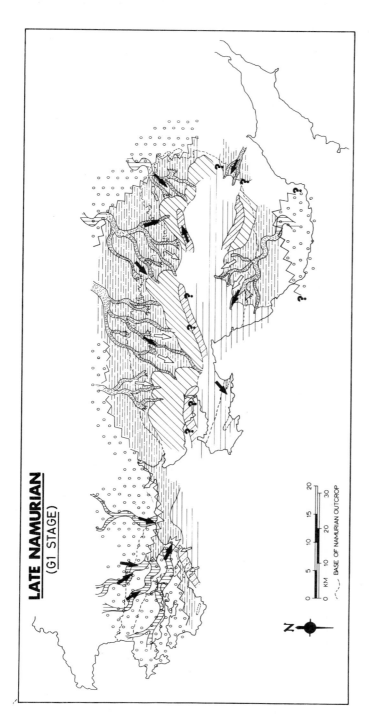

Fig. 46. Environmental reconstruction of South Wales in the late Namurian (G₁ stage), based largely on the work of G. T. George (1970) and R. O. Oguike (1969). For explanation of ornamentation, see fig. 45.

A complex pattern of overlapping environments is discernible in the late Namurian sediments of the northeastern portion of the basin, where sand-filled channels of fluvial aspect, indicating flow to the south or southwest, are incised into sheet-like bodies of littoral sand and silt. A principal locus of fluvial input appears to have existed in Llammarch Dingle, near Brymnawr.

On the South Crop, the delta front located near Aberkenfig continued to supply south-derived sand and silt to the Pontypridd-Margam trough and advanced first northwestwards then, during the final stages of Namurian sedimentation, in a westerly direction. The complex palaeocurrent patterns recorded from the tidal flat, lagoonal and deltaic late Namurian sediments of the Gower region may indicate the coalescence and modification of the southerly advancing Llandebie-Twrch delta system with the westerly prograding Aberkenfig lobe (Oguike, 1969, pp. 249–250).

In Pembrokeshire, deposition of the upper Namurian took place within the northwest-trending embayment described earlier (p. 191) and was controlled by the same factors (tectonic, eustatic and hydrologic) which operated in the main basin. Here, the basin-wide marine inundation represented in the *R. superbilingue* horizon led to a prolonged phase of marginal-marine sedimentation of muddy or silty sediments. Lagoonal, tidal flat, tidal channel and landward-migrating barrier-bar environments are recognised in this sequence, while a southwards prograding deltaic lobe gave rise to a thick upwards-coarsening association in the contemporaneous succession at Marros (G. T. George 1970, fig. 59). The succeeding *Gastrioceras cancellatum* marine incursion is preceded, in the Marros section, by a 9 metre thick transgressive sequence which records the landward (northward) advance of a brackish lagoon and accompanying beach, backed by high-energy outer barrier bar sands with a lag-pavement (G. T. George, 1970, pp. 122–123). As in the main basin (p. 194) the *G. cancellatum* marine transgression resulted in general deepening of the Pembrokeshire embayment and widespread deposition of muds, associated with thin beds of silt and fine sand which display grading and other attributes suggesting deposition from rather dilute density flows or turbidity currents. These units probably were formed at the base of prodelta slopes (G. T. George, 1970, p. 180) and an interesting feature of these beds is that they are finer in grade and more 'distal' in aspect in the northern outcrops than in the Tenby section. The shallower nature of the basin at Tenby is further attested by the abrupt upwards passage of the turbidites into wave-agitated sandy siltstones and cross-stratified distributary channel sands. The latter mark the intermittent advance of a delta lobe from the west (G. T. George, 1970, fig. 59). At Tenby this sequence is succeeded by basinal silts and distal turbidites which are erosionally overlain by the littoral barrier sands preceding the *G. cumbriense* marine transgression.

During the latter part of the G_1-stage and possibly as a result of regional uplift, there was a widespread renewal of fluvial activity in the Pembrokeshire area, which culminated in a resumption of coastal plain deposition in the north, gradually spreading southwards. The rivers cut deeply into earlier Namurian ('Middle Shales') sequences and ultimately formed the thick complex of fluviatile channel sands known locally as the Farewell Rock.

At least four major river systems appear to have existed during this period. Three were located in the north and flowed in a general southerly direction while the other river system was located in the southwest and flowed eastwards, producing the Farewell Rock sands of the Cresswell-Tenby area (G. T. George, 1970, p. 181). These fluvial complexes were separated by a zone of tectonic instability, trending east-west and located along the line of the Ritec Fault, a feature which also appears to have influenced sedimentation in Dinantian times (Sullivan, 1965; Kelling and Williams, 1966). By late G_1 time the prominent embayment of the earlier Namurian had been reduced to a small gulf, bounded to the south by the Ritec line. This gulf received most of the sediment being supplied by the northern rivers (fig. 46).

The uppermost Namurian is represented by argillaceous floodplain sediments which abruptly succeed the channel sands of the Farewell Rock member. The diachronous behaviour of the Farewell Rock (which occurs below *G. subcrenatum* in Pembrokeshire but appears at successively higher levels above that marine band as it is traced east and north into the main basin; cf. Leitch *et al.*, 1958) suggests that the river-systems responsible for the formation of these sands migrated in a north-easterly direction. However, this fluvial retreat was accompanied and interrupted by a major marine transgression which led to formation of the widespread *Gastrioceras subcrenatum* marine horizon.

III. LOWER WESTPHALIAN FACIES AND ENVIRONMENTS

The structural and physiographic framework within which the early Westphalian (Lower and Middle Coal Measures) sediments of South Wales were deposited appears to have been similar to that existing during the later Namurian stages (cf. George, 1962) and the *Gastrioceras subcrenatum* marine transgression ushered in paralic conditions comparable to those previously established in this area. Indeed, as Oguike (1969, p. 250) has pointed out, there is a sedimentological transition from the upper Namurian G_1 stage to the lower Westphalian G_2 stage in the North Crop sequences—a transition which is only briefly interrupted by the *G. subcrenatum* incursion. Moreover, the diachronous behaviour of the thick Farewell Rock fluvial sands makes it clear that the prime control of early Westphalian sedimentation resided not so much in the relative position of sea level, but rather in those factors, tectonic and climatic, which governed the location and effectiveness of the point-sources feeding sediment into the basin, and which thus determined the relative rates of supply and subsidence in different parts of the basin. In this context a eustatic rise in sea level might substantially alter the hydraulic gradient of a stream system by raising the base-level, but the pre-established energy relationships between the stream-nets would remain constant in spite of the base-level change. Thus the ratio of the amount and grade of detritus supplied by adjacent nets would remain constant, despite an overall diminution in the quantity and size of the transported sediment.

However, although the environmental conditions of the basal Westphalian in South Wales are substantially similar to those encountered in the upper Namurian, the decrease in the relative frequency of demonstrably marine units, the increase in the number and thickness of coal seams with their accompanying underclays and the predominantly argillaceous nature of the Lower and Middle Coal Measures all indicate a degree of balance between subsidence and supply and a relative stability of environments, both laterally and vertically, which is lacking in earlier formations. However, such conditions became widely established only after deposition of the M_5 marine band (Leitch *et al*, 1958).

In the main basin Bluck (1961) has demonstrated that the channelled proto-quartzitic sands of the post-*subcrenatum* Farewell Rock represent a fluvio-deltaic complex, accompanied by littoral silts and muds, which appears to have prograded in a broadly southward manner. At Pont-nedd-fechan in the Vale of Neath, the Farewell Rock rests directly upon the *G. subcrenatum* marine mudstones and at this locality the thick sand-body displays the typical coarsening-upwards attributes of a distributary mouth bar and passes up into a distributary channel development (Kelling, 1971). Elsewhere on the North Crop (e.g. in the Llech section), the sands include a basal mélange, comprising quartz pebbles, rounded ironstone nodules and angular chips of marine mudstone (Bluck, 1961, p. 33) and probably indicate more continuous reworking by distributary channels. On the South Crop the contemporaneous facies is more argillaceous, with small mudstone-filled channels, and the fine, muddy sands of northerly derivation which appear somewhat later are interpreted as the southern extension of the distributary channels seen in the north. The equivalent orthoquartzitic and pebbly sands of the East Crop occupy small, steep-sided channels and display west or northwest-directed transport vectors (Bluck, 1961, fig. 39; Bluck & Kelling, 1963, fig. 3). A similar coarse, but texturally and mineralogically mature lithofacies is developed intermittently in the basal Coal Measures of the South Crop and again provides evidence of transport towards the northwest. The contrast of these quartzose arenites with the interbedded proto-quartzites of northerly derivation is striking and indicated to Bluck (1961) a southerly source of pebbly quartzose detritus. However certain features of this orthoquartzite association suggest that these mature sands may have been deposited in a littoral zone, as barrier bars and tidal channel-fills, in which case their aberrant transport vectors indicate predominant onshore movement during phases of littoral reworking of the delta-front.

In Pembrokeshire the basal Westphalian sediments are mostly bioturbated muds and silts of pro-delta aspect, accompanied by sequences of intertidal character (mud- and sand-flats and tidal channels). Rheotropic transformation structures in these sediments indicate a regional palaeoslope inclined in a broadly southwards direction and palaeocurrents and facies relationships suggest the general southwards and eastwards advance of a prograding coastline of low relief (Williams, 1966, p. 183; 1968, p. 350) and the ultimate incursion of southeast-flowing distributary (or tidal) channels into the Saundersfoot region (Williams, 1968, fig. 17), Succeeding sequences up to the M_4 Marine Band record similar regressive phases interspersed with short-lived marine transgressions across the coastal plain (Williams, 1968, p. 359). However the M_4-Amman Marine Band interval is distinguished by a general increase in the proportion of sand and silt in the sequence, and by more pronounced and widespread channelling. In the south, this interval is represented by alluvial floodplain cycles in which both substratum (channel-fill) and topstratum deposits are preserved, whereas the northern areas are characterised by topstratum deposition of fine sediment, interdigitating with backswamp and basin-lake areas. Just prior to the Amman marine transgression there was an extension of the channelled floodplain regime across the earlier topstratum sediments of the northern area. The complex palaeocurrent pattern derived from the channel sands of this interval is attributed by Williams (1966, p. 185; 1968, p. 359) to deposition within a highly sinuous meander-belt subject to significant changes in orientation but generally flowing west or northwest.

Throughout South Wales, the period immediately following the Amman marine incursion is marked by what Williams (1968, p. 459) has termed 'proximal marine conditions' which resulted in the accumulation of predominantly fine-grained sediment on a low-lying coastal plain, adjacent to a muddy marine shelf. Numerous thick and extensive coals, underclays, sideritic ironstones and abundant 'non-marine' mussels point to the prevalence of swamps and fresh or brackish water lakes. Lenticular sands recur regionally in the middle part of the Amman-Cefn Coed interval. In the northwest portion of the main basin (west of Ammanford) fining upwards, channel-fill sands of fluvial aspect (distributary channels?) display north- and east-directed current-vectors (Thomas, 1967, fig. 43) while broadly contemporaneous channel sands in the Wiseman's Bridge and Broadhaven sections of Pembrokeshire were apparently transported to the northwest and north (Williams, 1968, figs. 24, 26). These sands furnish strong evidence for the existence of a fluvio-deltaic complex which prograded in a generally northwards direction at this time and presumably originated in an uplifted area to the south, on the site of the present Bristol Channel (fig. 47).

In the main basin several thin marine mudstones (Hafod Heulog, Brittanic etc.) record localised drowning of the coastal swamps and delta-lobes in the period immediately prior to the major Cefn Coed marine transgressions, and are associated with littoral muds and silts and widespread but thin units of clean sand deposited in low barriers or in tidal channels (Thomas, 1967, p. 151). However, renewed fluvial-deltaic activity is evident in the localised but overlapping sand-bodies which occur in the northwest (Ammanford-Glynneath), the northeast (Aberdare-Brynmawr), the east (Blaenavon-Risca), and the south (Maesteg-Kenfig-Cwmavan). The North Crop sands are characterised by southeast-directed flow-vectors and indicate revival of the northern deltaic complex which had dominated early Westphalian sedimentation (fig. 47). Moreover, quartzose sands reappear on the East Crop, where they are associated with a southwest-flowing distributary channel system (Thomas, 1967, fig. 42).

Areas of argillaceous backswamp deposition, centred around Gelligaer-Caerphilly in the east and Abernant-Neath in the west, separate the northern and eastern sandy facies from the thick wedge of channelled fluvial sand in the Maesteg-Cwmavon area (Thomas, 1967, figs. 9, 42). This sand development is of importance not only because it furnishes additional evidence of the east-west trending Margam-Maesteg downwarp whose effects are also detectable in geophysical studies (M. D. Thomas, 1967) and in studies of coal seam splitting and thickness changes (Parry, 1966), but also because sedimentary structures in the channel-sands consistently indicate northwards transport (Thomas, 1967, fig. 46). Northward progradation

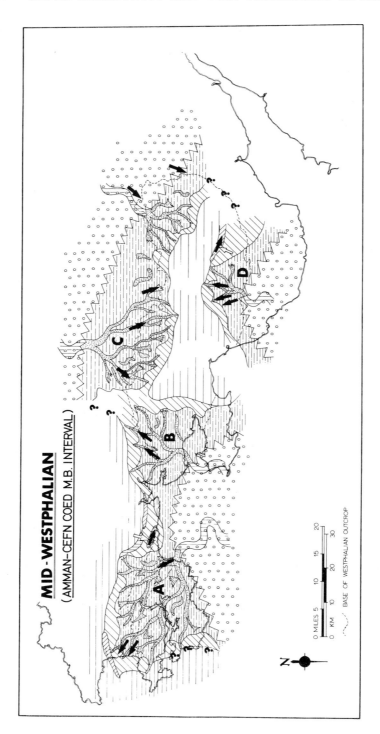

Fig. 47. Environmental reconstruction of South Wales in mid-Westphalian times (Amman to Cefn Coed Marine Band interval), based largely on the work of L. P. Thomas (1967) and P. F. Williams (1966, 1968). Note that whereas deltaic complexes A and B were most fully-developed in the early part of this time interval, delta-lobes C and D were more prominent in the period immediately preceding the Cefn Coed marine transgression. For explanation of ornamentation see fig. 45.

of a major delta-lobe is thus indicated, the sequence recalling the brief interludes of southerly derivation already noted in the upper Namurian and basal Westphalian. However, some sheet-like littoral sands in this area yield vectors indicating west northwest transport, probably related to longshore drift of sediment (Thomas, 1967, p. 103).

The period following the Cefn Coed marine transgression is again marked by widespread development of muddy backswamp and penemarine conditions, and sand-bearing distributary channels were active only in the Glynneath-Aberdare region, where they again indicate southwards flow.

Because of poor exposure and (especially in Pembrokeshire) structural complexity, sedimentological data are lacking from the upper part of the Middle Coal Measures (Gorllwyn Coal—Cwmgorse Marine Bands), but underground and borehole records indicate that centres of sand deposition during this interval were located along the eastern crop and southeastern margins of the coalfield (as far west as Llantrisant) and in a linear belt around the Llynfi Valley (Maesteg-Glyncorrwg).

These areas were separated by a north-trending zone of mud deposition which passed through the central parts of the Rhondda Fach and Rhondda Fawr valleys and the Gilfach Goch area. An important feature of this part of the Upper Carboniferous succession is the first appearance of 'pennant-type' lithic arenites, which occur spasmodically in the middle part of this stratigraphical interval (above the Pentre or Upper Yard coal) in the Maesteg district (see also Woodland & Evans, 1964, p. 159) and in the Gwendraeth Valley (Archer, 1968. p. 109). However, the overlying sandstones revert to the protoquartzitic and orthoquartzitic character typical of the Middle Coal Measures and the main development of the lithic arenite petrofacies did not take place until later (see below).

A final piece of evidence bearing on the environment of Lower Westphalian deposition in South Wales is provided by the lateral variations in faunal assemblage displayed by individual marine bands. Such variations are regarded as indicative of differing environmental conditions and are probably governed by factors such as salinity, sediment-supply, and water-depth (Calver, 1969; cf. Ferm & Williams, 1965). Careful study of these variations thus provides an estimate of the relative degree of 'marineness' under which specific sedimentary units accumulated, bearing in mind that at each location any marine band might be expected to exhibit a vertical sequence of faunas related to a gradual increase in the marine influence (and possibly a subsequent decrease) consequent on the marine invasion of the coastal environment and the ultimate retreat of the sea from that region (cf. Bloxam & Thomas 1969).

Three of the major early Westphalian marine bands (*Gastrioceras subcrenatum*, Amman and Cefn Coed) indicate the prevalence of fully marine conditions in the south and west of the South Wales region, with near-shore or even brackish conditions along the present North Crop and possibly along the East Crop (Calver, 1969, p. 246–251). However recent work by E. M. Chaplin (personal communication, 1970) contributes significantly to the environmental interpretations provided by the marine band variations.

According to Chaplin, regional variations in the Amman Marine Band faunas indicate a prevalence of shallow, muddy conditions in Pembrokeshire and the north (Gwendraeth-Ammanford-Ystradgynlais) and central (Cynon, Rhondda, Avan and Llynfi Valleys) parts of the basin but open, clear waters to the southeast are indicated by the rich and varied faunas obtained from localities east of Llanharan (cf. Moore, 1945, p. 175). Faunas from the Cefn Coed horizon suggest that clear water, open marine conditions prevailed in Pembrokeshire and in the central and southern region (Pontardawe-Maesteg-Aberbaiden-Margam). Further north, the fauna is impoverished and indicates shallow, muddy conditions (Cynheidre-Glynneath) while the equivalent horizon in the northeast (Aberdare-Taff Valley) carries a brackish water assemblage. In contrast to the Amman, the Cefn Coed marine band is very thin in the southeast part of the basin (Risca-Caerphilly) and carries a restricted linguloid fauna, suggesting the proximity of a local source of sediment, presumably to the south or southeast of these outcrops.

The behaviour of the Lower and Upper Cwmgorse Marine Bands is of particular interest. Chaplin maintains that these units are generally thickest and most marine in aspect in the north-central districts (Cynheidre-Rhondda Fach-Cynon valley) while impoverished faunas indicative of muddy, near-shore conditions are typical of the attenuated Cwmgorse sequences of the southern and southeastern outcrops (although a thick local development which includes an open-marine assemblage has been recorded at Aberbaiden by Woodland and Evans, 1964, p. 64). The general distribution of these faunas is consistent with the advent of an extensive southerly shore-line to the South Wales area and thus heralds the major changes in palaeogeography which are revealed in the overlying Pennant Measures.

IV.   UPPER WESTPHALIAN FACIES AND ENVIRONMENTS

The major features of upper Westphalian (Upper Coal Measures) lithostratigraphy which were outlined earlier (p. 188) clearly indicate that the South Wales area suffered a profound change in the character and environment of sedimentation during the interval following the Upper Cwmgorse marine invasion. The succeeding Llynfi Beds formation commences, almost everywhere in the main basin, with a coastal plain sequence of mudstones, siltstones, underclays and coals, but these are rapidly replaced upwards by thick sand bodies which record a major influx of lithic detritus. The lithic sands appear earliest in the west and southwest (Cynheidre, Margam-Maesteg) and spread gradually northeastwards. Significantly, although these subgreywacke sands represent the dominant lithotype of the upper Llynfi Beds, units of quartz-arenite aspect, sometimes pebbly, recur in this sequence, and can be traced from the East Crop and as far west as Maerdy and Caerau.

Sedimentological studies of this critical formation are not yet complete but preliminary investigations indicate that the salient features of the subgreywacke and orthoquartzite facies associations are similar to those displayed by the overlying Rhondda Beds (see below). Palaeocurrent data also reveal a high degree of complexity, with bimodal or polymodal vector distributions common, at both outcrop and regional levels. However, there is a significant preponderance of northerly-directed vectors within the subgreywacke facies, whereas westerly flow

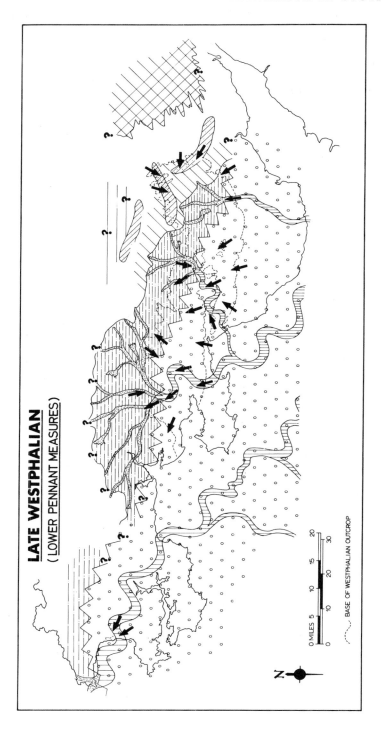

Fig. 48. Environmental reconstruction of South Wales in late Westphalian times (Lower Pennant Measures), based largely on the work of G. Kelling (1964, 1968, 1969) and P. F. Williams (1966, 1968). For explanation of ornamentation see fig. 45.

appears to be characteristic of the quartz-arenite association.

In general terms, the Llynfi Beds may be regarded as the deposits of a wide and active meander-belt, created by streams carrying immature detritus derived from a strongly elevated region to the south. This alluvial zone gradually encroached northwards and eastwards, advancing across pre-existing delta-plain and marginal marine environments (fig. 48).

The succeeding Rhondda Beds formation provides a more detailed example of this process of alluvial encroachment from the south (Kelling, 1964, 1968, 1969). The transport-pattern within the western subgreywacke facies is complex, with dominant northerly and northwesterly directed flow but with significant southerly vectors on the northern margin of the coalfield. Analysis of lithological sequence and vector variability in different parts of the basin demonstrates the alluvial character of the subgreywacke facies and indicates dominant deposition in braided or low sinuosity streams on the southern edge of the coalfield but deposition within a northwest-flowing meander-belt in the central and northern area (Kelling, 1968, 1969). Derivation from a source-area lying to the south is attested not only by the transport pattern and environmental association but also by the petrology of the sands. These carry a variety of rock-fragments and minerals reminiscent not only of the underlying paralic Carboniferous but also of the Devonian and older rocks of the geosynclinal terrain, including phyllite and spilite clasts.

In previous accounts (Kelling, 1964, 1968) the southerly transport vectors encountered on the North Crop were cited as evidence of a partial contribution from the northern barrier of St. George's land and some support for this view may be found in the broadly concentric pattern of isopachyte and isolith contours (see Kelling, 1968, figs. 2 and 3). However, the degree of petrographic distinction between sediments displaying south and north-directed vectors is so small that it indicates a remarkable (and unlikely) similarity in the geology of the northern and southern source areas. Alternatively, the petrographic similarity may result from derivation of the detritus from a single (southern) source: vectors indicating southerly transport then must be ascribed to local environmental factors, such as the gross variations in sediment transport direction to be anticipated in the highly sinuous stream-courses encountered in a lower alluvial valley or deltaic plain situation (fig. 48). Support for this interpretation is provided by the conspicuous bimodality of the North Crop vectors (Kelling, 1969, fig. 4) and the very high cross-bedding variance estimate for this region (Kelling, 1969, Table 5). Moreover, the fining-upwards fluvial sequences of subgreywacke sand which occur above the quartz-arenite facies in the eastern and northeastern parts of the coalfield display only westerly or northwesterly transport vectros.

Within the lower Rhondda Beds the alluvial lithic arenites of the western facies demonstrably pass eastwards and downwards into the pebbly quartz-arenite facies which is characteristic of the eastern margin of the South Wales coalfield throughout much of the Upper Carboniferous. Significantly, a zone of argillaceous deposition separates these two lithofacies over much of the eastern part of the basin (Kelling, 1968, fig. 2). Red, buff and green lutites occur sporadically within this argillaceous sequence and in the quartz-arenite belt to the east and have been

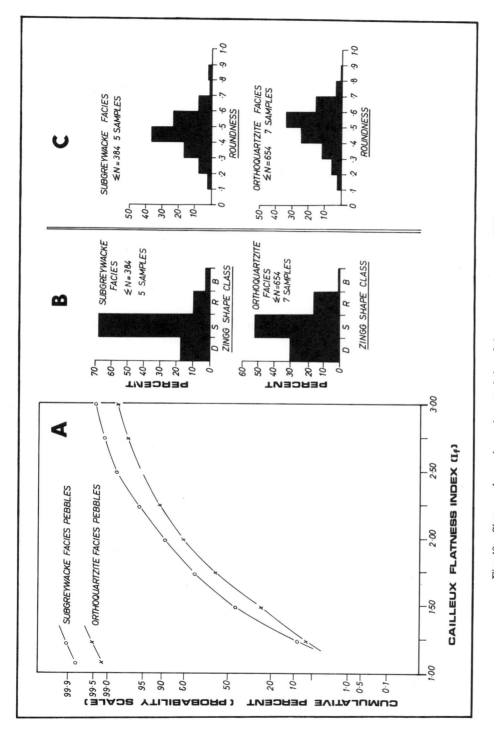

Fig. 49. Shape and roundness characteristics of the aggregate quartz pebble populations from the orthoquartzite and subgreywacke lithofacies in the Rhondda Beds (Lower Pennant Measures).

The combination of features enumerated above is, therefore, more readily compatible with a littoral or lagoonal situation rather than an alluvial environment, despite the absence of marine fossils from the upper part of the orthoquartzite sequence. Moreover, this re-interpretation accords well with relationships observed in the Pennsylvanian (Upper Carboniferous) sequences of the Appalachians by Ferm & Cavaroc (1969), Hobday (1969) and Ferm *et al.* (1971). In the American examples, the orthoquartzite sand-bodies intervene between alluvial/deltaic sequences dominated by 'pennant-type' lithic arenites, and deposits of the shallow marine shelf, mainly carbonates with red and green shales. The Pennsylvanian quartzose sands are interpreted as beach-barriers and the paucity of marine fossils is ascribed to the high hydraulic energy associated with this environment and to solution of calcareous debris in the acid conditions characteristic of the back-barrier lagoons which normally migrated across the bars. The quartz-sands are produced by marine re-working of lithic sands supplied by the delta distributaries and their compositional and textural maturity is a function of the intensity of the re-working processes.

Viewed in the context of a strand-line environment, the Lower Pennant pebbly orthoquartzites presumably represent beaches, barrier-bars and possibly tidal channel deposits, while the intervening bioturbated lutites, seatearths and coals were formed in lagoonal areas protected by the quartz-sand barriers and subject to periodic emergence and oxidation under appropriate climatic conditions leading to the formation (and preservation) of red argillaceous sediments. The discrepant current vectors of this facies association may then result from impinging processes of marine transport and erosion whose orientation is independent of the drainage-patterns on adjacent land-areas (see Klein, 1967; Selley, 1970).

The eastwards (and northwards) spread of lithic 'pennant' detritus is also attested by the Westphalian sequence of the Forest of Dean, where Upper Coal Measures sediments probably slightly younger than the Rhondda Beds of South Wales, rest with angular discordance on Dinantian and Devonian rocks. These basal beds (the Trenchard Group) include pebbly orthoquartzites, red and green mudstones, and display westerly transport vectors. They pass laterally (southwards) and upwards into a lithic arenite fluvial association dominated by northwesterly flow (J. T. G. Stead, personal communication). The provenance of these sediments is inferred to be similar to that of the eastern facies of the Rhondda Beds and they are believed to represent similar depositional environments.

The Pennant Measures of Pembrokeshire are exposed only in the coastal sections on St. Brides Bay, where they occur in a series of faulted blocks whose mutual stratigraphic and tectonic relationships are obscure. However, Jenkins (1962, fig. 5) has established a tentative succession, based on the floral and mussel evidence, which indicates the presence of the Lower Pennant Measures (*Anthraconauta phillipsi* mussel zone; floral zone G; Jenkins, 1962, p. 90).

Williams (1966) has demonstrated that the Pennant succession in Pembrokeshire is dominated by lithic arenites deposited in major fining-upwards sequences of alluvial character. The general substratum: topstratum sediment ratio is 3 : 1, indicating the relatively high energy of the local fluvial environment, and the overall facies development suggests the persistent operation of streams of moderate sinuosity, i.e. a braided-meandering transition, similar to the Rhondda Beds sequences of the Swansea-Maesteg area (fig. 48). However, thin coals and under-clays are abundant in some sequences, and especially in the Rickets Head section (? Rhondda Beds; cf. Jenkins, 1962, p. 94 who drew attention to the faunal affinities of the Rickets Head succession with the Gwscwm Vein of Carmarthenshire), where they are associated with thin coarsening-upwards cycles (Williams, 1966, fig. 54; Reading, 1971, p. 1405. Such successions result from protracted episodes of flood-basin infill and plant colonisation and indicate a degree of active channel confine-ment which is most commonly encountered within the subaerial portion of a large delta. Transport vectors derived from channels and cross-stratification in the channel-fill (substratum) sands conform to an aggregate unimodal distribution, but with a rather large spread (Williams, 1966, fig. 65). There is a statistically significant modal preference for transport towards the west.

Thus in terms of facies and inferred depositional environments the Upper Coal Measures of Pembrokeshire are broadly comparable with their stratigraphical equivalents in the main basin and represent a lateral extension of the Lower Pennant alluvial-deltaic wedge. The more westerly transport vectors encountered in the St. Brides Bay successions may reflect the diversity in orientation of the early Pennant drainage net or may represent a local environmental anomaly, similar to that which existed in the Margam-Maesteg area during deposition of the Rhondda Beds (Kelling, 1968, p. 2385). However, in view of the structural complexity of the Pembrokeshire Carboniferous, the possibility cannot be ignored of some tectonic re-orientation of the Pennant sequences by rotation of the fault-blocks (cf. Jenkins, 1962, p. 94). Unlike the effects of folding, such a rotation (of unknown magnitude), cannot be compensated for in the process of determining current vectors.

Following establishment of the broadly uniform northwest-flowing alluvial system recorded in the Rhondda Beds, later Pennant Measures indicate a general persistence of these environmental conditions. Although not yet studied in detail, preliminary vector data from the Brithdir and Hughes Beds indicate a regionally consistent pattern of stream flow towards the northwest, with indications of bimod-ality only along the North Crop and, locally, in the southeast. The average grade of sand in the Brithdir and Hughes Beds also appear to be slightly coarser than in the earlier Pennant formations. These vector and textural attributes are consistent with the increasingly alluvial character of the later Pennant sequences. In that portion of the alluvial wedge now preserved in South Wales there remains no hint of the marginal marine sediments typical of the earlier Pennant sequences and it appears that by late Westphalian C times the Carboniferous sea had retreated from the Southwest Province, presumably to the west or northwest.

Two factors modify this picture towards the end of the Westphalian period. First, recrudescent activity of the Usk-Cardiff axis apparently is attested by an eastwards-developing unconformity which is inferred to exist below the Mynydd-islwyn Coal of eastern South Wales, and is believed to result in the eastwards overstep of the Hughes and the Swansea Beds by the Grovesend Beds. However, there is no detectable discrepancy in the dips of the relevant strata, nor is there a conspicuous lithological break of the type which might be expected to follow a significant phase of uplift and erosion. Indeed, except in the extreme eastern area of the coalfield, the basal Grovesend Beds are dominantly argillaceous, with several economic coals and seatearths. Moreover it is significant that the strata immediately *beneath* the Mynyddislwyn seam become predominantly arenaceous as they are traced eastwards and lack the intervening mudstones and coals which are present in the west of the coalfield (cf. Squirrell & Downing, 1969, Plate VII). As an alternative to the concept of an unconformity developed by localised pre-Mynyddislwyn uplift and denudation, these observations suggest that the Swansea and Hughes Beds may be represented in the eastern outcrops by attenuated sequences of a sandstone facies from which the diagnostic coals are absent. Such a relationship is consistent with earlier Coal Measures depositional patterns and results from the persistently (rather then intermittently) positive character of the Usk Axis.

One aspect of Upper Carboniferous sedimentation in South Wales remains to be discussed: this concerns the nature and depositional environment of the Grovesend Beds, the youngest Carboniferous rocks preserved in this area (Westphalian D). This sequence presents something of a contrast to the subjacent formations not only in its dominantly argillaceous character and the common development of thick and economic coals, together with many seatearths and thin ironstones, but also because of the localised development of red, purple and green measures, especially common in the Caerphilly basin and the Gowerton syncline of the western area. (Coloured strata are also sparingly present in the Swansea Beds of the latter area, according to Archer, 1965). In environmental terms the prevalence of the mudstones, ironstones and coals probably indicates the re-establishment of deltaic plain conditions, in place of the alluvial valley environment represented in the underlying arenaceous Pennant Measures. On the other hand the sandy Grovesend sequences of the southeastern area of the coalfield probably mark the limit of retreat of the truly alluvial facies. The coloured strata in the Grovesend Beds have been interpreted as indicating ' . . . the onset of the arid climate that contributed to the change in conditions from Coal Measures to New Red Sandstone' (T. N. George, 1970b, p. 93). However, as indicated earlier, oxidising conditions recurred in earlier Coal Measures, both in South Wales and the Midlands (the Etruria Marls) and there is some indirect evidence that a considerable further thickness of late Carboniferous sediments (possibly including Stephanian) may have accumulated over South and Central Wales (Jones, 1956; Owen, 1964, 1970). Thus an assumption that alluvio-deltaic sedimentation ceased with the Grovesend Beds is probably not justified.

V. PETROLOGICAL CONSIDERATIONS

(a) Introduction

Previous sections have summarised the changes in sedimentary pattern and in the associated geographical/structural elements which have been detected in the Upper Carboniferous sequences of South Wales and only brief reference has been made to the composition of the sediments. However, knowledge of sediment composition is a useful and often a vital aspect of regional studies of sediment dispersal and is particularly important where other evidence suggests that major changes have occurred in the location and nature of the source-areas. The following section therefore provides a brief account of the main petrographic properties of the Namurian and Westphalian sandstones of South Wales, with particular reference to those features which may be of value in determining sediment-provenance.

Before proceeding to this account, it is necessary to interject a cautionary note respecting certain complicating factors. These factors fall into two basic categories, which may be termed *intrinsic* and *procedural*. *Intrinsic* complications derive from the proposition that the ultimate composition of a clastic sediment is the product of three inter-acting factors: source-area character, transport history, and diagenetic history (Pettijohn, 1957). Deciphering the relative role of these three factors in an ancient sediment is frequently difficult. Mineral composition broadly reflects the nature of source-rocks and the post-depositional history, while the texture reflects in some measure the nature and persistence of the various agencies responsible for transporting (and depositing) the detritus. However sediment composition can be profoundly modified during transportation by exposure to physical abrasion or chemical action, which not only produces compositional breakdown of the minerals but also promotes sorting of grains according to size, shape, or specific gravity, a process which may result in segregation and separation of different mineral species. Consequently, mineralogical as well as textural maturity is likely to be rapidly enhanced in an environment such as the littoral zone where sediment is being moved almost continuously and under conditions of high hydraulic energy. Thus assessment of stratigraphical or lateral changes in sandstone petrography must take due account of the environmental status of the sequences under scrutiny.

A related factor is the mean grain-size of the sediment, since this may affect the relative proportions of some petrological components. For example, the proportion of rock fragments recognised in a given sand population normally diminishes with a decrease in the mean grain size, mainly because of the difficulties inherent in distinguishing small rock-clasts from individual grains of the component minerals.

Procedural complications arise mainly through differences in the requirements and techniques of various operators, and, indeed, the variability of the same operator at different times (see Welsh, 1967). Operator differences are most pronounced (and most imponderable) in the identification of specific rock-types (especially of fine-grained types) among the clasts, and in the definition of matrix. Difficulty is often encountered in distinguishing between small fragments of certain fine-grained siliceous rocks such as sedimentary chert, rhyolite and some types of hornfels and, in the present case, this is an important factor in considering the relative role of igneous, metamorphic and sedimentary sources. Another problem concerns the classification of fragments of sheared and granulated igneous rocks, which have been included in an igneous category by some researchers and in a metamorphic class by others. With regard to definition of the matrix, although some attempt has been made to standardise the upper size-limit of matrix components at 5 ø (31 microns) the precise identification of matrix and its distinction from discrete clasts of mudstone or siltstone frequently presents difficulty, particularly in the lithic arenite petrofacies.

Thus, isolation of the provenance factor in the composition of the Upper Carboniferous sandstones considered in this discussion has been achieved by comparing sands of similar grade (medium sand: 250–500 microns mean size) and formed in similar environments, mainly fluvial. Moreover the limits of the categories of the various minerals and clasts have been maintained as wide as possible in order to minimise the variations in identification and classification to which allusion had been made.

(b) Sandstone composition

QLM diagrams (fig. 50B) reveal that Namurian sandstones are, in general, highly quartzose and relatively deficient in martix content. In Pembrokeshire the littoral sands of the Basal Grit formation are true orthoquartzites whereas the fluvial arenites of the Farewell Rock contain slightly enhanced proportions of matrix but a substantially greater content of labile components. The R_2-zone and G_1-zone sandstones in the main basin also appear to contain a higher proportion of matrix than the Pembrokeshire equivalents (fig. 50B) although this increase is partly accounted for by the choice of a different upper size limit for matrix (Oguike, 1969, p. 189). However a distinction again can be drawn between the dominantly littoral sands of the northwestern area (which are mainly true quartz-arenites, lacking matrix) and the fluvial sands of the eastern basin, most of which are quartz-wackes (Oguike, 1969, p. 238). Significantly, on a matrix-free basis most of the Namurian sandstones studied exhibit similar proportions of quartz, feldspar and rock-fragments and are mineralogically mature (fig. 50A). However, the Farewell Rock sandstones of Pembrokeshire are distinctly lithic in character, although the proportion of feldspar remains low.

The Namurian suite of rock-fragments includes a variety of clastic sedimentary rock-types of extrabasinal character, which are the most common clasts in the G_1-zone of the main basin (Oguike, 1969, p. 238), while rhyolitic and basic tuffs dominate among clasts in the Farewell Rock of Pembrokeshire (G. T. George, 1970, p. 156). Metaquartzites, phyllites, slates, and low-grade quartz-mica schists are the main metamorphic representatives, and again are more abundant in the Pembrokeshire sequences.

The Middle Coal Measures sandstones of the main basin are petrographically comparable with the Namurian Farewell Rock, possessing relatively little matrix but a substantial labile content, dominated by rock-fragments (fig. 50A and B). However, Thomas (1967) has detected significant geographical differences in the composition of the Middle Coal Measures arenites. Thus, sandstones from the south-derived fluvial sequences of the North Crop west of Ammanford contain an average of 10 per cent clay-matrix, with individual values as high as 17 per cent (Thomas, 1967, Table 5) and therefore exhibit reduced textural maturity compared with coeval fluvial (and littoral) sands from other areas. Moreover the Ammanford sandstones are clearly distinguishable from the arenites of other regions because they include generally higher proportions of feldspar and also on the basis of their

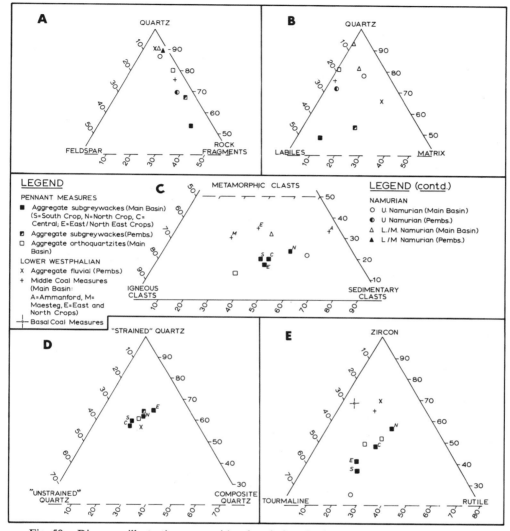

Fig. 50. Diagrams illustrating compositional variations in the Upper Carboniferous sandstones of South Wales. Since the compositional fields of the different formations overlap widely, only the mean composition is given, in order to avoid confusion. For full explanation, see text.

rock-fragment content, since they are characterised by a dominance of sedimentary chert clasts and lack igneous fragments (fig. 50C; cf. Thomas, 1967, figs. 57, 58). Sandstones from the East Crop and from the Aberdare-Ystradgynlais portion of the North Crop display similar rock-fragment assemblages which are, however, statistically separable from the clast-contents of Middle Coal Measures arenites in the Maesteg-Cwmavon area. These geographical differences are apparent both in the fluvial and the marine sandstones and evidently result from the admixing of three separate contributions of detritus (see below, p. 217).

The aggregate suite of rock-fragments present in the Middle Coal Measures sandstones appears more restricted than that found in the Namurian. Metamorphic clasts are almost entirely restricted to stretched and foliated metaquartzites while acid extrusives and felsites are the dominant igneous representatives. Cherts provide an overwhelming proportion of the sedimentary fragments.

The composition of fluvial sandstones in the Lower and Middle Coal Measures of Pembrokeshire has been described by Williams (1966). All these rocks contain sufficient matrix to be classified as 'wackes' on the Dott (1964) scheme (fig. 50B) but the content of labiles is uniformly low and the average composition on a matrix-free basis is similar to that of the Namurian arenites (fig. 50A). The Lower Coal Measures sands contain fewer rock-fragments than the Middle Coal Measures and the clasts are almost entirely of sedimentary origin whereas felsites, phyllites and metaquartzite are not uncommon as fragments in the younger formation (Williams, 1966, p. 159).

The Upper Coal Measures (Pennant Measures) of the main basin, as exemplified by the Rhondda Beds, display two main types of petrofacies, referred to in earlier sections as an orthoquartzite and a lithic arenite facies. The quartzose sands of the eastern and northeastern areas are marked by a notable deficiency of matrix and feldspar but they generally contain too high a proportion of rock-fragments to be classed as orthoquartzites *sensu stricto* (fig. 50A). This is particularly true of the coarser sands in the northeast which carry substantial proportions of siliceous clasts, mainly acid lavas and felsites. The East Crop quartzites are, on average, more quartzose than those of the northeastern outcrops and plot close to the position of the Namurian sands on a QFR diagram (fig. 50A).

The lithic arenites which stratigraphically succeed the quartzitic facies in the eastern areas are compositionally indistinguishable from Pennant sandstones sampled in other parts of the basin (South, Central, North Crop: see fig. 50C). They contain an average of about 10 per cent apparent matrix and a smaller content of feldspar, together with a high proportion (25 to 55 per cent) of rock-fragments (fig. 50A and B). This last category is dominated by the 'intrabasinal sediment' rock-types—mudstones, siltstones and fine-grained arenites identical in lithology with the interbedded sequences and the underlying Coal Measures and presumably derived mainly by contemporaneous erosion of these sediments. Clasts of igneous origin are also abundant in the Rhondda Beds arenites. These mainly comprise fine-grained siliceous volcanic and hypabyssal rocks such as rhyolites, keratophyres and felsites, some of which are distinctly porphyritic. Granitic fragments occur sparingly, together with more basic rock-types, including spilites, porphyritic basalts and badly weathered dolerites. Low-grade metamorphic rocks are well represented and are predominantly of metaquartzite aspect, although quartz-muscovite schists and quartz-chlorite schists are also common.

The relative proportions of the sedimentary, igneous and metamorphic rock-fragments occurring in Rhondda Beds lithic arenites from different parts of the basin are broadly similar, sedimentary and igneous rock-types providing roughly equal contributions and substantially exceeding the metamorphic components (fig. 50C). Lithic sands from the north and northwest provide an exception to that rule since on average they contain fewer igneous clasts and proportionately more sedimentary fragments of intrabasinal type, in addition to an increased proportion of metamorphic rock-types (mainly stretched metaquartzite). Significantly, all the Pennant sands of the main basin exhibit aggregate clast compositions distinctive from those displayed by the Middle Coal Measures and the Namurian (fig. 50B), which contain proportionately more metaquartzite fragments.

The Pennant arenites of Pembrokeshire, as described by Williams (1966), correspond petrographically to the lithic arenites or subgreywackes of the main basin but here there is no equivalent of the eastern quartz-arenite association. However, most of the Pembrokeshire Upper Coal Measures sandstones contain substantially more matrix and less labile components, including feldspar, than their main basin equivalents (fig. 50B). These changes may relate to the nature of the 'matrix' in the Pennant sands, much of which is probably derived from post-depositional alteration of unstable minerals and especially from mechanical and chemical breakdown of argillaceous and silty clasts (Williams, 1966). It is considered that these processes of matrix production (accompanied by elimination of labiles) were enhanced in the tectonically deformed Pembrokeshire rocks. At any rate, the Pennant sandstones of this area contain a suite of rock-fragments broadly similar to that seen in the Upper Coal Measures of the main basin.

(c) Heavy minerals

Although the indurated nature of most of the Upper Carboniferous sandstones restricts the value of quantitative data relating to heavy minerals, such results, together with the more qualitative data, may be used with caution as an additional indication of sediment provenance.

The sandstones of the Namurian and early Westphalian are characterised by a highly restricted suite of heavy minerals—the non-opaque assemblage is composed almost entirely of zircon, tourmaline and rutile. In the Namurian there is little evidence of consistent geographical variation in the proportions of these mineral species (Oguike, 1969, p. 224) but some variations are apparent in the basal Westphalian rocks (Bluck, 1961). In these sediments the East Crop is characterised by a general lack of tourmaline while the pink, well-rounded variety of zircon common in all the Upper Carboniferous sediments, is more abundant in the western outcrops. The Middle Coal Measures sandstones also yield a stable suite dominated by zircon, tourmaline and rutile although minor amounts of apatite, epidote, garnet, sphene and corundum occur in addition (Thomas, 1967, p. 140). In Pembrokeshire, the fluvial arenites of the Lower and Middle Coal Measures display similar ZTR suites (Williams, 1966, Table 12) and it is interesting to note that the mean Z-T-R ratios of the lower Westphalian are all zircon-dominated and are separable from the mean ratios derived from the upper Westphalian (Pennant Measures) assemblages both in the main basin and in Pembrokeshire (fig. 50E), which contain proportionately more tourmaline and less zircon.

However, although the lithic arenites of the Rhondda Beds of the main basin yield ZTR-dominated suites of heavy minerals, they also contain significant proportions of garnet and apatite and in that respect also differ from the earlier West-phalian and Namurian sediments. The quartz-arenites of the Rhondda Beds in the eastern and northeastern portions of the coalfield carry little garnet or apatite but the absence of these less stable species is ascribed to environmental factors (elimination by abrasion in a high-energy regime) rather than provenance, since these sands yield Z-T-R ratios indistinguishable from those characteristic of the coeval lithic arenites (fig. 50E). Moreover the Z-T-R ratios for this facies reveal significant regional variation, suites from the southern and eastern regions containing proportionately more tourmaline than the Central suites, which in turn are richer in tourmaline than the North Crop subgreywackes (fig. 50E). These changes might be ascribed to differences in source but the vectoral character of the variation, allied to the environmental interpretation offered earlier (p. 208) suggests the operation of sediment-modifying processes which produced a general down-stream decrease in tourmaline.

(d) Sandstone provenance

The petrographic data outlined above clearly indicate the importance of environmental factors in determining the detrital composition of these Upper Carboniferous arenites. However, a stratigraphic comparison of sands from identical environmental associations (such as the fluvial facies) reveals substantial residual differences in petrography which are ascribed to changes in the nature of the source-rocks. The main evidence for such provenance-controlled variations is provided by the less stable clasts and minerals, but the more resistant rock-fragments and mineral species also contribute evidence on source-area characteristics.

The quartzose Namurian rocks appear to have been derived largely from recycling of pre-existing sedimentary rocks, probably accompanied, at least in Pembrokeshire, by acid tuffs and lavas resembling those of the local Lower Palaeozoic sequence (G. T. George, 1970, p. 170). The northwesterly increase in plagioclase and uralite content observed in the fluvial Farewell Rock sandstones of Pembrokeshire (G. T. George, 1970, p. 171) is attributed to increasing proximity to a source in the Cambro-Ordovician terrain of North Pembrokeshire. In the main basin, despite the masking effects of environmental factors, Oguike (1969, pp. 238, 242) has also ascribed the provenance of the later Namurian sands to predominantly sedimentary source-rocks. Igneous clasts, which form a significant element in the earlier Namurian rocks (R_2-stage) of the main basin, are mainly siliceous in character and exhibit rounded to subrounded forms. Since these fragments are not accompanied by an increase in the content of detrital feldspar or the unstable heavy minerals, it is inferred that the igneous clasts are probably recycled grains derived from pre-existing lithic sandstones, possibly of Old Red Sandstone aspect. The petrographic identity of the north-derived Namurian arenites of the North and East Crops with the south-derived sands of the Aberkenfig-Margam area (see p. 212) is only explicable in terms of relatively uniform, if spasmodic upwarping of basin-bordering elements to the south, which exposed earlier Palaeozoic sediments to erosion.

The Lower and Middle Coal Measures sandstones reveal broadly similar suites of stable clasts and minerals indicative of a sedimentary provenance, although subordinate contributions from low-grade metamorphic sources can be identified (Williams, 1966; Bluck, 1961; Thomas, 1967). Bluck (1961, p. 84) has suggested that the basal Coal Measures sands may have been derived largely from rocks of Upper Old Red Sandstone age, but the lack of garnet and apatite and the virtual absence of sedimentary clasts exhibiting petrological features characteristic of the local Upper Old Red Sandstone throws some doubt on this interpretation. Moreover, the stratigraphic evidence indicates that the limit of deposition of the Upper Old Red Sandstone facies lay not far north of the present North Crop in the main basin and was actually located to the south of the Pembrokeshire North Crop (T. N. George, 1970, fig. 21; Allen, 1965, fig. 2). The potential area of Upper Old Red Sandstone outcrop is thus severely restricted and appears inadequate to account for the known volume of early Westphalian sediments in the South Wales area. A Lower Palaeozoic source-terrain is therefore envisaged for these sediments and the occurrence of occasional greywacke clasts and more common chlorite-bearing protoquartzite fragments provides corroborative evidence for this hypothesis. The increased abundance of metaquartzite and quartz-muscovite-schist clasts in the Middle Coal Measures, especially in Pembrokeshire (Williams, 1966, p. 159) may indicate partial un-roofing of the metamorphic basement, now represented in Anglesey. Significantly, the south-derived Middle Coal Measures arenites of the Maesteg area are relatively deficient in metamorphic fragments, compared to contemporaneous sands of northerly provenance (Thomas, 1967, p. 132) suggesting that unroofing of the southern source-area in the Bristol Channel had not yet proceeded to the extent of exposing basement rocks.

The late Westphalian sediments (Pennant Measures) reveal a different provenance which is reflected in the greatly enhanced amount and variety of rock-fragments. The dominant clasts in these south-derived sequences are of 'intra-basinal' character and reflect erosion both of contemporaneous sediments and of older Coal Measures and Namurian rocks. Progressive warping and cannibalisa-tion of the southern margin of the Westphalian basin is clearly indicated. However, the occurrence of abraded and recycled garnet and apatite grains and of limonite-cemented protoquartzite clasts suggests that significant erosion of Old Red Sand-stone formations also proceeded during later Westphalian time (cf. Heard & Davies, 1924; Heard, 1922). Moreover the abundant fragments of metaquartzite, slate, quartz-chlorite-schist and quartz-sericite-schist together with detrital flakes of mica and chlorite, represent additional contributions from low-grade meta-morphic sources and probably result from exposure of the metamorphic basement underlying the Palaeozoic successions of southernmost Britain. Some of the granitic and epidote-rich fragments found in the Rhondda Beds of the East and South Crops closely resemble rocks exposed in the metamorphic core of the Malvern Hills. However, direct derivation from this region is not necessarily implied by this petrology since the Malverns rocks probably represent a segment of an ancient cratonic block which palaeogeographic reconstructions indicate as extending from the Midlands of England towards the southwest during the Lower Palaeozoic (Wills, 1951, p. 12). Subsequent to late Carboniferous erosion, the remnants of this block in the Bristol Channel area have been concealed beneath the frontal thrusts of North Devon, associated with northward-advancing Armorican nappes (Bott, *et al.*, 1958; Anderson & Owen, 1970, p. 88).

VI. SUMMARY OF CONCLUSIONS

The Upper Carboniferous rocks of South Wales may be assigned to three inter-grading phases of sedimentation which together record a progressive environmental change from the carbonate marine shelf of the Lower Carboniferous to the alluvial valleys of the late Westphalian. This environmental evolution is paralleled by a change in geotectonic regime from a slowly subsiding basin dominated by older, mainly caledonoid structural elements in the Namurian, to a rapidly subsiding foredeep trough, flanked to the south by the rising Armorican structures, which dominates the late Carboniferous history of southern Britain.

The first phase of sedimentation is recorded in the Namurian sequences of South Wales, which reveal the progressive, if intermittent, encroachment of terrigenous detritus from the north across the warped surface of the carbonate shelf. Evidence cited earlier is consistent with the model of a low-lying, strongly embayed and scalloped coastline, characterised by extensive mudflats and low, sandy bars and spits. Streams traversing this coastal plain were small and sluggish and the few deltaic lobes which are preserved are also small and impersistent both laterally and vertically.

This complex of coastal plain sub-environments was subject to widespread and relatively rapid inundations by the sea but the frequency and duration of such marine invasions decrease throughout the successive stages of the Namurian, probably because sediment supply began to outpace the rate of subsidence as the pulse of uplift in the hinterland areas became quickened.

The late Namurian and early Westphalian rocks mark a second stage in the Upper Carboniferous development of the South Wales area. An early episode of fluvial deposition is reflected in the north-derived sands of the Farewell Rock lithology, which spread diachronously towards the south and east. This fluvial regime was superseded by a prolonged phase of deltaic sedimentation accompanied by recurrent emergence and vegetative colonisation of the delta-plain and the adjacent coastal swamplands. The relatively mature detritus being supplied to the coalescing and southwards-prograding delta-lobes was conveyed mainly from the northern source area (St. George's Land) by sluggish, serpentine streams and rivers. However, the localised northward and westward growth of a number of fluvio-deltaic complexes may be demonstrated in the Lower and Middle Coal Measures of Pembrokeshire and the main coalfield basin. These features probably represent the sedimentary response to renewed and spasmodic uplift of a southern source-area, sited in the region of the Bristol Channel. This conclusion is fortified by the distinctive and immature petrology of the south-derived sands (see p. 213 and Thomas, 1967, pp. 128–129).

The third and final phase of Upper Carboniferous sedimentation is recorded in the late Westphalian Pennant Measures of South Wales. These thick sequences of sandstones with their subordinate lutites and coals were deposited, under conditions of accelerating subsidence, in the channels and floodplains of a broad alluvial tract which advanced gradually to the north and east, as braided and meandering rivers conveyed vast quantities of lithic detritus from the rapidly rising positive areas lying to the south. Littoral and penemarine conditions may have persisted for a short time along the East Crop, if the revised interpretation of the Lower Pennant pebbly orthoquartzite facies is correct. Moreover the enigmatic North Crop sequences of Rhondda Beds displaying bimodal transport-vectors may represent a lower delta plain environment and thus suggest the existence of open marine conditions to the north and west of the South Wales area in early Pennant times. Later Pennant sequences, especially in Pembrokeshire, indicate further withdrawal of the sea and the establishment of a major drainage system, probably aligned west northwest.

The studies summarised here also have a bearing on the inter-relationship of tectonics and sedimentation. Thus, although the existence of certain structural elements was largely responsible for the establishment and delineation of the Upper Carboniferous basin of South Wales, the activity of these features appears to have had little direct effect upon the regional style of sedimentation. Local effects may be discerned and attributed to individual structures, for example the deviation of Rhondda Beds palaeocurrents in the vicinity of the Margam-Maesteg thickness culmination (Kelling, 1968, p. 2384). On the other hand, contemporaneous movement of the major marginal and intrabasinal elements is reflected primarily by marked changes in thickness across such features, whereas direct structural control of facies distribution seldom can be demonstrated (cf. Thomas, 1967, pp. 29 and 30). Moreover the tectonic movements associated with these elements occurred in a continuous rather than a periodic fashion, so that 'events' such as the intra-Morganian unconformity represent simply the local outstripping of sediment supply by the local rate of relative elevation (or, more accurately, the rate of absolute subsidence) of the Usk axis.

The importance of environmental controls in determining sandstone petrography has been demonstrated in several of the Upper Carboniferous formations. The fluvial sands are generally lithic, carry significant amounts of matrix and may be classed as lithic arenites or lithic wackes (subgreywackes and protoquartzites), whereas littoral and barrier-bar sands are quartz-rich and matrix-deficient (quartz-arenites or orthoquartzites). Sands deposited in the distributary channels associated with deltaic lobes are mainly quartzose or lithic arenites of protoquartzitic aspect. Thus the general decrease in compositional maturity which is observed in the sandstones of successive Upper Carboniferous formations (p. 212, fig. 50) may be attributed, at least in part, to the environmental evolution from marine to alluvial conditions, already described.

However there also exist residual petrographic differences which are assigned to factors of provenance. These indicate a general change in the source-rocks, from a local, mainly sedimentary suite in the Namurian and early Westphalian to the mixed sedimentary, igneous and low-grade metamorphic provenance, characteristic of the later Westphalian (Pennant Measures) rocks but also intermittently developed in the middle Coal Measures.

A final point concerns the ultimate derivation of the Upper Carboniferoue sediments of South Wales. The Namurian and early Westphalian sequences were supplied dominantly from an area (St. George's Land) lying to the north of the present limits of the basin. This area contributed muds and relatively fine-grained sands, with some quartz-pebbles, from a Lower Palaeozoic terrain of mixed sedimentary and acid volcanic aspect. However, commencing in the late Namurian there were spasmodic influxes of coarser, less mature sediment from the south, Much of the Middle Coal Measures sequence of Pembrokeshire and in the western and southern portions of the main basin may be attributed to the persistent progradation towards the west, northwest or north of a series of overlapping fluvio-deltaic complexes which must have flanked a southern source-area on the site of the present Bristol Channel.

The reality of this elusive and intermittently emergent southern terrain ('Sabrinia') is attested not only by the evidence in the South Wales Namurian and Coal Measures but also by the north derived wedges of paralic sediment which are intercalated within the Ammanian (mid-Westphalian) flysch-sequences of the Upper Culm of Devon and north Cornwall (Prentice, 1962; De Raaf et al., 1965). The minumum distance separating these rocks from their equivalents in South Wales is about 65 km at the present time. In view of the great crustal shortening indicated by the intense folding and major thrusting displayed by the Culm rocks, the original distance of separation was probably at least twice the present amount. Uninterrupted southwards progradation of the South Wales paralic belt over such a distance is considered to be highly improbable, especially in view of the observed frequency (and short-lived character) of the regional marine transgressions which are recorded in the Middle Coal Measures marine bands (see pp. 201–202). An intervening source-area, situated between South Wales and the Culm trough of North Devon and shedding sediment bilaterally, furnishes a more satisfactory explanation for the observations summarised above.

This terrain also may have contributed to the Upper Palaeozoic basin inferred on geophysical evidence to exist below the boundary thrust of the North Devon coastline (Bott et al., 1957; Webby, 1965; Brooks & Thompson, 1972). At any event 'Sabrinia' appears to have a venerable history, since the mid-Devonian Ridgeway Conglomerates of Pembrokeshire also reveal evidence of consistent southerly derivation from a nearby source (Williams, 1964).

However the south-derived fluvial sands of the Upper Coal Measures (Westphalian C) present a conspicuous contrast to the preceding sequences both in their environmental and petrological characteristics. These Pennant Measures are largely an alluvial facies, possibly passing northwards into delta-plain sediments. Moreover the nature of the rock-fragments and mineral grains contained in these sands clearly indicates that older Coal Measures and possibly Culm sediments were being exposed and re-worked northwards at this time. Such re-working presumably resulted from tilting of the basin margins and reflects the northward passage of the first waves of the 'Hercynian Storm' which had probably brought sedimentation to a halt in the Culm trough by late Ammanian times. One effect of this major change in geotectonic regime was to create a broad, north-facing palaeoslope over most of southern Britain and thus to warp off to the north and west the late Carboniferous sea which in earlier times had invaded South Wales from the south and southwest. Moreover the northward advance of this Hercynian front resulted in progressive extinction of the pre-existing framework structures, including the massif of St. George's Land, which by late Westphalian C times had ceased to furnish significant quantities of sediment to the South Wales—Forest of Dean basin and was probably no longer an effective barrier to sedimentation.

VII. ACKNOWLEDGEMENTS

The writer is glad to acknowledge his indebtedness to many past and present colleagues who have contributed in some measure to this study. Particular gratitude is expressed to T. R. Owen, whose stimulating advice and enthusiastic encouragement first sparked interest in this particular area of research, and to F. H. T. Rhodes and D. V. Ager who successively and generously supported the study. The basic data utilised here in part result from the researches of various former students of the Swansea Geology Department, notably B. J. Bluck, G. T. George, R. O. Oguike, L. P. Thomas, R. L. Thomas, and P. F. Williams while E. M. Chaplin generously made available some of the results of his studies. However the overall conclusions remain the responsibility of the writer alone. Stimulating discussions with J. T. G. Stead and J. C. Ferm have contributed substantially to the section on late Westphalian sedimentation. Part of the fieldwork for this study was financed by the Research Grants Committee of the University College of Swansea.

VIII REFERENCES

ALLEN, J. R. L. 1965a. A review of the original characteristics of recent alluvial sediments. *Sedimentology*, **5**, 91–180.

——. 1965b. Upper Old Red Sandstone (Farlovian) palaeogeography in South Wales and the Welsh Borderland. *J. sediment. Petrol.* **35**, 167–195.

——. 1967. Notes on some fundamentals of palaeocurrent analysis with reference to preservation potential and sources of variance. *Sedimentology*, **9**, 75–88.

ANDERSON, J. G. C., and OWEN, T. R. 1968. *The Structure of the British Isles.* Oxford (Pergamon), 162 pp.

ARCHER, A. A. 1965. Red beds in the Upper Coal Measures of the western part of the South Wales coalfield. *Bull. Geol. Surv. Gt. Britain*, **23**, 57–64.

——. 1968. The Upper Carboniferous and later formations of the Gwendraeth Valley and adjoining areas in parts of the Carmarthen, Ammanford and Worms Head sheets. *Mem. geol. Surv. Gt. Britain*.

BLOXAM, T. W., and THOMAS, R. L. 1969. Palaeontological and geochemical facies in the *Gastrioceras subcrenatum* marine band and associated rocks from the North Crop of the South Wales Coalfield. *Quart. J. geol. Soc. Lond.*, **129**, 239–82.

BLUCK, B. J. 1961. The sedimentary history of the rocks between the horizon of *G. subcrenatum* and the Garw Coal in the South Wales coalfield. *Ph.D. thesis*, Univ. of Wales (Swansea), 130 pp.

——. 1967. Sedimentation of beach gravels: examples from South Wales. *J. sedim. Petrol.*, **37**, 128–56.

—— and KELLING, G. 1963. Channels from the Upper Carboniferous Coal Measures of South Wales. *Sedimentology*, **2**, 29–53.

BLUNDELL, C. R. K. 1952. The succession and structure of the north eastern area of the South Wales coalfield. *Quart. J. geol. Soc. Lond.*, **107**.

BOTT, M. H. P., DAY, A. A., and MASSON-SMITH, D. 1958. The geological interpretation of gravity and magnetic surveys in Devon and Cornwall. *Phil. Trans. Roy. Soc.*, **215A**, 161–91.

BROOKS, M., and THOMPSON, M. S. 1972. The geological interpretation of a gravity survey of the Bristol Channel. *Geol. Soc. London Circular*, **169**, 11–12.

CAILLEUX, A. 1945. Distinction des galets marins et fluviatiles. *Bull. Soc. geol. France* ser. 5, **15**, 375–404.

CALVER, M. A. 1969. Westphalian of Britain. *Compte Rendu 6me Cong. Inter. Strat. et Géol. du Carbonifere* (Sheffield, 1967), **1**, 233–54.

DAVIES, D. K., ETHRIDGE, F. G., and BERG, R. R. 1971. Recognition of Barrier environments. *Bull. Amer. Assoc. Petrol. Geol.*, **55**, 550–65.

DE LA BECHE, H. T. 1846. On the formation of the rocks of south-western England. *Mem. Geol. Surv. Gt. Britain*, **1**, 296 pp.

DE RAAF, J. F. M., READING, H. G., and WALKER, R. G. 1965. Cyclic sedimentation in the Lower Westphalian of North Devon, England. *Sedimentology*, **4**, 1–52.

DOTT, R. H. 1964. Wacke, greywacke and matrix—what approach to immature sandstone classification. *J. sediment Petrol.*, **34**, 625–32.

DOWNING, R. A., and SQUIRRELL, H. C. 1965. On the red and green beds in the Upper Coal Measures of the eastern part of the South Wales coalfield. *Bull. Geol. Surv. Gt. Britain*, **23**, 45–56.

FERM, J. C., and CAVAROC, V. V. JR. 1969. A Field Guide to the Allegheny deltaic deposits in the upper Ohio Valley. *Ohio and Pittsburg Geol. Socs.* 21 pp.

—— and WILLIAMS, E. G. 1965. Characteristics of a Carboniferous marine invasion in western Pennsylvania. *J. sediment. Petrol.*, **35**, 319–30.

——, HORNE, J. C., SWINCHATT, J. P., and WHALEY, P. W. 1971. Carboniferous depositional environments in northeastern Kentucky. *Annual Spring Field Conference Guidebook, Geol. Soc. Kentucky*, 30 pp.

GEORGE, G. T., 1970. The sedimentology of Namurian sequences in South Pembrokeshire. *Ph.D. thesis*, Univ. of Wales (Swansea), 202 pp.

GEORGE, T. N. 1927. The Carboniferous Limestone (Avonian) succession of a portion of the North Crop of the South Wales coalfield. *Quart. J. geol. Soc. Lond.*, **83**, 38–95.

——. 1956. The Namurian Usk Anticline. *Proc. geol. Ass.*, **66**, 297–316.

——. 1962. Devonian and Carboniferous foundations of the Variscides in northwest Europe. In *Some aspects of the Variscan fold belt* (ed. Coe K.). Manchester (Manchester Univ. Press), 19–47.

——. 1969. British Dinantian Stratigraphy. *Compte Rendu 6me Cong. Inter. Strat. et Géol. du Carbonifere*, (Sheffield, 1967), **1**, 193–218.

——. 1970. *British Regional Geology: South Wales*. London (H.M.S.O.), 3rd ed., 152 pp.

HEARD, A. 1922. The petrology of the Pennant Series. *Geol. Mag.*, **59**, 83–92.

—— and DAVIES, R. 1924. The Old Red Sandstone of the Cardiff district. *Quart. J. geol. Soc. Lond.*, **80**, 489–519.

HOBDAY, D. 1969. Upper Carboniferous shoreline systems in northern Alabama. *Ph.D. thesis*, Univ. of Louisiana, 75 pp.

HOWELL, A., and COX, A. H. 1924. On a group of red measures or coloured strata in the east Glamorgan or Monmouthshire coalfield. *South Wales Inst. Eng. Proc.*, **43**, 139–74.

JENKINS, T. B. H. 1962. The sequence and correlation of the Coal Measures of Pembrokeshire. *Quart. J. geol. Soc. Lond.*, **118**, 65–101.

JONES, D. G. 1970. The base of the Namurian in South Wales. *Compte Rendu 6me Cong. Inter. Strat. Géol. du Carbonifere* (Sheffield, 1967) III, 1023–30.

JONES, O. T. 1956. The geological evolution of Wales and the adjacent regions. *Quart. J. geol. Soc. Lond.*, **111**, 323–51.

KELLING, G. 1964. Sediment transport in part of the Lower Pennant Measures of South Wales. In *Deltaic and Shallow Marine Deposits* (ed. Van Straaten, L.M.J.U.). Amsterdam (Elsevier), 177–84.

—— 1968. Patterns of sedimentation in Rhondda Beds of South Wales. *Bull. Amer. Assoc. Petrol.*

Geol., **52**, 2369–86.

—— 1969.. The environmental significance of cross stratification parameters in an Upper Carboniferous fluvial basin. *J. sedim. Petrol.*, **39**, 857–75.

—— and GEORGE, G. T. 1971. Upper Carboniferous sedimentation in the Pembrokeshire coalfield. In *Geological Excursions in South Wales and the Forest of Dean.* Bassett, D. A. and M. G. (eds.), Cardiff, 240–59.

—— and WILLIAMS, B. P. 1966. Deformation structures of sedimentary origin in the Lower Limestone Shales (basal Carboniferous) of South Pembrokeshire, Wales. *J. sedim. Petrol.*, **36**, 927–39.

KLEIN, G. de V. 1967. Paleocurrent analysis in relation to modern marine sediment dispersal patterns. *Bull. Amer. Assoc. Petrol. Geol.*, **51**, 366–82.

LEITCH, D., OWEN, T. R., and JONES, D. G. 1958. The basal Coal Measures of the South Wales coalfield from Llandybie to Brynmawr. *Quart. J. geol. Soc. Lond.*, **113**, 461–83.

MOORE, L. R. 1945. The geological sequence of the South Wales coalfield: the South Crop and Caerphilly Basin and its correlation with the Taff Valley sequence. *South Wales Inst. Eng. Proc.*, **60**, 141–52.

OGUIKE, R. O. 1969. Sedimentation of the Middle Shales (Upper Namurian) of the South Wales coalfield. *Ph.D. thesis*, Univ. of Wales (Swansea), 274 pp.

OWEN, T. R. 1964. The tectonic framework of Carboniferous sedimentation in South Wales. In *Deltaic and Shallow Marine Deposits* (ed. Van Straaten, L. M. J. U.). Amsterdam (Elsevier), 301–7.

——. 1971. The relationship of Carboniferous sedimentation to structure in South Wales. *Compt. Rendu 6me Cong. Inter. Strat. Géol. du* Carbonifere (Sheffield, 1967), **III**, 1305–16.

—— and JONES, D. G. 1961. The nature of the Millstone Grit/Carboniferous Limestone junction of a part of the North Crop of the South Wales coalfield. *Proc. geol. Ass.*, **72**, 239–49.

PARRY, C. 1966. The effects of differential subsidence on the principal coal seams of Ammanian age in the eastern part of the South Wales coalfield. *M.Sc. thesis*, Univ. of Wales (Swansea), 78 pp.

PETTIJOHN, F. J. 1957. *Sedimentary Rocks.* 2nd ed. New York (Harper), 728 pp.

POOLE, E. G. 1969. The stratigraphy of the Geological Survey Apley Barn borehole, Witney, Oxfordshire. *Bull. Geol. Surv. Gt. Britaln*, **29**, 1–104.

POTTER, P. E., and PETTIJOHN, F. J. 1963. *Paleocurrents and Basin Analysis.* Berlin (Springer Verlag), 330 pp.

PRENTICE, J. E. 1962. The sedimentation history of the Carboniferous in Devon. In *Some Aspects of the Variscan Fold Belt* (ed. Coe, K.) Manchester (Manchester Univ. Press), pp. 93–108.

RAMSBOTTOM, W. H. C. 1969. The Namurian of Britain. *Compte Rendu Cong. Inter. 6me Strat. et Géol. du Carbɔnifere* (Sheffield), 1967, **1**, 219–232.

READING, H. G. 1971. Sedimentation sequences in the Upper Carboniferous of north-western Europe. *Compte Rendu 6me Cong. Inter. Strat. et Géol. du Carbonifere* (Sheffield), 1967), **IV**, 1401–11.

ROBERTSON, T. M. 1948. Rhythm in sedimentation and its interpretation with particular reference to the Carboniferous sequence. *Trans. Edinburgh geol. Soc.*, **14**, 141–75.

SELLEY, R. 1967. Paleocurrents and sediment transport in the Sirte Basin, Libya. *J. geol.*, **75**, 215–23.

——. 1968. A classification of paleocurrent models. *J. geol.*, **76**, 99–110.

——. 1970. *Ancient Sedimentary Environments.* London (Chapman and Hall), 237 pp.

SQUIRRELL, H. C., and DOWNING, R. A. 1964. The attenuation of the Coal Measures in the south-east part of the South Wales coalfield. *Bull. Geol. Surv. Gt. Britain*, **21**, 119–32.

—— and ——. 1969. Geology of the South Wales coalfield: pt. I: the country around Newport (Mon.). 3rd ed., *Mem. geol. Surv. Gt. Britain.*

SULLIVAN, R. 1965. The mid-Dinantian stratigraphy of a portion of central Pembrokeshire. *Proc. geol. Ass.*, **76**, 283–99.

THOMAS, L. P. 1967. A sedimentary study of the sandstones between the horizons of the Four-

Foot Coal and the Gorllwyn Coal of the Middle Coal Measures of the South Wales coalfield. *Ph.D. thesis*, Univ. of Wales (Swansea), 176 pp.

THOMAS, M. D. 1967. Gravity surveys around the mouth of the Severn. *Ph.D. thesis*, Univ. of Wales (Swansea), 234 pp.

TRUEMAN, A. E. 1941. The periods of coal formation represented in the British Coal Measures. *Geol. Mag.*, **78**, 71–84.

——. 1946. Stratigraphical problems in the Coal Measures of Europe and North America. *Quart. J. geol. Soc. Lond.*, **102**, 49–86.

——. 1954. (Ed.). *The Coalfields of Great Britain*. London (Arnold), 396 pp.

WARE, W. D. 1939. The Millstone Grit of Carmarthenshire. *Proc. geol. Ass.*, **50**, 168–204.

WEBBY, B. D. 1965. The stratigraphy and structure of the Devonian rocks in the Quantock Hills, west Somerset. *Proc. geol. Ass.*, **76**, 321–48.

WELSH, W. 1967. The value of point-count modal analysis of greywackes. *Scot. J. Geol.*, **3**, 318–26.

WILLIAMS, B. P. 1964. The stratigraphy, petrology and sedimentation of the Ridgeway Conglomerate and associated formations in south Pembrokeshire. *Ph.D. thesis*, Univ. of Wales (Swansea), 245 pp.

WILLIAMS, P. F. 1966. The sedimentation of the Pembrokeshire Coal Measures. *Ph.D. thesis*, Univ. of Wales (Swansea), 259 pp.

——. 1968. The sedimentation of Westphalian (Ammanian) Measures in the Little Haven-Amroth coalfield, Pembrokeshire. *J. sedim. Petrol.*, **38**, 332–62.

WILLS, L. J. 1951. *A palaeogeographical atlas of the British Isles and adjacent parts of Europe*. London (Blackie), 64 pp.

WOODLAND, A. W., and EVANS, W. B. 1964. The geology of the South Wales coalfield: Pt. IV: Pontypridd and Maesteg. 3rd Ed. *Mem. geol. Surv. Gt. Britain.*

WORSSAM, B. C. 1963. The stratigraphy of the Upton Borehole, Oxfordshire. *Bull. Geol. Surv. Gt. Britain*, **20**, 107–62.

ZINGG, T. 1935. Beitrag zur Schotteranalyse, *Schweiz. mineralog. petrog. Mitt.*, **15**, 39–140.

THE GEOCHEMISTRY OF SOME SOUTH WALES COALS

M. M. Davies and T. W. Bloxam

I. INTRODUCTION

THE presence of anthracite in parts of the South Wales coalfield has posed many questions as to its origin and this study was undertaken to investigate geochemical differences between anthracite and the lower rank bituminous coals. All the coals are Carboniferous in age and selected elements were determined quantitatively from two coal seams that occur within the anthracite and bituminous zones over a wide area of the South Wales coalfield (Fig. 51). To get a clearer picture of the distribution of elements between the organic matter of the coal, and the associated mineral matter, the samples were divided into 'light' (organic) and 'heavy' (inorganic) fractions, and certain elements determined in each. Since correlation between coals in the main eastern part of the coalfield and westwards into Pembrokeshire is uncertain, the latter area was excluded from the study.

II. DESCRIPTION OF THE COAL SEAMS

The coal seams studied in South Wales were the Nine Feet and Six Feet seams from the Ammanian rocks of the Carboniferous Series which occur between the Amman and Cefn Coed marine bands. The vertical distance between them varies from 150 feet in the east to 300 feet in the west (Fig. 52), the Six Feet lying above the Nine Feet.

The Nine Feet seam can be traced continuously throughout the area where it varies in thickness from 7 feet (2·1 m) to 10 feet (3·0 m) and splits into two seams, the Upper and Lower Nine Feet seams. Throughout a considerable part of the coalfield the lower part of the Nine Feet seam converges with the top of the Bute seam in the south of the coalfield (Adams, 1956, and Fig. 53).

The Six Feet seam can also be correlated over the whole of the coalfield. In the south east it is a single seam, but the overlying Four Feet seam is mined with it in the extreme south, and in the vicinity of Pontypridd the underlying Red Vein is also mined with it. Towards the north and west the seam splits into two, the Upper and Lower Six Feet seams and the two seams join again on either side of the Neath Valley before splitting again further to the north west. In the extreme north west the overlying Four Feet seam is mined with it as the Big Vein (Adams, 1965, personal communication).

The shaft at Cynheidre (Carmarthenshire) reaches a depth of over 1800 feet (548 m) and 20 samples of coal seams cut by this shaft have been analysed. These seams include lateral equivalents of the Nine and Six Feet seams, although both seams have split into several additional seams (Big, Hwch, Stanllyd, etc.), many of which cannot be precisely correlated (Figs. 53 and 54).

III. ANALYTICAL METHODS

Sampling

The coals were all provided by the National Coal Board (South Western Division) and were sampled underground by their standard procedures (British Standards, 1960, No. 1017) and

225

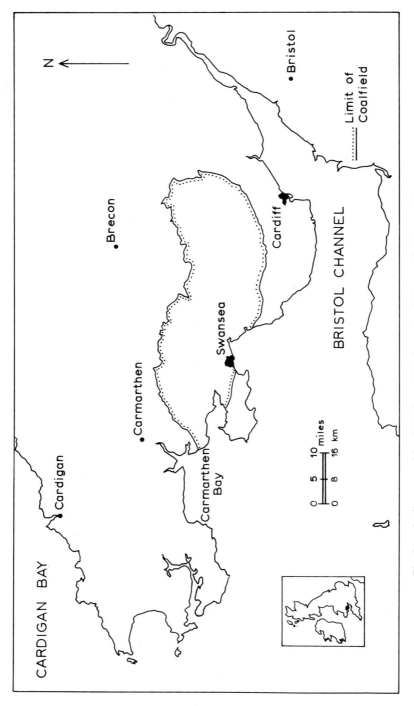

Fig. 51.   Outline map of South Wales and the boundary of the coalfield east of Carmarthen Bay.

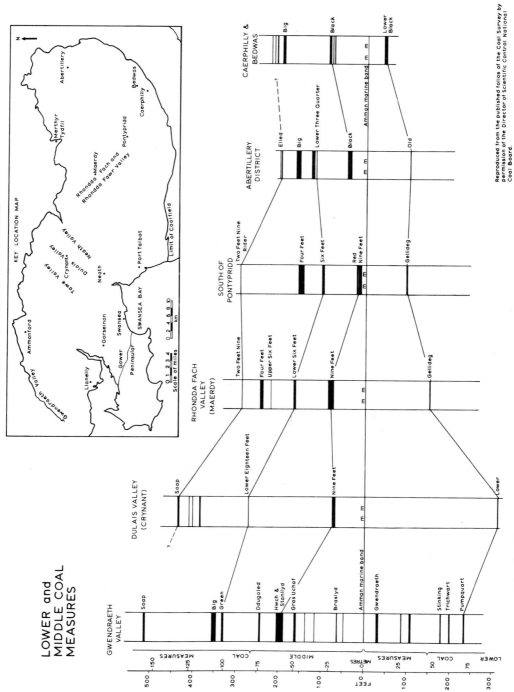

Fig. 52. Correlation diagram of Nine Feet, Six Feet, and associated coal seams of the Lower and Middle Coal Measures.

represent whole seam averages.

Separation of inorganic and organic fractions; Nine Feet and Six Feet seams

Before ashing, the heavy fraction (inorganic mineral matter) was separated from the light (organic) fraction using carbon tetrachloride. The specific gravity of carbon tetrachloride is 1·63 and coals vary between 1·39–1·50, the average being 1·45. Mineral matter is rarely less than 2·0. The mineral matter separates easily from the coal, although some organic matter forms a coating around individual mineral grains and, if this coating is thick, some organic matter may occur in the inorganic fraction. This cannot be entirely avoided but is minimized by fine grinding.

Ashing

After separation into organic and inorganic fractions the coals were weighed into silica dishes and ashed in a muffle furnace in a current of air. The temperature of the furnace was raised slowly to 400°C and was left at this temperature for 3–4 hours. They were then transferred to an electric furnace and, in a current of oxygen, the temperature was raised slowly to 450°C and the coals ashed to constant weight (5–6 hours). Since a certain amount of coal was lost during the separation process, the ash values are approximate. Accurate values for the unseparated coal (i.e. total ash content) were found using the standard method (British Standards, 1957, No. 1016).

Some carbon was left in all the samples, the amount ranging from 0·25–3·0%, averaging 1%. The values for the inorganic fractions are generally higher than for the organic fractions and may be attributed to the carbon contained in calcite. Since the amount of calcite was not known, it was impossible to correct the data to 100%, but for the minor elements this 1% is within experimental error, and for the major elements the data is used comparatively so that the error is largely eliminated.

Major and trace elements were determined in the ash obtained from the inorganic and organic separates (Tables 3 and 4).

Correction of Cynheidre samples

The samples from Cynheidre were not fractioned since they were smaller and would not have provided enough material. In these cases total ash was determined.

The total ash content of all the Nine Feet and Six Feet seam samples was found to vary with the percentage of the inorganic fraction (Fig. 55). Also, a fairly constant ratio was found for the concentration of certain elements between the inorganic and organic fractions. On this basis the *percentage distributions* of the elements in the Cynheidre coals were determined, which means that a better comparison can be made between coals of differing ash content. This was done for Cu, Pb, Sn, and Be. Other elements did not show a regular distribution ratio between the inorganic fractions (Table 5).

The average ratio of the Cu content in organic/inorganic fractions in the Nine Feet and Six Feet seams is 2·41 (s = ± 0·72). All ratios for Pb, Sn and Be are 2·25 (s = ± 0·94), 1·94 (s = ± 0·53) and 7·32 (s = ± 1·39) respectively. Assuming similar relationships apply at Cynheidre, the partition of Cu between organic and inorganic fractions may be estimated as follows:

Example   Total ash 6·07%   Cu content 660 p.p.m.
     Inorganic ash (from Fig. 55) = 1·95% (31·5% of total ash)
Therefore: Organic ash   = 4·12% (68·5% of total ash)
Since the ratio of Cu in organic to inorganic fractions averages 2·41:
Let 1% inorganic ash contain x p.p.m. Cu and
let 1% organic ash contain 2·41 p.p.m. Cu
     Then; 1·95x + (4·12 × 2·41x) = 660
          1·95x + 9·95x = 660   Therefore x = 55·5
     Hence; inorganic ash contains 1·95 × 55·5 = 108 p.p.m.
          organic ash contains   9·95 × 55·5 = 551·5 p.p.m.

$$\text{correcting to 100\% inorganic; } \frac{108 \times 100}{31\cdot5} = 343 \text{ p.p.m.}$$

correcting to 100% organic; $\dfrac{551 \cdot 5 \times 100}{68 \cdot 5} = 809$ p.p.m.

Determination of the major elements: Al_2O_3, Fe_2O_3, TiO_2, P_2O_5, MnO, CaO, MgO, Na_2O and K_2O (Table 3).

These elements were determined by wet chemical methods using colorimetric, volumetric and flame-photometric techniques. The samples were decomposed using HF and H_2SO_4; remaining traces of carbon were removed by oxidation with HNO_3. Na_2O and K_2O were determined by flame photometry using a strong lithium buffer, and CaO and MgO by titration with EDTA using a photoelectric titrator after the removal of 'R_2O_3'. Al_2O_3, Fe_2O_3 (total iron), TiO_2, P_2O_5 and MnO were all determined photometrically; Al_2O_3 as the oxinate after extraction in chloroform from Fe complexed as ferrous dipyridyl; Fe_2O_3 as the ferrous dipyridyl, (correcting for TiO_2); TiO_2 as the titano-peroxide complex (correcting for Fe); P_2O_5 as the molybdivanadic acid and MnO as the permanganate after oxidation with periodate (Shapiro and Brannock, 1956).

Determination of the minor elements: Co, Ni, V, Mo, Be, B, Zn, Cu, Ag, Ge, Ga, and Sn (Table 4).

The elements were determined spectrographically on a Hilger quartz spectrograph (E.742) under the following conditions: slit width $0 \cdot 01$ mm, focus on collimator, current 8 amps, 6 mm electrode gap. A rotating step sector with a transmission ratio of 2 gave a graded series of exposures recorded on Kodak B10 spectrographic plates. Cathode layer excitation was used and all burns were made in triplicate. The internal standards used were indium and palladium (neither of which was detected in the coal ash): indium for the volatile elements and palladium for the involatile elements. The Pd and In were introduced into the samples by addition to graphite powder ($0 \cdot 25\%$ of each), which was then mixed in a 1:1 ratio with the standards and samples to reduce the selective distillation of the volatile elements (Ahrens and Taylor, 1961).

In samples with < 30 p.p.m. B and a high iron concentration, Fe $2497 \cdot 8$ interfered with the boron line, but very few samples showed such low boron concentrations and there was no difficulty in separating and measuring the two lines.

The accuracy was checked using standard diabase Wl which has a similar, though not identical base composition. The standard deviations are as follows: Ga, Be, Cu, Co, V, B less than 10%; Ni, Sn, Pb, Mo 15%; Ag, Ge 20% and Zn 25%.

IV. GEOCHEMISTRY

(a) Distribution of the major elements over the Coalfield

Inorganic fractions (Table 3)

The major elements in the inorganic fractions reflect the environment in which the coals were formed. Ca-Mg, Fe-Mn, Al-Na-K all show fairly constant ratios and generally where Al, Na, and K (i.e. the clay fraction) decrease, Ca, Mg, Fe and Mn increase. Ti follows Al and shows no relationship to Fe (Fig. 56). This is a common feature of Ti in sedimentary rocks where it appears to be associated with the clay fraction (Goldschmidt, 1954; Rankama and Sahama, 1950).

Mn is associated with Fe in sedimentary rocks and the coal ash is no exception. The average Mn/Fe ratio of $0 \cdot 015$ is typical of shale (Goldschmidt, 1954) which is the commonest type of sediment present in the coals. The Ca and Mg values are also characteristic of a clay type material ranging from 1–13% CaO and $0 \cdot 7$–13% MgO. Where Ca and Al values are low, a more arenaceous sediment is inferred, although Si was not determined. The inorganic P may be contained in detrital apatite or precipitated inorganic phosphates. The bulk of P in clays is present as precipitated $Ca_3(PO_4)_2$ (Rankama & Sahama, 1950).

The Fe content of the coal ash is high and apart from the Fe contained in the

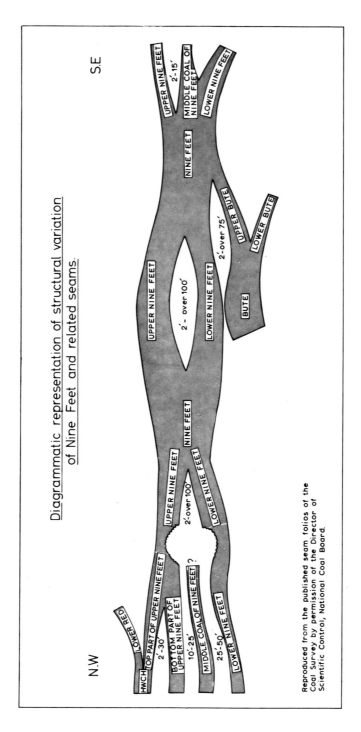

Diagrammatic representation of structural variation
of Nine Feet and related seams.

N.W

S.E

LOWER RED

HWCH

TOP PART OF UPPER NINE FEET

2'-30'

BOTTOM PART OF
UPPER NINE FEET

10'-25'

MIDDLE COAL OF NINE FEET ?

25'-50'

LOWER NINE FEET

UPPER NINE FEET

NINE FEET

2'-over 100'

LOWER NINE FEET

UPPER NINE FEET

2' - over 100'

LOWER NINE FEET

NINE FEET

2'-over 75'

UPPER BUTE

BUTE

LOWER BUTE

UPPER NINE FEET

2'-15'

MIDDLE COAL OF
NINE FEET

LOWER NINE FEET

Reproduced from the published seam folios of the
Coal Survey by permission of the Director of
Scientific Control, National Coal Board.

Fig. 53.   Diagrammatic representation of the splitting of the Nine Feet coal seam.

coal ash, both the Nine Feet and the Six Feet seams contain an ironstone band in the north west and centre of the area (National Coal Board, 1959; 1966). The source of the material forming these ironstones is not known but may be derived from iron-rich Old Red Sandstone rocks exposed around the perimeter of the coal basin.

There is a geographic variation of the elements and this very nearly coincides with the geographic variation of the rank of the coal (Fig. 57). The Cynheidre shaft coals were not analysed quantitatively for Fe but a comparison of Fe lines on the spectrographic plates showed no apparent increase with depth and thus the variation of Fe is thought to be geographic rather than a function of anthracitisation. The north west of the coalfield is richest in Fe and poorest in Al, and towards the east the Fe content decreases and the Al increases. The ironstone bands occur over the centre of the coalfield and die out towards the east. The Nine Feet seam has a further development of ironstone in the north west. Sedimentary studies of the Coal Measures in South Wales suggest a westerly flowing river system so that the ironstone band may be the result of a marine incursion affecting the seaward part of the delta. The increase of Al to the east represents the increase in clay fractions in that direction which also produces higher ash contents.

Organic fractions

In the organic fraction Al, P, and Ti show an enrichment when compared with the concentration of these elements in the inorganic fraction. P is an expected constituent of the organic fraction since it is present in living organisms. There is a high percentage of Al present and the amount is fairly constant whatever the Al content of the associated inorganic fraction. While it is possible that some of the Al represents clay material adhering to the organic particles, its amount suggests a more intimate association as clay-organic, or organic-metal chelates. Concentration of Al has been reported in the organic fractions of West Virginian coals (Headlee & Hunter, 1954) and Al is known to be accumulated by certain modern plants, notably Lycopodium (Hutchinson, 1945) which was an important member of the Coal Measure flora. The formation of metal-organic complexes by the interaction of plant organic matter and percolating waters is generally recognized together with displacement of more loosely held ions such as Ca^{2+} and K^+ (Robinson and Edgington, 1945).

The Ti enrichment shows no relationship with the amount of Ti in the inorganic fraction, neither is the correlation between Ti and Al as apparent as in the inorganic fraction. Ti is usually concentrated in the organic fraction of coals (Goldschmidt, 1954; Jones & Miller, 1939; Otte, 1953) and seems to be reactive with organic matter.

The sodium content is about the same as in the inorganic fraction and the K content lower but whereas the Na_2O/K_2O ratio of the inorganic fraction is approximately $0·8$, in the organic fraction it is nearer $1·1$.

The total Fe content is generally about 5% (as Fe_2O_3) but rises to $10–20\%$ where the inorganic content is highest due possibly to the availability of Fe in the solutions pervading the coal swamps and precipitation as Fe-organic complexes, generally secondarily as in the case of Al.

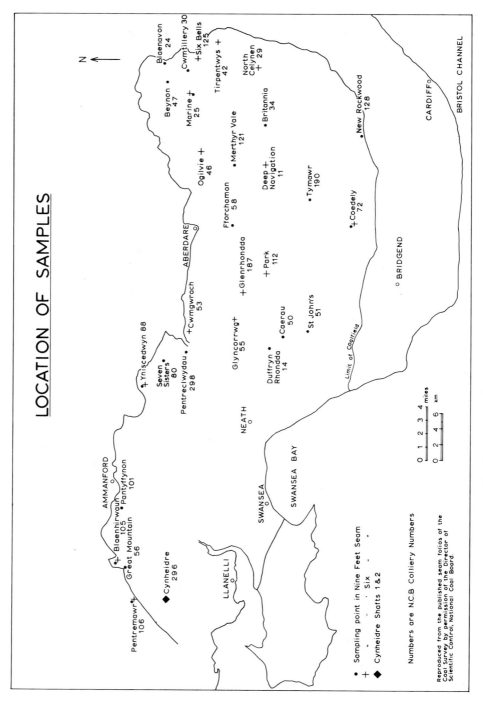

Fig. 54.   Sampling localities for Nine Feet and Six Feet seams.

TABLE 3

Nine Feet and Six Feet seam ash, and major element ranges and averages in the ash of organic and inorganic fractions

Weight %	Nine Feet				Six Feet			
	RANGE		AVERAGE		RANGE		AVERAGE	
*Volatile matter (d.m.m.f.)	4·8 –34·3		15·3		4·8 –29·9		15·5	
Total ash	2·59– 8·12		5·34		2·54–10·55		4·7	
Organic ash	1·85– 5·76		3·73		1·63– 7·40		3·6	
Inorganic ash	0·55– 3·39		1·65		0·39– 4·31		1·1	
	Organic	Inorganic	Organic	Inorganic	Organic	Inorganic	Organic	Inorganic
Al₂O₃	13·0–32·7	3·4–30·2	27·0	18·3	19·1–36·6	6·3–31·7	27·0	14·0
Fe₂O₃	3·5–15·8	5·9–29·7	9·6	15·9	2·0–15·9	1·3–31·9	6·9	11·9
MgO	1·1– 7·4	0·7–13·5	2·6	4·2	2·6– 9·0	0·8– 6·4	4·6	3·1
MnO	0·0– 0·8	0·0– 0·6	0·1	0·2	0·0– ·02	0·0– 0·4	0·1	0·2
CaO	1·4– 8·1	0·6– 3·8	4·8	1·6	6·6–17·2	1·7–10·7	10·2	6·2
Na₂O	0·4– 3·0	0·2– 4·5	1·1	1·3	0·4– 2·0	0·3– 1·0	1·0	0·6
K₂O	0·3– 2·2	0·2– 4·0	1·1	1·7	0·2– 2·0	0·4– 3·4	1·0	1·7
TiO₂	0·0– 1·9	0·1– 1·7	0·9	0·4	0·6– 1·5	0·1– 0·6	0·9	0·3
P₂O₅	0·3– 8·2	0·1– 3·7	3·0	1·3	1·0– 4·6	0·1– 1·6	2·1	0·7

*From N.C.B. data (with permission)

TABLE 4

Nine Feet and Six Feet seam trace element ranges and averages in the ash of organic and inorganic fractions

p.p.m.	Nine Feet				Six Feet			
	RANGE		AVERAGE		RANGE		AVERAGE	
	Organic	Inorganic	Organic	Inorganic	Organic	Inorganic	Organic	Inorganic
Be	13– 78	1·9– 10	35·5	4·3	3·5– 130	1– 7·5	39	4·4
B	75– 480	70 –270	244	128	75 – 440	60– 185	201	116
V	135– 475	35 –180	266	79	110 –1250	27– 540	345	124
Co	140– 500	20 –160	280	99	150 –1300	30– 160	448	66
Ni	360–1200	35 –460	741	195	440 –2000	35– 290	928	193
Cu	350– 710	120 –285	509	242	340 – 840	135– 340	492	215
Zn	100– 830	90 –970	287	323	100 –2300	100–1100	447	297
Ga	28– 71	11 – 41	50	20	23 – 120	10– 27	51	20
Ge	0– 22	<10	—	—	<10	<10	—	—
Mo	56– 565	9 –140	254	47	70 – 500	1– 320	211	53
Ag	—	1·7– 10	—	1·6	—	—	—	—
Sn	12– 73	16 –113	11	36	10 – 61	10– 70	28	29
Pb	100– 790	46 –532	338	186	205 –1310	60–1270	462	248

TABLE 5

Analyses of Cynheidre shaft coal ash (p.p.m.)

SEAM	Depth (Feet)	% Ash	B Total	Ni Total	Zn Total	Mo Total	Be Total	Be Org.	Be Inorg.	Cu Total	Cu Org.	Cu Inorg.	Sn Total	Sn Org.	Sn Inorg.	Pb Total	Pb Org.	Pb Inorg.
Lower Welsh	113 (34m)	1·95	145	760	870	1100	17·5	21·2	2·8	550	612	257	36	30·5	59·3	320	415	184
Red	157	6·07	80	750	345	100	19	26·4	3·6	660	780	405	32	24·5	48	465	564	251
Carway Fawr	548	6·09	100	535	325	140	19	26·5	3·6	360	550	182	40	—	—	365	442	191
Carway Fach	592	8·95	85	520	100	40	14·5	21·1	3·0	340	432	179	14	10·9	19·8	110	138	59
Drap	748	11·98	60	925	255	100	24	35·5	4·9	290	370	154	10	—	—	460	580	258
Soap	1032	4·78	74	470	195	200	13	17·3	2·1	435	520	216	28	22·2	42·5	180	216	95
Penny Pieces	1150	9·58	55	385	260	300	14	20·4	2·7	400	510	208	16·7	12·7	24	255	319	142
Big	1340	3·62	96	500	100	450	18	22·6	3·0	540	610	348	12	20·8	43	335	385	171
Green	1359	6·17	37	1100	210	85	24	33	4·4	520	635	263	3	24·5	48	200	241	108
Ddugalad	1427	5·29	51	800	900	100	14	18·9	2·6	540	660	270	10	—	—	390	467	208
Hwch	1437	10·05	50	340	275	240	18	24·4	3·5	400	505	210	75	56	109	160	199	89
Stanllyd	1551	7·45	48	700	150	630	15	22·9	2·9	480	620	207	10	—	—	220	274	121
Gras Uchaf	1566	2·60	27	620	300	300	52	60	7·9	700	770	320	17	10	—	880	965	427
Braslyd Fawr	1665	8·74	60	320	260	240	14	20·2	2·5	390	494	205	12	10	—	220	274	121
Braslyd Fach	1687	3·93	54	1500	450	260	90	117	15	710	840	349	36	14·8	28·7	1100	1283	574
Amman Rider	1748	24·10	48	450	220	40	10·5	—	—	210	—	—	14	—	—	240	—	—
Stinking Rider	1831	4·94	62	570	200	175	18	24·1	2·8	270	310	135	13	10·1	19·7	200	240	106
Gwendraeth	1850	5·03	37	550	620	155	26	35·2	4·0	770	930	385	27	22	41	620	740	331
Stinking Rider	1860	8·93	39	400	610	1100	24	34·9	4·7	540	690	230	130	91	192	610	760	341
Braslyd	1880 (573m)	3·86	62	1500	470	580	80	100	13·6	680	790	325	18	14·8	28·5	470	542	256

(b) Distribution of the minor elements over the Coalfield

Boron

B is usually enriched in coal ashes, although the amount varies. Goldschmidt

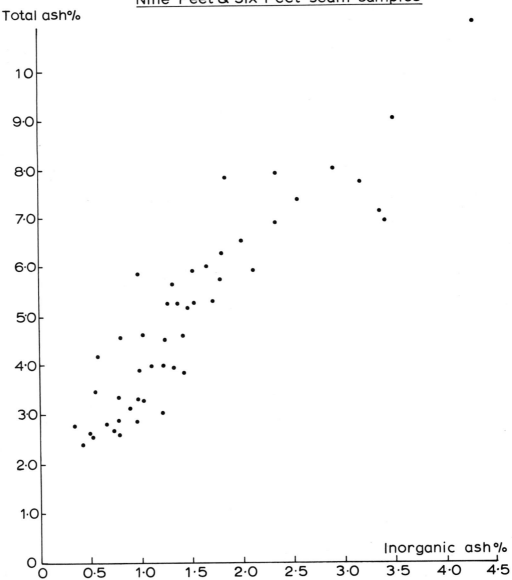

Fig. 55.   Inorganic *v*. total ash content of the Nine Feet and Six Feet seams.

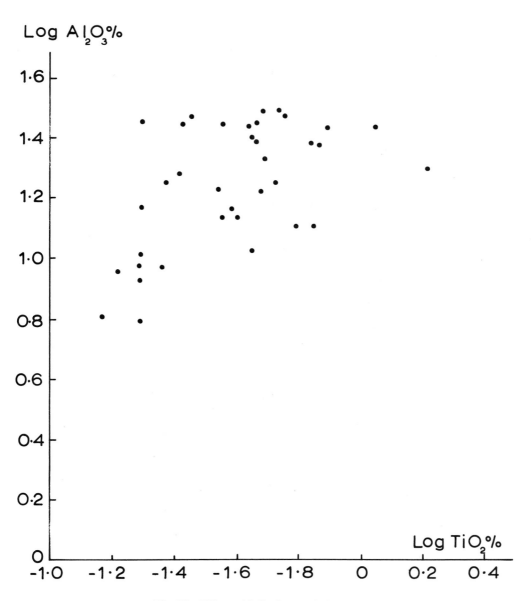

Fig. 56. TiO$_2$ v. Al$_2$O$_3$, inorganic fractions.

(1954) records an average of 600 p.p.m. but in South Wales coals values are well below this figure with a range of 60–27 p.p.m. in the inorganic, and 75–480 p.p.m. in the organic fractions. In many parts of the world B has been found to decrease in concentration with rank. This relationship has been reported by Roga (1958) in Polish coals, Kear & Ross (1962) in New Zealand and Zubovic, *et al.* (1960) in the United States who attribute it to the formation of organic esters which are lost as coalification proceeds. In South Wales B apparently decreases as the rank increases but the relationship is complicated by the fact that B follows Al (clay) in the inorganic fraction (Fig. 58) and the geographic distribution of the clay coincides with the distribution of rank.

In the organic fraction B can no longer be correlated with Al but there is still a decrease in B with increasing rank with the exception of a few samples of the Nine Feet seam from the anthracite zone which are associated with an ironstone band. Although this may be due in part to the availability of B, it is likely that the B in the organic fraction was driven off as coalification proceeded. This fact is supported by the Cynheidre coals which show a general decrease in B with depth (Fig. 59).

Although the Nine Feet and Six Feet seams contain similar amounts of B in their inorganic fractions, the organic fraction of the Nine Feet (average 244 p.p.m.) is richer than that of the Six Feet (average 201 p.p.m.). Assuming a constant supply of B to both, the Nine Feet seam *should* contain less B since it lies below the Six Feet seam and therefore has a higher rank at any locality. Since the values show the reverse there appears to have been more B available during Nine Feet seam times which implies a greater marine influence. Swaine (1962) found that the B content of New South Wales coals increased as the surrounding sediments changed from fresh water to marine in character, and sedimentary evidence from Pembrokeshire shows that the lower part of the Lower Coal Series below the Timber Vein (correlated with the Nine Feet seam by George & Trueman, 1925) contains more marine bands than the Upper part. It was a less stable area and therefore more likely to suffer marine incursions (Williams, 1966).

Ironstone bands occur in both the Nine Feet and Six Feet seams in the centre of the coalfield from the Taff valley to the Neath valley; the Nine Feet seam ironstone band extending further to the north west. This could indicate greater marine influence during Nine Feet seam times but there is no evidence to establish the fact since there is no marine band near either seam.

Beryllium

Be is greatly enriched in the organic fraction of the coals. It ranges from 3·5–130 p.p.m., average about 37 p.p.m. in the seam averages. The inorganic part rarely contains more than 7 p.p.m. Be may have been accumulated in the original environment either by the living fauna or flora or after death. The other alternative is the introduction of Be by mineralisation as may have occurred in the case of Cu, Pb, Zn and Sn. Since Be cannot be correlated with the rank of the coal the latter possibility is doubtful. However, in the Cynheidre coals the highest values occur in the highest rank coals in the lowest part of the sequence, although no direct correlation can be made with depth. Thus some Be may have been introduced this way,

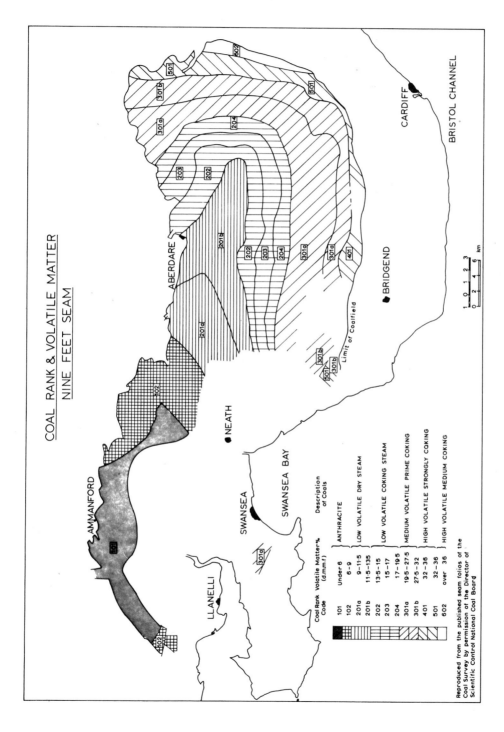

Fig. 57.   Map of the coal rank and volatile matter in South Wales coals.

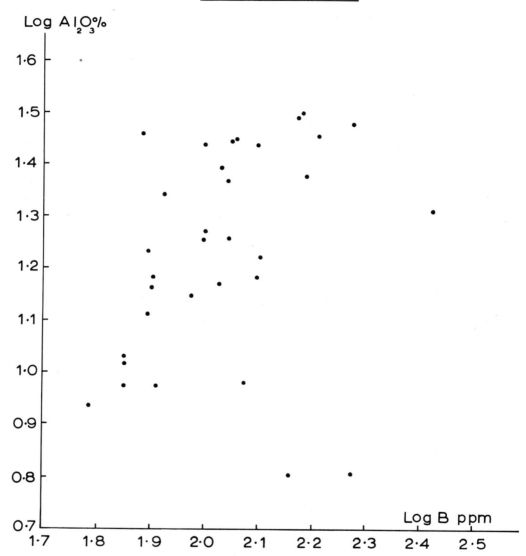

Fig. 58.   B v. Al_2O_3, inorganic fractions.

but the bulk of the Be was probably present in the original environment.

In the inorganic fraction Be shows some relationship to Al (Fig. 60). Since its chemical properties are similar this is not surprising but in the organic fraction Be no longer follows Al and instead shows some relationship with P (Fig. 61) and may occur as a phosphate. Vinogradov (1959) found Be in soils as the phosphate and Krauskopf (1955) notes that Be appears to be dependent on P in phosphorites, although the mode of occurrence is unknown. Al is known to fix phosphates in modern soils (Hutchinson, 1945) and Be may have precipitated by a similar mechanism although P does not correlate with Al in either the inorganic or organic fractions. The Be could therefore have been enriched by concentration in organic matter and

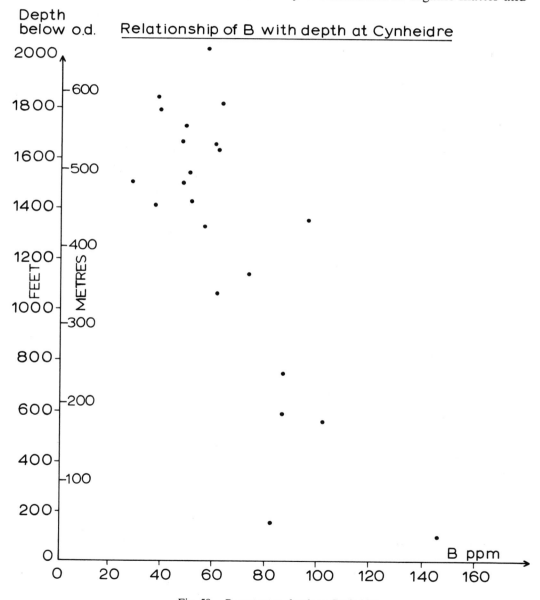

Fig. 59. B content v. depth at Cynheidre.

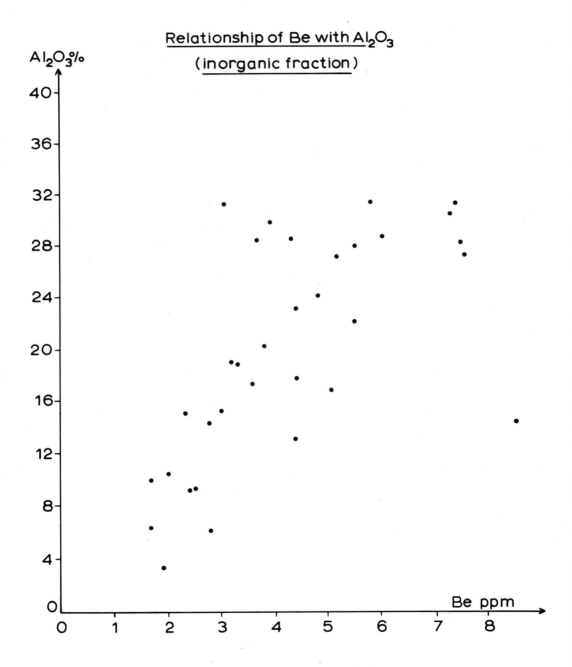

Fig. 60. Be *v.* Al₂O₃, inorganic fractions.

fixed by phosphorus or may have been present in original phosphorus-bearing organisms.

Germanium

Aubrey (1952) found very low concentrations of Ge in South Wales coals (average 2 p.p.m.) and this is confirmed for the Nine Feet and Six Feet seams by the present study. Ge concentrations below 10 p.p.m. could not be analysed. The few samples with higher concentrations than this were found in some organic fractions.

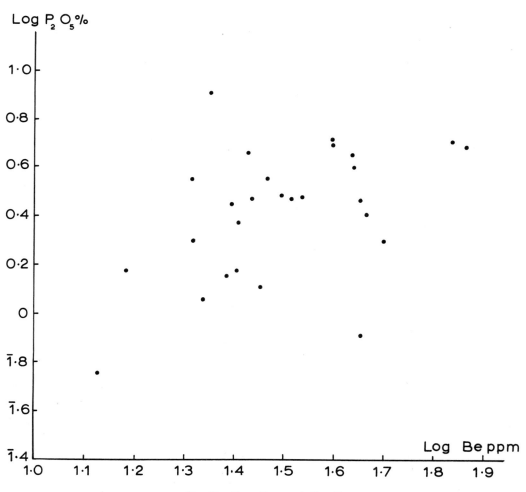

Fig. 61. Be *v.* P, organic fractions.

Ge is one of the elements which is consistently recorded in the organic matter of coals. Ge generally decreases with increasing rank as in the Soviet Union (Losomov, *In:* Roesler, 1963) and in Hungary (Szadeczky-Kardoss & Vogl, 1955). This could explain the low values in South Wales but samples with detectable amounts are not confined to the low rank coals, but occur in the anthracite zone as well. It seems likely that the whole coal-forming basin was extremely poor in Ge since the concentrations are well below the 8 p.p.m. in bright coals which is quoted as a world average (Bethall, 1962).

Gallium

Ga follows Al in both the inorganic and organic fractions of the coal ash (Fig. 62) and was undoubtedly introduced into the coal-forming basin along with Al so that the geographic distribution of these elements is identical. Ga has been found associated with organic matter by Dalton and Pringle (1962), Otte (1953) and Ryczek (1954), but since Al is itself associated with the organic fraction the pattern is rather complicated. Bethall (1962) suggested that the Ga/Al ratio would indicate whether the Ga was originally bound or not. In the present inorganic fractions the ratio averages $2 \cdot 15 \times 10^{-4}$ which is typical of a clay sediment but in the organic fraction the ratio increases to $3 \cdot 5 \times 10^{-4}$ so that the Ga is preferentially adsorbed or precipitated by organic matter relative to Al either primarily or by secondary mineralisation (compare Al). This is the opposite effect to that noted by Butler (1953) in Swedish coals where Al accumulated at the expense of Ga.

Molybdenum

Mo is greatly enriched in the organic fraction with up to fifty times the amount in the inorganic fraction and an average enrichment of about seven times. The range of values is 1–320 p.p.m. in the inorganic and 56–565 p.p.m. in the organic fractions.

The inorganic Mo can be related to the inorganic Fe (Fig. 63) and was probably precipitated as the sulphide MoS_3 on finely dispersed FeS_2 as suggested by Korolev (1958). This would explain the high values of Mo since averages for sedimentary rocks are usually lower. Mo shows little association with Fe in the organic fraction and may form various metal–organic complexes.

Mo shows some geographic variation with the highest concentrations in the west, both in the Nine Feet and Six Feet seams, although there are a few high values in the centre of the basin and in the south east. In the Cynheidre coals the values are high although there is no apparent increase or decrease with depth. The source of the Mo could either be from below by mineralisation as suggested for Cu, Zn, Pb and Sn (p. 256), or by the usual weathering processes, or both. Mo ores usually represent a very small part of a mineralisation field so that if the values for Cu and Pb are of the order of 800 p.p.m. and 1500 p.p.m. (Table 4), the concentrations of Mo are too high to consider them as having been introduced in the same way. The Mo may have come from the same source as the Fe and Mn which could be the Old Red Sandstone. According to Hauptmann and Bakoni (*In:* Goldschmidt, 1954) Mo accumulates especially in Mn-rich oxidate sediments.

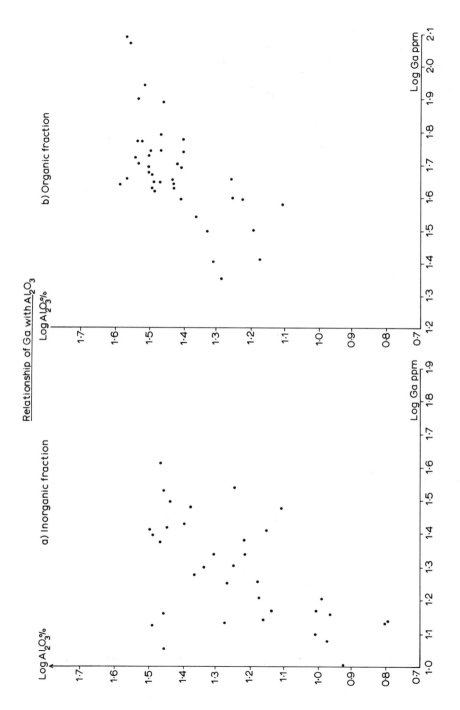

Fig. 62.   Ga *v.* Al$_2$O$_3$, inorganic fractions (a) and organic fractions (b).

Cobalt and Nickel

Both Co and Ni have been found concentrated in the organic matter by several authors (Otte, 1953) and South Wales is no exception. Ni always occurs in greater abundance than Co but the two elements are found in fairly close association (Fig. 64). The ranges of values and the Co/Ni ratio is typical of a hydrolyzate sediment (Rankama & Sahama, 1954) and suggests that the source of Co and Ni was from

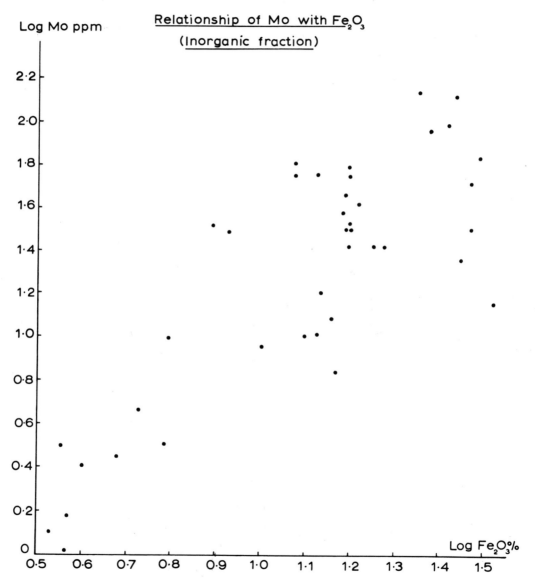

Fig. 63.   Mo *v.* Fe$_2$O$_3$, inorganic fractions.

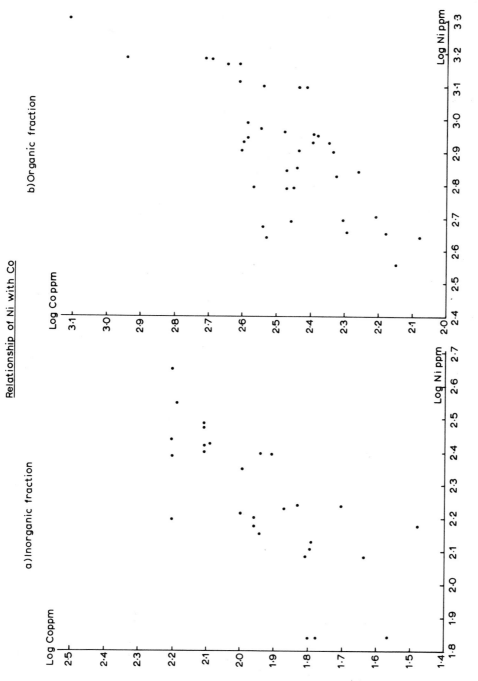

Fig. 64. Ni v. Co, inorganic fractions (a) and organic fractions (b).

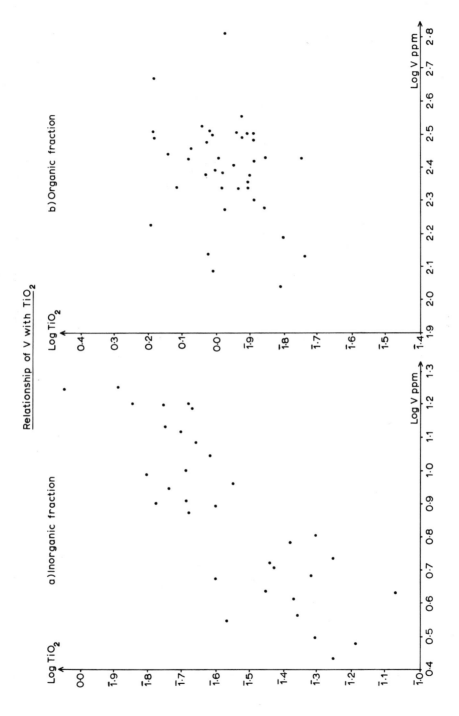

Fig. 65. V *v.* TiO$_2$, inorganic fraction (a) and organic fraction (b).

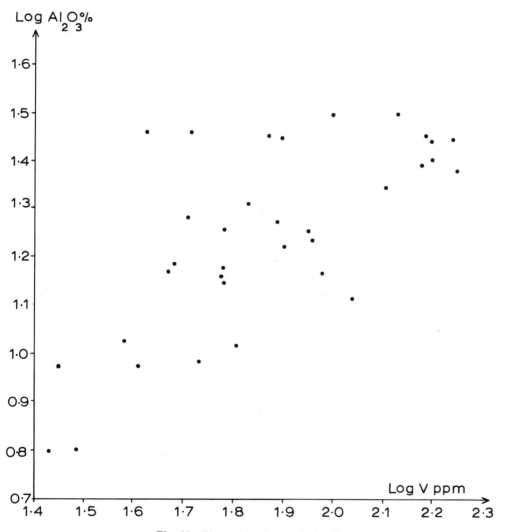

Fig. 66.   V v. Al$_2$O$_3$, inorganic fractions.

clays rather than oxidate sediments. Co and Ni show some geographic variation with highest values in anthracites of the north west, but elsewhere the concentration is sporadic.

Vanadium

Coals usually show high concentrations of V and average 60–100 p.p.m. (Bethall,

1962). In South Wales V closely follows Ti in the inorganic fraction and to a lesser extent in the organic fraction (Fig. 65) and, like Ti, can be related to Al in the inorganic fraction where it is probably incorporated in the clays (Fig. 66). In the organic fraction the Ti/V ratio is similar to that in the inorganic so that the enrichment factor is the same for both elements, V showing a very uniform distribution over the coalfield.

Copper, Lead, Zinc, Tin and Silver

These elements are considered together since they show similar distributions in the coal ashes and there seems to be a definite relationship between them. Cu and Pb are both enriched in the organic fractions and there is a fairly constant ratio between their concentrations in the organic and inorganic fractions; 2·4 for Cu and 2·25 for Pb. Sn is concentrated in the inorganic fraction but also occurs in the organic fraction, the ratio between the two being 1·9, but in samples from the south east the Sn content is usually below 10 p.p.m. Zn shows no preference for the organic or the inorganic fraction and can occur in high concentrations in either. Ag was found in a few samples but with a maximum concentration of only 5 p.p.m. and always in the inorganic fraction.

The highest concentrations of Cu, Pb, Zn, Sn and Ag occur in the N.W. in the anthracite zone. A graph of the concentration against the rank of the coal, as typified by the volatile matter (Fig. 57), shows a general although weak relationship between them. The scatter of points is to be expected since the concentration of these elements is probably due to more than one factor (Figs. 67 and 68). It may be related to primary accumulation by plants, or precipitation by percolating solutions at the peat stage while the geographical position of the plants themselves might affect the amounts of these metals available. Sn showed little relationship but samples from the north west have the highest values whereas those from the south east are often less than 10 p.p.m.

In a series of anthracite coals from the Cynheidre shaft the concentrations of these elements were plotted against depth and, in spite of some anomalies at the top of the sequence, there is some relationship with depth (Figs. 69 and 70). The values used to plot these graphs were corrected (p. 228) since the coals were not fractioned and better comparison could be made between coals of differing ash content. The increase of these elements with depth could be due to a decreasing ability of the plants to accommodate these elements as organo-metallic compounds, but, although there was a gradual evolution of the plants through Coal Measures times, there was no great change of the commonest forms in the assemblages (Dix, 1933).

Perhaps more important is the ratio between the pairs of elements Cu and Sn, Pb and Zn; both pairs showing a relationship in which Sn and Zn increase downwards (Fig. 71).

Ores of these elements outcrop in rocks of Ordovician age at the present time to the north of the coalfield near Llandovery. These are Pb and Zn ores normally found in the outer zones of mineralisation so that it is doubtful whether Cu and Sn, if they existed, were being eroded from this area during Carboniferous times.

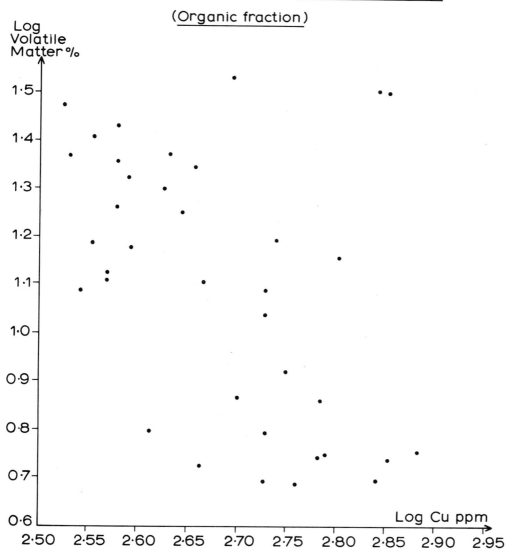

Fig. 67.   Cu *v.* volatile matter (d.m.m.f.), organic fractions.

Finlayson (1910) has suggested that South Wales was affected by Amorican mineral-isation and that the ore solutions utilised Caledonoid structures. The variation of Cu, Pb, Zn, Sn and Ag with coal rank in the Nine Feet and Six Feet seams could indicate a geographic variation coinciding with the distribution of rank, but in the Cynheidre coals the concentration of these elements would increase in coals higher in the sequence, and the ratios Cu/Sn and Pb/Zn would be reversed as the ores of the source area were eroded. This would be similar to the situation in Pennsylvania where a series of coals have been described by Zubovic, *et al.* (1960) in which the minor element concentration reflects the erosion of a deep seated igneous complex.

The Cu/Sn and Pb/Zn ratios can be compared with the Armorican mineralisation of South West England where two sets of mineral veins occur, Cu-Sn ores and Pb-Zn ores (Hosking, 1962) and here the ratios increase with depth as in the coals (Dewey, 1948). Armorican mineral deposits are also recorded from Central and North Wales and Cu-Pb mineralisation occurs in Carboniferous Limestone at Mynydd y Garreg near Kidwelly in Carmarthenshire. Pb ores are present in the south east of the coalfield between Rudry and Mynydd Machen in rocks of Keuper age (Thomas, 1961).

From the distribution of ore centres in South West England, Hosking (1962) suggested that emanative centres developed where well defined NE-SW trending folds were intersected by those with equally strong E-W trends. In South Wales a similar situation is present in which the Caledonoid Towy anticline trending NE-SW converges with E-W Armorican folds, providing a potential emanative centre just NW of the coalfield (Fig. 72).

Anthracites are generally found to contain smaller concentrations of minor elements than other coals since they are less reactive after anthracitisation due to an increased carbon content and decrease in hydrogen and oxygen. Roesler (1963) states that adsorption mechanisms are still possible at the beginning of the formation of bituminous coals, but it is doubtful whether adsorption can occur after this stage. The elements Cu, Pb, Zn, Sn and Ag may therefore have been introduced into the coal at a fairly early stage of development and not later than the formation of the bituminous coal.

It is concluded that the source of Cu, Pb, Zn, Sn and Ag may have been from below and that part of the coalfield lay in a zone of mineralisation which was effective during post-Carboniferous Armorican times when the coals were at an early stage in their diagenetic development.

Seam correlation

This study has not indicated any definite possibility of geochemical seam correlation, although there are some elements which may be useful. Average B in the organic fractions is higher in the Nine Feet seam than the Six Feet while the reverse is true of V, Co, Ni and Zn. The spread of values is large however and the lateral variations in element concentrations in the seams is greater than their differences between seams.

One exception is the Hwch seam from Cynheidre (equivalent to the Nine Feet) which has a high Mn content compared to the other seams at this locality. Using elements such as Cu which increase with depth at Cynheidre, it is possible to indicate whether seams are high or low in the sequence but it is impossible to isolate any particular seam.

V. METAMORPHISM OF THE SOUTH WALES COALS

The cause of anthracitisation in South Wales has not been explained satis-

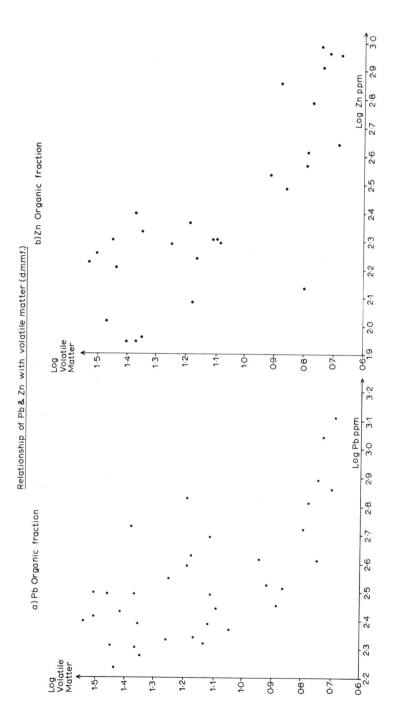

Fig. 68.   Pb and Zn *v.* volatile matter (d.m.m.f.), organic fractions.

factorily, although there are three main theories: (a) primary differences in the coal or its environment, (Strahan & Pollard, 1915; Fuchs, 1946; Mackenzie-Taylor, 1926, 1928), (b) pressure due to load or earth movements (Trotter, 1948), (c) heat due to pressure or magmatic intrusion.

The geological evidence suggests a relatively uniform type of environment over the whole coalfield, hence anthracite cannot be due to primary differences in the coal itself. Although Hilts Law holds at any point in the coalfield, it does not explain the regional distribution of coal rank, unless a thicker cover of sediments now eroded is postulated to the north west (Bailey, 1944). Indeed, substantial differences in thickness of the Coal Measures between north and south show the reverse effect. Furthermore, studies of coals in Germany show that pressure opposes the chemical processes of anthracitisation rather than promoting them, and many examples occur along the thrusted areas of the Hercynian and Alpine forelands in which the coal rank has been unaffected by strong earth movements (Teichmuller & Teichmuller, 1965). Anthracite is undoubtedly produced by heat, but whether the heat generated by earth movements is sufficient to convert coals to anthracite grade is questionable. In the case of the German coals frictional heat would be dissipated too quickly to have any effect on the coals. The earlier nineteenth century theories of heat from magmatic intrusions have been discarded from lack of proof.

Although pressure can cause changes in low rank coals by the elimination of moisture and reduction of pore space, it opposes the chemical processes which are more important in coals which have passed the lignite stage, (Teichmuller & Teichmuller, 1965). For this reason it is generally accepted that coals are thermally metamorphosed either by heating increased by geothermal gradient, earth movements, magmatic intrusions, or depth of burial (Francis, 1961).

The coals of the Pennsylvanian coalfield become gradually more anthracitic as the Appalachian tectonic region is approached, also the rocks become thicker in this direction, and the anthracites are thus attributed to an increased geothermal gradient (Francis, 1961). The heat due to earth movements is not always great enough to produce the highest rank in coal, as is shown in the German coalfields of the Ruhr and Osnabruck (Teichmuller, 1948). Here the anthracites are not associated with the areas of intense folding and faulting.

As in the Pennsylvanian coalfield, the South Wales coals become more anthracitic towards the area where the earth movements were most intense. But it must be pointed out that in Pennsylvania the change from bituminous grade to anthracite occurs over a distance of 250 miles, whereas in South Wales this same change occurs in less than 10 miles. Furthermore, although the thickest strata of the Upper Coal Measures lies in the west rather than in the east, it thickens to the south west rather than to the north west. Yet there is a rapid decrease in rank from north west to south west, the values for the Nine Feet seam ranging from less than 6% near Ammanford to more than 20% near Gower, a horizontal distance of only 10 miles. The thickness of strata thus offsets the effect of earth movements. If the earth movements were intense enough for the heat over a period of time to form anthracites in the north west of the area, it is difficult to understand why the south and south east were virtually unaffected.

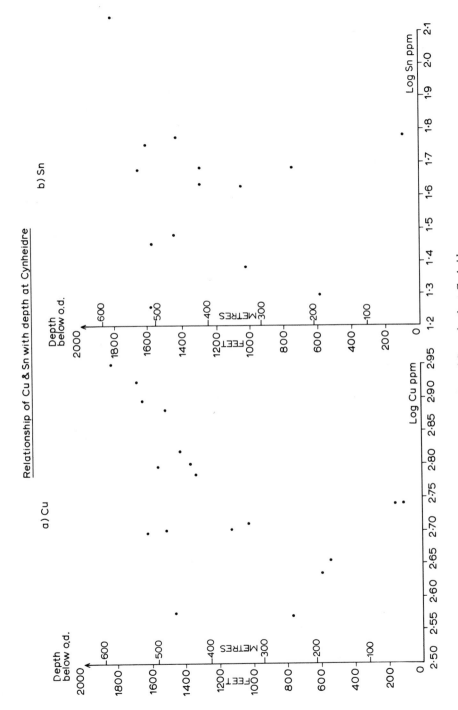

Fig. 69.   Cu and Sn v. depth at Cynheidre.

Coals are very susceptible to thermal metamorphism as the organic material is very reactive. Opinion varies as to the minimum temperature required to convert peat to anthracite but temperatures as low as 150°C have been quoted, (Francis, 1961) while the duration of heating is also important (Teichmuller & Teichmuller, 1965). Loss of volatiles such as Ge and B are indicative of heat increases. In this area Ge is very much lower than average and this may be due to the influence of heat; also a relationship has been shown between the rank of the coal and B showing loss of B in areas of high rank.

Since the organic matter of coal may absorb metals to form organic complexes, a regional mineralisation might be detectable in coals even though there were no indications in the surrounding sediments. The suite of metals that shows a correlation with rank, both vertically and horizontally, suggests a regional mineralisation from below. The timing of this mineralisation was important since if the coals were already converted to their present rank before mineralisation started, coals with the high carbon content would not be expected to react with metals as readily as coals with a higher percentage of hydrogen and oxygen. Hence the concentration of metals would be greatest in low rank coals (Szadecsky-Kardoss & Vogl, 1955). The mineralisation must therefore have occurred fairly soon after burial of the coals and can thus be correlated with the Armorican mineralisation of South West England and Mid Wales, and the local deposits of Pb/Zn ores which occur in the Carboniferous Limestone.

The source of the metals is questionable. The contours of mineralisation indicate a focus somewhere to the north west of the coalfield. The metals present indicate a granitic source comparable to the South West England granites, but if it occurs it is very deep since it is not detected by geophysical evidence (Bullerwell, 1965, personal communication) and a siltstone taken from a borehole near Cynheidre, Carmarthenshire, 3,000 feet (914 m) from the surface appears to show no sign of metamorphism, so that any existing magmatic intrusions would be below this. On the other hand numerous mineral deposits occur in Wales with no detectable granite below them.

Recent geophysical surveys of the area are inconclusive. The gravity anomaly that exists can be accounted for by the Carboniferous sediments, and the magnetic anomalies are associated with the tectonic disturbances at Carreg Cennen and the Vale of Neath (the granites of South West England show no magnetic anomaly). Even the increased temperature gradient can be explained as indicating that the heat flow has not returned to equilibrium after the last Ice Age (Bullerwell, 1965, personal communication).

The temperatures involved in the development of Cu, Pb, Sn, Zn and Ag in the coals are unknown, but the correlation between metals and the distribution of coal rank seems to be significant. It seems likely that the geothermal gradient in the area was above normal since there were earth movements in progress which were probably the primary cause of the anthracitisation, but heat accompanying the mineralisation seems to have been responsible for the final conversion into anthracite.

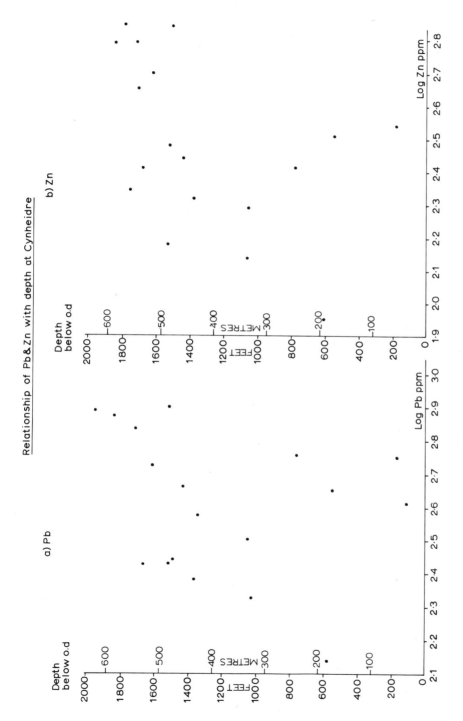

Fig. 70.   Pb and Zn v. depth at Cynheidre.

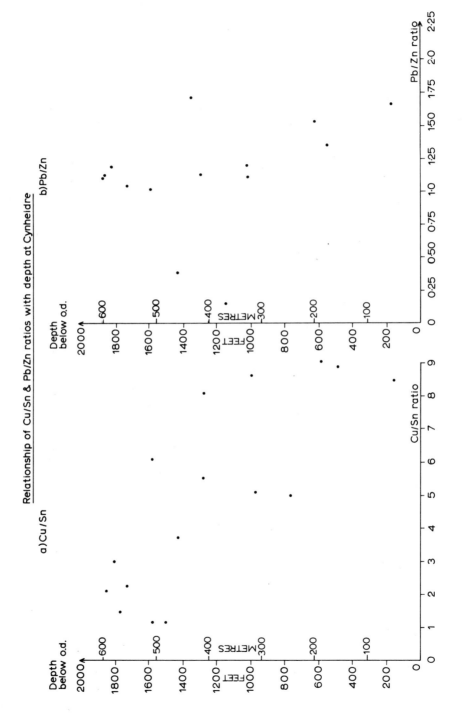

Fig. 71. Cu/Sn and Pb/Zn ratios *v.* depth at Cynheidre.

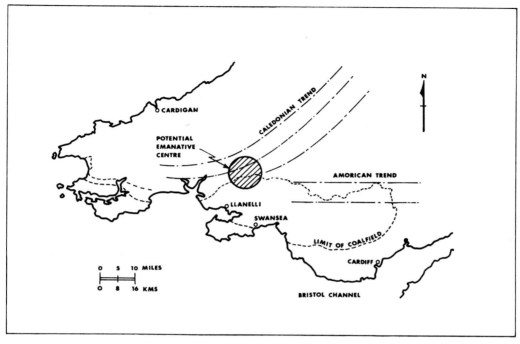

Fig. 72.   Position of a possible emanative centre on the N.W. edge of the coalfield.

VI.   ACKNOWLEDGMENTS

The authors wish to acknowledge the encouragement and help of Professor F. H. T. Rhodes and Mr. T. R. Owen. The samples on which the study was based were kindly provided by the National Coal Board (South Western Division) and we would like to thank in particular Mr. H. F. Adams, and Mr. S. R. Crook for the information they have supplied and for their generous co-operation and interest. Several diagrams are reproduced from the published seam folios of the Coal Survey by kind permission of the Director of Scientific Control, National Coal Board. Thanks are also due to Mr. T. R. Rees and other members of the technical staff of the Department of Geology, University College, Swansea, and to Mrs. J. Waring for typing the manuscript. The study was made possible by a grant from the Science Research Council for which the senior author is most grateful.

VII.   REFERENCES

ADAMS, H. F. 1956. Seam structure and thickness in the South Wales Coalfield. *Trans. Inst. of Min. Eng.*, **115**, II.
AHRENS, L. H., and TAYLOR, S. R. 1961. Spectrochemical Analysis (Pergamon Press, London).
AUBREY, K. V. 1952. Germanium in British Coals. *Fuel*, **31**, 429.
BAILEY, E. 1944. The natural resources of Great Britain: Minerals. *J. Roy. Soc. Arch.*, **42**, 538.
BETHALL, K. V. 1962. The distribution and origin of minor elements in coal. *Brit. Coal Util. Res. Ass. Monthly Bull.*, Pt. II, Review No. 217, 26, No. 12.
BRITISH STANDARDS. 1957. Proximate analysis of coal. 1016, Pt. 3.

BRITISH STANDARDS. 1960. The sampling of coal. No. 1017, Pt. 1.

BUTLER, J. R. 1953. Geochemical affinities of some coals from Svalbard. *Norske Polarinst. Skrifter*, **96,** 1.

DALTON, I. M., and PRINGLE, W. J. S. 1962. Gallium content of some Midland coals. *Fuel*, **41,** 41.

DEWEY, H. 1948. South West England. *Brit. Reg. Geol.*, 2nd Edn., H.M.S.O.

DIX, E. 1933. The sequence of floras in the Upper Carboniferous with special reference to South Wales. D.Sc. thesis. University College, Swansea. (University of Wales).

FINLAYSON, A. M. 1910. Metallogeny of the British Isles. *Quart. J. geol. Soc. Lond.*, **66,** 281.

FRANCIS, W. 1961. Coal—its formation and composition. (Edward Arnold, London).

FUCHS, W. 1946. Origin of coal and change of rank in coalfields. *Fuel in Sci. and Pract.*, **25,** 132.

GEORGE, T. N., and TRUEMAN, A. E. 1925. Notes on the Coal Measures of East Pembroke-shire. *Proc. S. W. Inst. of Engs.*, **12,** 409.

GOLDSCHMIDT, V. M. 1954. Geochemistry. (University Press, Oxford).

HEADLEE, A. J. W., and HUNTER, R. G. 1955. Characteristics of minable coals of West Virginia. *West Virginia Geol. Surv.*, **13A,** Pt. V, 36.

HOSKING, K. F. G. 1962. The relationship between the primary mineralisation and the structure of S. W. England. *In:* Some Aspects of the Variscan Fold Belt (Manchester University Press).

HUTCHINSON, G. E. 1945. Al in soils, plants and animals. *Soil Sci.*, **60,** 29.

JONES, J. H. and MILLER, I. M. 1939. Concentration of trace elements in vitrains. *J. Soc. Chem. Ind.*, **58,** 237R.

JONES, J. H., and MILLER, I. M. 1939. Concentration of trace elements in vitrains. *J. Soc. Chem. Ind.*, **58,** 237R.

KEAR, D. and ROSS, J. B. 1961. Boron in New Zealand coal ashes. *N.Z. Journ. Sci.*, **4,** No. 2, 360.

KOROLEV, D. F. 1958. Role of iron sulphides in the process of accumulation of Mo in sedimentary rocks of the reduction zone. *Geokhimiya*, 359.

KRAUSKOPF, K. B. 1955. Sedimentary deposits of rare elements. *Econ. Geol.*, 50th Anniversary Vol., Pt. 1, 411.

MACKENZIE-TAYLOR, E. 1926. Base exchange and its bearing on the origin of coal. *Fuel in Sci. and Pract.*, **5,** 195.

MACKENZIE-TAYLOR, E. 1928. Base exchange and the formation of coal. *Fuel in Sci. and Pract.*, **7,** 230.

NATIONAL COAL BOARD. 1959. South Wales Coalfield—seam maps, Nine Feet seam. (Britannic Litho. Co. Ltd., London).

NATIONAL COAL BOARD. 1966. South Wales Coalfield—seam maps, Six Feet seam. (Britannic Litho. Co. Ltd., London).

OTTE, M. 1953. Trace elements in some German mineral coal. *Chem. Erde.*, **16,** 287.

RANKAMA, K., and SAHAMA, Th. 1950. Geochemistry (University of Chicago Press).

ROBINSON, W. O., and EDGINGTON, G. 1945. Minor elements in plants, and some accumulator plants. *Soil Sci.*, **60,** 15.

ROESLER, H. J. 1963. Neurere Arbeiten und Erkenntrisse uber. *Bergakademi*, **15,** No. 2, 77.

ROGA, B. 1958. Tests on B content in Polish coals. *Prace Glownego Inst. Gornictwa*, Ser. B, No. 212.

RYCZEK, M. 1959. Occurrence of Ge and Ga in Bituminous Coal. *Przeglad Gorniczy*, **9,** 420.

SHAPIRO, L., and BRANNOCK, W. W. 1956. Rapid analysis of silicate rocks. *U. S. Geol. Surv. Bull.*, **1036**—C.

STRAHAN, A., and POLLARD, W. 1915. The coals of South Wales, with special reference to the origin of anthracite. *Mem. geol. Surv. Gt. Britain*.

SWAINE, D. T. 1962. Trace elements in New Zealand. *C.S.I.R.O. Tech. Comm.*, No. 45, Pt. II.

SZADECZKY-KARDOSS, E. and VOGL, M. 1955. Geochemical investigation on ashes of Hungarian coals. *Foldtani Kozlony*, **85,** 7.

TEICHMULLER, M., and TEICHMULLER, R. 1965. The diagenesis and metamorphism of coal. Preprints of 13th Inter-University Geol. Congress, Newcastle-upon-Tyne.

THOMAS, T. M. 1961. The mineral wealth of Wales and its exploration. (Oliver and Boyd, Edinburgh and London).

TROTTER, F. M. 1948. The devolatilization of coal seams in South Wales. *Quart. J. geol. Soc. Lond.*, **104**, 389.

VINOGRADOV, A. P. 1959. The geochemistry of rare and dispersed chemical elements in soils (Consultants' Bureau, New York).

WILLIAMS, P. F. 1966. Unpublished Ph.D. thesis. University College, Swansea (University of Wales).

ZUBOVIC, P., STADNICHENKO, T. and SHEFFEY, N. B. 1960. Relation of the minor element content of coal to possible source rocks. *U.S. Geol. Surv. Bull.*, **400—B**, 87.

GEOCHEMICAL PARAMETERS FOR DISTINGUISHING PALAEOENVIRONMENTS IN SOME CARBONIFEROUS SHALES FROM THE SOUTH WALES COALFIELD

T. W. Bloxam

I. INTRODUCTION

A study of shales belonging to the *Gastrioceras subcrenatum* marine band and associated rocks from the South Wales Coalfield has provided mineralogical and geochemical parameters which can be correlated with changing depositional and faunal environments. The initial investigation (Bloxam & Thomas, 1969) commenced with a faunal study, the results of which established four distinct faunal assemblages or phases representing the effects of changing environments related to marine transgression and regression. Details of sampling and faunal descriptions will be found in Bloxam & Thomas (1969).

The faunal phases are similar to those recognised by Calver (1968) in the Pot Clay marine band (*G. subcrenatum*) of the Pennines and East Midlands coalfields, although our system of numbering are in the reverse order:

Phase:	of Bloxam and Thomas (1969)	Phase:	of Calver, 1968
1	Unfossiliferous—probably freshwater	–	—
2	Marine brackish—*Lingula B, Nuculopsis*	3	*Lingulas-Serpulids*
3	Marine inshore—*Lingula A, Goniatites*	2	*Goniatites-Pectinoids-Bivalves*
4	Marine offshore—*Goniatites, Bivalves* (deepest water)	1	*Goniatites-Pectinoids* (deepest water)

The significance of these four phases is confirmed by the fact that they can be recognised (faunally and geochemically) in *G. subcrenatum* shales from three widely separated localities (Bloxam & Thomas, 1969). The *Lingula* in phase 2 and 3 differ, the larger form (*Lingula A*) being associated with a marine fauna in phase 3 and the smaller form (*Lingula B*) with a shallow or brackish fauna in phase 2. Hence phases 1 to 4 progress in order of their increasing distance from the shoreline.

The shales are very similar mineralogically and contain detrital quartz, clay, collophane, pyrite and organic material. Quartz varies from 11% in offshore shales of phase 4 to 56% in some silty shales of phase 1. The clay mineralogy is relatively constant and dominated by illite (about 85% of total clay) which is reflected in the rather uniform value for K_2O in all samples. Unless element *ratios* are used, chemical comparisons between shales require correction for the varying amounts of 'inert' quartz present. The quartz corrected means and standard deviations of chemical analyses of phases 1 to 4 are given in Table 6 for a total of 63 samples from all localities.

Table 6: Quartz corrected means and standard deviations for the elements grouped into their respective phases.

Phase	1		2		3		4		Total	
Wt. %	Mean	Standard deviation	Mean	Standard deviation	Mean	Standard deviation	Mean	Standard deviation	Mean	Standard deviation
Fe^{3+}	1·65	0·39	1·74	0·41	1·70	0·72	1·31	0·91	1·62	0·70
Fe^{2+}	3·89	0·67	4·93	0·90	3·89	1·54	5·12	0·58	4·26	1·29
Mg	1·57	0·15	1·58	0·15	1·39	0·21	1·67	0·57	1·50	0·32
Ca	0·88	0·55	1·16	0·40	0·42	0·30	2·67	1·72	1·05	1·18
Na	1·09	0·34	0·84	0·35	0·58	0·20	0·35	0·15	0·68	0·36
K	4·29	0·52	4·37	0·37	4·61	0·22	3·93	0·44	4·39	0·45
P	0·07	0·01	0·09	0·01	0·07	0·02	0·26	0·19	0·11	0·11
CO_2	0·74	0·92	1·37	0·88	0·17	0·46	3·34	3·39	1·04	1·99
C (organic)	0·85	0·30	0·91	0·25	1·03	0·20	5·31	2·67	1·79	2·08
S	0·32	0·25	1·41	0·70	0·81	0·58	1·95	0·46	1·00	0·76
Quartz	35·37	8·02	29·34	4·81	25·44	5·61	19·05	3·45	26·77	7·86
U ppm	4·6	1·8	6·0	1·4	6·3	1·5	15·2	5·4	7·6	4·6
Th ppm	14·3	2·6	14·3	1·4	13·4	1·9	11·2	2·1	13·3	2·3
F ppm	948·0	139·0	1030·0	45·0	960·0	77·0	1403·0	425·0	—	—
Th/U	4·0	2·8	2·5	0·5	2·2	0·7	0·8	0·4	2·3	1·7
Q/C	85·5	46·6	48·0	12·0	35·0	11·5	6·4	4·1	41·7	34·5
Na/K	0·26	0·10	0·20	0·08	0·13	0·04	0·09	0·03	0·16	0·09
Fe^3/Fe^2	0·45	0·15	0·37	0·14	0·66	0·64	0·26	0·18	0·50	0·48
F/P_2O_5	0·55	—	0·47	—	0·62	—	0·31	—	—	—

Note: All elements except C (organic) are corrected for quartz. Phase 1 = 13 samples; phase 2 = 8 samples; phase 3 = 30 samples; phase 4 = 12 samples. Total of 63 samples.

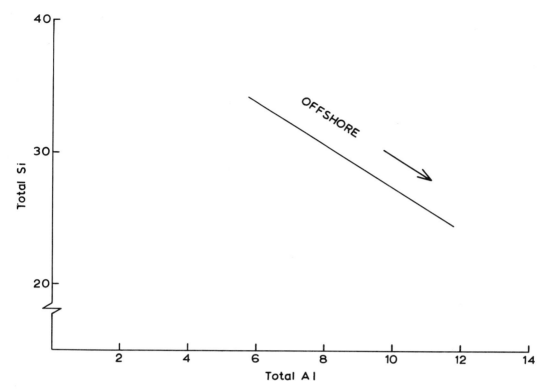

Fig. 73. Curve to illustrate decreasing Si/Al ratios offshore.

II. DISTANCE FROM SHORELINE, SEDIMENTATION RATE AND PALAEOSALINITY

(a) General statement

Distance from shoreline is the primary control of parameters such as depth, salinity, sedimentation rate, land derived detritus, grain size, fauna and composition. Hence an estimation of any one or more of the above will establish several others in normal situations where the offshore environment becomes deeper and more marine in character. However, in sediments exhibiting evidence of disturbances such as slumping and submarine scouring, many distance from shoreline parameters might give misleading results.

Increasing distance from shoreline will also influence the physical-chemical environment on the sea bottom and in the sediment just beneath the water interface. In particular, redox potentials may decrease due to the accumulation and bacterial decay of organic material in deeper, less oxygenated water.

A complete lack of oxygen and production of H_2S on the sea bottom are uncommon, even in the deepest oceans, and the presence of bottom-dwelling bivalves in phase 4 confirm that this deepest water phase was oxygenated (Sverdrup, et al., 1942, p. 871).

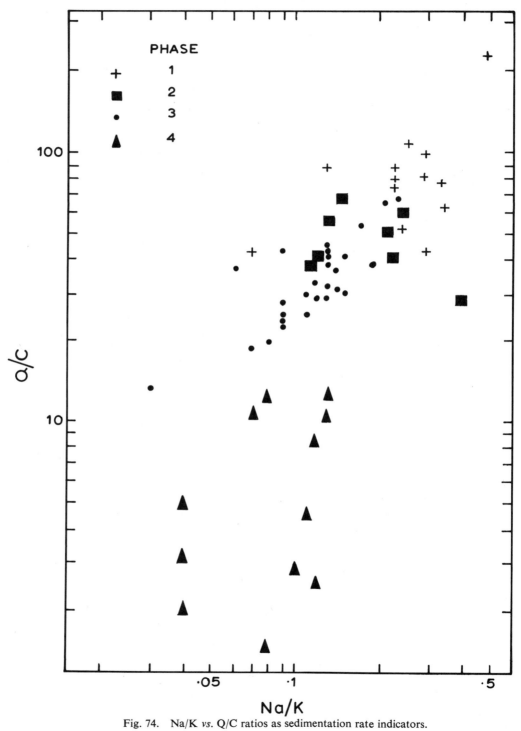

Fig. 74.   Na/K *vs.* Q/C ratios as sedimentation rate indicators.

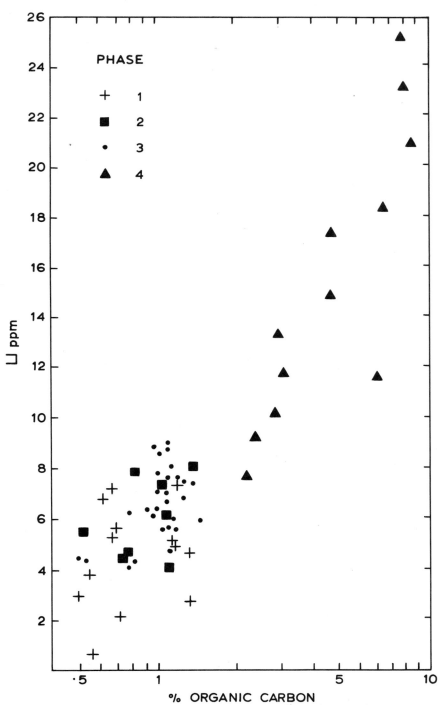

Fig. 75. Plot illustrating covariance of U and C-organic.

(b) Mineralogical and textural parameters

Quartz and carbon

Angular detrital quartz provides a useful indicator of distance from shoreline and decreases progressively from a maximum of 56% in phase 1 to a minimum of 11% in phase 4 (Table 6). It also decreases in grain size from a mean value of 0·05 mm in phase 1 to 0·01 mm and less in phase 4.

Organic carbon (expressed as total non-carbonate carbon) exhibits an inverse relationship to quartz; increasing progressively from phase 1 to phase 4 (Table 6). Correlation of both organic carbon and quartz with the faunal phases is good and the ratio of quartz to organic carbon (Q/C) forms a useful parameter for relative distance from the shoreline and sedimentation rate. This does not mean that the abundant organic material in deeper offshore phases is entirely autochthonous. A considerable proportion is land-derived (p. 280) and its abundance in deeper offshore waters is partly a result of its extremely fine grain size and preservation in less oxygenated environments (p. 282).

The mean values of Q/C ratios are included in Table 6 and range from 85·5 in phase 1 to 6·4 in phase 4.

The Lamination. Phases 3 and 4 exhibit a distinct fine lamination, although the laminae are frequently disturbed by bioturbation. The lamination is defined by dark organic-rich and lighter inorganic (quartz-rich) laminae which are probably annual. Counts on a phase 4 slide gave about 500 sets per inch representing a period of 18,000 years for the deposition of the observed thickness of this shale unit (Bloxam & Thomas, 1969, p. 280).

Calvert (1964) reports laminated sediments from the Gulf of California at depths of 900–1400 meters where the ocean floor intersects the water oxygen minimum (less than 1ml/1). Here lamination is preserved by virtue of the lack of burrowing benthos in these poorly oxygenated zones. In deeper parts of the basin (down to 2000 meters) where the bottom is more oxygenated, lamination is rapidly destroyed by organisms.

In the present samples evidence of an abundant benthic marine fauna in phase 4 does not support an oxygen minimum nor great depth, since these shales pass rapidly into *Lingula*-bearing shallow-water facies. A partially enclosed marine basin with oxygenated bottom water, even in the deepest parts, is their most probable depositional environment.

(c) Inorganic geochemical parameters

Si/Al ratios. The ratio of Si/Al is in effect a measure of the quartz and clay content since the clay mineralogy is relatively constant in these shales. Like Q/C ratios, their ratio should decrease offshore. The plot of Si *v.s.* Al in fig. 73 has a negative slope representing mutual interdilutions of clay (Si/Al) and quartz (Si).

Na/K ratios. Clays transported to a marine environment undergo changes in composition as the result of ion exchange reactions with sea water. Generally these changes are reflected in the Na/K ratio which increases due to the uptake of K (Goldschmidt, 1958; Hirst, 1962). Assuming suitable lattice sites are available in the clay, the amount of exchange will also depend upon how long the clay was in

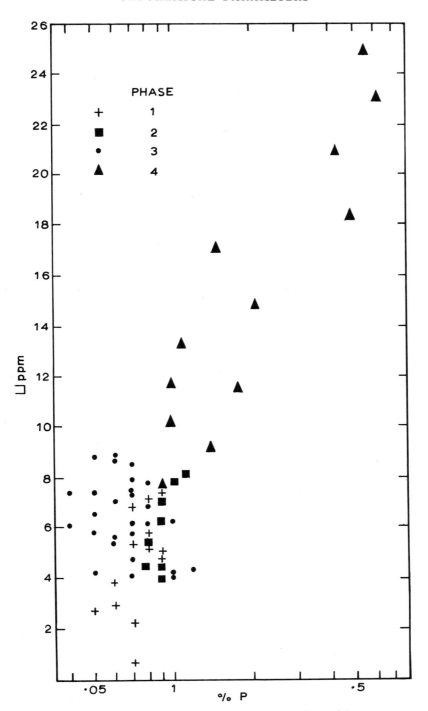

Fig. 76. Plot illustrating covariance of U and P.

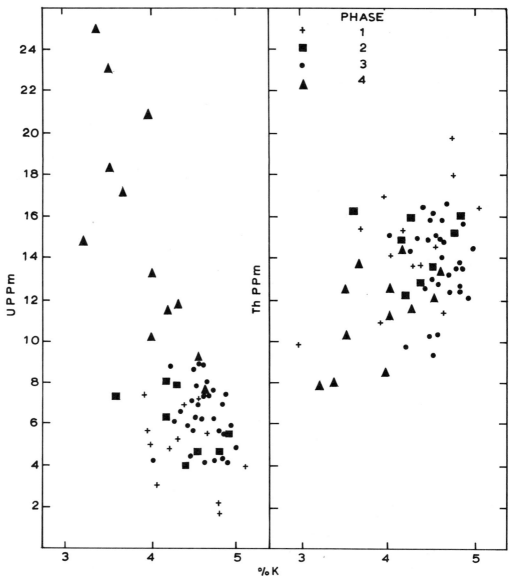

Fig. 77.   U and Th plotted against K showing weak correlation of Th with K.

contact with sea water (sedimentation rate).

Mean values for Na/K cover a range varying from 0·26 in phase 1 to 0·09 in phase 4 although the standard deviations are rather large (Table 6). To test whether or not the Na/K ratios are related to sedimentation rate the values are plotted against the Q/C ratios in fig. 74. There is a good separation of phases 1 and 4 but some overlap of phases 2 and 3. A perfect separation can hardly be expected since the same clay may have been exposed to the influence of more than one environment and carries, to varying degrees, the chemical imprint of them all. However, the consistent separation of phase 4 suggests that slower sedimentation rates in deep water enabled clay-water interactions to proceed more completely over longer periods of time.

Statistically there is little variation in the K content of the clays, changes in Na/K being largely the result of different Na contents (Bloxam & Thomas, 1969). The relatively high K in the illites from these rocks suggests that the detrital clays entering the basin were 'mature'—very little degradation of the source material seems to have occurred and a temperate, rather than tropical or subtropical, climate is indicated (Strakhov, 1969, vol. 3, p. 360).

U, Th and Th/U ratios. Uranium exhibits a strong positive correlation with organic carbon (Beers & Goodman, 1944; Bloxam, 1964) and phosphorus in the form of collophane (Figs. 75 and 76). The source of uranium may be from one or more of the following:

(i) an original constituent of the organic and phosphatic material.
(ii) a component of the aqueous environment (including interstitial water trapped in the sediment) adsorbed and complexed by organic matter, phosphates, and clays under appropriate conditions,
(iii) material introduced into the sediment as epigenetic uranium by subsequent near-surface processes.

The distribution of uranium is not sufficiently random for (iii) to have operated. With respect to (i), while certain marine organisms are known to concentrate uranium, in the present rocks most of it is considered to be the result of process (ii) involving adsorption from seawater. This is the mechanism favoured by Swanson (1961) and Landis (1962) for uranium concentrated in the Chattanooga and other black shales. The association of uranium with organic carbon may therefore be only a physical one—slow sedimentation rates favouring the concentration of both, although lower redox potentials might also assist (p. 276).

Thorium, which is essentially a hydrolyzate element (Kovalev, 1965) varies less than uranium and displays a weak correlation with K (Bloxam, 1964 and Fig. 77). Samples with large amounts of detrital quartz show no enrichment in thorium so that the latter cannot be due to the presence of thorium-bearing detritus.

Decreasing Th/U ratios from inshore to offshore sediments have been noted by Koczy (1949); Baranov, Ronov & Kunasheva (1956); Adams & Weaver (1958); and Bloxam (1964). Similar results have been obtained for the present rocks, (Table 6 and Fig. 78) which, like the Na/K ratios, separate phase 4 but show overlap of phase 2 and 3.

Phosphorus, fluorine and sulphur. The presence of calcium phosphate is evidenced in the development of collophane (? faecal material) in phase 4. The mean concentration of P is nearly the same in phases 1 to 3 but increases by about four times in phase 4 (Table 6).

Phosphates have been used as a palaeosalinity indicator by Nelson (1967) but in the present investigation the presence of simple aluminium and iron phosphates,

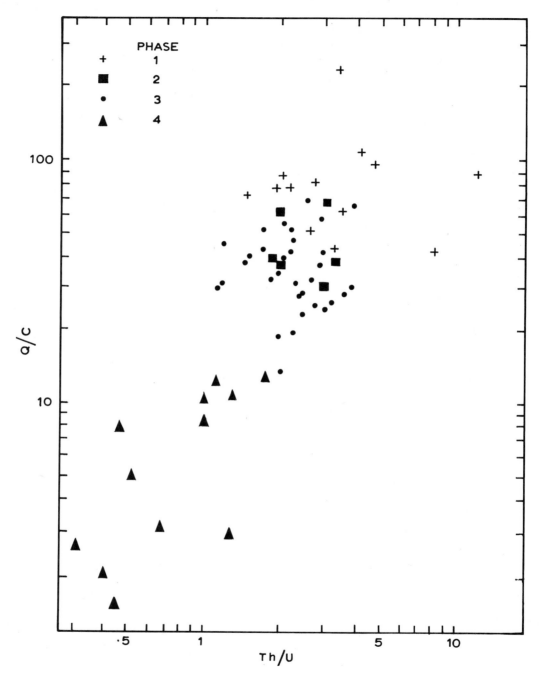

Fig. 78. Th/U *vs.* Q/C ratios as sedimentation rate indicators.

which are the bases of the method, could not be established. High calcium phosphate indicates a thoroughly marine environment and like uranium, the association of phosphorus with organic carbon may be a physical one—a consequence of slow sedimentation in the offshore environment of phase 4.

The fluorine content of all the samples averages 779 ppm and exhibits a close relationship to phosphorus since both are constituents of fluorapatite (collophane). Fluorine is highest in phase 4, decreasing in near-shore sediments except for indications of a reversal of this trend in phase 2 (Table 6). Since phase 2 is the first marine phase, these near-shore sediments may contain detrital apatite although the mineral has not been definitely identified (Bloxam & Thomas, 1970). The mean value for phosphorus is also slightly higher in phase 2 than in the adjacent phases 1 and 3.

Keith & Degens (1959) noted that among the major elements, sulphur might be an indicator of marine environment. In the present samples, sulphur is highest in the marine offshore phase 4 with a possible secondary maxima in phase 2, although the standard deviation is well outside the differences of the means (Table 6). The possible distribution of sulphur between sulphates and sulphides has not been investigated but it is unlikely that significant amounts of original sulphates are present, although they have been identified as later alteration products of pyrite. The relationship of sulphur to iron and organic material is significant and will be described subsequently (p. 274).

Boron. This element has found some favour as an indicator of palaeosalinity although its application is more complex than originally envisaged (Frederickson & Reynolds, 1960; Lerman, 1966). Palaeosalinity based upon boron in the less than 0.5μ clay extracts from the present rocks shows a rather poor correlation with inshore to offshore phases. Indeed it is the only criteria which is out of line with all the other parameters. In Table 7, the observed values for boron are given, together with the 'adjusted' boron values (Walker, 1964):

$$\text{adjusted boron} = \frac{\text{observed boron}}{\% \, K_2O} \times 8.5$$

The relatively 'mature' state of the illite being deposited in the basin has already

TABLE 7
Means and standard deviations for boron and potassium in clay fractions.

Phase	Boron (ppm, observed)		Boron (ppm, adjusted)	% K₂O	
	Mean	Standard deviation	Mean	Mean	Standard deviation
1	55	26·87	77	5·93	0·59
2	83	22·63	111	6·38	0·35
3	52	12·45	78	5·77	0·91
4	79	19·77	126	5·32	0·72

been noted (p. 271), and it seems likely that further boron uptake in the marine environment was small and largely obscured by the inherited boron.

III. BOTTOM CONDITIONS DURING AND FOLLOWING SEDIMENTATION

(a) General statement

Although the bottom conditions were initially non-reducing the sediments, shortly after burial, would pass into zones of reduction (ZoBell, 1946; Emery & Rittenberg, 1952). Negative redox potentials are characteristic of organic-rich fine-grained sediments and are the result of bacterial decomposition of the organic matter. Hence the amount of reduction will be controlled by the composition of the sediment in terms of its organic content, interstitial water, and the availability of suitable redox-coupled elements.

(b) Iron

The mean Fe^{3+}/Fe^{2+} ratios for the four phases (Table 6), show a general decrease offshore. However, phase 3 exhibits some reversal of this trend, although the standard deviation is almost as great as the mean. This is the result of local oxidation (which was not obvious in hand specimens) produced by weathering of some of the phase 3 shales which have also lost sulphur by oxidation and solution of pyrite. Although the iron appears to be easily affected by weathering, the remaining constituents are not.

Iron and sulphate in the original sediment might be reduced by a process such as: $2Fe_2O_3 + C \rightleftharpoons 4FeO + CO_2$; and also $SO_4 + 2C \rightleftharpoons S + 2CO_2$ (Strakov and Zalmanzon, 1955; Golyaeva, 1956). The amount (and type) or organic matter may therefore exert a considerable control over the sediment's redox history.

Following the methods adopted by Strakov and Zalmanzon, (1955) iron in the samples is sub-divided into five chemical divisions:

$$\text{Total Fe} = \underset{\text{Fe }total}{\underbrace{\text{Fe }det}} + \underset{\text{Fe }inert}{\underbrace{Fe^{3+}HCl}} + \underset{\text{Fe }reactive}{\underbrace{Fe^{2+}HCl + Fe^{2+}S}}$$

$Fe^{3+}HCl$ and $Fe^{2+}HCl$ are extracted by boiling in HCl and represent iron in clay and carbonate. $Fe^{2+}S$ is the iron in pyrite (determined from the sulphur content of the rock) and Fe det is the iron introduced in heavy detrital minerals, etc. This is also designated as the $inert$ Fe as opposed to the $reactive$ Fe available for redox reactions (Thomas & Bloxam, 1971). Means and standard deviations are given in Table 8. Maximum total iron (Fe $total$) and Fe $reactive$ occurs in phase 4 which also contains the minimum amount of Fe $detrital$ and maximum $Fe^{2+}HCl$ and $Fe^{2+}S$ (Fig. 79).

Using reactive iron Slavin (1959) has sub-divided redox fields into four based on the following criteria:

I Oxidizing:	$Fe^{3+}HCl > Fe^{2+}HCl + Fe^{2+}S$	
II Weakly reducing:	$Fe^{2+}HCl > Fe^{3+}HCl > Fe^{2+}S$	
III Reducing:	$Fe^{2+}HCl > Fe^{2+}S \quad > Fe^{3+}HCl$	

Table 8: Means and standard deviations for the various forms of iron in the phases.

Element	PHASE 1 Non-marine		PHASE 2 Transgressive Brackish to Inshore marine		PHASE 3 Inshore marine		PHASE 4 Offshore marine		TOTAL	
	Mean	Standard Deviation	Mean	Standard Deviation	Mean	Standard Deviation	Mean	Standard Deviation	Mean	Standard Deviation
Fe total %	3·59	0·42	4·68	0·47	4·21	1·02	5·21	0·94	4·33	1·01
Fe reactive %	2·08	0·38	3·06	0·43	2·71	1·05	3·93	0·63	2·86	1·01
Fe cly %	3·43	0·40	3·55	0·59	3·68	0·54	4·02	0·66	3·68	0·58
Fe det %	1·51	0·19	1·62	0·23	1·50	0·35	1·28	0·46	1·48	0·35
$Fe^{3+}HCl$ %	0·63	0·20	0·51	0·24	0·76	0·38	0·65	0·35	0·68	0·34
$Fe^{2+}HCl$ %	1·25	0·38	1·69	0·24	1·41	0·75	1·92	0·49	1·51	0·64
$Fe^{2+}S$ %	0·20	0·13	0·85	0·38	0·55	0·39	1·36	0·26	0·67	0·51
Fe det ⎞ As	42·4	5·2	34·1	5·4	37·7	13·5	24·2	5·8	36·6	11·8
$Fe^{3+}HCl$ ⎬ Percent	18·0	6·3	11·2	5·8	18·8	10·1	12·2	5·7	16·4	8·8
$Fe^{2+}HCl$ ⎮ Fe total	34·5	8·0	36·2	3·5	31·4	12·5	36·6	3·0	33·6	9·8
$Fe^{2+}S$ ⎠	5·0	3·3	17·8	7·2	12·2	8·7	26·9	6·1	14·2	10·2
$Fe^{3+}HCl$ ⎞ As	32·3	13·6	17·8	10·6	33·9	23·2	15·9	6·6	28·1	19·4
$Fe^{2+}HCl$ ⎬ Percent	59·3	10·1	55·6	2·2	48·4	13·4	48·6	5·5	51·6	11·6
$Fe^{2+}S$ ⎠ Fe reactive	8·5	4·9	26·6	9·8	17·7	11·7	35·5	8·0	20·3	13·3
Fe^{3+}: Fe^{2+} (reactive)	·556	·409	·240	·184	·878	1·091	·196	·096	·601	·832

IV Very reducing:　　$Fe^{2+}S$　$> Fe^{2+}HCl$ (No $Fe^{3+}HCl$)

The positions of the shales are plotted in fig. 80 in relation to these four redox fields. Most of the samples have a *present* redox state varying from weakly reducing to reducing (Fields II and III), the extent of reduction being directly related to the content of organic material. It is unlikely that these redox states were established on the bottom but rather just beneath the sediment-water interface.

(c) Sulphur and organic carbon

An interesting relationship exists between sulphur and organic carbon (Thomas & Bloxam, 1971). They show covariance in phases 1 to 3 (Fig. 81, A) up to $1 \cdot 2\%$ organic carbon. However, in phase 4 shales, which contain organic carbon ranging from $1 \cdot 7$ to 8%, the correlation is negative (Fig. 81, B). This is unusual since the sympathetic variation of sulphur (pyrite) with organic carbon is the general rule. It has been pointed out by Zobell (1946) that the more highly reducing the sediment the lower the redox *capacity*. However, an alternative explanation is preferred here. The organic material in phase 4 includes large amounts of land-derived non-extractable kerogenous (lignin and coaly) constituents (p. 282). This material was probably already denatured and non-reactive when finally deposited in the offshore sediments and was unable to play any significant part in redox reactions.

(d) Th/U ratios

A strong negative correlation exists between organic material and Th/U ratios (p. 271) The concentration of U in the offshore phases may be due in part to lower redox potentials attained by these organic-rich shales. In addition to absorption from seawater by organic matter, reduction of the soluble U^{6+} ion to the less soluble U^{4+} ion also allows its incorporation into collophane where it substitutes for Ca (Altschuler, *et al.*, 1958). Iron may act as a half cell in the reduction of uranium:

$$2Fe^{2+} + U^{6+} \rightleftharpoons 2Fe^{3+} + U^{4+}$$

(e) Conclusions

In sedimentary rocks, estimates of *absolute* redox potential of the original depositional environment based upon the *present* oxidation state of iron and other elements are more likely to represent the redox state attained shortly after burial when redox potentials are generally at their lowest. *Relative* differences in the present oxidation states of sedimentary rock sequences similar in lithology and general composition may be significant however, and reflect original differences in the depositional environment.

Hence, although the presence of a benthic fauna suggest that the waters were oxygenated during the deposition of the sediments, upon burial the final stamp of a more reducing environment was impressed upon them. Their geochemical response was largely conditioned by their composition, particularly with respect to organic matter.

IV. ORGANIC GEOCHEMICAL PARAMETERS
(a) Extractable hydrocarbons

In the previous sections organic carbon has been expressed in terms of total non-carbonate carbon. However, since the organic material in the shales appears to

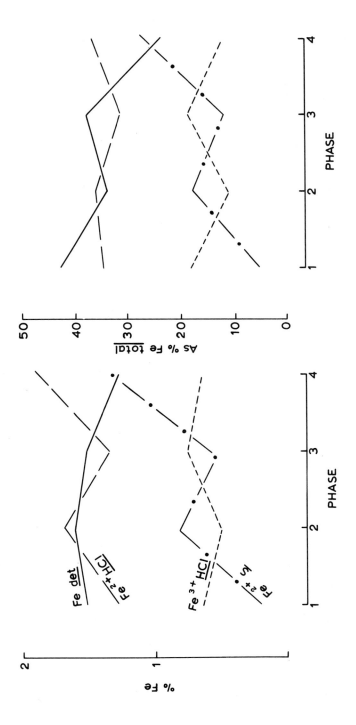

Fig. 79. The main forms of Fe in the four phases.

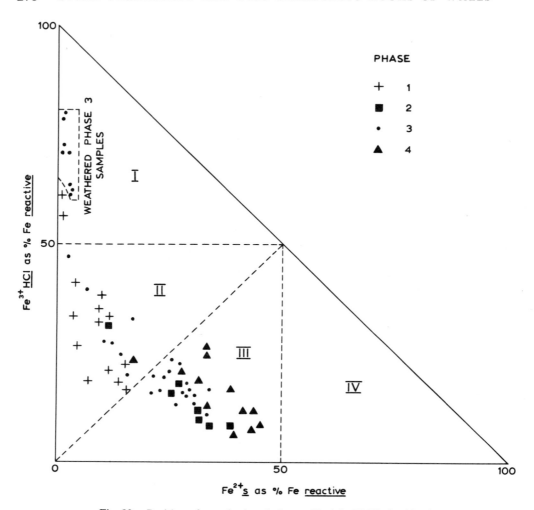

Fig. 80.   Position of samples in relation to Slavin's (1959) classification.

play an important part in their geochemical history, more data on its composition has been obtained. The methods employed have been described elsewhere (Bloxam, 1971) and the organic constituents quantitatively determined were nitrogen, hydrocarbons, and *n*-alkanes.

Benzene-acetone-methanol extractable hydrocarbons were fractionated into the following groups: heptane insoluble asphaltenes (As), paraffins, aromatic hydrocarbons and non-hydrocarbons (O-N-S compounds).

The end of organic maturation and diagenesis of marine sapropel are frequently saturated and aromatic hydrocarbons, together with kerogenous material derived from organically denatured humic and other compounds (Forsman, 1963). The hydrocarbon content (particularly paraffins) in sediments, generally increases with age but since all the present samples have similar ages, lithologies, and hence diagenetic histories, their present differences in organic composition must represent original differences related to source material plus the effects of the environment.

The results are plotted in Figs. 82 and 83, from which it is evident that the four phases exhibit differences in their hydrocarbon compositions. Phases 1 and 2 are

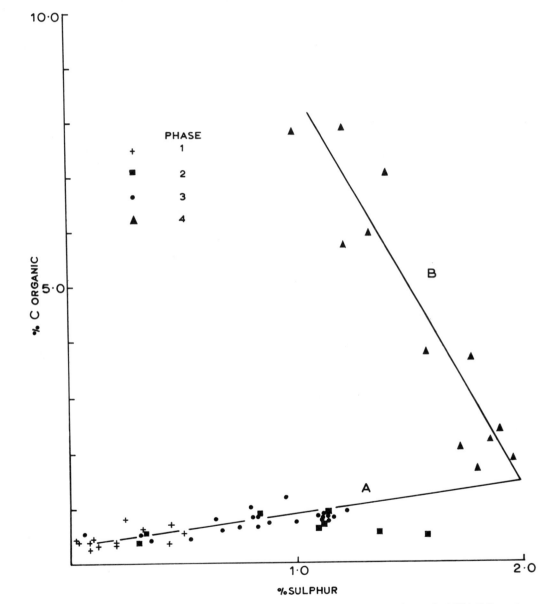

Fig. 81. Plots of S *vs.* C-organic. Curve A shows covariance up to approixmately 1.2% C-Organic; curve B is antipathetic from 1.7 to 8% C-organic.

high in O-N-S compounds while phase 3 shows enrichment in asphaltenes (As) and aromatic compounds. The offshore marine phase 4 contains a higher proportion of paraffinic hydrocarbons. These variations are similar to those noted by Shimada (1963) from Modern shelf sediments which ranged from lagoonal to offshore marine. The former were characterised by O-N-S compounds, the latter by paraffinic and aromatic hydrocarbons.

(b) n-alkanes

The distribution of *n*-alkanes is of considerable significance in environmental interpretation (Bray & Evans, 1961). Land plants tend to show maximum concentrations of *n*-alkanes over the range C_{27} to C_{29} with a strong odd over even preference (Eglinton, *et al.*, 1966). Marine organisms are a less likely source material for high molecular weight alkanes although they may provide maxima over the range C_{12} to C_{17} (Oro, *et al.*, 1967; Mathews, *et al.*, 1970).

The *n*-alkane distribution in a phase 4 shale shown in Fig. 84 has a broad maxima around C_{17} to C_{19} and another at C_{23} with general preponderance of odd over even carbon numbers. These distributions suggest a large contribution of organic material derived from land-plants, although *n*-alkanes in the region of C_{11}

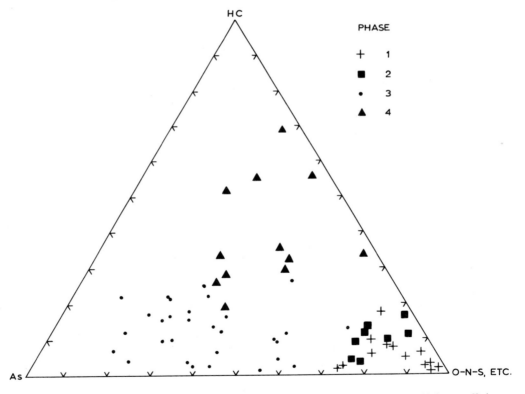

Fig. 82. Plot of chromatographically separated "heptane insoluble asphaltenes (As), paraffinic + aromatic hydrocarbons (HC), and non-hydrocarbons (O-N-S compounds, etc.)

upwards are also present and represent the contribution made by autochthonous marine products (Mathews, *et al.*, 1970).

(c) *Carbon/Nitrogen ratios (C/N)*

Investigation of Recent sediments suggests that the C/N ratio is related to the nature and origin of the organic matter. Generally loss of nitrogen accompanies burial and diagenesis of sediments which will increase the C/N ratio of the rocks, but high C/N ratios also occur in unconsolidated Modern sediments. Trask (1932) found C/N ratios of 5·3 to 13·6 (average 8·4) in Recent marine sediments which is close to that of plankton which provide the source material. Land plants and lake sediments generally exhibit higher C/N ratios. Recent sediments from the northern Caspian Sea and elsewhere have C/N ratios of 29 to 40 indicating large contributions from landplants and related detritus of humic and lignin-type (Bordovskiy, 1965, p. 48). In the present samples C/N ratios of phases 1 and 2 average 3, increasing to 5 in phase 3 and rising to over 30 in phase 4 which is close to that for oil shales (Arrhenius, 1951).

(d) *Conclusions*

The benzene-acetone-methanol extractable hydrocarbons indicate the progressive marine character of the offshore organic material (phases 1 to 4), but data from *n*-alkane distributions and C/N ratios show that a large proportion of the organic material in the offshore shales is non-marine in character. The marine organic hydrocarbon components are relatively easily extracted by organic liquids whereas

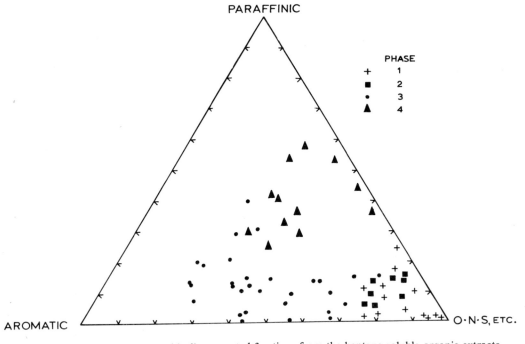

Fig. 83.   Plot of chromatographically separated fractions from the heptane-soluble organic extracts.

the non-marine components (lignin and coaly material) are not so readily removed except by thermal decomposition which releases them in the gas chromatograph (n-alkanes) and in total carbon and nitrogen determinations.

It had been assumed earlier (Bloxam & Thomas, 1969 p. 254) that the organic material was autochthonous and entirely marine in origin. But, despite its marine fauna, organic constituents include large proportions of land-derived plant detritus, much of which is probably spores. Its concentration in offshore waters may have resulted from plant detritus swept offshore into deeper water where it accumulated in less oxygenated sediments, while faster sedimentation in shallower water phases, coupled with some oxidation of organic material, lowered the C/N ratios.

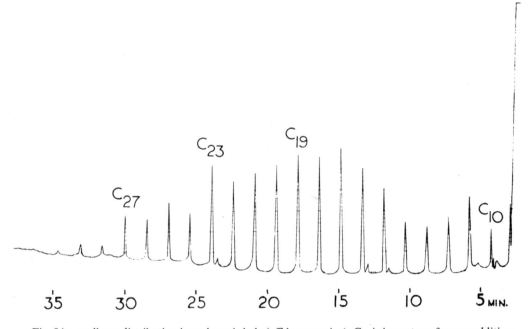

Fig. 84.   *n*-alkane distribution in a phase 4 shale (offshore marine). C_{10} is in part a reference addition.

V.   REFERENCES

ADAMS, J. A. S. and WEAVER, C. E. 1958. Thorium to uranium ratios as indicators of sedimentary processes: an example of geochemical facies. *Bull. Am. Assoc. Petrol. Geol.*, **42**, 387–430.

ALTSCHULER, R. S., CLARKE, R. S. JR., and YOUNG, E. J. 1958. Geochemistry of uranium in apatite and phosphorite. *U.S. Geol. Surv. Prof. Pap.* **314-D.**

ARRHENIUS, G. 1951. Carbon and nitrogen in subaquatic sediments. *Geochim. Cosmochim. Acta*, **1**, 15–21.

BARANOV, V. I., RONOV, A. B., and KUNASHEVA, K. G. 1956. On the geochemistry of the dispersed thorium and uranium in the clays and carbonaceous rocks of the Russian platform. *Geochem. Int.*, **3**, 3–8.

BEERS, R. F., and GOODMAN, C. 1944. Distribution of radioactivity in ancient sediments. *Bull. Geol. Soc. Amer.*, **55**, 1229–53.

BLOXAM, T. W. 1964. Uranium, thorium, potassium and carbon in some black shales from the South Wales coalfield. *Geochim. Cosmochim. Acta*, **28**, 1177–85.

——. 1971. Organic geochemistry of some Carboniferous shales from the South Wales coalfield. *Rep. No. 71/5, Inst. Geol. Sci.*, 6 pp.

—— and THOMAS, R. L. 1969. Palaeontological and geochemical facies in the *Gastrioceras subcrenatum* marine-band and associated rocks from the North Crop of the South Wales coalfield. *Quart. Jl. Geol. Soc. Lond.*, **124**, 239–81.

—— and ——. 1970. Fluorine in Carboniferous shales from the South Wales coalfield. *Chem. Geol.*, **5**, 179–93.

BORDOVSKIY, O. K. 1965. Accumulation of organic matter in bottom sediments. *Mar. Geol.*, **3**, 33–82.

BRAY, E. E., and EVANS, E. D. 1961. Distribution of *n*-paraffins as a clue to recognition of source beds. *Geochim. Cosmochim. Acta*, **22**, 2–15.

CALVER, M. A. 1968. Distribution of Westphalian marine faunas in Northern England and adjoining areas. *Proc. Yorks. Geol. Soc.*, **37**, 1–72.

CALVERT, S. E. 1964. Factors affecting distribution of laminated diatomaceous sediments in Gulf of California. *In:* van Andel, T. H. and Shor, G. G. (editors). *Marine geology of the Gulf of California. Amer. Assoc. Petrol. Geol.*, Mem. 3, 311–330.

EGLINTON, G., SCOTT, P. M., BELSKY, T., BURLINGAME, W. R., and CALVIN, M. 1966. Occurrence of isoprenoid alkanes in a Precambrian sediment. *In:* Hobson, G. D. and Louis, M. C. (editors). *Advances in organic geochemistry.* Oxford (Pergamon), 41–71.

EMERY, K. O., and RITTENBERG, S. C. 1952. Early diagenesis of California basin sediments in relation to the origin of oil. *Bull. Amer. Assoc. Petrol. Geol.*, **36**, 735–806.

FORSMAN, J. P. 1963. Geochemistry of kerogen. *In:* Breger, I. A. (editor.) *Organic geochemistry.* Oxford (Pergamon), 148–182.

FREDERICKSON, A. F. and REYNOLDS, R. C. 1960. Geochemical methods for determining salinity. *In:* Swineford, A. (editor.) *Proc. 8th. Nat. Conf. on Clays and Clay Mins.* Oxford (Pergamon), 202–13.

GOLDSCHMIDT, V. M. 1958. Geochemistry (Muir, A., editor). Oxford (Clarendon Press).

GOLYAEVA, L. A. 1956. Oil and gas-bearing deposits in the Ural-Volga region. *Izv. Akad. Nauk, SSSR*, 242–45.

HIRST, D. M. 1962. The geochemistry of modern sediments in the Gulf of Paria –1. *Geochim. Cosmochim. Acta*, **26**, 309–34.

KEITH, M. L., and DEGENS, E. T. 1959. Geochemical indicators of marine and fresh-water sediments. *In:* Abelson, P. H. (editor.) *Researches in geochemistry.* New York (Wiley), 38–61.

KOCZY, F. F. 1949. Thorium in sea water and marine sediments. *Geol. För. Stockh. Förh.*, **71**, 238–42.

KOVALEV, V. A. 1965. Geochemical aspects of the investigation of the Th/U ratio in sedimentary rocks. *Geochem. Int.*, **2**, 861–64.

LANDIS, E. R. 1962. Uranium and other trace elements in Devonian and Mississippian black shales in the Central Midcontinent area. *U.S. Geol. Surv. Bull.*, 1107–E.

LERMAN, A. 1966. Boron in clays and estimates of palaeosalinities. *Sedimentology*, **6**, 267–86.

MATHEWS, R. T., BURNS, B. J. and JOHNS, R. B. 1970. Comparison of hydrocarbon distributions in crude oils and shales from Moonie Field, Queensland, Australia, *Bull. Amer. Assoc. Petrol. Geol.*, **54**, 428–38.

NELSON, B. W. 1967. Sedimentary phosphate method for estimating palaeosalinities. *Science*, **158**, 917–20.

ORO, J., TORNABENE, T. G., NOONER, D. W., and GELPI, E. 1967. Aliphatic hydrocarbons and fatty acids of some marine and freshwater micro-organisms. *Journ. Bact.*, **93**, 1811–18.

SHIMADA, I. 1963. Extractable organic constituents in sediments with reference to the relation between composition of the chromatographic fractions in extracts and depositional environment. *Sci. Rep., Tohoku Univ., 3rd Ser.*, **8**, 421–82.

SLAVIN, P. S. 1959. The forms of iron in Cretaceous and Tertiary deposits of Turkmen. *Dokl. Akad. Nauk, SSSR*, **130**, 871–74.

STRAKHOV, N. M. 1967–1970. Principles of lithogenesis, vols. 1–3. Edinburgh (Oliver and Boyd).
—— and ZALMANZON, E. S. 1955. Distribution of authigeno-mineralogical forms of iron in sedimentary rocks and its significance to lithology. *Izv. Akad. Nauk, SSSR. (Ser. Geol.)*, *No.* 1, 34–51.
SVERDRUP, H. U., JOHNSON, M. W., and FELMING, R. H. 1942. The oceans. New Jersey (Prentice-Hall).
SWANSON, V. E. 1961. Geology and geochemistry of uranium in marine black shales: a review. *U.S. Geol. Surv. Prof. Pap.*, **356–C**, 67–112.
THOMAS, R. L., and BLOXAM, T. W. 1971. Some observations on the distribution of quartz, organic carbon, sulphur, and iron in the *Gastrioceras subcrenatum* marine band of the South Wales coalfield. *6me intern. Congr. Carb. Strat. Geol. Sheffield*, 1967 **IV**, 1555–70.
TRASK, P. D. 1932. Origin and environment of source sediments of petroleum. Houston, Texas (Gulf Pub. Co.).
WALKER, C. T. 1964. Palaeosalinity in Upper Viséan Yoredale Formation of England—geochemical method for locating porosity. *Bull. Amer. Assoc. Petrol. Geol.*, **48**, 207–20.
ZOBELL, C. E. 1946. Studies on redox potential of some marine sediments. *Bull. Amer. Assoc. Petrol. Geol.*, **30**, 447–513.

THE VARISCAN OROGENY IN WALES

T. R. Owen

I. INTRODUCTION

THE major orogenies have each spanned a considerable range of geological time. The Variscan (Armorican or Hercynian) Orogeny can be said to have ranged from at least late Devonian times to the middle Triassic. Radiometric dating suggests crustal unrest in South Cornwall in the Upper Devonian. Appreciable block-faulting occurred in central and northern areas of England during the Permian and Triassic periods. In any one area affected by an orogeny, there are minor preliminary phases of unrest leading up to the main pulses which are subsequently followed by waning movements (often of a tensional character and usually accompanying uplifts and contemporaneous erosion). As far as Western Europe is concerned, this orogenic climax appears to have moved north with time. In France, less-folded, limnic-type Coal Measures (of high Westphalian or of Stephanian age) rest unconformably on intensely deformed, and even granitised, earlier Carboniferous or Devonian rocks. In Britain however, Coal Measures of paralic type are involved in the main climax, usually placed within, if not even at the close of, Stephanian times. The northward time-rise of the Variscan climax is explained by the contraction of the Mid-European ocean as an African plate gradually closed up from the south. South West England probably experienced its climax before Wales. No Westphalian sediments younger that the Lower Coal Measures have been proved in Cornubia and the Upper Coal Measures (Pennant Measures) of South Wales are largely of southern derivation (Kelling, 1964).

II. PRELIMINARY MOVEMENTS

Proto-Variscan movements in Wales occurred at the beginning of the Dinantian, mid-way through the Dinantian, late in the Dinantian and mid-way through the Westphalian (more especially in North Wales). Lesser movements occurred at other times in some regions. The Usk area, for example, may have experienced fairly continuous unrest during parts of Carboniferous time.

Early Dinantian movements are difficult to locate in areas where Upper Devonian strata are absent through non-deposition following on mid-Devonian uplift and erosion. Nevertheless some discordance occurs between the Lower Limestone Shales and the Devonian on the northwestern border of the South Wales Coalfield and one can assume some upward pulse of areas like northern Pembrokeshire before the advance of the Dinantian sea. In North Wales, too, the stratigraphical gap below the Carboniferous Limestone is too great to locate any early Dinantian unrest. Devonian rocks are represented only locally in Anglesey. Neverthless the coarse character of South Wales and Welsh Borderland Upper Devonian sediments implies some appreciable uplift of North Wales at least in the late Devonian.

Mid-Dinantian movements can be located particularly along the North Crop of the South Wales and Pembrokeshire coalfields. George (1954) has demonstrated the relationship of mid-Dinantian breaks to early unrest along the Neath Disturbance. Marked marine regression in mid-Dinantian times over Carmarthenshire and the northern areas of Pembrokeshire results in the unconformable relationship of Viséan to Tournaisian or older strata in those areas. Early definition of a Ritec line at this time can be demonstrated by the remarkable lithological and thickness changes that occur in Pembrokeshire from Pendine to Tenby (Sullivan, 1966). Even as far south as Caswell Bay in Gower, there are signs of mid-Dinantian unrest.

The absence of Dinantian strata below high S_2 horizons in North Wales and above Z in areas like Titterstone Clee makes detection of mid-Dinantian movement in those areas difficult. If any Tournaisian strata are discovered in the Irish Sea then the possibility of extensive lower Dinantian deposition over northern areas of the Principality could become likely, with the removal of those pre-S_2 horizons by mid-Dinantian movements. If Devonian deposition was restricted to Anglesey in North Wales, then it is remarkable that pre-S_2 Dinantian and Devonian erosion (lasting some 60 million years) could not erode beds deeper than Ludlovian before the deposition of the thin basal pebble beds in the Dinantian of northern Denbighshire.

That late Dinantian unrest was widespread is evidenced by the great lithological change which marks the Carboniferous Limestone-Millstone Grit boundary. This change does not however occur everywhere at the same time in Wales. On the flanks of the Berwyn Dome and the Llangollen Syncline, the highest limestones are sandy and pebbly. In the Forest of Dean, higher S_2 deposition becomes arenaceous and this sandy influx spread southwards with time in the D zones. A similar sandy and pebbly spread extended southwestwards into the northeastern areas of the South Wales Coalfield, reaching the Brynmawr area by later S_2 times and the head of the Neath valley by D_2-D_3 times (Owen & Jones, 1961; Jones & Owen, 1966). The source rock of much of this debris would appear to be higher Devonian grits and conglomerates although the emergence of older (Malvern-type) areas in the southern Welsh Borderland cannot be ruled out. A marine regression westwards over the eastern portion of the South Wales Coalfield in late Dinantian times was followed by a new advance in the Namurian though that deposition was not to be renewed in the eastern margin areas until well through Namurian times, and in the Blorenge district not until early Westphalian times (Jones, 1971). Similar regressions and later Namurian renewal of deposition occurred in Pembrokeshire. Near St. Bride's Bay the Limestone was completely removed by pre-Namurian erosion. One cannot say how much pre-Trenchard deposition followed the Limestone in the Forest of Dean. The survival of a thin Ammanian remnant near Mitcheldean (Sullivan, 1964) suggests that during later Namurian and earlier Westphalian times the Dean Forest area was the scene of intermittent, much-interrupted deposition, with sediments being eroded as soon as they were deposited.

During the late Namurian and the lower half of Westphalian times in South Wales, widespread movements did not occur, though local unrest has been cited for developing structures like the Neath Disturbance (Bluck, 1958), Johnston Thrust (Owen, 1971) and the Ritec Fault (Jenkins, 1962). It is in the middle of the Westphalian that a main sedimentary change occurs in South Wales with active erosion somewhere to the south of Wales flushing in the Pennant molasse to South Wales. The presence of westward-carried gravels to the early Pennant on the East Crop (Kelling, 1964) is indicative of further growth of the Usk fold.

In North Wales, the absence of true Namurian in Anglesey is indicative of movements between the end of the Dinantian and the deposition of 'millstone Grit' containing Lower Westphalian goniatites. Some early Namurian unrest in the Prestatyn area is indicated by the erosional contact of Millstone Grit and the Limestone. Fairly important movements seem to have occurred in N.E. Wales during the deposition of the Coal Measures with flexuring during the deposition of the Productive Coal Measures (George, 1961), doming of Flintshire during the early part of the Upper Coal Measures and elevation in the neighbourhood of Oswestry. Still further southwards, of course, intra-Westphalian earth movements became much more intense, resulting in the absence of the Lower Carboniferous and much of the Coal Measures (below the Keele Beds) over the Longmynd Massif. Whilst it is difficult to unravel post-Triassic movements from Variscan effects in that area, it seems as if a broad anticlinal effect during the deposition of the Carboniferous over the Longmynd area preceded a downwarping along a Cheshire-Longmynd line prior to the deposition of the Permo-Trias.

III. THE MAIN MOVEMENTS

The main structural elements produced during the main climax of the Variscan Orogeny in Wales are shown in fig. 85. It at once becomes apparent that whereas in the extreme south of Wales the Variscan structures are typically east-west trending, that trend is virtually absent in the rest of the country. Elsewhere the main fold and fracture trend is a caledonoid one on the whole with malvernoid to charnoid elements in the southern Welsh Borderland and N.E. Wales. Many of the caledonoid structures shown for Central and North Wales are of Caledonian age, these elements being rejuvenated by the preliminary and climax movements of the Variscan Orogeny (see, for example, Owen, 1971). Structures like the Church Stretton line and the Bala Fault have had a long Palaeozoic history of unrest (see George, 1963; Bassett, Whittington and Williams, 1966). Their growth in fact even post-dates the Variscan Orogeny (post-Oligocene faults off the Cardigan Bay coast are shifted by the Bala Fault) (Dobson, 1971, fig. 3).

(a) Pembrokeshire

It is in Pembrokeshire that the most intense Variscan structures occur. South of Milford Haven, the Ordovician to Namurian successions are thrown into sharp folds with steep limbs. The upfolds expose horizons down to Llanvirn whilst the synclines preserve high Dinantian or even Namurian (as at Lydstep and Bullslaughter Bay). The folds trend about 10 N. of W. and most of them pitch eastwards, the exceptional cases being the Bullslaughter and Angle synclines and the Freshwater East upfold. Axial planes incline both northwards and southwards. Axial and limb buckling is common, especially in the case of the complex Orielton Anticline. Several minor thrusts replace minor fold limbs. Some incipient cleavage is developed in the thick Red Marls: the Lower Limestone Shales become quite brittle and cleaved south of Milford Haven. The fold axes are displaced by numerous cross-faults. Those trending NNW-SSE have dextral shifts, those trending NNE-SSW have sinistral shears. Much of the horizontal movements along these fractures is post-folding as displaced structures can be matched (for example in Stackpole Quay).

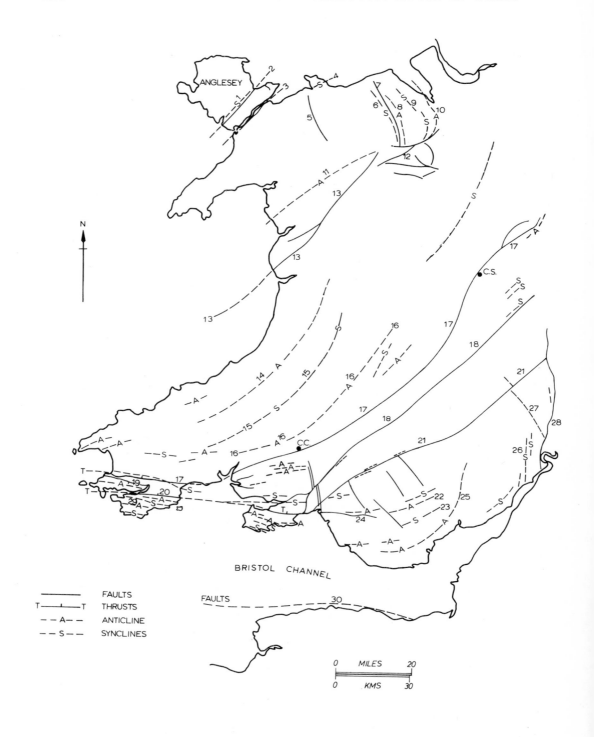

ANGLESEY

N

C.S.

CC

BRISTOL CHANNEL

——————— FAULTS

T——+——T THRUSTS

– – A – – ANTICLINE

– – S – – SYNCLINES

FAULTS   30

0   MILES   20

0   KMS   30

Fig. 85.   Variscan structures in Wales. Key: 1 Malldraeth Syncline; 2 Berw Fault; 3 Dinorwic Fault;
4 Llandudno Syncline; 5 Conway Fault; 6 Vale of Clwyd Syncline; 7 Clwyd Fault; 8, 9 and 10 Inner Crescentic Anticline, Outer Crescentic Anticline, etc.; 11 Derwen Fold. 12 Bryneglwys Fault, Llanelidan Fault,
Aqueduct Fault, etc.; 13 Bala Fault; 14 Teifi Anticline; 15 Central Wales Syncline; 16 Towy Anticline;
17 Church Stretton-Careg Cennen Disturbance; 18 Swansea Valley Disturbance; 19 Benton Fault; 20 Ritec
Fault; 21 Vale of Neath Disturbance; 22 Pontypridd Anticline; 23 Caerphilly Syncline; 24 Moel Gilau
Fault; 25 Usk Anticline; 26 Forest of Dean Syncline; 27 Mayhill-Woolhope Anticline; 28 Malvern Fault;
29 Folds in South Pembrokeshire; 30 Folds in Gower.

Even some Tertiary movement is likely in view of Dearman's findings along wrench faults in SW England.

Milford Haven has been eroded along the important Ritec Fault. There are many splays in the Tenby area. The northerly downthrow of the chief fracture is 500 metres near Flemington and double that amount near Llanreath. Shear planes in the vicinity of the Ritec belt would indicate a general 50° southward dip of the main fractures (cf. Jones, 1956, Plate XV). Many thrusts break the continuity of the sequences in the Namurian and basal Westphalian of the Tenby-Saundersfoot coastline. Much of the Namurian at North Sands, Tenby dips southwards yet thrusting results in a general younging of structural blocks in a northward direction. Important thrusts occur at Saundersfoot where much higher Ammanian Coal Measures occur in the cliffs north of that resort.

Mid-Pembrokeshire is dominated structurally by the Johnston-Benton fault belt. The Benton fracture was largely active in Lower Palaeozoic and mid-Devonian times. The Johnston fracture (or a differently-hading ancestor) was also an important line in the structural and depositional history of Central Pembrokeshire (see Owen et al., 1971, fig. 15). George (1963) has clearly shown the importance of this Johnstone-Benton block during the pre-Upper Llandovery movements. During Variscan thrusting, Silurian and underlying Precambrian have been thrust on to Upper Carboniferous along much of the Johnston Thrust. Slices of Lower Carboniferous are caught up in the thrust on Goultrop Roads (see Jones, 1956, 346). The eastward continuation of this fracture can be seen in the cliffs at Amroth; from here the belt becomes the Llandyfaelog-Careg Cennen Disturbance, north of the main coalfield (17 in fig. 85). Crustal shortening resulting from the Ritec and Johnston thrusting must be fairly considerable. Sanzen-Baker has recently (1972) drawn attention to the widely different Silurian and Devonian successions and sedimentary environments on either side of these fractured blocks. Very different Upper Carboniferous successions occur across the coastal districts on the western side of Carmarthen Bay. Very old, fundamental, basement fractures are involved in this Ritec—Johnston block. The Variscan climax finally broke the rock sheets and thrust forward slices over the site of the earlier controlling fractures. The easily yielding Coal Measures and Namurian of the St. Bride's Bay area were appreciably contorted as the brittle Precambrian thrust upwards and over along the Johnston Thrust.

The southward tilt of the Coal Measures at Newgale, on the north side of St. Bride's Bay suggests some arching during the Variscan movements of the St.

David's area. It is possible that Carboniferous rocks could therefore occur below the water (and below younger strata) northwest of the St. David's peninsula.

(b) Gower and the Vale of Glamorgan

The Variscan folds of Gower trend WNW-ESE but turn more eastwards in the east of that peninsula. This change becomes even more apparent in the Vale of Glamorgan where a northeastward swing sets in by the Cardiff area. The folds of Gower are not as closely packed as Pembrokeshire, though near vertical strata are not uncommon on South Gower's cliff line. The major upfolds, Cefn Bryn, Rhosili Down, Llanmadoc Hill and Harding's Down) expose Devonian while the synclines of Oystermouth, Oxwich and Port Eynon preserve Namurian. Parallel thrusts occur, particularly on Cefn Bryn and near the Langland upfold. Numerous cross-faults, trending between NNW and NNE, break the continuity of the folds but George (1940) has shown that a contrast of fold pattern frequently occurs across these fractures and that folding was, at least in part, accompanied by wrench faulting.

The Variscan folds of the Vale of Glamorgan are not as numerous and are more gentle. The main anticlines are the Candleston, Cowbridge and Pen-Llin folds. The Cowbridge upfold probably continues northeastwards into the Usk Anticline.

Before leaving these southern peninsulas of the Principality, it is worth pausing to consider the widely held view that the 'northern front' of the Variscan Orogeny runs through South Wales from Pembrokeshire to the Vale of Glamorgan. This surely cannot be so, as the folds of the Vale of Glamorgan are too gentle. If a well marked 'front' must be defined for the Variscan movements in the British Isles then it must lie south of Gower and the Vale of Glamorgan. The postulated thrust (Cannington Park?) lying near the Exmoor coast would seem to be a better position, comparing well with the thrust complex running across southernmost Eire.

(c) The main South Wales Coalfield

The main coalfield is a major basin extending for 60 miles from Pontypool to the Burry Estuary. The main structure has an east-west axis that trends near to Llanelli in the west and Glyncorrwg further east. Dips on the north-eastern portion of the coalfield and along the North Crop as far west as the Black Mountains are gentle but then increase into Carmarthenshire. On the East Crop dips are high as they are also on the Gower South Crop and around Briton Ferry. The major downfold includes a number of minor folds arranged an echelon. These include the Pontypridd and Maesteg anticlines and the Caerphilly, Pengam, Gowerton and Llanelli downfolds. Within these downwarps are preserved the highest Pennant Measures in South Wales. As with the folds of the Vale of Glamorgan, there is a gradual northeastward turning of the minor folds as they are traced eastwards across Glamorgan into Monmouthshire. Many smaller flexures are developed in the incompetent Lower and Middle Coal Measures, compressed as they were during the major folding between the more competent Millstone Grit and the Pennant Measures. Many such structures, often overfolded, have been detected by opencast working in areas like Glynneath, Llandybie and at Carway (near Kidwelly). Examples are photographed in the Gwendraeth Valley memoir (Archer, 1968, Plate IV).

A large number of dip or cross-faults affect the coalfield. These range NNW-SSE in western and central areas but become more NW-SE in the eastern parts. They include the Dyffryn, Glyncorrwg, Pen-y-castell, Gardeners, Bettws, Merthyr Church and Tredegar faults. Throws in some cases exceed 300 metres. Some can be shown to separate areas of differing tectonic pattern. Trotter (1947) has described examples in the Ammanford and Cross Hands areas. They grew in form and slip as the strata were being buckled. Other cross faults appear to have had a tensional origin and to post-date the folds. During this tension, troughs and horsts have formed using one of the earlier mentioned type of cross faults as a plane of movement. The Tredegar Trough is an example.

Strike-thrusts are particularly a feature of the South Crop, especially in the Margam and Llanharan areas. Many are underthrusts (Woodland and Evans, 1964) and pre-date the cross faults. Many minor thrusts and lag faults affect the incompetent portions of the Ammanian measures in the coalfield, especially in the anthracite belt from Ammanford to the Gwendraeth valley and again on the South Crop. Throws range from 1 to 100 metres. Good examples occur in the Tondu Brickpit at the horizon of the Amman Marine Band (see fig. 27 in Owen *et al.*, 1971). In the Newport portion of the coalfield, the incompetent disturbances trend W.20 N and E.20 N (Squirrell and Downing, 1969). A more important E-W fracture near the South Crop of the coalfield is the Ty'n-y-Nant and Moel Gilau fault-system. The southerly downthrow of this system reaches 1200 metres near Cwmavon. The dip of the fault varies from about 25° near Baglan (probably here a low tectonic level of the fracture—Woodland and Evans, 1964) to almost 70° further east. The fracture has had a long history, probably being initiated before folding. Even a sinistral shift of about 400 metres occurs along it. Tertiary movement is likely to have occurred.

The main coalfield is crossed obliquely by two belts of folded and fractured strata—the Neath and Swansea Valley disturbances. They trend NE-SW into the coalfield from the wide Devonian tracts to the north and east. Both belts are narrow but both involve quite intense folding of Carboniferous Limestone and Millstone grit, at Penderyn and Craig-y-dinas (Neath Disturbance) and at Cribarth in the Swansea Valley. The main structural element of both belts is a persistent NE-SW aligned fracture, largely of wrench origin but which has almost certainly suffered many later vertical movements, some as late as Tertiary times. The prolonged history of unrest along the Neath belt and the narrowness of each belt strongly suggests that the disturbances are sited over lines of fracture in the basement. A third disturbed belt runs to the northwest of the coalfield from Ferryside through Llandyfaelog, Careg Cennen and probably on to Old Radnor and Church Stretton (17 in fig. 85). Over the southwestern portion of this disturbed belt, the main movement appears to have been thrusting of the Upper Palaeozoic cover on to the Central Wales foreland (Trotter, 1948). This thrusting may have at least in part, been responsible for the conversion of the Carmarthenshire coals into anthracite. It is likely that some element of sinistral wrench enters this disturbance further to the northeast.

(d) The southern Welsh Borderland

The main Variscan structures here include the Usk Anticline, the Forest of

Dean basin and the Woolhope anticline. All have peculiar trends for Variscan elements. The Usk upfold has a NNE-SSW trend and is a pericline revealing Silurian rocks. Minor folds occur on its south-eastern limb (Walmsley, 1959), their axes being displaced by faults. Important fractures define the western limb. Sudetic and Malvernian phases preceded the Variscan movements. Earlier phases are very apparent on the southern side of the Forest of Dean basin, where the Upper Coal Measures cut sharply across the much more sharply folded Devonian and Lower Carboniferous rocks of the Lydney Syncline and adjacent folds. These earlier structures and the main Variscan downfold of the coalfield trend N-S. The main basin is sharply monoclinal towards its eastern edge. Further west is a highly faulted zone—the Cannop belt of N-S fractures with mostly westerly downthrows.

The Woolhope inlier has been described by Squirrell and Tucker (1960). It too is an elongate pericline fold but its axis trends NW-SE and moreover is shifted *dextrally* by the continuation of the Neath Disturbance into the Woolhope-Shucknall gap. This could be an early Variscan wrenching, the result of the strong E-W compression which was operating in this eastern area of Wales and its Gloucester-Hereford neighbourhood in mid-Carboniferous times. There is a possibility that the main folding of the Woolhope fold took place before the deposition of the Coal Measures.

(e) Central and Northwest Wales

It is difficult to locate and detect Variscan effects in areas of Lower Palaeozoics and Precambrian. Many of the major structures in Central and Northwest Wales are undoubtedly of Caledonian origin. It is however likely that some of these structures flexed a little more in the Variscan Orogeny, thereby determining the sites of more gentle warps in any Upper Palaeozoic cover. It is reasonable to assume that there was some buckling along the line of the Towy-Builth belt and possibly also along the Teify line (into North Pembrokeshire, as suggested earlier). The part ringing around of the Carboniferous fringe from near Oswestry through Colwyn Bay to Anglesey strongly suggests some upwarping on a major scale (and probably along a caledonoid line) of the Harlech Dome. The Derwen Anticline could therefore have had some Variscan growth too. This upwarping was one of the modifications of the N-S aligned Variscan regional stress in this area. The other, accompanying, modification was the sinistral shear induced along a reactivated Bala Fault, sending its Variscan splays into the Wrexham and Llangollen areas (see below). The possibility of Merionethshire upwarping by Variscan stresses should be borne in mind when it comes to interpret sub-Permian layers in Cardigan Bay.

In Anglesey, Carboniferous rocks are preserved along two NE-SW Variscan downfolds. The Malldraeth Syncline, preserving Coal Measures, is cut off to the southeast by the Berw Fault (2 in fig. 85). The Dinorwic Fault of the nearby Caernarvonshire mainland brings Lower Carboniferous against Arvonian volcanics, this southeasterly upthrow being the reverse of pre-Arenig movement (Shackleton, 1954). Post-Variscan movements are almost certain to have occurred along these North Wales faults (the Dinorwic fracture moved as recently as 1906). Further south, the Ystwyth Fault (with a downthrow of 1000 metres) could have suffered some Variscan slip, as mineral veins (Variscan?) are cut by such fractures in that area.

(ƒ) Northeast Wales

Here the main warps curve around the northeastern nose of the Harlech-Derwen upwarp. Upwarping along NNW-SSE lines along the Clwyd Range was accompanied by a parallel downwarp along the Vale of Clwyd. Some movement probably began along the Vale of Clwyd faults at this time, to be accentuated in Permo-Triassic times. Renewed slip along the northeasterly splays of the Bala Fault affected the Minera-Llangollen area on a fairly intense scale. Sinistral slip occurred along the Llanelidan, Bryneglwys, Llangollen and Glyn-Ceiriog faults, resulting in the rotation of intervening blocks, thereby producing great curved fractures, like the Aqueduct and Minera faults. Numerous horst and graben structures characterise the neighbouring coalfield. The striking ridge of Cefn Mawr, near Ruabon, is the result of faulting isolating strong sandstones in the Coal Measures.

IV. LATE OROGENIC EFFECTS

Associated with each orogeny there is usually magmatic activity, manifested by actual igneous intrusion or by associated mineralisation. In Devon and Cornwall both effects are to be seen. In Wales, no igneous intrusion can be proved. Yet there was fairly widespread mineralisation. In Northeast Wales, mineral veins are abundant along the Carboniferous tracts of Halkyn and Minera. Zinc blende, galena, copper pyrites, fluorspar and barytes occur, the ores being restricted to the Limestone and the overlying calcareous sandstones. The ores are post-fracturing but pre-Triassic (George, 1961, 74). Magmatic solutions from deep-seated sources carried the ores into the country rock.

In mid-Wales, zinc and lead occurs in the Lower Palaeozoics, especially in Cardiganshire. This mineralisation, too, may well be Variscan (Jones, 1954). The mineralisation in the Shelve Ordovician could date to the same time. Actual igneous activity can be demonstrated for the Lower Carboniferous (near the Wrekin) and the Coal Measures (Little Wenlock, Clee Hills) in nearby areas. In South Wales, some mineralisation occurs near Carmarthen. Near Kidwelly, a vein of copper pyrites occurs in the Millstone Grit. Studies of South Wales coals suggest that the present distribution of coal rank is related to the influence of mineralisation during Armorica times (Davies, 1971).

V. REFERENCES

ARCHER, A. A. 1968. Geology of the Gwendraeth Valley and adjoining areas. *Mem. geol. Surv. Gt. Britain.*

BASSETT, D. A., WHITTINGTON, H. B. and WILLIAMS, A. 1966. The stratigraphy of the Bala District, Merionethshire. *Quart. J. geol. Soc. Lond.*, **122**, 219–71.

BLUCK, B. J. 1958. The sedimentary history of the rocks between the *G. subcrenatum* Marine Band and the Garw Coal in the South Wales Coalfield. Unpublished Ph.D. thesis of the University of Wales.

DAVIES, M. M. 1971. Geochemistry of some South Wales Coals. In *Mineral Exploitation and Economic Geology.* (Proceedings of the Univ. of Wales Intercollegiate Colloquium, Cardiff, 1970.) 64.

DOBSON, M. R. 1971. A review of the economic potential of part of the Welsh continental shelf. (As Davies, M. M., above). 49–56.

GEORGE, T. N. 1940. The structure of Gower. *Quart. J. geol. Soc. Lond.*, **96**, 131–98.

——. 1954. Pre-Seminulan Main Limestone of the Avonian series in Breconshire. *Quart. J. geol. Soc. Lond.*, **110**, 283–322.

——. 1961. North Wales. *Brit. Reg. Geol.*, 3rd ed.

——. 1963. Palaeozoic growth of the British Caledonides. *In* Stewart, F. H. and Johnson, M. R. W., (Eds.): *The British Caledonides*, 1–33. Edinburgh.

JENKINS, T. B. H. 1962. The sequence and correlation of the Coal Measures of Pembrokeshire. *Quart. J. geol. Soc. Lond.*, **118**, 65–101.

JONES, D. G. 1971. The base of the Namurian in South Wales. *Compt. Rend. 6me. Congr. Strat. Carb.* **III**, 1023–29.

—— and OWEN, T. R. 1966. The Millstone Grit succession between Brynmawr and Blorenge, South Wales, and its relationship to the Carboniferous Limestone. *Proc. Geol. Ass. Lond.*, **77**, 187–98.

JONES, O. T. 1954. The trends of geological structures in relation to directions of maximum compression. *Adv. Sci.*, **11**, 102.

——. 1956. The geological evolution of Wales and the adjacent regions. *Quart. J. geol. Soc. Lond.*, **111**, 323–57.

KELLING, G. 1964. Sediment transport in part of the Lower Pennant Measures of South Wales. In *Developments in Sedimentology*, **1**, 177–84.

OWEN, T. R. 1971. The relationship of Carboniferous sedimentation to structure in South Wales. *Compt. Rend. 6me. Congr. Strat. Carb.* **III**, 1305–16.

—— and JONES, D. G. 1961. The nature of the Millstone Grit—Carboniferous Limestone junction of a part of the north crop of the South Wales coalfield. *Proc. Geol. Ass. Lond.*, **72**, 239–49.

——, BLOXAM, T. W., JONES, D. G., WALMSLEY, V. G. and WILLIAMS, B. P. 1971a. Summer (1968) Field meeting in Pembrokeshire, South Wales. *Proc. Geol. Ass. Lond.*, **82**, 17–60.

—— *et al.*, 1971b. Excursion 2: South Wales. *Compt. Rend. 6me. Congr. Strat. Carb.* **IV**, 1669–98.

SANZEN-BAKER, I. 1972. Stratigraphical Relationships and Sedimentary Environments of the Silurian-Early Old Red Sandstone of Pembrokeshire. *Geol. Ass. Lond.*, Circular No. 741.

SHACKLETON, R. M. 1954. The structural evolution of North Wales. *Lpool and Manchr Geol. J.*, **1**, 261–97.

SQUIRRELL, H. C., and DOWNING, R. A. 1969. Geology of the country around Newport (Mon.). *Mem. geol. Surv. Gt. Britain.* 3rd Ed.

—— and TUCKER, E. V. 1960. The geology of the Woolhope Inlier, Herefordshire. *Quart. J. geol. Soc. Lond.*, **116**, 139–

SULLIVAN, H. J. 1964. Miospores from the Drybrook Sandstone and associated measures in the Forest of Dean Basin, Gloucestershire. *Palaeontology*, **7**, 351–92.

SULLIVAN, R. 1966. The Stratigraphical Effects of the Mid-Dinantian Movements in South-West Wales. *Palaeogeogr., Palaeoclim., Palaeoecol.*, **2**, 213–44.

TROTTER, F. M. 1947. The structure of the Coal Measures in the Pontardawe-Ammanford area, South Wales. *Quart. J. geol. Soc. Lond.*, **103**, 89–133.

——. 1948. The devolatilization of coal seams in South Wales. *Quart. J. geol. Soc. Lond.*, **104**, 387–437.

WALMSLEY, V. G. 1959. The geology of the Usk Inlier (Monmouthshire). *Quart. J. geol. Soc. Lond.*, **114** (for 1958), 483–522.

WOODLAND, A. W., and EVANS, W. B. 1964. Geology of the South Wales Coalfield, Part IV. Pontypridd and Maesteg. *Mem. geol. Surv. Gt. Britain.* 3rd ed.

THE PERMIAN AND TRIASSIC DEPOSITS OF WALES

H. C. Ivimey-Cook[1]

I. INTRODUCTION

EROSION of the post-Armorican landmass continued throughout the Permian and much of the Triassic. Deposits regarded as late Permian in the Vale of Clwyd and in the western flank of the Cheshire basin are the only deposits of this system known in Wales. The former owe their preservation to their situation within a down-faulted block. Triassic rocks are represented by the deposits proved below the coast of Merionethshire, in the western margin of the Cheshire basin and in South Wales and Monmouthshire.

Mainly as a result of recent palynological work on the Trias of England it is now clear that the conventional division of the British Trias, based primarily on gross lithology, is not satisfactory for chronostratigraphic correlation. Proposals are being made to formally distinguish lithostratigraphic units in the Trias to conform with the principles of the Geological Society of London's 'Stratigraphic Code' (George and others 1969). Published names which apply to the main units present in South Wales are shown below.

Stage	Current Nomenclature mainly after Richardson 1905b and 1911			Nomenclature based on George and others 1969, Kent 1970			
R H A E T I A N	Upper Rhaetic	{	Watchet Beds	Watchet Member	} Lilstock Formation	}	} Penarth Group
			Langport Beds	Langport Member			
			Cotham Beds	Cotham Member			
	Lower Rhaetic	{	Westbury Beds		Westbury Formation		
			(*bristovi* limestones) } Sully Beds				
			Grey Marls				
	Keuper	{	Tea Green Marls	Proposals not published			
?							
NORIAN		{	Red Marls				

The recommendations of the Geological Society Code have been further developed here with regard to the 'Rhaetic Series' (Richardson 1911) as South Wales contains one of the finest exposures of these beds in Great Britain and the type area of the Penarth Group (p. 309). New lithostratigraphic names are under consideration for rock units lower in the Triassic but these are not used as they have not been defined or published.

---

[1] Published by permission of the Director, Institute of Geological Sciences.

Fig. 86.   The distribution of Permian and Mesozoic rocks (stippled) in Wales and neighbouring counties.

Historical

The maps published in 1815 by William Smith showed a generalized picture of the geology of South Wales. The first systematic survey of the area was by de la Beche (1846) who discussed a number of exposures of the marls, conglomerates and limestones of the 'New Red Sandstone'. Later the Rhaetic was examined in detail by Bristow (1864) and Etheridge (1872), both of whom paid particular attention to the Penarth area. Tawney (1866) and Moore (1867) on the other hand described the Rhaetic of the west of the Vale of Glamorgan and discussed the age of the Sutton Stone in relation to the position of the base of the Jurassic. The Rhaetic of both Glamorganshire and Monmouthshire were later re-examined by Richardson (1905a, 1905b, 1911). The major contribution to the knowledge of the early Mesozoic rocks and their environments was however that made by A. Strahan and T. C. Cantrill. The results of their detailed surveys appeared as the series of Geological Survey Memoirs on the South Wales Coalfield.

II.   PERMIAN

(a) The New Red Sandstone of the Vale of Clwyd

Within the Vale of Clwyd are two extensive areas of red aeolian sandstones which are correlated with the Lower Mottled Sandstone (Dune Sandstone) of the Bunter facies (Shotton 1956). However no palaeontological evidence has been found for the age of these beds. They are lithologically similar to the Lower Mottled Sandstone of the western margin of the Cheshire basin which is known to underlie the Bunter Pebble Beds. Wills (1948), Audley-Charles (1907a), amongst others, have regarded them as of Permian age though the lack of positive evidence of age is recognized in the 'Permo-Triassic' classification used in recent Geological Survey maps of these areas.

The most northerly outcrop occurs around Abergele, Rhyl and Rhuddlan. The sandstones overlap to the west across Coal Measures onto the Carboniferous Limestone. The eastern margin of the basin is marked by the Vale of Clwyd Fault and the western margin is cut by smaller faults which can be traced northwards into the sandstones. This forms the inland part of a basin extending into the Irish Sea.

The borehole at Foryd (SJ 993 806), west of Rhyl, penetrated 499 ft (152 m) of beds, mainly red sandstones but with some paler and white sandstones with thin bands of red shales near the base. They were all attributed to the New Red Sandstone (Bunter Sandstone) by Strahan (1885) though Smith (in Wedd, Smith and Wills 1928) considered that the lower part may have included some Lower Coal Measures. The borehole is close to the deepest part of the inland extent of this basin.

In the Denbigh-Ruthin area, further south in the Vale of Clwyd, exposures are scarce although some occur where small streams have cut through the cover of glacial drift (Strahan 1890, Reade 1891, Double 1926). Although generally loosely coherent, the bright and pale red sandstones contain better cemented horizons which have been used as building stone. Quarries near Fron-ganol yielded the stone used in the construction of Ruthin Castle (Wedd & King, 1924). The sandstones are usually strikingly false bedded and the direction of foreset dips enabled Shotton (1956) to conclude that, at least in the southern part of the Vale, the unit had been deposited as crescentic dunes. The orientation of these indicated remarkably constant winds blowing from the east across the 'Permian sand sea'. Generally the sandstones contain very few pebbles although George (1961) refers to the presence of coarser breccias and conglomerates near the base.

Double (1926) examined the petrography of the sandstones exposed in some of the small streams. He recorded that the quartz grains were subangular and normally coated with limonite. They occasionally showed secondary outgrowths of quartz. Feldspar was a ubiquitous accessory mineral, ilmenite dominated the heavy residues and could be found encrusting deep red rutile crystals; tourmaline and zircon were frequent whilst garnet was present in subsidiary amounts. He contrasted the variety of minerals present with the paucity of accessory minerals in the Carboniferous rocks forming the local basement. The only mineral derived from this source appears to have been silica in the form of chert.

Wilson (1959) concluded from geophysical surveys that several 'Permo-Triassic' basins existed within the Vale and that their position had been influenced by the partly pre-Triassic north-north-easterly trending faults which cut the western margin of the basin. These faults have throws estimated at between 400 and 1000 ft (120–305 m) and can be traced northwards out of the Carboniferous into the 'Permo-Triassic' deposits. Along the east of the Vale lies the Vale of Clwyd Fault with a downthrow to the south-west estimated at 5000 ft (1500 m). Three main basins were indicated. The Foryd borehole lies within the northern basin where a thick glacial drift sequence covers 'Permo-Trias' which extends to a depth of at least 500 ft (152 m). In the southern part of the Vale the outcrop contains two deep basins. The more northerly is centred between Denbigh and Bodfari and also contains drift on about 725 ft (220 m) of 'Permo-Trias'. The most northerly basin is deeper and situated south-east of Denbigh. It may contain up to 1250 ft (380 m) of 'Permo-Trias' below the drift. This compares with Shotton's (1956) estimate of between 700 and 1000 ft (213–305 m) for the thickness of Dune Sandstone resting on the pre-Permian surface.

(b) East Denbighshire and Flintshire

The Permian and Triassic rocks of the western margin of the Cheshire basin continue, below the extensive cover of drift deposits, into the north-eastern counties of Wales. The Lower Mottled Sandstone is known to overlap westwards onto the Carboniferous rocks of eastern Denbighshire and the eastern parts of Flintshire. It is possible that west of the River Dee it may exceed 1000 ft (305 m) in thickness. The sandstones are similar to those of the Vale of Clwyd. The soft, red, well rounded, wind-polished sands with some bedded micaceous mudstones are proved in boreholes and wells (Poole & Whiteman 1966).

III.   THE TRIAS OF NORTH WALES

(a) The Upper Triassic of the Llanbedr (Mochras Farm) Borehole

This borehole, situated about 2 miles (3 km) west of Llanbedr, near Harlech, Merionethshire (SH 5533 2594), was completed at a depth of 6361 ft (1938·83 m). Below 1974 ft 6 in (601·83 m) of Recent, Pleistocene and Tertiary, resting on about 4282 ft (1305 m) of Lower Jurassic strata, it entered rocks of Triassic age (Wood & Woodland 1971).

The 105 ft 2 in (32·05 m) of Triassic rocks penetrated are divided (Harrison 1971) into an older, Terrigenous Formation, unbottomed after 49 ft 11 in (15·21 m) and a younger, Carbonate Formation 55 ft 3 in (16·84 m) thick.

The Terrigenous Formation consists dominantly of pale red-brown calcitic, calcarenitic and oolitic sandstone with fan gravels and coarse sand. Both above and below this occur rhythmic alternations of pink-brown bedded calcitic sandstone, siltstone, and dark red-brown marly shales. The lowest unit also contains thin conglomerates and breccias. The Carbonate Formation contains a variety of grey-green dolomitized carbonates with local spreads of arenaceous material and some hematitic sandy beds. The highest part of the unit is a pelletty dolomitic pseudo-breccia passing down into arenaceous calcitic dolomite.

These two facies differ in lithology from most other deposits of about this age known elsewhere in Great Britain and it was suggested that they represent part of a playa environment which was subject to flash floods. The area was however gradually subsiding below sea-level and the evidence of increasing salinity heralded the marine calcareous mudstones and limestones of the overlying Lower Jurassic (*planorbis* Subzone) which follow unconformably (Ivimey-Cook, 1971).

The age of these beds was established by Warrington (1971) from an examination of sparse and poorly preserved miospore assemblages. Samples from the Terrigenous Formation were mainly barren but the inconclusive results suggested that the lowest strata might be of Norian age. The lowest part of the Carbonate Formation between 6292 ft (1917·8 m) and 6308 ft 11 in (1923·14 m) yielded miospore assemblages suggestive of a late Rhaetian age whilst samples from the upper part were inconclusive but could be as young as Hettangian.

The composition of the miospore assemblages from the Carbonate Formation invites comparison with part of the *Heliosporites* (miospore) Zone (Orbell 1971), which extends from about the base of the Cotham Member into the *planorbis* Zone. As the *planorbis* Subzone was proved in this borehole, close above the top of the Carbonate Formation, the miospore evidence suggests that the Carbonate Formation is yet another of the calcareous facies of the Upper Rhaetian Lilstock Formation which are developed in South Wales and England.

Wood & Woodland (1971) consider the evidence for the thickness and extent of the Permo-Triassic rocks which underlie Cardigan Bay; Wright and others (1972) review the submarine geology of the Irish Sea area to the north of Wales.

(b) The Trias of east Denbighshire and Flintshire

The eastern extremities of the counties of Denbigh and of Flint, with its detached areas, extend across the edge of the thick Triassic sequence of the Cheshire basin. The basin contains beds ranging in age from the (?early Triassic) Bunter Pebble Beds up through the Upper Mottled Sandstone, Keuper Sandstone and Keuper Water-stones. The overlying Keuper Marl is divided by two major saliferous sequences and overlain by the Tea Green Marls and the 'Rhaetic' (Penarth Group).

Most of this succession may be present below the thick glacial deposits covering the large detached area of Flintshire west of Whitchurch. Detailed knowledge of the sequence has been gained from a series of boreholes in Cheshire and Shropshire (Poole & Whiteman 1966). One of these, the Plattlane Borehole, lies in Shropshire only a short distance east of Flintshire and proved glacial drift on Lower Jurassic and the higher parts of a Triassic sequence including 44 ft 3 in (13·5 m) of Penarth Group (Ivimey-Cook 1969). Another, the much deeper Wilkesley Borehole, was drilled close to the axis of the Wem–Audlem Syncline about 5 miles (8 km) east of Whitchurch. It proved part of the thick Lower Jurassic sequence of this syncline and went on to penetrate formations in the Keuper to a depth of 5531 ft 7 in (1686·03 m). The western parts of the basin contain sandstones, mudstones and conglomerates, referred by Poole & Whiteman (*op. cit.*) to the Keuper Waterstone and Keuper Sandstones. They overlie thick developments of the red, current-bedded micaceous sandstones with thin shale bands of the Upper Mottled Sandstone. These in turn rest on the red-brown sandstones and sandy mudstones of the Bunter Pebble Beds.

A general account of the Trias of the Midlands is given by Hains and Horton (1969); parts are discussed by Audley-Charles (1970a, 1970b), whilst detailed surveys are given in the Memoirs describing the Geological Survey maps of the area.

IV. THE TRIAS OF SOUTH WALES

The distribution of the two principal facies (Red Marls and marginal[1]) of the Keuper and of the Penarth Group (Rhaetic) in South Wales is shown in Fig. 87. The changes in lithology and palaeogeography which took place as the Keuper, and later the Rhaetic and Lower Jurassic deposits, gradually overlap and overstep across the eroded Palaeozoic floor of the Vale of Glamorgan have been graphically described by Strahan & Cantrill (1902, 1904, 1212), Richardson (1905b), Miskin (1922), Cox & Trueman (1936) and George (1970). Strahan & Cantrill (1904) examined the evidence for the location of the Triassic 'mainland' postulated to lie close to the Pennant scarp of the South Wales coalfield. They could find firm evidence for this only in the vicinity of Llantrisant and Llanharan. Within the Vale the steady submergence of the chain of islands which formed the western extension of Richardson's 'Mendip Archipelago' (1901) is clearly marked. The deposits of conglomerates, breccias and redeposited limestones forming on or closely adjacent to the 'basement', rose up the sides of the islands as these sank below the level of deposition. As each island in turn was inundated the successive shorelines contracted around the remaining islands until at the end of the Triassic only the Cowbridge Island remained. This later became further reduced to form the St. Brides and Cowbridge islands in the Hettangian (Robinson 1971).

(a) The Red Marls

The dominant lithology of the Keuper Red Marls is a brownish-red, often calcareous mudstone. It commonly shows uniformly thick, laterally-persistent bedding although the lower beds are often more massive.

The relatively high calcium carbonate content of some of the red marls used for brick-making in the outcrops west of Cardiff was commented on by Thomas (1961). Most of the mudstones are only moderately hard and they may contain patches and bands of grey-green mudstones or calcareous mudstone. Occasional ripple marks, mudcracks and raindrop imprints are recorded (Klein 1962) although many of these were in or close to the marginal deposits. Miskin (1922) commented on the lack of salt pseudomorphs and salt in Glamorgan but other evaporite minerals do occur. Celestine is found in the Red Marls in Cogan quarry and Thomas (1968) foun nodules and films of it in the extensive area of marginal dolomitic conglomerate about $1\frac{1}{2}$ miles ($2 \cdot 5$ km) south of Llantrisant. Bands of white and pale pink gypsum occur near the top of the Red Marls and also within the Tea Green Marls. Where these outcrop along the coast near Penarth they form nodular beds up to 1 ft ($0 \cdot 3$ m) thick and were once dug on a small scale as "Penarth alabaster" (Cox and Trueman 1936). Miskin (op. cit.) also recorded copper carbonate on the under surface of three of the gypsum beds.

---

[1] Willey (1966b) has urged the use of the terms 'proximal' and 'distal' (deposits) for the contrasting facies of the Keuper in South Wales, where a purely geometrical rather than an environmental description is to be conveyed. In this account 'marginal' has been used for the facies he termed 'proximal'. They are regarded as deposits formed sub-aqueously close to the fringes of a large area of water. Littoral is used for deposits of the marginal zone of a marine sea.

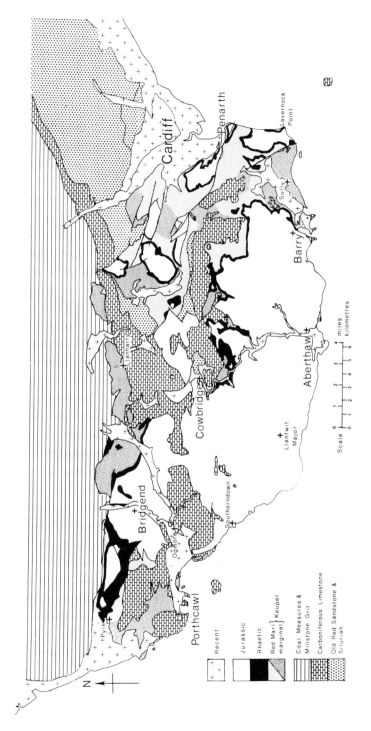

Fig. 87. Simplified map of the geology of the Vale of Glamorgan (Based on Geological Survey maps, by permission of the Director, Institute of Geological Sciences.)

The Red Marls display considerable variation in clay mineral content. In one sample, a light brown calcareous marl from near Newport, analysis gave 52% carbonate, 43% illite, 4% corrensite, 2·5% hematite and 1·5% quartz. A second, from near Cardiff, contained 62% illite, 38% corrensite, 3·5% quartz and 2·2% hematite (Perrin 1971). The mineralogy has been interpreted as a phase equilibrium of a complex system including illite-corrensite-calcite-dolomite-hematite-quartz (Davis 1967). The particles making up the Red Marls are concentrated into unusually large aggregates as, although 50% to 90% of the rock consists of clay minerals, only between 15% and 35% are actually of clay grade (i.e. $< 2\mu$) (Dumbleton 1967).

The thickest sequences of Red Marls occur in east Glamorgan but estimation of the maximum thickness present is difficult as the beds cover a buried topography in an area also affected by major north-north-west trending faults. Good sections of the higher parts of the Red Marls are seen along the coast near Penarth, in the brickworks quarries near Cogan and around the north of the Leckwith plateau. In the main Cogan quarry [ST 177 721] these upper beds can be seen for over 75 ft (23 m) and the junction to the Tea Green Marls may be at about 100 ft (30 m) above sea level. A borehole in the quarry was completed at 317 ft (96·6 m) below sea level in 9 ft (2·7 m) of conglomerate believed to be at the base of the Keuper. This would indicate a thickness of over 408 ft (124·3 m) of Red Marls (Strahan and Cantrill 1902). In the Roath Borehole, Cardiff there were 23 ft (7 m) of superficial deposits overlying Trias. The junction of the Trias and the Old Red Sandstone was taken at a depth of 342 ft (104·2 m). However the Keuper 'Red Marls' here contained numerous conglomeratic, gritty and sandy beds which were also often dolomitic (North 1916). Another borehole, at Roath Dock, was interpreted by Strahan and Cantrill (1902) to show over 367 ft (111·9 m) of Keuper. It is difficult to estimate how far below the top of the Red Marls these boreholes commenced.

The basal conglomeratic beds encountered in these boreholes were considered by Audley-Charles (1970b, p. 86) to represent a thin development of his 'division 4' of the Trias. This would correspond to the Lower Keuper Marl of the English Midlands. In the absence of palynological or other dating they are here regarded as a basal, marginal, facies of his next higher unit (division 5), the upper part of the Keuper Marl.

The marginal facies of the Red Marls

A wide variety of sediment types occur between the characteristic red mudstone facies of the Red Marls and the basement of the main area of deposition. The Keuper sediments rest on an irregular basement topography consisting largely of outcrops of Old Red Sandstone and Carboniferous Limestone. Parts of the basement had been eroded to leave, by the middle of the Triassic, a number of isolated masses or "islands" rising above the floor of the basin. The character of the marginal facies and the extent of its development at the expense of the Red Marls depended largely upon the nature and topography of the rocks forming the laterally adjacent basement. Erosion of the Old Red Sandstone could produce a coarsely arenaceous or rudaceous facies which can locally be very thick. At Radyr Quarries the Radyr Stone con-glomerate reaches 60 ft (18·3 m) whilst near Capel Llaniltern they reach 120 ft

(36·6 m) in thickness. This contrasts with the Newport region where the poorly calcareous Red Marls have a local marginal facies of light buff sandstone against the Old Red Sandstone (Squirrell & Downing 1969). Further east near Portskewett up to 12 ft (3·7 m) of sandstone occur above a thin dolomitic conglomerate at the base of the main sequence of Red Marls (Welch & Trotter 1961). Similar very thin basal beds occur at Michaelston-le-Pit, in the Vale south-west of Cardiff, where a small outcrop of Old Red Sandstone [ST 146 733] is overlain by the Keuper. There a thin conglomerate of mixed Old Red Sandstone quartz pebbles mixed with Carboniferous Limestone fragments passes rapidly upwards through a thin sandy facies into the Red Marls.

The series of islands eroded out of the Carboniferous Limestone of the southern flank of the Cardiff-Cowbridge anticline provide a wide variety of conditions for the deposition of the marginal facies. These islands are now partially re-exposed by erosion and can be seen to have been fringed, particularly around their eastern and southern sides and in the bays and channels between, with a series of dominantly clastic Triassic deposits. These range from coarse breccias and conglomerates (the Dolomitic Conglomerate of authors) to fine grained reprecipitated limestones with partings of calcareous mudstone. The junction of these beds with the basement can be very clear cut as at Sully Island and Hayes Point (Plate 2), or be extremely intricate where the breccias fill in against the eroded sides of the limestone islands. The finer grained limestones may contain remanié Carboniferous Limestone crinoids, corals, brachiopods and, rarely, trilobites as in the Westra Quarry, Dinas Powis. The junction of these limestones against the Carboniferous Limestone can be extremely difficult to place accurately as the Triassic limestone fills crevices, penetrates amongst fallen blocks and banks up against the old shoreline. It can be most readily distinguished by its more argillaceous nature and tendency to become thinner bedded and more argillaceous upwards as the supply of available carbonate for redeposition decreased when the island was nearly submerged (Plate 1).

The remains of an algal development are known from some laminated limestones which lie between the Keuper marginal conglomerate and the overlying massive grey marginal limestones on the fringe of the Carboniferous Limestone island north-west of Dinas Powis. The material contains laminations of reddish silt and limestone with a distinct vertical arrangement of calcite suggesting pre-existing tubes or threads. These were some $10-20_\mu$ in width and closely packed together. No organic matter is now present. Penecontemporaneous minor slumping, micro-brecciation and mud cracks together with later recrystallization obscure the structure but it was attributed by Professor T. M. Harris (personal communication) to the binding effects of blue-green (Cyanophyte) algae. These algae would probably occupy a fresh to brackish water environment. This illustrates the degree of contrast which could be present locally within the more widespread lacustrine (inland sea) environment found at this time in the Triassic when minor evaporite deposits and other indicators of dessication were also being formed.

Within many of the Carboniferous Limestone outcrops occur fissure or cavern infillings. These occupy what were initially solution phenomena, described by Robinson (1971) as slot fissures and underground watercourses. They later became

Plate 1.   Marginal Triassic limestones (1), with thin beds of mudstone and conglomerate, banked against and over Carboniferous Limestone (2).   Slot fissure at (3).   Westra Quarry (now infilled), Dinas Powis, Glamorgan.

filled with mudstone and marl at least some of which was of Triassic age. In the Cardiff area very early slot fissures later filled with bedded red marl were formerly well seen in the Westra Quarry, (ST 146 711) (Plate 1). They occur below a cover of marginal deposits lateral to Red Marls. Horizontal and vertical fissures of this type also occur in the Beachley inlier near Chepstow. In the middle and west of the Vale fillings of more mature cavernous fissures have yielded late Norian vertebrate faunas (Robinson 1971) near Ruthin and at Pant-y-ffynnon Quarry west of Bonvilston. Fissures containing red marly clay with the remains of slightly later vertebrates, including *Morganucodon watsoni*, described by Kühne (1949) as an early mammal, are known from the Duchy Quarry, Ewenny (Robinson 1957). *Kuehneotherium praecursoris* was described from Pontalun (Kermack and others 1968) and Pant Quarries (Kermack & Mussett 1959). Thomas (1952) described the Triassic marls, fine breccias and marly sandstones with scattered reptilian fragments, contained in joints and solution hollows in the Llanharry Quarries. In the Carboniferous Limestone quarries in Mynydd Ruthin occur mature underground watercourses now partially filled with later sediment. This ranges from angular blocks of oolitic Carboniferous Limestone up to 8 ft (2·4 m) in length set in a matrix of Triassic red marls and breccia; elsewhere the fill consisted of finer red breccia with small angular limestone fragments whilst in another cave the infill consisted of 3 to 5 ft (0·9 to 1·5 m) of fine grained shale, red marly sandstone and grey inorganic limestone seen extending over a distance of about 20 ft (6·1 m). The

Plate 2. Marginal Triassic limestones and calcareous mudstones (1) overstepping Carboniferous Limestone (2). Hayes Point, Barry, Glamorgan.

marly sandstones have yielded comminuted reptile bones including the small jaw bones of *Tricuspidosaurus thomasi* Robinson.

In addition to the Triassic fissure deposits found within the Carboniferous Limestone of the Vale of Glamorgan other types of marginal deposits occur in the west adjacent to the larger islands. Bluck (1965) described the alluvial fan deposit containing an alternation of stream flood and stream sediments in the area around Porthcawl, Ogmore and Stormy Down. The sequences are particularly well seen near Sker Point and at Newton, north and east of Porthcawl respectively, and also to the south-east on the coast at Ogmore. They include badly sorted stream flood (or mudflow) conglomerates overlain by better sorted stream conglomerates. The bases of the fans are usually channelled and some can be seen to rest unconformably on Carboniferous Limestone (Plate 3). They may extend for at least 500 to 1000 ft (150 to 300 m) in length and some can be seen to thin in the inferred downstream direction. The Ogmore fans contain particularly large boulders with subordinate amounts of red mud and marl. Structures within the fans include trough cross-bedding whilst imbrication is well shown by disc shaped pebbles and small cobbles.

Between the fans occur reddish sandstone and marls frequently containing angular limestone and chert fragments up to 50 mm diameter. It was this kind of bed which yielded the footprints of *Brontozoum thomasi* Sollas near Newton.

The fan deposits represent screes accumulating in deep canyon-like watercourses which were then flushed out by torrential streams as a result of storms. At least one of the canyons has itself become plugged by similar material. Thomas (1969) also studied the Ogmore breccias and the Trias filled fissures in the quarries of the Alun valley north of Ogmore Down.

In the north of the Vale of Glamorgan the Keuper oversteps onto the Lower and Middle Coal Measures and comes to rest against an almost vertical 'cliff' of Pennant Sandstone east of Llanharan. Here, even when overlying Coal Measures, the basal beds of the Keuper are coarse angular limestone breccias and conglomerates. Over the Carboniferous Limestone of the Llanharry hematite mining area these breccias are often 100 to 150 ft (30 to 45 m) thick, rising to over 230 ft (70 m) west-north-west of Llanharry (Thomas 1961). Further west bands of conglomerate are overlain by red marls with green mottling. The conglomerate facies recurs as a 20 ft (6 m) bed near the top and the total thickness reaches several hundred feet. To the east the total thickness decreases and the coarse marginal facies is relatively more important; it contains breccias interbedded with red to buff-cream porcellaneous limestones and red marly micaceous sandstones (Woodland and Evans 1964).

Still further west, beyond the main part of the Vale, a small outlier of reddened Triassic limestone conglomerate rests, almost horizontally, on steeply dipping Carboniferous Limestone near Port Eynon in Gower; and a dark-red sandy marl

Plate 3.   Coarse Triassic breccia and conglomerate resting on Carboniferous Limestone, Ogmore-by-Sea, Glamorgan.

occurs in a chasm in Mewslade Bay (Strahan 1907). Major breccia infillings, known as 'gash breccias' and also containing clay, sandstone, stalagmite and calcite occur within the Carboniferous Limestone of the South Pembrokeshire coast near Bosherston, south-west of Tenby between Giltar Point and Shrinkle Haven, and at several inland localities (Dixon 1921, Thomas 1970, 1971). These features have been attributed to cavern collapse beneath the former Triassic surface (Dixon 1921) but Thomas (1971) advocates a primarily tectonic origin with secondary modification by preferential solution and a limited amount of collapse during the Trias. Thomas has pointed out that the blocks within the breccias do not show evidence of being waterworn and that the sides of the structures are gradational, not abrupt. Many of the structures can be related to Armorican shatter belts whilst others are related to mid-Tertiary earth movements.

Hematite Deposits

Between Llanharry and Rudry occur deposits of hematitic iron ore lying in the northerly dipping limestones of the southern limb of the South Wales Coalfield syncline (Sibly 1927).

The detailed occurrence and mineralogy of these deposits are described by Gayer & Criddle (1970). The ore bodies lie mainly within the limestones just below the base of the Namurian although they also extend up to 40 ft (12 m) into the overlying marginal Keuper breccias and conglomerates. The adjacent, unaltered, *Dibunophyllum* Zone limestones are typically crinoidal, coraliferous and shelly biosparites with horizons of shelly micrites and oolites. Secondary dolomitization has commonly produced a coarsely crystalline sugary dolomite. The ore-bodies are controlled by the intersections of north-north-west (Hercynian) faults with beds of grey shale which occur at several levels in the upper part of the limestone sequence.

The iron deposits consist of interbanded dark and light brown layers ranging between 1 mm and 1 m in thickness and containing both colloidal and crystalline hematite and goethite. Numerous cavities, ranging from microscopic sizes up to 0·5 m in diameter also occur. They are often lined with crystalline hematite, goethite, calcite and bipyramids of quartz.

Gayer and Criddle consider in detail the introduction of the ore in solution. The minor flunctuations between successive solutions produced the different mineral phases during a cycle involving the initial solution of the limestone, flocculation, deposition within the cavities, dehydration and further solution. These workers favour the overlying marginal Keuper as the source of the iron, considering it to be a late Keuper deposit as no similar mineralization is known in Jurassic or younger strata. They point out that derived fragments of hematite and broken bipyramids of quartz are known from the Rhaetic of St. Mary Hill (Crampton 1960).

Elsewhere in east Glamorgan small hematitic bodies are known in the Carboniferous Limestone at Garth near Taffs Well, Rudry, Cwrt-yr-ala and South Cornelly. They occur as replacements following joints, faults and bedding planes in finely crystalline dolomites and dolomitic limestones. Other similar bodies may exist south of Pencoed (Thomas 1961).

The origin of these ore-bodies and also of the red staining common in the

Carboniferous Limestone, is disputed. Strahan (1909), Sibly & Lloyd (1927) considered that the iron originated from the Keuper but O. T. Jones (1930) and Trotter (1942) believed that a more probable origin lay in the iron compounds derived from the Coal Measures, then being eroded further north in the coalfield. If this iron was carried by streams draining south into the Keuper basin it might be deposited when the streams entered the limestone sequence. The origin of similar hematite deposits near Bristol was considered by Kellaway (1967).

Tea Green and Grey Marls

These grey and greenish-grey, often calcareous, mudstones contain subordinate darker grey shales and thin, sometimes dolomitic, limestones. They are the youngest of the Keuper facies mudstones and are well seen in the Penarth area, particularly along the coast, and also to the west of Cardiff. Etheridge (1872) referred all these beds to the Tea Green Marls. Richardson (1905b) discussed the introduction, by Etheridge, of the Grey Marls as a distinct deposit lying between the Tea Green Marls and the Westbury Beds. Richardson also introduced, at this time, the Sully Beds for a part of the Grey Marls ('the topmost 14 ft or so—which contain Rhaetic fossils'), together with the overlying conglomerate and the limestones with *Liostrea* [*Ostrea*] *bristovi* (Richardson). The type locality was St. Mary's Well Bay, west of Lavernock. The conglomerate and *bristovi* limestones are here considered to form the basal unit of the Westbury Formation (Strahan & Cantrill 1912, Ivimey-Cook, 1962, 1970 p. 288). The top of the Tea Green and Grey Marls is taken at the irregular erosion surface and non-sequence below the conglomerate, this marks the change to the marine sediments of the Westbury Formation.

Whilst 'Tea Green and Grey Marls' is an apt description of the colour, though not of the lithology, of these beds it is felt that a separate unit of 'Grey Marls' is not useful in south Glamorgan. The gradual change of colour and lithology upwards can be ascribed to the increasingly marine condition spreading into the area. The earliest evidence of these is known from the thick Tea Green and Grey Marls of north Somerset which contain thick beds of grey shale with some organic remains including bivalves and miospores (Richardson 1911, Stevenson 1971).

Between Lavernock and Penarth the sequence is about 54 ft (16·4 m) thick and is dominated by grey, frequently calcareous, mudstones with some beds of pale grey, reddish-brown and greenish yellow mudstone. The base is taken at a horizon of lenticular gypsum resting on the top of the main mass of Red Marls. The mudstone around these lenticles is usually brecciated and displaced and the associated joint planes are lined with gypsum. Above this occurs 14 ft (4·3 m) of massive grey calcareous mudstone and another gypsiferous horizon. Both reddish-brown and grey beds are present in the overlying 8 ft (2·4 m) of locally massive but generally thin-bedded mudstone with rare gypsum lenticles. Above this level the bedding in the greyish mudstones is usually thin and often laminated. Locally irregular partings of fine breccia and conglomerate, slump bedding and 'suncracks' occur. At the top of the sequence at Lavernock are a pair of conspicuous, slightly dolomitic limestones, often divided by a shale parting. The base of each limestone may contain mudstone pellets and fine pebbles.

These beds are very poorly fossiliferous. Near Lavernock the former presence o invertebrates is shown by burrow infillings of about 10 mm diameter and up to 35 mm long in the calcareous mudstone 20 ft 6 in (6·25 m) below the Westbury Formation. In addition some of the fine conglomeratic partings yield rare scales of *Gyrolepis* and other fish fragments. The pyritized teeth of the amphibians *Masto-donsaurus* and '*Trematosaurus*', associated with bones and teeth of the reptile *Palaeosaurus*, *Sphaerodus* teeth and pebbles were recorded by Storrie (1894) from a 'pocket' about 6 ft (1·8m) down in the Tea Green and Grey Marls near Lavernock. Plant remains are also known and include *Otozamites sp.* and *Hirmerella* [*Cheirolepis*] *sp.* about 13 ft (4 m) below the top of the unit.

To the east, in the Newport area, only the Tea Green Marls are recorded. They consist of pale-yellowish and greenish-grey marls between 12 and 20 ft (3·7 to 6·1 m) thick. At Goldcliff an ossiferous sand fills runnels in the marl about 5 ft (1·5 m) below the top of the Marls. (Richardson 1905a). The sharp junction with the overlying Westbury Formation, as well as the lithology of the marls, suggests that the upper part of the Tea Green and Grey Marls is missing.

To the west of Cardiff, along the Vale of Glamorgan, the Tea Green and Grey Marls may thin and also pass laterally into a marginal facies. This frequently appears as a breccia or yellow dolomitic limestone (Strahan & Cantrill 1902). A particularly hard yellow dolomitic limestone occurs near St. Nicholas where it is known as the Cromlech Bed. The marginal facies south of the Cowbridge anticline often contains thickly bedded yellowish, sandy calcareous mudstones with numerous limestone pebbles. In the Bridgend area Strahan indicated that the marginal beds became more sandy whilst the marl facies is also seen again in the red and yellow marls near Pyle and North Cornelly (Strahan & Cantrill 1904). Willey (1966a) studied the Triassic of the Cowbridge area in detail and drew conclusions on the sequence in the area.

(b) The Penarth Group (Rhaetic)

Introduction

Exposures of beds proved to belong to the Rhaetian Stage of the Triassic are confined to the Vale of Glamorgan and to a few scattered localities in south Monmouthshire. They are also now known at depth in the Llanbedr (Mochras Farm) Borehole and may occur below the drift in north-east Wales (p. 299).

It was recognized from the earliest studies that the 'Rhaetic', as developed in the Penarth area, was one of the finest developments of these rocks in Britain. Bristow (1864) defined the 'Rhaetic or Penarth Beds' for the Geological Survey. It was primarily a lithostratigraphic unit and included all the beds between the top of the Red Marls and the base of the thin paper-shale horizon at the base of the Lower Lias limestones and clays with *Psiloceras* [*Ammonites*] *planorbis*. The proposed Penarth Group is also primarily lithostratigraphic and contains two Formations— given the names Westbury Formation and Lilstock Formation (Kent 1970), see also p. 295. The Group extends upwards from the base of the conglomerate and limestones with *Liostrea bristovi* (the upper part of the Sully Beds); through the Westbury Shales and the Cotham, Langport and Watchet Beds of Richardson (1911) to the same horizon of 'paper-shale' as was used by Bristow. This concept of the 'Rhaetic' was later developed by Etheridge (1872), Strahan & Cantrill, and Richardson. The upper boundary, chosen originally on lithostratigraphic grounds, approximates closely to the definition of the boundary between the Rhaetian and Hettangian Stages as established by Renevier (1864 p. 41, 89–92), and consequently to the boundary between the Triassic and Jurassic Systems. The base of the Penarth Group is a lithostratigraphic boundary at a horizon within the Rhaetian Stage. For this reason, and for its similarity with the stage name, the term 'Rhaetic' is considered inappropriate (George and others 1969, Pearson 1970). Although it is clear from faunal and palynological evidence that at least part of the Tea Green and Grey Marls are also Rhaetian in age there is no clear evidence as to the position of the Rhaetian-Norian Stage boundary in any British exposure.

South-east Glamorgan

Westbury Formation

The Westbury Formation consists primarily of medium to dark grey fairly fissile mudstones. These typically contain the bivalve *Rhaetavicula contorta* (Portlock) so receiving the name '*contorta* Shales'. Within the Formation occur horizons of calcareous silt, calcareous sandstone and argillaceous limestone which often contain shell and bone fragments. In the Penarth area each of these marks a slightly more turbulent phase and indicates the base of one of a series of cycles or rhythms of deposition (Ivimey-Cook 1962). Within each cycle the sediment tends to become finer upwards, with some fluctuations, and there is a concurrent change in the fauna. The most varied fauna is found in the lower half of each cycle whilst in the higher part the bivalves, which are the dominant element in the preserved fauna, often decrease in number and variety until they are very sparse. Some of the cycles have eroded tops but where they are more complete a thin transition back through silty shale is present. Eight primary cycles are developed in this area though two of these can be divided into sub-cycles. Within the Formation the marine fauna becomes gradually more diverse upwards; in the lowest cycle only a few genera, mainly bivalves and vertebrates, are present but in the highest (incomplete) cycle the fauna is considerably more diverse. Cycles have also been recorded in the 'Rhaetic' in the Bristol region (Hamilton 1962).

The base of the Westbury Formation is best seen at St. Mary's Well Bay (ST 176 677), where it is faulted against the marginal Triassic limestones, breccia and siltstones of the Sully area. East of this fault about 5 ft (1·5 m) of alternating grey-brown calcareous mudstones and grey-black shales of the Tea Green and Grey Marls can usually be seen above beach level. The top of the mudstone is channelled to a depth of about 4 in (0·1 m), the channels and irregularities being filled by 3 to 9 in (0·08–0·23 m) of mudstone conglomerate in an argillaceous matrix. The finer, upper part is overlain sharply by grey coquinoidal limestone rich in *Liostrea bristovi* (Richardson). This bed is often divided into two by a lenticular grey shale. The upper limestone is more massive and argillaceous than the lower and the oysters are broken and often bored. *Rhaetavicula contorta*, rare specimens of a large *Modiolus*, *Sphaerodus* teeth and wood fragments also occur. The top of the limestone is argillaceous and silty and has been eroded into channels later infilled with mudstone pellets and quartz grains. The *bristovi* limestones are confined to an area south and west of Penarth and represent the earliest phase of the marine transgression in Glamorgan.

Higher parts of the Penarth Group occur in the landslipped and overgrown cliff above this exposure; the base of the Lilstock Formation, just over 13 ft (4 m) thick here, is well seen close to the ramp leading to the beach.

A clearer section of the higher beds is seen just to the north of Lavernock Point (ST 187 682) where the Penarth Group is 33 ft 4 in (10·16 m) thick. The Westbury Formation, which is 23 ft 10 in (7·26 m) thick at Lavernock is also well exposed both here and at intervals along the coast to the north and around to Penarth Dock. At Lavernock the equivalent of the *bristovi* limestones of St. Mary's Well Bay is a thin wedge of shale capped by limestone, resting in the channelled top of the Tea Green and Grey Marls. The limestone is 2 to 4 in (0·05–0·1 m) thick on the foreshore but thins northwards to a line of nodules in the cliff. It is also extensively dissected by the channels which contain the overlying coarse, ossiferous, quartz conglomerate of Storrie's Fish Bed. This, unlike the Aust Bone Bed, contains only rare specimens of *Ceratodus* but a wide variety of other fish remains are present. These include the spines of *Hybodus* and *Nemacanthus*, teeth attributed to *Acrodus*, *Birgerta, Dalatias, Hybodus*, and *Saurichthys* and the scales of *Gyrolepis*, in association with the remains of other fish and reptiles (*Ichthyosaurus, Palaeosaurus* and *Plesiosaurus*.

Higher in the sequence the erosive phase at the base of the cycles is often less evident and deposition of shell-detrital, silty limestones marks the base of several cycles. The top of the nearly barren fine-grained grey mudstones of the cycle which started with Storrie's Fish Bed is only slightly eroded. It is overlain by a complex alternation of arenaceous, ossiferous and pyritic limestones and mudstones with thin lenses of calcareous siltstone. This sequence includes Storrie's Bone Bed, a locally decalcified, thin arenaceous limestone densely packed with comminuted fish remains. The thin mudstones contain a fauna of ostracods assigned by Anderson (1964) to the *Rhombocythere penarthensis* Zone. They also contain fragments of the cirripede *Eolepas rhaetica* (Moore) and bivalves including *Eotrapezium* and *R. contorta*.

The calcareous bases of the succeeding three cycles are characterized by the presence of *Chlamys valoniensis* (Defrance), *Placunopsis alpina* (Winkler), comminuted *R. contorta* and echinoid fragments. The overlying mudstones also contain *Eotrapezium concentricum* (Moore), *Lyriomyophoria postera* (Quenstedt), *Protocardia rhaetica* (Merian), *Tutcheria cloacina* (Quenstedt) and the pyritized remains of the small gastropod '*Natica*' *oppelii* (Moore). This gastropod sometimes becomes abundant in the otherwise nearly barren mudstones high in each cycle. Complete specimens of *Ophiolepis* occur in the siltier mudstones together with rare *Discinisca*, the polymorphinid *Eoguttulina* and fish remains.

About 3 ft (0·9 m) below the top of the Formation the lithology changes to pale grey calcareous silty mudstone. *Rhaetavicula contorta* is however still present, in abundance, to within a few inches of the 'intra-Rhaetic' unconformity which marks the top of the Formation. In the Penarth area these calcareous beds can be seen to occupy extensive channels cut in the underlying grey mudstone. The basal parts of the infilling can contain such quantities of mudstone fragments as to suggest a transition from one lithology to the other. A very diverse fauna of bivalves occurs in this basal siltier part and it also contains abundant specimens of the foraminifer *Eoguttulina liassica* (Strickland). The higher parts have a restricted fauna and become laminated.

The top of these silts is deeply cleft with vertical cracks extending down for over 1 ft (0·3 m) below the eroded surface. This erosional phase of the 'intra-Rhaetic unconformity' is most evident at Lavernock and decreases in significance to the north-west. This is away from the Lower Severn Axis postulated by Kellaway and Welch (1948). An unknown amount of the Westbury Formation is missing though, as the change to calcareous sedimentation is found over a wide area of Southern England within the top few feet of the top of the Formation and in beds containing *R. contorta*, the thickness of beds missing here is probably small.

Lilstock Formation

The characteristic pale grey-green rather soapy textured mudstones of the 'Cotham facies' are almost entirely missing at Lavernock. This is partly due to erosion, or non-deposition, and partly to a change in lithology to calcareous silt and siltstone. The Cotham Member at St. Mary's Well Bay contains 4 ft 6 in (1·37 m) of laminated pale grey calcareous silt in contrast to the 1 ft 6 in (0·46 m) present at Lavernock. At Lavernock the ripple marked silts contain thin lenticular calcilutites at the top and rest on a hard calcareous grit or sandstone with pseudo-ooliths. The latter forms a basal bed about 10 mm thick but also infills the deep cracks in the underlying Formation. The overlying Langport Member is represented by 10 in (0·27 m) of limestone, in three beds of which the highest is the thickest, and partings of calcareous silt. The very pale grey-white, buff weathering limestones are finer in grain size and purer than the Blue Lias (Hydraulic Limestones) of the Jurassic. A rich microfauna includes a variety of ostracods (Anderson 1964) both polymorphinid and lagenid foraminifera and numerous comminuted echinoid fragments. *Montlivaltia rhaetica* Tomes occurs at Penarth Head and recrystallized coral fragments are not uncommon. Only occasional bivalves of which *Liostrea*, *Modiolus* and *Plagiostoma* are the most common can be extracted from the limestones. The Watchet Member is represented by about 7 ft 2 in (2·18 m) of blocky grey calcareous silts with frequent small calcareous siltstone lenticles in the lower part. The higher beds are more argillaceous. Fine cross-bedded sandy lenticles and ripple marks occur. The microfauna is again abundant, particularly near the base and is similar to that of the Langport Member; it becomes scarcer upwards. The macrofauna, except where scattered lenses are rich in *Liostrea* or *Modiolus*, is generally sparse. The Lilstock Formation totals 9ft 6in. (2·9m) in thickness at Lavernock (Plate 4).

The Watchet Member is capped conformably and without any obvious break by an 11 in (0·28 m) bed of dark grey laminated calcareous silt (paper-shale), rich in comminuted fragments of *Liostrea* and *Modiolus* and frequent echinoid fragments. This, and the overlying alternation of grey-brown limestones and shales of the pre-*planorbis* Beds, form the lowest part of the *planorbis* Subzone of the Hettangian

Plate 4.   The Rhaetian and lowest Hettangian north of Lavernock Point, Glamorgan. The *planorbis* Subzone of the Hettangian (1) overlies beds of the Watchet Member (2), Langport Member (3) and Cotham Member (4). The Westbury Formation (5) is mainly covered by debris. The limestones of the Tea Green and Grey Marls (6) stand out near the base of the cliff. The rule is marked in 1 ft. (0.3m) bands.

Monmouthshire and the littoral facies of west Glamorgan

Eastwards from Cardiff the Penarth Group is not seen again until the Newport area is reached (Richardson 1905a, Squirrell & Downing 1969). Few of the recorded sections are still visible but the sequence is known to thin eastwards and the lithology changes, particularly in the 'Cotham Beds' towards Aust, Sedbury Cliff and Westbury-upon-Severn (Welch & Trotter 1961). At Goldcliff, Richardson (1905a) records 12 ft 3 in (3·7 m) of grey-black shales with four beds of limestone forming the 'Westbury Beds'. The 'Upper Rhaetic' is only 5 ft 6 in (1·68 m) thick and is composed mainly of 'Cotham-type' greenish-grey shales including a Cotham Marble equivalent and a thin '*Estheria*' Bed limestone (but without *Naiadita*). The top of the upper limestone is bored and overlain non-sequentially by the *planorbis* Zone. The Penarth Group is similarly developed at Sedbury Cliff, Gloucestershire, near Chepstow (Richardson 1903). The continued reduction in thickness and presence of pronounced non-sequences can be attributed to activity of the Lower Severn Axis.

To the south-west of the Penarth area, Banner and others (1971) recovered core samples locating the Triassic outcrops south of Barry beneath the Bristol Channel. In mid and west Glamorgan the facies of the Group changes in response to the proximity of the remaining emergent Carboniferous Limestone islands—principally the Cowbridge-Southerndown island. To the north of this island the sequence contains substantial amounts of terrigenous sediment, attributed by George (1970) to the erosion and reworking of Millstone Grit and Coal Measures in the area north of Bridgend. In that area the Group is now found faulted against the Upper Carboniferous and the old shore may have been composed of arenaceous rocks in contrast to the earlier, dominantly limestone, shore. The littoral beds also differ from those of the marginal Keuper in containing both oolitic and bioclastic limestones in addition to derived limestone debris. Crampton (1960) discussed the heavy mineral assemblages of samples from the Keuper and Rhaetic in the Vale of Glamorgan.

Most of the exposures near Cowbridge lie on the south of this last remaining major island. Those around St. Hilary and to the south and west were recorded by Strahan & Cantrill (1904), Richardson (1905b) and recently by Willey (1966a). As the grey mudstones of the Westbury Formation transgress westwards they become more arenaceous with thin conglomerates and sandstones; whilst the overlying beds include white limestones reminiscent of the Sutton Stone. In the southern part of the Cowbridge-Aberthaw railway cutting 19 ft 8 in (6 m) of marls, limestones and thin conglomerates occur; these thin northwards to arenaceous limestones and sandy oolites near St. Mary Church and to a thinner arenaceous sequence at Ty Ganol, near Bonvilston. The latter occurrence reflects the drift of arenaceous sediment around the eastern end of the island from the northern shoreline. Francis (1959) correlated the arenaceous beds as part of the Westbury Beds. Harris (1957) described the abundant carbonized fragments of the wood, bark, leaves and seeds of *Hirmerella* [*Cheirolepis*] from the Rhaeto-Liassic clay-filled fissures in the Cnap Twt and Ewenny quarries in this island. Associated with these fragments are carbonized beetle elytra of *Metacupes* (Gardiner 1961) and a few dispersed mega-spores and miospores including *Bacutriletes*, *Classopollis* and *Heliosporites*. Harris (*op. cit.*) attributed this combination to periodic charring, during forest fires, of vegetation dominated by the conifer *Himerella*. Another local plant was *Otozamites bechei* Brongniart, fragments of whose leaves occur in the Cowbridge railway cutting (Harris 1961) whilst the conifer *Mesembrioxylon rhaeticum* McLean was found near Penarth.

North of the Cowbridge island the arenaceous beds not only dominate the sequence but also have a striking effect on the topography. Francis (1959) describes two thick horizons of sandstone. The Lower Sandstone, which reaches 20 ft (6·1 m) in thickness west of Bridgend, is a white or yellowish, medium grained sandstone resting unconformably on the Keuper. Near Coity (SS 929 806) it contains *Eotrapezium sp.*, '*Gervillia*' *praecursor* (Quenstedt) and *Lyriomyophoria postera*. Between this and the Upper Sandstone are up to 6 ft 7 in (2 m) of black shales. A thin pebbly bed at their base contains poorly preserved fish and reptile remains whilst at Pyle Inn Quarry the black-green shales contain two nodular limestones and *R. contorta*. The Upper Sandstone (Quarella Sandstone) reaches 35 ft (10·7 m) in thickness and may overlap the Lower Sandstone east of Bridgend. The basal part is a fine-grained, thin-bedded, soft, brown sandstone often with abundant '*Natica*' *oppelii* (= *N. pylensis* Tawney) and has intercalated brown and green shales with *R. contorta*. The middle is coarser grained and locally calcareous, whilst the topmost few feet of sandstone are hard, mottled and calcareous. These beds probably all correlate with the Westbury Formation. Overlying them are up to 20 ft (6·1 m) of soft greenish marls, concretionary limestones, thin detrital limestones and pockets of reddish and green sandstone. All of these were attributed to the Cotham Beds by Francis (1959). A coarse littoral facies of pebbly sandstones is found on Stormy Down and at Tythegston. The quarries for silica brick at Stormy Down yielded the massive jaw, with curved teeth, belonging to the teratosaurid dinosaur '*Zanclodon*' *cambrensis* Newton. The Upper Sandstone was also extensively quarried for building stone at Quarella, on the north of Bridgend, where about 23 ft (7 m) of sandstone were worked (Strahan & Cantrill 1904).

Although Bristow (1867) and others showed that Tawney's (1866) view of the order of the Sutton, Southerndown and *contorta* Series was untenable, Tawney did record (p. 77) a section at Cwrt(-y-) Coleman, north-west of Bridgend which included 7 ft (2·13 m) of white shelly limestone with '*Myophoria postera*' overlain by 4 ft (1·22 m) of 'mottled green and yellow marls' and 2 ft (0·61 m) of red marls. At Laleston (churchyard) he found 'Sutton type' white limestone beneath indurated marl with *Meleagrinella* [*Monotis*] *decussata*, (Münster), well preserved plants and ostracods. These beds all lie below the '*Ostrea*' limestones of the Lias. At Stormy Lime and Cement Works he saw 9 ft (2·74 m) of dark and pale green marl overlain by 1½ ft (0·46 m) of *contorta* shales with '*Pecten valoniensis*' and then an alternation of grey smooth limestones and dark brown-grey shales. The latter were described in detail by Woodward (1893, p. 114). Bristow (1867) recorded 12–15 in (0·30–0·38 m) of pale argillaceous limestone below 2 ft (0·61 m) of siliceous shelly conglomerate (?Sutton Stone) below Liassic limestones in the Stormy Works. These pale limestone were assigned to the 'White Lias as the uppermost member of the Rhaetic series'. Thus above the main group of greenish marls there seems to be evidence of local thick developments of white limestones, shales and marls belonging to the highest parts of the Penarth Group.

V. PALAEOGEOGRAPHY AND ENVIRONMENT

The palaeogeography of the Triassic of the British Isles has recently been reviewed by Audley-Charles (1970b) and Warrington (1970). The occurrence of late Triassic rocks in the Llanbedr Borehole and the probability of their occurrence elsewhere in the Irish Sea somewhat modifies the picture then presented. In particular, the correlation of the thick Lower Jurassic sequence in this borehole with those in Cheshire and Glamorgan cast doubt on the long continued existence of a 'Welsh massif' into the Jurassic. It is reasonable to conclude that it may have been of only low relief in the late Triassic, particularly in the Rhaetian.

Little can be deduced as to the relative disposition of areas of erosion and deposition during the early and middle Triassic within Wales, though it seems probable that the area was the source for substantial amounts of the sediments deposited elsewhere. More evidence is available to provide some detail to the palaeogeography of Wales in the Norian and Rhaetian ages of the late Triassic.

Stretching northwards across the South Wales Coalfield through central and north Wales lay a moderately low-lying, and now dominantly sinking, land area. To the east lay the extensive, largely landlocked, shelf 'sea' covering central England. This was linked to the south, through a strait, to another shelf 'sea'. This southern 'sea' covered south Glamorgan, the Bristol Channel, most of Gloucestershire and extensive areas of the south of England. Near its northern margin lay the islands of the Mendip archipelago and south Glamorgan. To the south a deep channel stretched through central Somerset to the English Channel. This marked the site of an earlier river system bringing sediment northwards to the Midlands. These very extensive lakes or small seas received torrential and stream deposits from the remaining uplands and also wind blown detritus. They may also have been supplied by other rivers from the south and east. The lack of halite in the evaporites of Glamorgan may reflect the diluting effect of fresh water draining south from the main Welsh island, whilst to the south, in Somerset, salt deposits were being formed in the Keuper. By the end of the Norian a marine influence may have become established in the west and south-west as Stevenson (1970) noted coastal type sabka environments in the early Rhaetian of north Somerset. Marine conditions were soon to extend rapidly to the north and east as the marine transgression of the Rhaetian. This could have been produced by the opening of a North Atlantic seaway in the late Triassic so opening up the extensive Keuper lacustrine complex to marine conditions.

Within the Vale of Glamorgan the islands of Carboniferous Limestone developed underground drainage systems controlled by the distribution of jointing, faults and dip surfaces. In the east of the Vale the islands were soon flooded and only immature slot fissures are found, these became filled with red mudstone during the deposition of the Keuper Red Marls. In the centre and west of the Vale these underground drainage systems existed for sufficient time to become much larger and mature; some were filled with Norian detritus including vertebrate fragments. In parts of the more westerly islands other watercourses developed whose mature systems were filled with Rhaeto-Liassic debris (Robinson 1971).

Around the islands accumulations of debris built up as a result of their erosion and dissolved carbonates were also reprecipitated to produce new limestones. The deposits of mudstone formed in the main body of the Keuper 'sea' appear to be laterally extensive suggesting deposition in one or more very large bodies of water rather than a complex of fluctuating lakes and intermittently flooding rivers. In particular the even bedded, red calcareous mudstone facies of the Red Marls were laid down under water without important breaks in deposition. They only rarely contain beds (e.g. evaporites) which suggest a reduced water cover. Most of the structures which suggest exposure (mudcracks, rainprints) are either in the marginal facies or in the Tea Green and Grey Marls which may have been deposited in conditions very close to wave base and with intermittent exposure.

Davis (1967, 1968) suggested that the Keuper Marl was deposited in a broad, shallow, epicontinental seal into which rivers supplied mud and the wind transported dust; these were then deposited in the basic, saline to hypersaline, waters. He estimated the water temperature at between 30° and 40°C and considered that the carbonates present were formed immediately after deposition in association with the formation of corrensite, an expanded chlorite. The local concentration of magnesium in the pore spaces of the sediments was suggested to favour dolomitization especially in non-turbulent conditions. Temperatures of this order would certainly restrict land based animals to the vicinity of fresh-water whilst aquatic forms would be deterred by the high salinity.

The lack of hematite coated particles in the Tea Green and Grey Marls contrasts with the Red Marls and is evidence for a major change in the environment. This change is also shown by the occurrence of rare marine fossils, including *Rhaetavicula*, in the upper part of these beds in north Somerset (Richardson 1911). In Glamorgan the only known marine forms are vertebrates.

The Westbury Formation was laid down in a normal, though shallow, marine environment. The fluctuating level of the water in this sea is however evident from the local development of channels, winnowed 'bone-bed' conglomerates, laminae of comminuted shell fragments and the cyclic deposits developed in response to minor movements of the Lower Severn Axis. The fauna was dominated by bivalves and contained both burrowing and surface-living forms. The remains were generally endemic though often sorted and locally transported by current or wave action. The fauna also contained foraminifera, one genus of brachiopod, gastropods, cirripedes, ostracods and marine vertebrates. The lack of ammonites can be attributed to the general rarity of Rhaetian ammonites especially in shallow water mud environments. Where particularly quiet water conditions developed, as towards the tops of the cycles, the fauna was reduced in variety and numbers. Although the fauna of these beds is often reputed to consist of dwarfed or stunted forms, evidence from the Lavernock area (Ivimey-Cook 1962) does not support this suggestion. Comparison of the size distributions within several species in the fauna, from different horizons localities, indicate normal patterns of growth. Slight changes in aeration or and may however account for the fluctuations of species abundance within the salinity series of cycles.

During the deposition of the Lilstock Formation the widespread occurrence of minor non-sequences, small channels, ripple marks and small lenses of sediment in a carbonate sequence suggests fairly shallow and warm waters. Within these conditions a variety of environments appeared. These supported faunas which range from local areas of possibly brackish-water into the predominant marine sequences. The extensive microfauna contains a variety of both foraminifera and ostracods, whilst the frequent presence of echinoid fragments again points to the marine nature of the environment. The bivalve fauna is in part inherited from species present in the upper part of the Westbury Formation. Of the 19 genera of fairly common bivalves which are first seen in the Westbury Formation, 3 do not range outside it (*Lyriomyophoria*, '*Pleurophorus*' and *Rhaetavicula*), of the remainder 7 persist into the *planorbis* Subzone. Four new genera (*Meleagrinella*, *Plagiostoma Pleuromya* and *Plicatula*) first become common in the Lilstock Formation and all of these persist into the Hettangian. The *planorbis* Subzone also records the appearance of 12 new genera of bivalves as well as the first ammonites in this area. On land the fauna included small reptiles and early mammals whose remains have been preserved because they were washed into fissures in the conifer clad islands.

At the close of the Rhaetian, marine deposition over much of east and southeast England was abruptly, but temporarily, terminated by a short period of very gentle uplift. Regionally a different pattern appeared for the areas of deposition in the early Hettangian. In the Mendips the thickly developed 'White Lias' facies of the Lilstock Formation was succeeded by a condensed Lower Jurassic sequence, whilst around the western end of the Oxfordshire-Ardennes positive area the successive zones of the Lower Jurassic can be seen to overlap and overstep the Lilstock Formation. In Glamorgan, the Bristol Channel, the Severn valley and Cheshire, the basins began to subside again on a regional scale and this change was marked by a sudden change in facies. The lithological change to alternating fissile mudstones and limestones suggest both deeper waters and a greater supply of terrigenous material. The new environment gradually acquired a distinctive fauna and its appearance marks the second phase of the marine transgression which soon submerged the whole of Wales beneath the Jurassic sea.

VI. REFERENCES

ANDERSON, F. W. 1964. Rhaetic ostracods. *Bull. geol. Surv. Gt. Br.*, No. 21, 133–174.
AUDLEY-CHARLES, M. G. 1970a. Stratigraphical correlation of the Triassic rocks of the British Isles. *Quart. J. geol. Soc. Lond.*, **126**, 19–47.
——. 1970b. Triassic palaeogeography of the British Isles. *Quart. J. geol. Soc. Lond.*, **126**, 48–89 (including discussion of both above papers).
BANNER, F. T., BROOKS, M. and WILLIAMS, E. 1971. The Geology of the approaches to Barry, Glamorgan. *Proc. geol. Ass.*, **82**, 231–47.
BLUCK, B. J. 1965. The Sedimentary History of some Triassic Conglomerates in the Vale of Glamorgan, South Wales. *Sedimentology*, 4, 225–45.
BRISTOW, W. H. 1864. On the Rhaetic or Penarth Beds of the neighbourhood of Bristol and the south-west of England. *Geol. Mag.*, **1**, 236–39.
——. 1867. On the Lower Lias or Lias-Conglomerate of a part of Glamorganshire. *Quart. J. geol. Soc. Lond.*, **23**, 199–207.
COX, A. H., and TRUEMAN, A. E. 1936. The Mesozoic Rocks, 19–59. *In* TATTERSALL, W. M.

(Editor). *Glamorgan Country History*, **1**, *Natural History*. Cardiff.

CRAMPTON, C. B. 1960. Petrography of the Mesozoic succession of South Wales. *Geol. Mag.*, **97**, 215–28.

DAVIS, A. G. 1967. The mineralogy and phase equilibrium of Keuper Marl. *Quart. J. engng. Geol.*, **1**, 25–38.

——. 1968. The structure of Keuper Marl. *Quart. J. engng. Geol.*, **1**, 145–53.

De la BECHE, H. T. 1846. On the Formation of the rocks of South Wales and South Western England. *Mem. geol. Surv. Gt. Britain*, **1**, 1–296.

DIXON, E. E. L. 1921. The Geology of the South Wales Coalfield, Pt. XIII, The Country around Pembroke and Tenby. *Mem. geol. Surv. Gt. Britain*.

DOUBLE, I. S. 1926. The Petrography of the Triassic rocks of the Vale of Clwyd. *Proc. Lpool. geol. Soc.*, **14**, 249–62.

DUMBLETON, M. J. 1967. Origin and mineralogy of African red clays and Keuper Marl. *Quart. J. engng. Geol.*, **1**, 39–45

ETHERIDGE, R. 1872. On the Physical Structure and Organic remains of the Penarth (Rhaetic) Beds at Penarth and Lavernock, also with description of the Westbury-on-Severn section. *Trans. Cardiff Nat. Soc.*, **3**, 39–64.

FRANCIS, E. H. 1959. The Rhaetic of the Bridgend District, Glamorganshire. *Proc. geol. Ass.*, **70**, 158–70.

GARDINER, B. G. 1961. New Rhaetic and Liassic beetles. *Palaeontology*, **4**, 87–9.

GAYER, R. A., and CRIDDLE, A. J. 1970. Mineralogy and genesis of the Llanharan iron ore deposits, Glamorgan. *Proc. 9th Commonw. min. metall. Cong.* 1969, Paper 15, 1–22; 605–25 of Volume.

GEORGE, T. N. 1961. North Wales. 3rd Edition. British Regional Geology.

——. 1970. South Wales. 3rd Edition. British Regional Geology

—— and others. 1969. Recommendations on stratigraphical useage. *Proc. geol. Soc. Lond.*, No. 1656, 139–166.

HAINS, B. A., and HORTON, A. 1969. Central England. 3rd Edition. British Regional Geology.

HAMILTON, D. 1962. Some notes on the Rhaetic sediments of the Filton By-Pass substitute, near Bristol. *Proc. Brist. Nat. Soc.*, **30**, 279–85.

HARRIS, T. M. 1957. A Liasso-Rhaetic flora in South Wales. *Proc. R. Soc.* (*B*), **147**, 289–308.

——. 1961. *Otozamites bechei* Brongniart from the Rhaetic of County Antrim and its comparison with British specimens. *Proc. R. Irish. Ac.*, **61(B)**, 341–344.

HARRISON, R. K. 1971. The Petrology of the Upper Triassic Rocks in the Llanbedr (Mochras Farm) Borehole. 37–72 *in* WOODLAND, A. W. (Ed.). 1971. The Llanbedr (Mochras Farm) Borehole. *Rep. No.* 71/18. *Inst. Geol. Sci.*, 115 p.

IVIMEY-COOK, H. C. 1962. On the relationship between the fauna and the sediments of the Rhaetic in south-east Glamorgan. Ph.D. thesis, University of Wales, Cardiff.

——. 1969. *In A. Rep. Inst. Geol. Sci. for* 1968, p. 98.

——. 1970. *In* General discussion on Symposium 'Triassic rocks of the British Isles'. *Quart. J. geol. Soc. Lond.*, **126**, 288.

——. 1971. Stratigraphical Palaeontology of the Lower Jurassic of the Llanbedr (Mochras Farm) Borehole, 87–92; *In* WOODLAND, A. W. (Ed.) 1971. The Llanbedr (Mochras Farm) Borehole. *Rep. No.* 71/18, *Inst. Geol. Sci.* 115 p.

JONES, O. T. 1930. Some aspects of the geological history of the Bristol Channel region. *Rep. Br. Ass. Adv. Sci.*, 57–82.

KELLAWAY, G. A. 1967. The Geological Survey Ashton Park Borehole and its bearing on the geology of the Bristol district. *Bull. geol. Surv. Gt. Br.*, No. 27, 49–124.

—— and WELCH, F. B. A. 1948. Bristol and Gloucester District. 2nd Edition. British Regional Geology.

KENT, P. E. 1970. Problems of the Rhaetic in the East Midlands. *Mercian Geologist*, **3**, 361–72.

KERMACK, D. M., KERMACK, K. A. and MUSSETT, F. 1968. The Welsh Pantothere *Kuehneotherium praecursoris. J. Linn. Soc.* (*Zool.*), **47**, 407–23.

KERMACK, K. A., and MUSSETT, F. 1959. The first Mammals. *Discovery*, **20**, 144–51.

KLEIN, G. de VRIES. 1962. Sedimentary Structures in the Keuper Marl (Upper Triassic). *Geol. Mag.*, **99**, 137–44.

KÜHNE, W. G. 1949. On a triconodont tooth of a new pattern from a fissure-filling in south Glamorgan. *Proc. zool. Soc. Lond.*, **119**, pt. II, 345–50.

MISKIN, F. F. 1922. The Triassic rocks of South Glamorgan. *Trans. Cardiff Nat. Soc.*, (for 1918), **52**, 17–25.

MOORE, C. 1867. On abnormal conditions of Secondary deposits when connected with the Somersetshire and South Wales Coal-Basin; and on the age of the Sutton and Southerndown Series. *Quart. J. geol. Soc. Lond.*, **23**, 449–568.

NORTH, F. J. 1916. On a boring for water at Roath, Cardiff. *Trans. Cardiff Nat. Soc.*, **48**, 36–49.

ORBELL, G. 1971. Spores and pollen grains from three sections in the south of England and their relationship to the Swedish Rhaeto-liassic macroflora. *Proceedings* In *J. geol. Soc. Lond.*, **127**, 299–300.

PEARSON, D. A. B. 1970. Problems of Rhaetian stratigraphy with special reference to the lower boundary of the stage. *Quart. J. geol. Soc. Lond.*, **126**, 125–50.

PERRIN, R. M. S. 1971. *The clay mineralogy of British sediments.* Mineralogical Society, London.

POOLE, E. G., and WHITEMAN, A. J. 1966. Geology of the country around Nantwich and Whitchurch. *Mem. geol. Surv. Gt. Britain.*

READE, T. M. 1891. The Trias of the Vale of Clwyd. *Proc. Lpool geol. Soc.*, **6**, 278–89.

RENEVIER, E. 1864. Infralias et Zône à Avicula contorta (Ét. Rhaetien) des Alpes Vaudoises. *Bull. Soc. Vaud. Sci. Nat.*, **8**, 39–97.

RICHARDSON, L. 1901. Mesozoic geography of the Mendip Archipelago (with appendix). *Proc. Cottes. Nat. F. C.*, **14**, 59–73.

——. 1903. The Rhaetic and Lower Lias of Sedbury Cliff, near Chepstow (Monmouthshire). *Quart. J. geol. Soc. Lond.*, **59**, 390–95.

——. 1905a. The Rhaetic rocks of Monmouthshire. *Quart. J. geol. Soc. Lond.*, **61**, 374–84.

——. 1905b. The Rhaetic and contiguous deposits of Glamorganshire. *Quart. J. geol. Soc. Lond.*, **61**, 385–424.

——. 1911. The Rhaetic and contiguous deposits of West, Mid and part of East Somerset. *Quart. J. geol. Soc. Lond.*, **67**, 1–72.

ROBINSON, P. L. 1957. The Mesozoic fissures of the Bristol Channel area and their vertebrate faunas. *Proc. Linn. Soc. (Zool.)*, **43**, 260–82.

——. 1971. A problem of faunal replacement on Permo-Triassic continents. *Palaeontology*, **14**, 131–53.

SHOTTON, F. W. 1956. Some aspects of the New Red Desert in Britain. *Lpool and Manchr geol. J.*, **1**, 450–65.

SIBLY, T. F., and LLOYD, W. 1927. Special reports on the Mineral Resources of Great Britain, Vol. 10, Iron ores, the Heamatites of the Forest of Dean and South Wales, 2nd Edition. *Mem. geol. Surv. Gt. Britain.*

SQUIRRELL, H. C., and DOWNING, R. A. 1969. The geology of the South Wales Coalfield. Part 1. Geology of the country around Newport (Monmouthshire). 3rd Edition. *Mem. geol. Surv. Gt. Britain.*

STEVENSON, C. R. 1970. In discussion, p. 81. *In* AUDLEY-CHARLES, M. G. 1970b. (*op. cit*).

STORRIE, J. 1894. Notes on the tooth of a species of *Mastodonsaurus*, found with some other bones near Lavernock. *Trans. Cardiff Nat. Soc.*, **26**, 105–6.

STRAHAN, A. 1885. The geology of the coasts adjoining Rhyl, Abergele and Colwyn. *Mem. geol. Surv. Gt. Britain.*

——. 1890. The geology of the neighbourhood of Flint, Mold and Ruthin. *Mem. geol. Surv. Gt. Britain.*

——. 1907. The geology of the South Wales Coalfield. Part IX. West Gower and the country around Pembrey. *Mem. geol. Surv. Gt. Britain.*

——. 1909. The geology of the South Wales Coalfield. Part 1. The country around Newport, Monmouthshire. *Mem. geol. Surv. Gt. Britain.*

—— and CANTRILL, T. C. 1902. The geology of the South Wales Coalfield. Part III. The country around Cardiff. *Mem. geol. Surv. Gt. Britain.*

——. 1904. The geology of the South Wales Coalfield. Part VI. The country around Bridgend. *Mem. geol. Surv. Gt. Britain.*

——. 1912. The geology of the South Wales Coalfield. Part III. The country around Cardiff. 2nd Edition. *Mem. geol. Surv. Gt. Britain.*

TAWNEY, E. B. 1866. On the western limit of the Rhaetic Beds in South Wales and on the position of the 'Sutton Stone'. *Quart. J. geol. Soc. Lond.*, **22**, 69–89.

THOMAS, T. M. 1952. Notes on the structure of some minor outlying occurrences of Littoral Trias in the Vale of Glamorgan. *Geol. Mag.*, **89**, 153–62.

——. 1961. *The Mineral Wealth of Wales and its exploitation.* Oliver and Boyd; Edinburgh and London.

——. 1968. A new occurrence of celestite, near Llantrisant, Glamorgan. *Geol. Mag.*, **105**, 185–86.

——. 1969. The Triassic rocks of the west central section of the Vale of Glamorgan with particular reference to the 'Boulder' Breccias at Ogmore-by-Sea. *Proc. geol. Ass.*, **79**, 429–39.

——. 1970. Field Meeting of the South Wales Group on the Stack Rocks to Bullslaughter Bay section of the South Pembrokeshire Coast; Report by the Director. *Proc. geol. Ass.*, **81**, 241–48.

——. 1971. Gash breccias of South Pembrokeshire: fossil karstic phenomena? *Trans. Inst. Br. Geog.*, No. 54, 89–100.

TROTTER, F. M. 1942. Geology of the Forest of Dean Coal and Iron-ore field. *Mem. geol. Surv. Gt. Britain.*

WARRINGTON, G. 1970. The stratigraphy and palaeontology of the 'Keuper' Series of the central Midlands of England. *Quart. J. geol. Soc. Lond.*, **126**, 183–223.

——. 1971. Palynology of the Upper Triassic strata in the Llanbedr (Mochras Farm) Borehole, pp. 73–85. *In* WOODLAND, A. W. (Ed.) 1971. The Llanbedr (Mochras Farm) Borehole. *Rep. No.* 71/18, *Inst. Geol. Sci.*, 115 p.

WEDD, C. B., and KING, W. B. R. 1924. The geology of the country around Flint, Hawarden and Caergwrle. *Mem. geol. Surv. Gt. Britain.*

——, SMITH, B. and WILLS, L. J. 1928. The geology of the country around Wrexham. Part II. Coal Measures and Newer Formations. *Mem. geol. Surv. Gt. Britain.*

WELCH, F. B. A., and TROTTER, F. M. 1961. Geology of the country around Monmouth and Chepstow. *Mem. geol. Surv. Gt. Britain.*

WILLEY, E. C. 1966a. On the Mesozoic strata of the Cowbridge District, Glamorgan. Ph.D. thesis. University of Wales, Cardiff.

——. 1966b. Nomenclature of the Keuper-Lias facies of the Bristol Channel Region. *Nature*, **211**, 398–399.

WILLS, L. J. 1948. *The palaeogeography of the Midlands.* London, Liverpool.

WILSON, C. D. V. 1959. Geophysical investigation in the Vale of Clwyd. *Lpool and Manchr geol. J.*, **2**, 253–70.

WOOD, A., and WOODLAND, A. W. 1971. Introduction, pp. 1–10 *In* WOODLAND, A. W. (Ed.) 1971. The Llanbedr (Mochras Farm) Borehole. *Rep. No.* 71/18 *Inst. geol. Sci.*, 115 p.

WOODLAND, A. W. and EVANS, W. B. 1964. The Geology of the South Wales Coalfield, Part IV, Pontypridd and Maesteg. 3rd Edition. *Mem. geol. Surv. Gt. Britain.*

WOODWARD, H. B. 1893. The Jurassic Rocks of Britain, Vol. III, The Lias of England and Wales (Yorkshire excepted). *Mem. geol. Surv. Gt. Britain.*

WRIGHT, J. E., HULL, J. H., McQUILLAN, R. and ARNOLD, S. E. 1972. (for 1971). Irish Sea Investigations 1969–70. *Rep. No.* 71/19, *Inst. geol. Sci.*, 55 p.

THE JURASSIC PERIOD IN WALES

D. V. Ager

I. INTRODUCTION

TO write about the Jurassic rocks of Wales seems almost a contradiction. In fact the schools in Wales have, until recently, excluded the Mesozoic from their A level syllabus. But the Jurassic rocks of Wales are both gloriously exposed and highly deserving of attention. What is more, paradoxically, when considering the Jurassic Period in Wales, the areas without sediments are even more interesting than those with, since the crucial question is whether or not the unfortunately named 'St. George's Land' ever existed at this time. It is therefore proposed to consider first the actual exposures in the Vale of Glamorgan, then the evidence from surrounding areas, including the remarkable evidence of the Mochras borehole, and finally to try to bring this all together in considering the palaeogeography of Wales during the Jurassic Period.

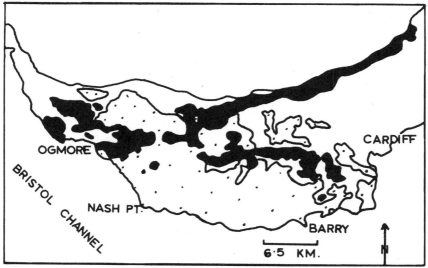

Fig. 88. St. David's Archipelago.

Sketch map showing the outcrops of Carboniferous Limestone (black) and Lower Jurassic (stippled) in South Wales (after Wobber, 1968a).

II. THE GLAMORGAN COAST

The magnificent Liassic cliffs of Glamorgan are probably the most neglected of all the much-studied Jurassic cliffs of Britain. They extend for 25 miles (c. 40 km) from Cardiff to Ogmore with 14 miles of high cliffs from Barry to Nash Point and Dunraven (fig. 88). Inland exposures are few but do exist and add to the standard story a third dimension which has commonly been overlooked.

323

Until recently, these rocks were known chiefly through the classic work of Trueman (1920, 1922a, 1930). Chronostratigraphically these beds extend from the so-called 'pre-planorbis Beds' through the *planorbis, liasicus, angulata, bucklandi* and *semicostatum* Zones of the Hettangian and Sinemurian (see Table 9). No later rocks, other than the Quaternary and the doubtful Tertiary pocket deposits in Pembrokeshire, Flintshire and Denbighshire are known *in situ* anywhere in Wales. There is, however, a strong probability that later Liassic strata and even post-Liassic Jurassic strata once covered considerable areas of South Wales. This is suggested by the presence of later Liassic strata of considerable thickness in the Mochras (Llanbedr) borehole in Merionethshire and of an almost complete Jurassic succession in the Bristol Channel. Both of these occurrences are discussed below, but on the mainland of South Wales the Sinemurian sediments have an eroded top and the rest of the Jurassic is sadly lacking.

TABLE 9

Zonal succession in the Lower Jurassic of Glamorgan (from Hallam 1960)

zones	sub-zones	thickness in metres
semicostatum	*gmuendense*	–
	bucklandi	2·70
bucklandi	*rotiforme*	3·90
	conybeari	23·40
angulata	*angulata*	18·00
	laqueus	18·00?
planorbis	*johnstoni*	6·60
	planorbis	6·60

The eastern part of the outcrop, that is east of Trwyn-y-Witch (*c.* 3 km south-south-east of Ogmore-by-Sea) consists of what may be regarded as the 'normal' Lower Lias, as seen on the Dorset coast and across the other side of the Bristol Channel from Glamorgan in north Somerset. This is the well-known 'Blue Lias' with its alternations of shales and argillaceous limestones. It reaches a thickness of nearly 100 m and follows conformably on top of Rhaetian strata near Cardiff. There seems to be no doubt that the classical interpretation of this as an off-shore shelf facies is correct. Hallam's study of the Blue Lias of Dorset and Glamorgan (1960) led him to the conclusion that it was deposited in a tropical or sub-tropical sea, by today's standards, at a depth of not more than 150 m. Bottom currents he thought were weak, with periodic stagnation of the bottom waters. The primary or secondary nature of the limestone bands has long been a matter of dispute, but Hallam came down strongly in favour of a primary origin, as he had in an earlier paper (1957), though he showed that there had been a limited amount of early diagenetic segregation of calcium carbonate producing the nodular appearance of many of the limestones.

Hallam also noted that there were two distinct types of limestone in the Dorset and Glamorgan sections. By far the most common type may or may not be nodular but is never laminated and contains abundant shelly material. The other type is never nodular, but always laminated and contains very little fossil material. Hallam's conclusion was that the laminated limestones (and the occasional finely laminated shales or 'paper shales') were laid down in anaerobic conditions, when there were no organisms on or in the bottom sediments to destroy the lamination by their burrowing activity.

The limestones in general were thought to have been formed by simple re-crystallization of an original lime mud and the alternations of limestones and shales or mudstones were produced by small-scale oscillations in sea-level every few thousand years or so. Nearly all the carbonate present (in both limestones and shales) proved to be calcite and the clay mineral content was almost entirely illite, with subordinate kaolinite. Bottom conditions were interpreted as quiescent and remarkably constant over a very wide area, though sedimentation was probably more rapid in Glamorgan than in Dorset coupled with a higher rate of subsidence, but even so did not exceed a fraction of a millimetre per year. The sort of evidence that is readily seen in the coast sections to confirm this interpretation is, for example, the presence of beds of the triangulate bivalve *Pinna hartmanni* Zieten in life position. In life we know that these molluscs live partly buried in the sediment and only slight disturbance or reworking of the bottom sediments is needed to lay them on their sides.

Trueman showed that the 'normal' facies of the Hettangian in the cliffs between Lavernock and Nash Point passes laterally into a shallow water facies westwards towards Southerndown and northwards towards Cowbridge. He also showed that the successive Liassic strata overlap one another on a steeply shelving floor of Carboniferous Limestone or supposedly Triassic breccia. This contrasts with the conformable relationship to the Rhaetian away to the east. The 'littoral' or 'near-shore' facies had long been divided into two lithological formations: the light-coloured, conglomeratic or at least coarse-grained Sutton Stone and the darker, comparatively well-stratified and finer-grained (though still locally conglomeratic) Southerndown Beds.

These three lithologies: 'normal' Blue Lias, Southerndown Beds and Sutton Stone were shown by Trueman (*op. cit.*) to be diachronous through the three chronostratigraphical zones then recognised. This became a classic example of diachronism and of the lateral transition of strata into a near-shore facies. Though ammonites are rare in the coarser sediments, this classic at least has not so far suffered from the iconoclasts and still seems to be generally accepted. Both Hallam (1960) and Wobber (1965) accepted Trueman's explanation and the earlier ter-minology, though the distinction between the Sutton Stone and the Southerndown Beds is not always very obvious, as might be expected. The rich fauna described by Duncan (1867) from the Sutton Stone at inland quarries has not been seen again for many years and Hallam's figures for the number of species present were shown by the present writer (Ager 1963, p. 236) to indicate an increasing diversity with time. This accords with the general picture of a marine transgression slowly submerging the old land-masses and/or islands and the gradual establishment of a full marine fauna. It should be pointed out that though the Blue Lias is famous for its fossils, both in Glamorgan and in south-west England, the richness is in abundance rather than diversity and there are really rather few species present. Particular phyla such

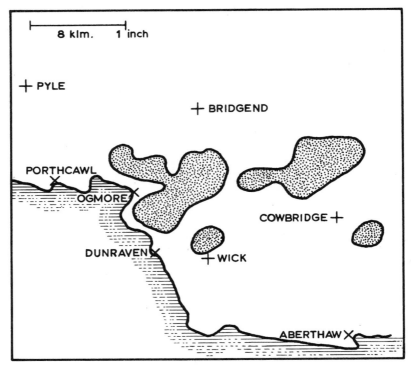

Fig 89.   Reconstruction of "St. David's Archipelago" showing the probable extent of the islands of Carboniferous Limestone (stippled) as they were in early Jurassic times (after Trueman, 1922).

as the Echinodermata and the Brachiopoda are represented by very few species indeed. The abundance of corals in the near-shore facies has often been commented upon. Glamorgan is remarkable for its early Jurassic corals, not only in Britain but also in Europe as a whole.

Though Trueman's diagram of the progressive submergence of islands of Carboniferous Limestone has often been reproduced, comparatively little attention has been paid to ancient coastline features preserved beneath the Glamorgan Lias. Thus there are fossil placer deposits of galena at the base of the Lias in places, though most of this mineral in the Lias seems to be authigenic (Wobber 1965). Thomas (1970) has commented on the occurrence of minor veins of galena and barytes near the base of the Sutton Stone when it rests on the Carboniferous Limestone. Near the blow-hole at Pant-y-Slade the top surface of the Carboniferous Limestone can be seen to be encrusted with what are probable Jurassic coral growths that have been highly and selectively dolomitised.

Hallam's brilliant survey of the sediments and fauna of the Blue Lias (1960) was concerned with both Glamorgan and the Dorset coasts and it is difficult to separate the evidence from each. His sedimentological conclusions have already been discussed. His palaeoecological conclusions relate both to the autecology of the various groups present and to the synecology of the whole. Some of his comments are well worth re-mentioning and merit further discussion.

Among the bivalves Hallam noted that some, such as pectinids, are almost always disarticulated, whilst others such as *Modiolus* are not. All the common burrowing bivalves such as *Pholadomya*, *Pleuromya* and *Mactromya* are commonly found in life position as are forms such as *Lima gigantea* (J. Sowerby) which evidently lived with its broad lunules adpressed to the sea floor. The burrowing forms have usually lost their thin shells in diagenesis, as is general in the Jurassic, and an interesting point is the way oysters tend to be cemented to the posterior ends of these molluscs, which were presumably protruding above the mud surface.

The larger ammonites all lie parallel to the bedding, though juveniles less than about 5 cm. in diameter seem to have been preserved in the attitudes in which they sank into the soft bottom ooze.

Brachiopods unfortunately do not seem to be so common in Glamorgan as they are in age-equivalent beds in Somerset and Dorset. By far the most common is the small rhynchonellid *Calcirhynchia calcaria* Buckman, which is very closely related to its late Triassic ancestors of central Europe. This species (like many other Mesozoic rhynchonellids) tends to occur in clusters or 'nests' which have been reasonably interpreted as original colonial associations. Hallam seemed to suggest in his 1960 paper that they may have lain loosely on the muddy sea floor as does the modern form *Neothyris*, in which the slender pedicle does no more than prevent the shell from drifting about. Hallam later suggested that *Calcirhynchia* may have lived attached in groups to anchored seaweed and the author suggested (1962) that it might have been one of a group of small, thin-shelled Jurassic rhynchonellids that lived an epiplanktonic life attached to floating weed. An item of evidence that may support either or both hypotheses is that *C. calcaria* characteristically occurs in intensely burrowed limestones, as though it were associated with an abundant benthonic fauna, though the sea floor may not have been stable enough for its attachment.

Other members of the shelly fauna are of minor importance though some, such as stem ossicles of crinoids, occur abundantly in local drifts. Trace fossils do not seem to be so common in the Glamorgan Lias as elsewhere, though it is nearly everywhere evident that there has been intensive bioturbation. Those that are observed, such as *Chondrites*, *Rhizocorallium* and *Thalassinoides* do not approach the sizes they achieve in later Jurassic strata and whatever made them may not have been living at its ecological optimum.

Plant remains are locally abundant in the offshore facies and Thomas (1970) commented on the presence of streaks of lignite and even of coal, of anthracite-like appearance, in the *angulatum* Zone. He related this anthracitisation to later fault and fold movements within the Lias.

Hallam's general conclusions on the synecology of the Glamorgan Lias were that true 'life assemblages' of popular parlance are rare and that mixtures of 'life' and 'death' assemblages are the rule. Of the latter, the vast majority may be termed 'indigenous death assemblages'; that is to say, the constituents have been a little transported but still remain in the general sedimentary environment in which they lived.

Hallam found some evidence of relative stunting of species in the shales compared with the same forms in the limestones and he also commented that one of the most remarkable features of the 'normal' Lias was the progressive increase in size of many of the elements as one climbs the succession. Thus the species *Pinna hartmanni*, already referred to, increases from a maximum dimension of 11 cm. in the *johnstoni* subzone up to 33 cm. in the *gmuendense* Subzone. This may be related, like the increasing diversity mentioned earlier, to the progressive nature of the transgression and to the gradual establishment of optimum conditions for normal marine invertebrates. It must be commented, however, that the phenomenon of size increase through the Hettangian and Sinemurian seems to be general in western Europe and is not confined to particular regions.

The tropical or subtropical aspect of the Liassic fauna, with its compound scleractinian corals and large thick-shelled molluscs, is more obvious in Glamorgan than elsewhere in Britain and is testified to by all workers on the strata. In this connection it is important to consider the palaeomagnetic evidence concerning the probable latitude of South Wales at this time. Unfortunately the British Lias seems to be rather lacking in magnetism so far as the physicists are concerned. The only relevant estimate known to the present author is that of Girdler (1959) who worked on the Toarcian sands of western England. He concluded that the pole at this time was in the region of the Caspian Basin. Assuming no post-Jurassic drifting of Wales relative to England, this would put South Wales at a latitude comparable to that of the north coast of Spain at the present day. However, such polar wandering is difficult to reconcile with the known distributions of early Jurassic faunas unless one postulates a tropical belt that was very wide indeed.

What is more, Girdler's results do not appear to fit in with those of other geomagnetists. Most of these do not attempt a pole position for Jurassic times, or only do so on the basis of evidence from far away places such as Argentina and Antarctica, which can have no bearing on the matter in hand. Thus Runcorn (1962) did not plot a Jurassic pole, but his Triassic pole (on European evidence) lies near Vladivostok and his Cretaceous pole lies in the Arctic Ocean north of Siberia. Irving (1964) plotted a wide scatter of Jurassic poles ranging from the Middle East to the Bering Straits, but his synthesis would seem to place the Jurassic north pole in the north-east corner of Siberia. Similarly, Meyerhoff's recent critical review of palaeomagnetic data (1970) and his plots of 'reliable' polar estimates has Permian poles widely scattered in eastern Asia, the north Pacific and western North America, whilst the Cretaceous points show a much more condensed scatter around the Bering Straits. At the same time Meyerhoff casts considerable doubts on the validity of these results and suggests that the main conclusion must be that the remanent magnetism vector in rocks became less stable with time. The recent description by the Russians of Tethyan type faunas in the Upper Triassic and Lower Jurassic of this region makes such a pole position even less likely.

Certainly the evidence of fossil distributions would place the Welsh Jurassic firmly in the northern European, extra-Tethyan belt. The faunas known from Glamorgan contain no 'Tethyan' or 'Alpine' elements, though such forms are known in Somerset. So far as hermatypic corals are concerned, these are known as far north as Sutherland and the Queen Charlotte Islands (near the southern tip of Alaska) in later Jurassic times.

Crampton's short paper (1960) concerned itself solely with the heavy mineral suites of the Mesozoic sediments of South Wales. He concluded that the deposits were laid down at sublittoral depths in a sea that was bordered to the north, in early Liassic times, by a coastline composed mainly of Carboniferous Limestone. Perhaps more significantly, however, was the evidence he found of currents bringing heavy minerals such as kyanite and staurolite from Brittany and an assemblage of andalusite, spinel and sillimanite from England. The detritus so derived was thinly scattered through mainly locally derived material and certainly could not have come from a land-mass to the north. Crampton found evidence of some sorting of the minerals, notably the evident retention of such light constituents as garnet and tourmaline in the littoral zone and of the heavier zircon and rutile in deeper water. This led to the conclusion that there was probably a rather steeply inclined coastal sea-bed. This is what one would also conclude from the actual sections through the basal unconformity seen along the coast.

The most recent work on the Liassic rocks of Glamorgan is that contained in a series of papers by an American visitor to the University College of Swansea, Frank J. Wobber (1965, 1966, 1967, 1968a, 1968b, 1968c). He was concerned chiefly with the sedimentation and sedimentary petrography of the Welsh Lias. In the first of these papers, Wobber revised the petrography of the Sutton Stone, the Southern-down Beds and the 'normal' off-shore limestone-shale sequence. He confirmed the idea of the first two having been deposited around islands of Carboniferous Lime-stone, and showed that they are formed chiefly of poorly-sorted lithoclasts of limestone and chert in a calcilutitic matrix. He showed that the Sutton and Southern-down Beds are most easily distinguished by the characters of their matrices, following a modified version of the Folk limestone classification. Rock types present in the nearshore facies include pseudo-oolitic lithosparites, lithosparites, lithomicrosparites, fossiliferous lithoclast-bearing micrite, fossiliferous lithomicrite and oosparite. The facies as a whole ranges from the white polymictic conglomerates with cobble and boulder beds, characteristic of the Sutton Stone, to grey, clay-bearing gravel conglomerates, characteristic of the Southerndown Beds. These features are indicative of active environments with much current agitation. The fauna of the Sutton Stone is normally fragmented, as one might expect in very shallow, high energy conditions. Hallam (1960) commented particularly on the abundance of the ribbed pectinids *Chlamys valoniensis* (Defrance) and *Terquemia arietis* (Quenstedt) in the Sutton Stone, both of which are only minor elements in the off-shore facies. Thick-shelled bivalves and the patellid gastropod *Acmaea schmidti* (Dunker) are also significant members of the fauna. On the other hand, in the Southerndown Beds, both Hallam (1960) and Wobber (1965) concluded that the fauna resembled that of the off-shore facies, though Hallam did draw attention to the relative abundance and large size of several species of gastropods.

Wobber found that the dense blue-grey limestones of the offshore facies were mostly micrites, pelmicrites, fossiliferous micrites and biomicrites. Occasional shallowing is indicated by bands of winnowed fossil debris.

With the recent great interest in trace fossils, the Liassic rocks of Glamorgan have received their fair share of attention and both Hallam (1960) and Wobber (1965) emphasized the importance of burrowing organisms, not only as elements in the fauna, but also as agents in breaking down shell and rock fragments. Rarely preserved laminations are indicative of a local absence of burrowing and other organic activity rather than of special conditions of sediment supply.

In a later paper, Wobber (1966) concerned himself with the specific problems of the depositional area of the Welsh Lias. Using orietentation studies of echinoid spines, sponge spicules and small sedimentary structures, he came to the conclusion that the principal source area of the Lias of the Vale of Glamorgan lay to the south and not to the north as had previously been supposed. He drew particular attention to the absence of fragments of Millstone Grit and Coal Measure sandstones in the Lias and deduced that the South Crop of the South Wales coalfield, most notably the Pennant escarpment, was not emergent during early Jurassic times. He thought that the northern shoreline, if it existed at all, lay far to the north of the present Liassic outcrops.

In a discussion of Wobber's paper, Owen (1967) stated firmly his view that Jurassic deposits must have extended right over the South Wales coalfield. He pointed out that the old argument was that a projection of the base of the Chalk from the Chilterns would only just clear the Welsh uplands, leaving little room for the Jurassic. But this took no account of the Malvern Fault, the easterly downthrow of which would leave room for hundreds of feet of pre-Chalk Mesozoic. Another argument was that all that remains now of the Lias in the Vale of Glamorgan is probably only a third of the original thickness and this coupled with mid-Tertiary warping in the Brecon Beacons area, would allow the Jurassic strata to cover the Pennant scarp and the Old Red Sandstone scarp beyond.

In his 1967 paper, Wobber concerned himself with post-depositional structures in the Glamorgan Lias. These are particularly well displayed in the alternating limestones and shales of the normal 'Blue Lias' facies and consist of load structures due to overburden and interstratal sliding features due to differential subsidence. Interesting cases are where fossils such as *Pinna* in life position are 'smeared' and broken off along bedding planes due to post-depositional sliding.

In two of his later papers on the Welsh Lias (1968a, 1968b) Wobber attempted a detailed faunal analysis related to the observed variations in both fossils and sediment. He concluded that the most important factor controlling the distribution of benthonic organisms was the texture of the sea-floor, whilst bottom turbidity was only of secondary importance. He followed Hallam in his approach of considering first the autecology of the different fossil groups and then synthesizing the synecology of the whole assemblage. His conclusions are not, however, fully in accord with the earlier work.

The whole subject is entangled with 'hen and egg' arguments, as Wobber himself clearly demonstrated. Thus the stability of the substratum increased with increasing amounts of organic debris and increasing stability lead to a greater abundance of bottom organisms. Similarly he suggested that burrowing activity by organisms led to a loss of stability through the exposure of bottom sediments to current winnowing, whilst a soft, shifting substratum became unfavourable to burrowers.

Parallel orientation of fossils and lithoclasts was taken as evidence of current action strong enough to transport and mix bottom-living communities. On the other hand, the distribution of most fossils through rocks of different lithologies was thought to indicate a high degree of faunal tolerance.

Wobber's general palaeoecological conclusions were that most early Jurassic organisms in this area could withstand limited changes in aeration, sedimentation and faunal competition. The gastropods were said to show a greater degree of tolerance than the bivalves, occurring as they do in relative abundance in both the near-shore and the offshore facies. Less tolerant organisms were (most obviously) the corals, the brachiopods and some of the pectinid bivalves.

The way in which similar Liassic faunas occur in both the limestones and the shales led to many of the earlier arguments about the primary or secondary nature of the former, but careful examination shows that the uniformity is more apparent than real. Thus the most famous of the Glamorgan fossils—*Gryphaea arcuata* Lamarck—certainly seems to occur more commonly in the shales than in the limestones.

It is impossible to leave the Liassic faunas of Glamorgan, however, without saying something more about this notorious fossil *Gryphaea*. Trueman, the founder of the Swansea department, was one of the few real geniuses of geology. His early study on the evolution of *Gryphaea* in the cliffs of Glamorgan (1922b) was one of the classics of our science. *Gryphaea* became, through Trueman, the *Drosophila* of palaeontology. Years later, another student of this hoary lineage observed that more papers had been written on *Gryphaea* than on any other fossil. He then used this as an excuse to write yet another paper on the overworked oyster.

In fact at least twelve important papers could be mentioned, but this is not the place to turn over the midden of old oyster shells yet again. In the paper that started it all, Trueman was concerned with demonstrating how *Ostrea* and *Gryphaea* could be used stratigraphically in the Glamorgan Lias. At the same time he claimed to show that the flat *Ostrea* (now *Liostrea*) of the lower beds evolved into the tightly coiled *Gryphaea arcuata* of the upper beds. His paper soon became a classic and stimulated a whole generation of evolutionary palaeontologists. It was typical of the brilliant ideas that adorned so many of Trueman's papers. However, and one is tempted to say 'unfortunately', his brilliant idea in this case appears to have gone beyond the facts, and like several other evolutionary classics (e.g. Carruther's *Zaphrentis*, Lang's Cretaceous bryozoa and Trueman's own *Androgynoceras-Liparoceras*) Trueman's *Ostrea-Gryphaea* lineage must now certainly be abandoned. The original theory based on field collecting was supported with some statistical evidence by Swinnerton (1939) and was not challenged until Hallam (1959) who found no evidence of the transition from *Liostrea* to *Gryphaea* and maintained that Trueman had been misled in his postulation of an evolutionary increase in tightness of coiling in *Gryphaea* by allometric growth and increase in size. Hallam briefly summarized the controversy that followed in his latest paper on *Gryphaea* (1968), but since the matter now goes far beyond the oysters of the Glamorgan cliffs, it need not be considered further here.

III. THE BRISTOL CHANNEL

The first definite information published on Jurassic rocks in the floor of the Bristol Channel was in a brief paper by Donovan and others (1961). This mentions Liassic sediments in a broad belt of Mesozoics in the eastern part of the Bristol Channel extending down to about Ilfracombe.

Two years later Lloyd (1963) published a note recording foraminiferal evidence of Kimmeridgian sediments between Morte Point and Worm's Head and of Lower Oxfordian 15 miles east of this. Lloyd concluded that a large syncline was present containing strata ranging from at least the *angulatum* Zone of the Hettangian to the *rotunda* Zone of the Kimmeridgian. It should also be mentioned that undated Mesozoic rocks had been reported earlier by Day (1958) off north Cornwall. The Oceanography Sub-department at Swansea has in the meantime found considerable areas of Jurassic rocks, of various ages, in the Channel south and east of Swansea (F. T. Banner and D. J. Evans, personal communication 1969). This was confirmed by Lloyd's report to a Geological Society of London meeting at Bristol in 1970 on the work of the University College of London/National Institute of Oceanography study in the Channel 10 years before. Lloyd reported the presence of all stages of the Jurassic, with the possible exception of the Portlandian/'Purbeckian'. Most of them were in a clay facies with no indications of a shoreline to the north. The only sandy developments were in the Upper Bathonian and Upper Oxfordian. The whole of the Aalenian, Bajocian and Bathonian amounted to no more than 20 m and presumably indicated some breaks in sedimentation. The sands of the Upper Oxfordian thin and disappear to the north and presumably imply a southerly source. The whole of the Jurassic, together with the Upper Triassic, has been estimated to reach a thickness of some 1,700 m in the Bristol Channel, and other estimates have put the Jurassic thickness alone at about 1,000 m.

IV. THE MOCHRAS (LLANBEDR) BOREHOLE

Five years ago there was published the first note (Wood and Woodland 1968) about what the present writer has called . . . 'the most important discovery in Welsh geology since the heroic age of Sedgwick and Murchison' (Ager, 1970 p. 14). This was the account of the first 1,000 metres of a borehole put down on the famous shell 'island' of Mochras, about three miles south-south-west of Harlech in north Wales.

Mochras is the most westerly projection of the coastal sand hills of Morfa Duffryn between Harlech and Barmouth and was well-known for the abundance and diversity of its recent shell accumulations. Inland, one is in the middle of the

Harlech Dome of Lower Cambrian sediments and the off-shore submarine ridge of Sarn Badrig had been interpreted as the underlying Precambrian. A study of the geological map might therefore lead one to the conclusion that a borehole here would reveal basal Cambrian and Precambrian rocks at a shallow depth. In fact what the cores revealed was thick unconsolidated sediments, probably of Tertiary age, resting on the thickest Lower Jurassic known anywhere in the British Isles.

Some years ago, O. T. Jones (1952, p. 217) drew attention to the marked similarities (both in shape and in drainage) between Cardigan Bay and the Cheshire Plain. He suggested that Cardigan Bay and the southern Irish Sea might be a comparable basin filled with Mesozoic sediments. This was argued on purely geomorphological grounds and began an epoch when almost all the arcuate stretches of British coastline were interpreted as the rims of Mesozoic basins. O. T. Jones had long been a protagonist of a Mesozoic sedimentary cover over much of Wales, and was particularly influenced in this case by the convergent radial drainage of this part of Wales.

Powell (1956) found marked gravity and magnetic anomalies on a traverse seaward from the Harlech Dome just south of Mochras. The model he used to explain these anomalies, founded on Jones' idea, was of a considerable fault throwing down Pleistocene drift and supposed Triassic sediments against the denser Cambrian rocks of the Dome.

Three later geophysical papers, by Griffiths, King and Wilson (1961), Blundell, Davey and Graves (1968) and Blundell, King and Wilson (1964) confirmed both the coastal fault and the deep basin, containing as much as 6,000 m of sediment, though differing views were expressed as to the probable age of that sediment.

Mochras was chosen because of its position as the dry land site farthest out towards the centre of this basin. Boring was begun in November, 1967 as a joint project by the University of Wales and the Institute of Geological Sciences. The first accounts were published by Wood & Woodland (1968) and by the Institute (1969) while drilling was still going on. Subsequently there appeared a short note by Lewis (1970) and the core and other data were exhibited on various occasions. The main publication, with complete logs has only just appeared (Woodland 1971). A condensed version of the Lower Jurassic part of the core is shown in fig. 90. The boring as a whole may be summarised as follows:

Quaternary sands, till, etc.	c.	55 m
Tertiary clays, lignites, etc.	c.	370 m
Lower Jurassic calcareous mudstones and siltstones	c.	1300 m
Triassic (probably Rhaetian)	c.	32 m

There is evidence of a sharp break at the bottom of the Lias, where it rests on dolomites of the Upper Trias (Ivimey-Cook in Woodland, 1971). The Lower Jurassic succession is also eroded at the top but is nevertheless the thickest known anywhere in the British Isles and includes representation of every zone of all four stages: Toarcian, Pliensbachian, Sinemurian and Hettangian. The *planorbis* zone at the bottom and the *davoei* zone higher up are thinner than elsewhere, but the total thickness of the Lias is 4284 ft 11 in (c. 1,305 m) which is between 2½ to 3 times as

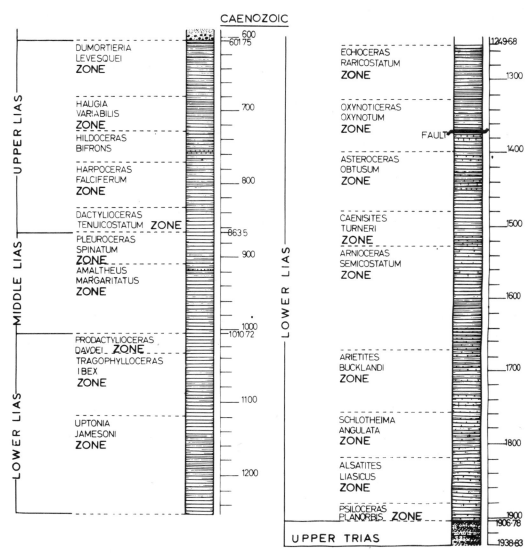

Fig. 90. The zonation, thickness and lithology of the Lower Jurassic in the Mochras borehole (after H. C. Ivimey–Cook *in* Woodland, 1971).

thick as any other Lower Jurassic deposits in the British Isles. It might be noted, however, that thicknesses up to 5,000 ft (*c.* 1,538 m) are known in troughs in north-west Germany and thicker developments are also found in the French Sub-Alps. The Lias consists almost entirely of mudstones and siltstones with only occasional thin beds of other lithologies, such as shelly limestones. Most of the succession is unlaminated and has apparently been reworked by burrowing organisms, though there is evidence of carbonate rich/poor rhythms as in the 'Blue Lias' and like them probably of primary origin. The Upper Lias is very like that of Yorkshire with jet bands. There is no trace of any sandy development in the Middle Lias such as characterize that part of the succession elsewhere in Britain. The Lower Lias contains limestone and ironstone nodules and there is some indication of hydrocarbons.

The most interesting feature of the Mochras Lias, therefore, apart from its location and great thickness is that it displays an 'offshore' mudstone facies throughout. The shelly fauna is dominantly pelagic, consisting largely of ammonites and belemnites. *Calcirhynchia calcaria* was found at one level and a few benthonic bivalves and crinoids occur in places as does drifted plant debris. The general bioturbation implies aerobic conditions, and although the facies is clearly an offshore one, it does not suggest very deep water. An influx of arenaceous foraminifera at the top of the Lias at Mochras may suggest a change towards shallower water deposition and perhaps the final silting up of the basin, though there is no indication of this in the general lithology. The only suggestion of possible shallowing in the whole monotonous Liassic sequence is one thin conglomerate band in the Upper Lias.[1]

V. THE COTSWOLDS

The author's long-term interest in the Jurassic rocks of the Cotswolds has always directed his eyes towards Wales. The famous diachronous sands of Toarcian and Aalenian age reach their lowest stratigraphical level in Gloucestershire and seem to require a westerly source.

Similarly, the sands recorded by the present writer in the most westerly outcropping Upper Pliensbachian (Ager 1955, p. 360) seemed to imply that the supposed Welsh land mass was more extensive in late Middle Lias times than in the usual palaeogeographical reconstructions founded on Lower Liassic evidence.

In the Inferior Oolite Formation (or Group according to taste) of Aalenian and Bajocian age, the thick oolitic and pisolitic limestones of the Cotswold scarp imply (on the basis of modern analogies) shallow water, very close inshore sedimentation. The presence of rolled coral fragments at several levels led to the romantic notion of a fringe of coral reefs along the line of the Malvern Hills, the geological (if not political) border of Wales.

However, more recent work has led to some reassessment of these ideas. The Liassic sands have retained their famous and surprising diachroneity (Davies, 1969) but are usually nowadays referred to a supposed Bath axis or (as by Davies) to a meta-igneous complex in the western approaches to the English Channel, rather than to a Welsh land mass. Particular interest attaches to the prominent cross-bedding in the Lower Freestone member of the Inferior Oolite. At least in the Cheltenham district, my own observations and those of Dr. Peigi Wallace seem to show a preferred orientation towards the east and south-east. Conventional geological attitudes might have presumed from this a land area to the west as suggested in such palaeogeographical maps as those of Wills (1951). However, Dr. Graham Evans suggested to the author, on the basis of his observations of modern oolites along the Trucial coast of the Persian Gulf, that the fore-set cross-bedding was more likely to be

---

[1] For some of the above information the author is deeply indebted to Dr. M. Dobson of Aberystwyth.

directed onshore rather than offshore. This suggested a land mass, to the east (where the London 'island' was known to be surrounded by coral reefs at this time) and open sea to the west.

Several other items of information in the Cotswold area lead to similar conclusions, such as the observed thickening of the Upper Lias towards Cleeve Hill and the outlier of Bredon Hill in the extreme north-west of the outcrop (Arkell, 1933). McKerrow's observation of westerly-directed slumps in the Inferior Oolite near Bourton-on-the-Hill (in McKerrow, Ager and Donovan, 1964, p. 4) may also be significant, since this is probably the only indication of original sea floor slope in the Jurassic rocks of this region.

My work on the geographical distribution of brachiopods in the British Middle Lias (Ager, 1956) showed close connections between the faunas of late Pliensbachian age in south-west England, the Cotswolds, the Prees outlier and western Scotland. Subsequently (Ager, 1967, p. 158) I was able to confirm that brachiopods of this age found in the drift of Northern Ireland were also similar and presumably came from an outcrop in the Irish Sea. A notable and easily-recognised species common to all the above areas is *Homoeorhynchia acuta* (J. Sowerby) which is virtually unknown in the English Midlands east of the Cotswolds and in Yorkshire. This requires a shallow marine connection not only along the eastern borders of Wales, but also across it.

VI.   THE PREES OUTLIER

The small Prees outlier in Cheshire provides the only evidence of Jurassic rocks close to the north-easterly border of Wales. It has long been known to include Lower and Middle Liassic sediments with the Marlstone Rock-bed at the top of the latter only normally exposed when graves are dug in the churchyard of Prees church at the top of the hill. Recent temporary exposures have revealed a complete record of the lower beds, though exact estimates of thicknesses are not yet available. The Lower Lias has long been said to amount to more than 400 ft (*c.* 123 m) of shales and calcareous mudstones, followed by some 100 ft (*c.* 30 m) of Middle Lias sandy beds and capping ferruginous limestone.

The Nantwich and Whitchurch memoir (Poole *et al.*, 1966) covered most of the Prees outlier and provided the very important new evidence found in the Wilkesley borehole (about 8 km east of Whitchurch). In that borehole, the Hettangian and lower part of the Sinemurian alone amounted to some 460 ft (*c.* 140 m). In both this and another borehole at Plattlane the Lias consisted almost entirely of fine-grained terrigenous sediment in the form of calcareous mudstones. There is very little coarse material, and the whole sequence implies steady subsidence and uniform sedimentation.

The facies is said to be exactly similar to that of the nearest main outcrop, in Leicestershire, nearly 70 miles to the east. There is no evidence whatever of thinning or change in facies towards a supposed Welsh land mass.

VII.   CONCLUSIONS

The main conclusion that must emerge from this review must be that Wales was largely, if not wholly, submerged beneath the sea in early Jurassic times. This is not altogether a new suggestion. T. R. Owen (1967) maintained that marine Jurassic sediments had formerly extended over the highlands of South Wales and M. Dobson (personal communication 1970) came to the same conclusion with respect to North Wales on the evidence of the Mochras borehole. It was also the conclusion of several speakers dealing with different parts of the period at the International Field Symposium on the British Jurassic in April 1969 (Ager 1969, p. 216). Perhaps the Welsh will not, however, altogether lament the passing of that tactlessly named land mass 'St. George's Land'.

Certainly the classic picture of St. George's Land, either as an island or as a peninsula from a western land mass throughout Jurassic times, no longer stands up to critical examination. The sediments in Glamorgan and the Bristol Channel seem to have been derived either from small, immediately adjacent islands, or, more generally, from the south or south-west. The Jurassic sediments of the Cotswolds and other areas along the Welsh Borderland seem to have come from the east rather than from the west; there is no evidence of a land mass to the west. The Jurassic now known on the coast of North Wales provides evidence of a deepish sea and certainly no derivation of sediments from a local source.

What then remains? Certainly not the extensive land mass of St. George's Land as supposed by earlier workers. Instead we must think of Wales, through most of Jurassic times, as covered by a shelf sea of moderate depth, with (at least in its earliest stages) an island or group of islands in what is now Glamorgan. Comparable islands across the Bristol Channel in Somerset are usually called the 'Mendip archipelago', but clearly a name is needed to include all these islands, both Welsh and English. Since the evidence is so much clearer, and so much more easily demonstrated on the Celtic side, I propose to name this palaeogeographical feature 'St. David's Archipelago' (fig. 89). So the English saint has sunk beneath the waves and the Welsh saint has appropriately risen to take his place.

VIII.   REFERENCES

AGER, D. V. 1955. Field meeting in the central Cotswolds. *Proc. Geol. Ass.*, **66**, 356–65.
——. 1956. The geographical distribution of brachiopods in the British Middle Lias. *Quart. J. geol. Soc. Lond.*, **112**, 157–88.
——. 1962. The occurrence of pedunculate brachiopods in soft sediments. *Geol. Mag.*, **99**, 184–86.
——. 1963. *Principles of Paleoecology*. McGraw-Hill (New York), 371 pp.
——. 1967. A Monograph of the British Liassic Rhynchonellidae. IV. *Palaeontogr. Soc. Monogr.*, 137–72.
——. 1969. International field symposium on the British Jurassic. 9th–24th April, 1969. *Proc. geol. Soc. Lond.*, No. 1658, 215–27.
——. 1970. *Geology as an Environmental Science*. Univ. Coll. Swansea, 23 pp.
ARKELL, W. J. 1933. *The Jurassic System in Great Britain*. Oxford Univ. Press, 681 pp.
BLUNDELL, D. J., DAVEY, F. J. and GRAVES, L. J. 1968. Sedimentary basin in the South Irish Sea. *Nature*, **219**, 55–6.
——, KING, R. F. and WISON, C. D. V. 1964. Seismic investigations of the rocks beneath the northern part of Cardigan Bay, Wales. *Quart. J. geol. Soc. Lond.*, **120**, 35–50.
CRAMPTON, C. B. 1960. Petrography of the Mesozoic succession of South Wales. *Geol. Mag.*, **97**, 215–28.

DAVIES, D. K. 1969. Shelf sedimentation: an example from the Jurassic of Britain. *J. sediment. Petrol.*, **39**, 1344–70.

DONOVAN, D. T., SAVAGE, R. J. G., STRIDE, A. H. and STUBBS, A. R. 1961. Geology of the floor of the Bristol Channel. *Nature*, **189**, 51–2.

DUNCAN, P. M. 1867. On the Madreporaria of the Infra-Lias of South Wales (chiefly from the Sutton and Southerndown Series). *Quart. J. geol. Soc. Lond.*, **23**, 12–28.

GIRDLER, R. W. 1959. A palaeomagnetic study of some Lower Jurassic rocks of N.W. Europe. *Geophys. J. roy. astronom. Soc.*, **2**, 353–63.

GRIFFITHS, D. M., KING, R. F. and WILSON, C. D. V. 1961. Geophysical investigations in Tremadoc bay, North Wales. *Quart. J. geol. Soc. Lond.*, **117**, 171–91.

HALLAM, A. 1957. Primary origin of the limestone-shale rhythm in the British Lower Lias. *Geol. Mag.*, **94**, 175–76.

——. 1959. On the supposed evolution of *Gryphaea* in the Lias. *Geol. Mag.*, **96**, 99–108.

——. 1960. A sedimentary and faunal study of the Blue Lias of Dorset and Glamorgan. *Phil. Trans. roy. Soc.*, (B), **243**, 1–44.

——. 1968. Morphology, palaeoecology and evolution of the genus *Gryphaea* in the British Lias. *Phil. Trans. roy. Soc.*, (B), **254**, 91–128.

INSTITUTE OF GEOLOGICAL SCIENCES. 1969. *Annual Report for* 1968. H.M. Stationery Office, 195 pp.

IRVING, E. 1964. *Paleomagnetism and its application to geological and geophysical problems.* Wiley (N.Y.), 399 p.

JONES, O. T. 1952. The drainage systems of Wales and the adjacent regions. *Quart. J. geol. Soc. Lond.*, **107**, 201–25.

JOYSEY, K. A. 1959. The evolution of the Liassic oysters *Ostrea-Gryphaea*. *Biol. Rev.*, **34**, 297–332.

LEWIS, B. J. 1970. Liassic sediments of the Mochras borehole. *Abstracts 4th Irish Sea Colloquium, Aberystwyth.*

LLOYD, A. J. 1963. Upper Jurassic rocks beneath the Bristol Channel. *Nature*, **198**, 375–76.

McKERROW, W. S., AGER, D. V. and DONOVAN, D. T. 1964. Geology of the Cotswold Hills. *Geol. Assoc. Guide* No. 36, 1–26.

MEYERHOFF, A. A. 1970. Continental drift: implications of paleomagnetic studies, meteorology, physical oceanography and climatology. *J. Geol.*, **78**, 1–51.

OWEN, T. R. 1967. 'From the South': a discussion. *Proc. Geol. Ass.*, **78**, 595–601.

POOLE, E. G., WHITEMAN, A. J. and others. 1966. *Geology of the country around Nantwich and Whitchurch. Mem. geol. Surv. Gt. Britain.* H.M. Stationery Office, 154 pp.

POWELL, D. W. 1956. Gravity and magnetic anomalies in North Wales. *Quart. J. geol. Soc. Lond.*, **111**, 375–97.

RUNCORN, S. K. 1962. Palaeomagnetic evidence for continental drift and its geophysical cause. In '*Continental Drift*' (Runcorn, S. K. edit.), N.Y. Acad. Press, 1–40.

SWINNERTON, H. M. 1939. Palaeontology and the mechanics of evolution. *Quart. J. geol. Soc. Lond.*, **95**, xxxiii–lxx.

THOMAS, T. M. 1970. Field notes on the coastal section from Ogmore-by-Sea to Dunraven, Glamorgan. *Welsh geol. Quart.*, **5**, 46–58.

TRUEMAN, A. E. 1920. The Liassic rocks of the Cardiff district. *Proc. Geol. Ass.*, **31**, 93–107.

——. 1922a. The Liassic rocks of Glamorgan. *Proc. Geol. Ass.*, **33**, 245–84.

——. 1922b. The use of *Gryphaea* in the correlation of the Lower Lias. *Geol. Mag.*, **59**, 256–68.

——. 1930. The Lower Lias (*bucklandi zone*) of Nash Point, Glamorgan. *Proc. Geol. Ass.*, **41**, 148–59.

WILLS, L. J. 1951. *A palaeogeographical atlas of the British Isles and adjacent parts of Europe.* Blackie (London and Glasgow), 64 pp.

WOBBER, F. J. 1965. Sedimentology of the Lias (Lower Jurassic) of South Wales. *J. sediment. Petrol.*, **35**, 683–703.

——. 1966. A study of the deposition area of the Glamorgan Lias. *Proc. Geol. Ass.*, **77**, 127–37.

——. 1967. Post-depositional structures in the Lias, South Wales. *J. sediment. Petrol.*, **37**, 166–74.

——. 1968a. *A faunal analysis of the Lias (Lower Jurassic) of South Wales, Great Britain.* I.B.M. Space Systems Center, Gaithersburg Maryland, 84 pp.

——. 1968b. A faunal analysis of the Lias (Lower Jurassic) of South Wales (Great Britain). *Palaeogeogr., Palaeoclimatol., Palaeoecol.*, **5**, 269–318.

——. 1968c. Microsedimentary analysis of the Lias in South Wales. *Sediment. Geol.*, **2**, 13–49.

WOOD, A., and WOODLAND, A. W. 1968. Borehole at Mochras, west of Llanbedr, Merionethshire. *Nature*, **219**, 1352–54.

WOODLAND, A. W. (Editor). 1971. The Llanbedr (Mochras Farm) borehole. *Rep. No.* 71/18, *Inst. geol. Sci.*, 115 pp.

THE CENOZOIC EVOLUTION OF WALES

T. Neville George

I. CENOZOIC HERITAGE

WALES, in a traditional 19th-century stratigraphy, has always lacked a significant development of Cenozoic rocks. In Cenozoic times, its evolution, unlike the evolution of southern England or of Ulster or of western Scotland, has correspondingly been regarded, until very recent years, as in major part almost eventless except in slight warping and in the slow removal of whatever Mesozoic rocks may have rested on the Palaeozoic foundation. 'Subaerial denudation', river dissection, peneplanation, were the main agents controlling landform until Glacial times; and the basic profile of the present landscape was to be interpreted in the two-fold terms of the physique and rock-sequence of end-Mesozoic Wales, and of the manner in which the sub-Mesozoic floor was exhumed and remoulded from Eocene times onwards.

The traditional stratigraphy is embedded in a hypothesis that, for nearly a century, has revolved with a surprising vitality about a few postulates which over the years have hardened into declaratory assumptions—notably of the dominant and persistent influence of a Chalk cover, and (less emphatically) of the survival in present-day landforms of the major elements of a very ancient (Triassic) physique. The hypothesis goes back to Ramsay (1878), who suggested that the Welsh massif, except perhaps in its highest summits, was drowned before the end of Cretaceous times to be overlain by a veneer of Chalk. The elevation of the cover into an asymmetrical dome in Palaeogene times, and the initiation of a radial river system, were recognised by Lake (1900, 1934) as accordant elements in the hypothesis: he added a postulate of mid-Cenozoic deformation along concentric structural belts to explain the occurrence of 'transverse' rivers that diverted the primary radial drainage.

Strahan (1902) took up the theme, placing it in mid-Cenozoic times: the uplift of the Chalk cover and its radial tilt he thought to be coincident with the imposition of the transverse crumpling that brought into being the structurally determined 'transverse' streams; and in correlation with the similar crumpling in the Chalk of south-eastern England he gave a post-Oligocene date to the association of events. The complex drainage pattern of South Wales he thus ascribed to direct super-imposition from the Chalk cover, whose form it reflected. But he made no comment on the Palaeogene history of Wales, which (it may be inferred) he considered to be without significant incident, the Chalk presumably being at such low altitudes as to escape any appreciable erosion (despite its known pre-Eocene erosion on the flanks of the London and Hampshire basins).

O. T. Jones (1930, pp. 69ff.; 1952, pp. 204ff.) accepted in principle the Cenozoic inheritance of a Cretaceous cover, and with it the pattern of drainage evolution proposed by Lake and Strahan; but he remained attracted by Ramsay's view, supported by Jukes-Browne, that some of the highest present-day mountain summits —highest in relative altitude, above a notional sub-Cretaceous floor—might have stood up as islands above the Chalk sea in faint reflection of a much older landscape largely buried under Chalk. Impressed by the intensity of post-Hercynian erosion recognisable in the relations of the Keuper sediments to the underlying Palaeozoic rocks in South Wales, and the delineation there of a sub-Triassic oldland by Cenozoic exhumation, he was also prepared to extrapolate a Triassic peneplain over much of Wales, the monadnock ranges standing above it including the mountain-tops that emerged as islands above the Chalk sea. Correspondingly, between the Chalk cover and the sub-Triassic floor he postulated an intercalation of variably thick Triassic and Jurassic rocks that filled the hollows at the foot of the mountain ranges; and to him the present landscape was then a relic in part of a sub-Cretaceous, in part of a sub-Triassic, surface, its geomorphic frame of residual mountain masses of the oldland massif—Snowdonia, the Merioneth summits, Cader Idris, the Berwyns, Plynlimon, the Brecon Beacons, the Black Mountains, Mynydd Preseli—and the depressions between them being thus in broad pattern of Triassic origin, revealed by a stripping of the Mesozoic rocks and not as yet mutilated beyond recognition. (Fig. 91.)

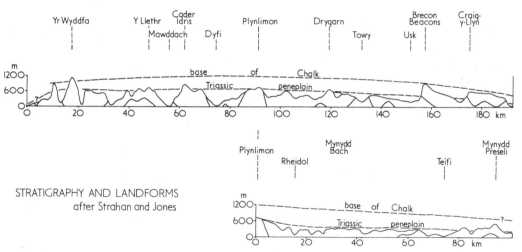

Fig. 91.   Interpretations of the geomorphological evolution of Wales. Both Strahan and O. T. Jones regarded the primary river system to be superimposed from a Chalk cover. Jones in addition considered much of upland Wales to be a re-exposed Triassic peneplain from which the Mesozoic rocks had been stripped (except in Glamorgan). The upper section, approximately north-to-south, is uncertainly reconstructed from the vague allusions to be found in Jones's writings, the Chalk capping the Brecon Beacons (but perhaps not the Snowdonian summits) and being some hundreds of metres above the crest of Plynlimon. The lower section, approximately northeast-to-southwest, illustrates more precisely Jones's conception of the Triassic peneplain, Mynydd Preseli being a geomorphic homologue of Plynlimon despite its being lower in altitude by more than 200 m.

The simple hypothesis of a drainage system inherited from a Chalk cover and of the progressive exposure of a sub-Mesozoic oldland floor, despite its time-honoured place in much assumptive stratigraphy and geomorphology, has nevertheless had increasing doubt cast upon it during recent years notably in the shadow of the Cenozoic history of Ulster (see George 1955, pp. 292ff; 1961, pp. 245ff.; 1967). The doubt has been suddenly and amply confirmed by the rock sequence proved in the Mochras borehole (Wood & Woodland 1968; Woodland 1971), whose record of thick Lias immediately beneath Palaeogene sediments, little more than a kilometre from the Cambrian rocks of the Harlech dome, has transformed Welsh stratigraphy in novel and wholly unsuspected ways, and added to the Tertiary evolution of Wales a chapter radically separating Eocene heritage from Neogene terminus. It has no less been confirmed by the recent geophysical evidence of a complexity of structure and sequence in the sea-floors surrounding the Welsh massif, whose thick deposits of Mesozoic, Palaeogene, and Neogene rocks contrast with the barren enptiness of the post-Mesozoic Welsh landscape. (See Fig. 92.)

II. THE MESOZOIC COVER
(a) Trias and Lias

Few direct signs remain of the nature of the Mesozoic cover in Wales at the close of Cretaceous times, and fewer still of the stages of Cenozoic erosion of the cover. The outcrops of New Red Sandstone now to be seen are ambiguous in their implications. On the one hand, thick Triassic accumulations contain the oldland to east, north, and west. They reach more than 2000 m in the Cheshire basin whose range westwards into Powys includes Bunter sandstones overstepping transgressively onto Carboniferous and Ordovician rocks. A tongue of the comparably deep basin in Liverpool Bay, continuous around the Clwydian range with the Cheshire basin, occupies the Vale of Clwyd in a residual thickness of over 600 m, also in overstep onto Carboniferous rocks. Triassic development in Cardigan Bay is not so certainly known, but it is probably very great, perhaps of the order of several thousand metres thick. (See Pugh 1960; Powell 1956, p. 283; Griffiths and Others 1961; Blundell, Davey, and Greaves 1971, p. 367.)

On the other hand, Triassic development in South Wales is insignificant. The Somerset basin, in which a thickness of not much less than 500 m occurs, may well continue westwards beneath the Jurassic rocks of the Bristol Channel (see Audley-Charles 1970, pls. 9–13); but in the Vale of Glamorgan, in Gower, and (if the gash breccias are a sign) in Pembrokeshire, marginal Trias is locally reduced to nil, being followed by Lias that overlaps onto a Palaeozoic oldland precisely defined by the Rhaetic feather-edge (see George 1970, fig. 34, p. 112). There can be no doubt of the complete absence through non-deposition of New Red cover in parts of hinterland South Wales, as in parts of the Hercynian Mendips.

How far elsewhere there was complementary thinning of the Trias centripetally remains unknown. Bunter rocks are thick along the borders of Flintshire, as they are in the Clwyd tongue that penetrates for over 30 km into the heart of the massif; and, present outcrops being late-Cenozoic residuals, it is certain that thick Trias, perhaps with some internal overlap, spread widely if not completely over North Wales in early Palaeogene times—unless it is gratuitously supposed that present outcrops approximately define the pattern of the sub-Cretaceous outcrops. The hypothesis put forward by O. T. Jones (1930, pp. 58ff.) that oldland Wales is now geomorphically a reflection of the sub-Triassic floor from which the New Red Sandstone has been removed, has some justification in being partly based on extrapolation from the peripheral outcrops.

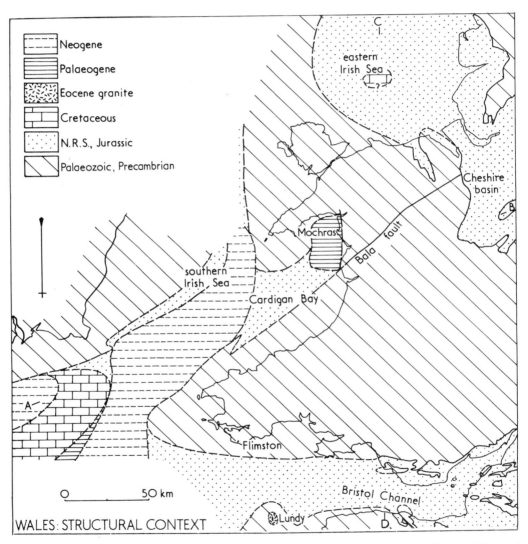

Fig. 92. Rock outcrops in Wales and the adjacent sea-floors. The map is much generalised and simplified, especially in the details of the southern Irish Sea and Cardigan Bay, where the precise locations of feather-edge or faulted margains are mostly unknown, but there is no doubt of the occurrence of deep basins of Mesozoic and Cenozoic rocks above which the Welsh massif of Palaeozoic rocks emerges. After several authors.

Within the Cheshire basin the Trias shows much variation in thickness, a product of contemporaneous sagging and faulting; and farther south the Worcester basin shows even stronger evidence of acute differential movement (see Audley-Charles 1970, pls. 4, 5). It is possible, perhaps probable, that the Malvern fault zone locally defined the westward limit of Keuper sedimentation, at least on the flanks of an archipelagic Malvern terrain, the mid-Welsh massif then being a continuation of the Keuper hinterland recognisable in South Wales. It is less likely, perhaps improbable, that the Clwydian range was a comparable island system upthrown along the Vale of Clwyd fault. At the same time, the western margins of the present outcrops in the Vale, as in the Cheshire basin, extend with simple overstep onto the Palaezoic rocks.

The fragments of evidence provided by the residual Lias is even less directly informative. In Glamorgan, northward overlap in variant detail of Sinemurian (and no doubt later) Liassic rocks continued indefinitely into the coalfield, as it did into Gwent on the flanks of the Usk anticline, with perhaps pocketed Hettangian and Keuper rocks beneath (see O. T. Jones 1956, p. 348; Wobber 1966, p. 136; Owen 1967, p. 598). The Prees outlier in Shropshire, of some 150 m of Lower Lias capped by Marlstone with thick Trias conformably beneath, points strongly to a former westward continuation into Powys. The Lias at Mochras, exceptionally thick at nearly 600 m, leaves no doubt of the original extension of the series in strength into Gwynedd, where also there may well have been a significant thickness of underlying Trias. On such a foundation, with its implications of sustained early-Mesozoic subsidence, later Jurassic rocks, more particularly Upper Jurassic clays, may also have been deposited in North Wales, as O. T. Jones surmised; and a Jurassic sequence in the Bristol Channel, some 1600 m thick, including Kimmeridge beds, has similar implications for South Wales (See Donovan, Lloyd, and Stride, 1971).

As the Liassic rocks are without signs (unlike those of the Kimmeridge beds of Helmsdale) of the proximity of a cliffed coast at time of sedimentation, the Llanbedr fault is not to be regarded as a contemporary barrier to deposition of the Lias; and a probable inference is that the Lias, like the Trias, covering in thickness much or all of North Wales, continued southwards to link with the rocks of Glamorgan and Gwent, the only interruptions to the continuity being the remains of the Hercynian mountains in South Wales. Correspondingly, as in southern England, there was no necessary correlation between the structures in the Palaeozoic rocks and the basins of Mesozoic sedimentation.

It does not follow that the thick Mesozoic blanket was preserved until Palaeogene times, for if the Llanbedr fault fractured the blanket during or at the close of Mesozoic times, separating areas at very different erosional levels, the Jurassic and Triassic rocks of Merioneth may have been stripped, or nearly so, before Cenozoic times began. A reconstruction of the Palaeogene environment is thus dependent on the nature and age of the fault, and on what hypotheses of fault growth are preferred (see p. 358).

(b) Chalk

The continued influence of a Cretaceous cover on the development of changing landform in Cenozoic times having been a theme persistent in the geomorphology of Wales (as of other parts of Britain) for many decades, doubts of the validity of the theme have not been eagerly received. The doubts, however, have deepened, not only as analogies of drainage and platform evolution between parts of Wales and southeastern England have been seen to imply a Neogene origin of the regional landscape, but also as the more completely preserved stratigraphical sequence in the dominantly oldland terrain of western Scotland and of Ulster makes its indirect comment on the tectonics of Wales. The highly revealing record of the Mochras borehole (in which Cretaceous rocks are not found) and the geophysical signs of deep basins of young rocks in the seas flanking Wales, are a final confirmation of the complex stages of Cenozoic change that concede only an insignificant place in a present geomorphology to a Cretaceous cover.

In a lack of any direct evidence, it is no more than speculation to suppose that the development of post-Liassic Jurassic rocks in Wales was inconsiderable. Nevertheless, the lithology and the thinning of the Oolites as they are seen in their nearest outcrops in Wessex and Mercia are without strong suggestion that they extended far across the borders into Wales. The Upper Jurassic clays may have been more widespread, but they also reveal thinning, facies change, and internal nonsequence as they are traced in their present outcrops towards the north-west; and it may be assumed that Portland and Purbeck rocks were not deposited in Wales, and equally, that Lower Cretaceous rocks were limited to southern and eastern England and contributed no protective cover to the Welsh massif. (See George 1962.)

The 'Cenomanian transgression', sign of mid-Cretaceous deformation followed by a drowning of a great part of the British area, has on the other hand been the cited evidence that Greensand and Chalk once covered Wales to form the foundation on which both Cenozoic history and Cenozoic geomorphology have been based; and a critique of derived theory must take the Cretaceous cover as a primary datum. A sceptical reservation concerns the range and nature of the cover. Even at their most widespread the Cretaceous rocks may not have completely blanketed the hinterland of the massif, despite the locus of the massif in the core of a vaguely defined anticline between the Wessex outcrops and the Chalk of Ulster and the Celtic Sea. The Chalk in its nearest residual escarpment of the Wiltshire downs is over 350 m thick: at a not-improbable rate of thinning it formerly rose (in up-dip projection) over the earlier Mesozoic rocks (and whatever Palaeozoic rocks were exposed) of south-eastern Wales to blanket them completely. But in further extension it may not have had the same continuity.

The signs of repeated non-sequence in the greensands and Chalk of Ulster and the Hebrides, and less radically (but perhaps more significantly) of the eastern flank of the Cornubian massif as the rocks are followed westwards from Dorset, are pointers to a Cretaceous sea floor that may have continued in subdued changes to reflect a tectonic frame—a pre-Albian dome or swell of Gwynedd—of ancient establishment. Like the East Anglian platform, the core of the Welsh massif may have suffered pulsed emergence above Jurassic and early Cretaceous sea levels, any remaining Jurassic and Triassic sediments (including inferred thick Lias and Keuper) being repeatedly eroded, locally perhaps down to the Palaeozoic base; and, as in Ulster, the core of the Welsh massif may have been a restless foundation for Upper Cretaceous sediments, which then formed an incomplete cover, and a cover thin and broken in sequence by overstep and internal overlap. That is, the massif, conceptually in early origin a Caledonian–Hercynian St. George's Land, retained its influence as a positive block intermittently throughout Mesozoic times, and the Cretaceous rocks behaved as the marginal Keuper and Jurassic deposits are now seen to behave on the flanks of the Mendips. In short, even if the Chalk formed a continuous cover, it may well have transgressed in the heart of the massif whatever Jurassic and Triassic rocks had survived (even if they were still thickly present in Merioneth) to rest as a thin veneer on the Lower Palaeozoic and Precambrian rocks. Above the Chalk a few incompletely degraded monadnock summits may conceivably have escaped submergence.

This theorising is so much conjecture, based on analogy and slender peripheral evidence. What is certain, as later Cenozoic events show, is that the form and the nature of the Welsh surface-rocks in immediately post-Cretaceous times find no recognisable counterpart in the present landscape: extinction by Palaeogene erosion and earth-movement appears to have reached a stage at which the Chalk cover, which may safely be supposed to have been not at all uniform, had few if any surviving remnants; and the determining and continuing control of the cover on the Cenozoic history of Wales, for so long a central hypothesis of geomorphological evolution, can be discounted. (See George 1961, pp. 247ff.)

III. PALAEOGENE EVENTS

(a) Igneous dykes

Dolerite dykes of early Palaeogene age, the oldest Cenozoic rocks so far identified, have long been known in North Wales. There are scores of them, planed at the present level of the land surface, in Anglesey, Arfon, Lleyn, and Bardsey, where Greenly (1919, pp. 684ff.; 1942) and Matley (1913, p. 525; 1928, p. 486) first distinguished them from the similar and similarly alined Palaeozoic dykes, and where they are unusually well represented perhaps because of the depth of later-Cenozoic erosive penetration in the coastal platforms. A few are known at higher altitudes in the mountainous hinterland (Williams 1924; Greenly 1938; Archer & Elliott 1965): the Marchllyn dyke on Moel Perfedd is now truncated at nearly 750 m, the Bwll-y-cywion dyke a mile to the south-east at 780 m. (See Fig. 93.)

The dykes have isotopic ages proving them to be Eocene (Fitch and others 1969, p. 27): an earlier phase of emplacement at about $61 \pm$ m.y. was followed by a later at 50–55 m.y. ago. The ages compare with those of the Ulster and Hebridean rocks. Their relations with the underlying rocks may be hypothesised as comparing with those in Ulster—where the associated basalts locally overstep the Chalk and any underlying Mesozoic rocks and rest discordantly on a variety of deformed Palaeozoic rocks down to Precambrian (see George 1967, fig. 3, p. 420). While the dykes lack

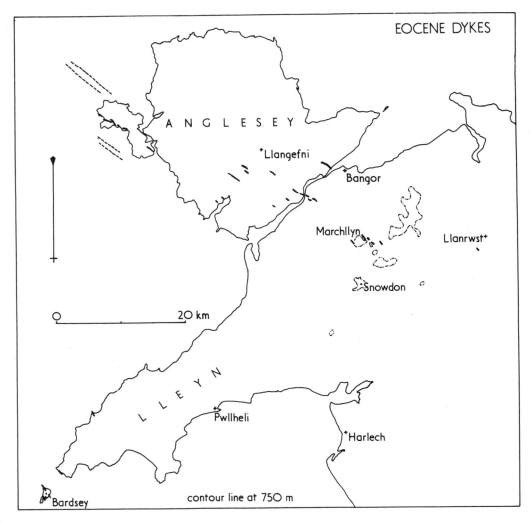

Fig. 93.   Map of the known Eocene dolerite dykes in north-west Wales.

the impressive significance of their counterparts in Ireland and Scotland, they, especially the Marchllyn and Bwlch-y-cywion dykes, no less imply an Eocene terrain whose surface was at an unknown level above the present summits of at least some of the North-Welsh mountains.

The dykes have a general north-westerly alinement. Off-shore there are geophysical signs of similarly alined dykes both west and north-west of Anglesey (Bullerwell, 1965; Al-Shaikh 1969, p. 898). In projection the dykes trend towards the Mourne igneous centres, and they may be regarded as distant representatives of the Mourne dyke swarm, with which as olivine dolerites they offer close petrological comparison. In Ulster, however, the dykes are cognate with the thick Antrim basalts;

and although no contemporary extrusives are known in Wales it is possible, perhaps probable, that a basaltic cover was formerly a blanket over parts of Gwynedd. The notional cover is now gone, and with it whatever Mesozoic rocks that lay beneath; and in the unroofing of the dykes there is the sign that the present land surface, at altitudes above the 'high plateau', is not of Eocene, still less of late-Cretaceous, age. Analogy of the present landforms in Gwynedd is thus closely with those of the Sperrin Mountains in Ulster, where in neighbouring outcrops of basalts and Chalk there are clearly displayed the alternations of rock formation and rock destruction from early Mesozoic into mid-Palaeogene times, and where basalts and Chalk have been completely extinguished by erosive stripping to lower tectonic and geomorphic levels. (See George 1967, pp. 422, 437.)

The importance of the dykes in North Wales, while in part intrinsic, thus lies also in their allusion to an Eocene environment at no great distance away in northern Ireland whose essential characteristics may with high probability be transferred to Wales in a restoration that gives a geological unity to the whole region of the Tertiary igneous province. Similarly, while there are no known Eocene plutons in Wales, like the Mourne granite in Ireland and the Red Hills granite in Skye, that in residual high summits prove the erosive removal of a thick roof in the evolution of the present landscape, the Lundy granite, 45 km from Gower and south Pembrokeshire and intruded about 52 m.y. ago (J. K. Miller & Fitch 1962; Dodson & Long 1962), is a comparable allusive sign, even if the island is scarcely a high summit but is benched with outliers of what appears to be a homologue of the '400-foot' platform.

Greenly (1938), in a perceptive awareness of the significance of the dykes, despite his not knowing their precise age, drew attention to the improbability of hypotheses then current on the ancient origin of the Snowdonian landscape. Since the dykes imply a cover at time of intrusion, erosion of which has now exposed their planed surfaces, and since none of the deeply excavated Snowdonian valleys displays any signs either of Triassic sediments or of extrusive lavas, he concluded that the present geomorphic pattern of North Wales was evolved in post-Eocene times, the valleys being no older, and the coastal platforms notably in Anglesey not being inherited from Triassic ancestors (doubts of which even O. T. Jones conceded when he thought the 'high plateau' of North Wales to be a sign of a Triassic floor). Greenly was prepared to conjecture, however, that the cover into which the dykes were intruded was a cover of Chalk, and with Fearnsides (1910, p. 822) that the present skyline profile of Snowdonia defined a stripped sub-Cretaceous floor gently domed in mid-Tertiary times, on the surface of which the dykes, presumably feeding lavas, came to outcrop.

(b) Palaeogene sediments

The thick Mesozoic sediments proved in the Mochras borehole were at first discovery startling in their proximity to the Cambrian rocks of the Merioneth dome; but the possibility of their occurrence had been anticipated notably by O. T. Jones, and in generalised tectonic relationships they do not depart widely in the frame of their occurrence from the rocks of the deep Worcester and Cheshire basins on the eastern flanks of the Welsh massif. Thick Palaeogene sediments, probably of Oligocene but perhaps in their highest members of Miocene age (Herbert-Smith in Woodland 1971, pp. 95 ff.), were on the other hand altogether unexpected. They throw a flood of light both on early Cenozoic palaegeography and on mid-Cenozoic deformation, and they give a complexity to a geomorphology of Wales that for too long was over-simplified into a gradual evolution of the present-day physique by a mildly imposed reduction of an early-Eocene physique (see Linton 1951, p. 69).

The sediments at Mochras have a residual thickness of 550 m. In general appearance they compare with the Bovey Tracey Beds (more than 200 m, perhaps as much as 750 m, thick) and the Petrockstow Beds (more than 660 m thick) of Devon, and with the Lough Neagh Clays (at least 350 m thick) of Ulster. Their lithological characteristics have been summarised by Wood & Woodland (1969) and O'Sullivan (in Woodland 1971, pp. 14 ff.). They are a suite of yellow, buff, and grey clays, sometimes red and mottled, many of them 'plastic', with silty and sandy layers. Plant debris is common, and there are carbonaceous and lignitic partings occasionally thickening to a lignite bed. A few massive conglomerates, one reaching 10 m, interrupt the generally fine-grained sequence.

The manner of deposition of the Palaeogene rocks, and the palaeogeographical environment in which they accumulated, are not easily reconstructed. They are non-marine at least in greatest part, and were almost certainly isolated if not by a physical barrier then by distance from the contemporary seas of south-eastern England. While they are mostly fine-grained and suggest gentle transport and deposition in a region of subdued relief, they are thick and had their provenance in a hinterland that was either extensive or repeatedly revived. They may be considered to occupy a 'basin' presumably tectonic; but the basin must have been large, for there is strong geophysical evidence that they continue in thickness westwards from Mochras into and beyond Cardigan Bay and are widespread in the southern Irish Sea; and at time of sedimentation they may have linked with the Lough Neagh Clays and the Devon deposits, as the pocket at Flimston in Pembrokeshire hints. Although they are no longer to be seen at surface over much the greatest part of Wales, the signs of them in Powys point to a regional extension of the 'basin', or of collated 'basins', running into the English midlands.

The lithological uniformity of the sediments against the Llanbedr fault, and their great thickness, establish the inference that (as the Lias shows) the fault was no hindrance to sedimentation, and there can be little doubt of the eastward continuation of the Palaeogene 'basin' into the Merioneth heartland. The Palaeogene rocks rest directly on Lias at Mochras, on the downthrown side of the Llanbedr fault. It is to be presumed that they originally rested on rocks no younger—rocks probably older—on the eastern upthrown side of the fault. In Tremadoc Bay they transgress the Lias northwestwards onto Trias and Lower Palaeozoic rocks; and in Powys they originally overlay eroded gently warped structures in whatever early-Mesozoic rocks survived beneath. (See Figs. 96 and 97.)

The contents of the conglomerates in the Mochras rocks add strong confirmation. In some of them the pebbles, which may be of the order of centimetres in diameter, are of mudstone lithologically similar to the yellow or buff matrix in which they lie, and they appear to be the product of contemporaneous erosion of temporarily desiccated mudflats—signs of a local environment but not otherwise greatly informative. In others, however, the pebbles in abundance, giving a grey tone to the rocks, are of vein quartzes and tough gritty rock-fragments of Lower Palaeozoic aspect, pointing to sources in exposed oldland, all Mesozoic rocks removed. None of the pebbles yet identified is of flint or greensand chert. (See O'Sullivan in Woodland 1971, pp. 20 ff.)

An inevitable conclusion makes radical comment on traditional hypothesis. The direct evidence at Mochras of Lias immediately overlain by Palaeogene rocks, Cretaceous absent, and the collateral inferences of the effect of the Llanbedr fault (see p. 357), match the regional evidence in Ulster (George 1967, p. 424): if ever the Chalk covered the Welsh massif, with or without earlier Mesozoic rocks beneath, it had been removed in great part, perhaps altogether, by Oligocene times; and where erosion had not penetrated to the Palaeozoic floor there remained only a residue of early Mesozoic rocks between the Palaeogene clays and the Palaeozoic rocks. A sub-Cretaceous landform vanishes as a primary datum in an explication of the geomorphic emergence of oldland Wales.

The Flimston clays of the Castlemartin peninsula (Dixon 1921, p. 167), also probably of Palaeogene (or early Miocene) age, are pocketed in Carboniferous Limestone. They cover a small area and are estimated to be only some 14 m thick. Their precise relations with the limestone are uncertain—conceivably they may occupy a solution cavity, or be faulted down—but they add, at a distance, to the evidence that neither Chalk nor other Mesozoic rock was locally present at the time of Palaeogene deposition.

The similar pockets in north-east Wales (Walsh & Brown 1971), some of which exceed 70 m in depth, some of which lie at relatively high altitudes, occupy solution cavities of uncertain age in Carboniferous Limestone. Their Palaeogene sediments—lignitic clays and sands, with some pebble beds—are uncontaminated by identifiable Cretaceous (or other Mesozoic) rocks, and are without derived flints or cherts. In resting directly on a Carboniferous floor they match the Flimston clays in proving the removal of any Chalk from the neighbourhood before their deposition.

(c) Contemporary earth-movement

The effects of Palaeogene earth-movements on stratigraphical sequence are to be recognised, if not in every detail, in the relation of the Palaeogene sediments to the rocks beneath. The Cretaceous cover, actual or notional, was uplifted, presumably domed and possibly faulted, in early Palaeogene times, and with it whatever Jurassic and Triassic rocks were underlying, over the whole of the Welsh massif between Tremadoc Bay and the Cheshire basin. The intensity of Eocene erosion was dependent on the amount of uplift and the warps of the uplifted surface: if Ulster is a guide (the base of the Eocene Antrim basalts being used as a datum), the pared-down surface may locally—in Anglesey, on the Bangor ridge—have reached in cumulative removal erosional levels penetrating to Precambrian rocks (compare George 1961, pp. 248 ff.; 1967, fig. 3, p. 420). The amplitude of the composite dome in North

Wales may then have been of the order of hundreds, even thousands, of metres.

A second datum is available at Mochras—the base of the Palaeogene sediments in contact with thick Lias—and by extrapolation it is also available less directly farther east, where inferentially the pipe-clays rested on the Palaeozoic floor. Westwards in Tremadoc Bay there is a complementary gentle rise of the Lias beneath the Palaeogene rocks, which overstep onto Trias towards Lleyn; and a sub-Palaeogene syncline in the residual Mesozoic rocks, between Lleyn and Merioneth, has an amplitude of not less than 1290 m (the measured thickness of the Mochras Lias) and probably of much more if thick Trias was also transgressed, the Palaeogene sediments coming to rest on Palaeozoic and crossing into Lleyn (see Fig. 97, p. 358). The age of the syncline is not proved by the overstep, for, Chalk not being preserved, the break may be wholly or mainly of pre-Albian age, as comparable structural relations are in Ulster. To the south-west, however, in Cardigan Bay and the southern Irish Sea, and in the Celtic Sea, the geophysical evidence shows the Triassic and Jurassic rocks to return in thickness, perhaps to 5000 m or more, and the Palaeogene sediments to display offlap to rest unconformably on Chalk; and a major component of the synclinal fold in Tremadoc Bay, and of the dome to the east, may be regarded as Eocene. (See Griffiths, King, and Wilson 1961; Bullerwell & McQuillin 1969; Eden, Wright, and Bullerwell 1971, pp. 133ff,; Blundell, Davey, and Graves 1971, fig. 14, p. 368.)

There may also have been movement along the Llanbedr fault in Palaeogene times, but there is no certain proof of it, and the evidence can be variously interpreted (see p. 358).

IV. MID-CENOZOIC EARTH-MOVEMENTS

(a) The folds of South Wales

Evidence being imperfect, conjecture and hypothesis have tended to be based on analogy and correlation in the recognition of 'Miocene' structures in Wales, the only direct sign being the Llanbedr fault and the Flimston pocket. To see the folds in the Lias and Trias of Glamorgan as evidence of 'Alpine' deformation was a natural consequence of assumptions of widely pulsed orogenic events—'Caledonian', 'Hercynian', 'Alpine'—derived from a relatively elementary 19th-century analysis of rock structures in southern Britain. Thus the first explicit discussion of the occurrence of mid-Cenozoic folds and faults in Wales was introduced by Strahan (1902) when he attributed the 'tranverse' direction of flow notably of the Towy and the Tawe to corrugations on the tilted Chalk surface on which he supposed the river system to have been initiated: the analogy was clearly with the comparable structures and the simply interpreted river pattern in the Chalk terrain of southern England, demonstrably (apart from mid-Eocene warping) including the Palaeogene sediments of the London and Hampshire basins and therefore of 'Miocene' age. The argument was cyclical: it inferred the Chalk cover from the pattern of the drainage, and it explained the pattern of the drainage by inferences of the form of the Chalk cover. It fails when, dependent on a hypothesis of drainage evolution, the 'tranverse' rivers—'tranverse' to Strahan only in being discordant with the 'longitudinal' south-flowing streams but like them no less 'primary' consequents—are in fact recognised to be 'subsequents', late products of river capture and secondary diversion (see p. 367).

It was left to O. T. Jones (1931, pp. 16ff., pl. II) to analyse in detail the form of the Jurassic folds in the Vale of Glamorgan, not merely in comparing them with 'Miocene' folds farther east, but in attempting to trace confluent or homologous structures from the Jurassic rocks into Chalk terrain. Thus he saw the depression of the Bristol Channel to be in broad continuity an extension of the Bridgwater syncline, and the posthumous Cowbridge anticline (of Hercynian inception) to be in approximate alinement with the Pewsey anticline on whose flanks Chalk is preserved. The more recent discovery of Upper Jurassic sediments (but not yet of Chalk) in the Bristol Channel strengthens Jones's highly plausible hypothesis in its material evidence and in emphasising the magnitude of the deformation (Lloyd, 1963; Donovan, Lloyd, and Stride 1971; Owen 1971).

Nevertheless, a notable weakness in Jones's argument is a neglect of the possibilities of intra-Mesozoic movements of both mid-Jurassic and mid-Cretaceous age. The amplitude of the Cowbridge anticline in the Rhaetic and Lias is of the order of 60 to 80 m: the 'Bajocian transgression' in the not-far-distant Cotswolds is a measure of mid-Jurassic folding of amplitude exceeding 100 m; and along the nearer Bath axis there was comparable if less explicit contemporaneous warping. Analogy or homology of the Cowbridge anticline with the Pewsey anticline similarly requires a recognition that folding of the Albian-Cenomanian rocks along the Pewsey axis was superimposed on already-deformed Jurassic rocks, Purbeck, Portland, and Kimmeridge formations being completely overstepped on the flanks of the fold (and it is to be noted that the correlated overstep in south Dorset is marked by discrepant dips of over thirty degrees and by pre-Albian faults in the Jurassic rocks several hundred metres in throw).

If the structures in the Jurassic rocks are regarded as 'Alpine', then the Alpine orogeny occupied a prolonged interval from mid-Mesozoic times onwards, and while no doubt there was structural revival in mid-Cenozoic times a reference of the folding in the Glamorgan Lias to the Miocene period can be at best tentative.

There is a likeness to the relations in Ulster and Scotland, where mid-Mesozoic deformation of the same order of magnitude resulted in the overstep of Upper Cretaceous rocks onto Precambrian from Lias, and where sub-Eocene deformation of the Chalk in folds over 100 m in amplitude preceded the sub-Oligocene folding of the basalts in amplitudes exceeding 500 m, which in turn preceded the Miocene folding of the Lough Neagh Clays in amplitude of over 1000 m.

The further elaboration by Jones of Cenozoic structural growth in South Wales, in his assumption of two stages of movement—an Eocene fall of the Chalk cover east of south to establish the primary drainage, and a Miocene tilting to the south-west to explain the secondary 'transverse' rivers—was a gratuitous alternative to Strahan's hypothesis, having no stronger basis than conjecture.

The only structure in South Wales that demonstrably is mid-Cenozoic ('Miocene') is the Flimston pocket of Palaeogene rocks. It may be a solution pocket, but if it is a gentle fold, perhaps faulted, whose dimensions as estimated by Dixon approximate to its present form, it has flanking dips of the order of 1 in 10 or less. It is truncated by the surface of the low coastal platform at about 50 m above sea level, of late Pliocene or early Pleistocene age (see p. 363), and its sediments are planed flush with the Carboniferous Limestone surrounding. Its floor, projected northwards, notionally overtopped the summits of the Ridgeway, and, farther afield, of Mynydd Preseli; and, projected southwards, did not (so far as knowledge goes) descend over the crest of an intervening anticline to the floor of the present Bristol Channel. Despite its very small residual size, it thus implies Miocene deformation on a scale commensurate with the folding of the Pewsey and Wardour structures. It also adds weight to an expectation that comparable deformation is to be found in the Jurassic rocks farther east, and thus to the view that at least in part the folds in the Vale of Glamorgan, and more considerably the syncline of Jurassic rocks in the Bristol Channel (with an inferred marginal fault off the Devon coast), may incorporate Miocene components. (See Fig. 94.)

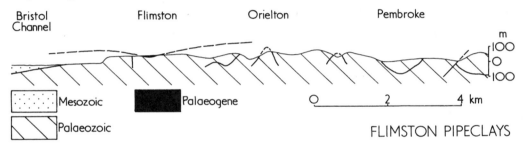

Fig. 94. Section, approximately south-west to north-east, across the south-Pembrokeshire platform. The relations of the Palaeogene rocks at Flimston to the Carboniferous floor are interpreted as gently synclinal, without intervening Mesozoic. Palaeogene outliers are not known to the north on a Palaeozoic floor, or to the south on a Mesozoic floor in the Bristol Channel.

(b) The origin of the geomorphic massif

The structural Welsh massif, of ancient origin in Devonian times as a swell or complex anticline and continually renewed by uplift throughout the Carboniferous period and enlarged by Hercynian accretions in the Permian, was reduced to an erosionally subdued uniformity in early Mesozoic stages when Trias and Lias (interpolation between Cardigan Bay, the eastern Irish Sea, and the Cheshire basin being regarded as valid) were deposited over the greater part if not all of it. By mid-Jurassic times its degraded and blanketed surface was negligibly ruffled. Less certainly it was again submerged, with a remnant Jurassic cover variably preserved, beneath Upper Cretaceous seas and sediments, to be uplifted, more or less completely veneered by Chalk, in an early-Eocene episode. Between Mochras and the Powys outcrops of Trias, the Mesozoic rocks were substantially removed before the third major post-Hercynian burial again obliterated, under Palaeogene sediments, any roughnesses etched or moulded in the landscape during Cretaceous and Eocene times. Post-Palaeogene uplift, partly by faulting, was a final phase in the sequence of tectonic events that transformed the ancient Caledonian structural swell into the

Neogene geomorphic massif.

In complement, the flanking basins of thick sediments now beneath the sea show that the emergence of the Miocene faulted dome was a late episode in a long and complex tectonic history the events of which add comment on differential movements between massif and flanks, and at the same time give some coherence to a unified structural evolution of massif and flanks.

The revealing geophysical work carried out in recent years in Cardigan Bay and the southern Irish Sea, amplified and integrated by Eden, Wright, and Bullerwell (1971, pp. 123ff.) and by Blundell, Davey, and Graves (1971), may be questioned in the derived geological interpretation of sequence and structure; but it can leave no doubt of a long-sustained deep synclinal basin of caledonoid trend lying between Palaeozoic Ireland and Palaeozoic Wales. The growth of the basin was nevertheless not unbroken. Mesozoic and Cenozoic deposits, in the deeper parts of the basin approaching or even exceeding 7000 m in thickness, are interrupted by unconformities that are a comment on vicissitudes of subsidence and show the basin to have been at times sufficiently uplifted to emerge into a zone of erosion, probably as land above sea level; and there is no reason to suppose that a present contrast between basin and massif has been consistently maintained in differential subsidence of the basin, uplift of the massif.

Thus thick Chalk (proved in dredged high-zonal samples), overlying thick Triassic and Jurassic rocks, is widely developed in the Celtic Sea off the south coast of Ireland from Carnsore Point almost to the edge of the continental shelf; but elsewhere, both in the southern Irish Sea and in Tremadoc Bay, it is absent beneath transgressive Palaeogene rocks. The Palaeogene rocks are themselves limited in outcrop to synclinal pockets isolated by culminating swells, and along their feather-edge margins they are overstepped by Neogene sediments that locally spill onto a Palaeozoic basement.

These structural relations—Jurassic rocks emergent above sea level before the deposition of Cretaceous, Chalk emergent and removed before the deposition of Palaeogene, Palaeogene uplifted and eroded before the deposition of Neogene—are not significantly different from those inferred to have existed on mainland Wales, the contrasts between Cardigan Bay and the massif being attributable to pulses of differential elevation rather than to independence in stratigraphical theme. Minor folds, their axes swinging about a caledonoid trend, ripple the main structures; and faults, of which the Llanbedr fault and an extended Bala fault are members, give the syncline a rifted form in complement to the horst element in the massif. (See Blundell, Davey, and Graves 1971, figs. 6, 14.) (See Fig. 95.)

The sediments of the eastern Irish Sea are not so informative. They include thick accumulations of New Red Sandstone, perhaps reaching 5000 m, in a broad synclinal structure continuing westwards from the Lancashire basin (see Bott in Donovan 1968, pp. 93ff.; Bott & Watts 1971, p. 96). No younger rocks have been directly identified, but it is appropriate to recall that the distribution of flint erratics in the Glacial drift of Anglesey led Greenly (1919, p. 777) convincingly to suppose that an outlier of Chalk occupied a small area on the sea bed a short distance to the

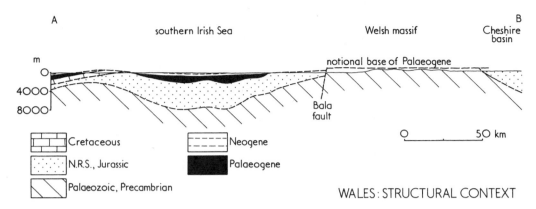

Fig. 95.   Section along the A-B of Fig. 92. As in Fig. 92, the stratigraphical boundaries on the southern Irish Sea are imprecisely located; the distinction between one system and another is mostly not certain; and the thicknesses of the systems are at best approximate; but there is no doubt of the contrast between massif and basin. As drawn, the notional base of the Palaeogene on the mainland is horizontal to reflect the inferred relations in Merioneth (see Fig. 97): it may well have been deformed by 'Alpine' movement, the model being intended to show only its overstep onto Palaeozoic rocks.

north-east, with possible signs of thin Jurassic rocks beneath. The presumption is that pre-Albian erosion was more intense or prolonged—pre-Albian uplift was more considerable—in the eastern Irish Sea than in Tremadoc Bay; and during early Cretaceous times the floor beyond the present Welsh coast was effectively part of the massif (an extended massif, linking with Cumbrian and Manx elements). The general absence of Cretaceous rocks, and everywhere of known Palaeogene and Neogene, is also a sign of prolonged Cenozoic exposure, if not continuingly at least intermittently, to intense erosion and perhaps to Neogene non-depositions it reflects early-Eocene and Miocene uplift and only slight Neogene subsidence in relation to the sequence in Cardigan Bay.

A feature of the sea floor around Anglesey and Lleyn is the wide development of Palaeozoic and Precambrian rocks at the surface, over an area apparently not fault-bounded but forming a dome-like swell, flanked by Mesozoic rocks to east and south, and in elongated form continuing northwards to the Isle of Man (see Al-Shaikh 1969; Eden, Wright, and Bullerwell 1971, fig. 2; Wright and others 1971, fig. 14). If the proposed regional palaeogeography is acceptable, this empty area in its size and location can scarcely be regarded as never having received any deposits of Triassic or Jurassic or Cretaceous or Palaeogene age: on the contrary, it is ground from which a multiple cover has been stripped. It has thus all the features of the mainland massif, it is indeed a structural extension of the massif; and, in being submarine, lying between deep basins of younger sediments, it offers direct comment on the stratigraphical evolution of the massif.

The Cheshire basin in its Bunter outcrops is confluent with the Trias of Liverpool Bay, and in its Keuper outcrops is separated only by the erosive planing of a gentle arch. Structurally it is a counterpart of the Irish Sea basin to the north-west, and differs only in not being submerged. The Prees outlier of Lias is a sign of a more complete stratigraphical sequence than is found to the north-west, a sign that may justifiably be extrapolated. The floor of the Cheshire basin now lies perhaps 2500 m below sea level (the base of the Keuper is known to lie at over 1500 m); it rises to or towards sea level on the eastern shore of the Dee estuary, and then descends to depths of several thousand metres in the Irish Sea. This is a range of altitude, the product of gentle warping, that dwarfs the present-day elevation of the highest Snowdonian mountains; and it diminishes to weak conjecture the importance of present-day landform as a guide to early-Mesozoic or late-Cretaceous or Palaeogene landform.

In combination, the evidence of the synclinal basins surrounding the Welsh massif points to a mid-Cenozoic origin of the massif by uplift in relation to the synclinal basins, the geomorphic evolution of the structural massif to its present landform being a product of the erosion by Neogene agents of the rocks exposed in consequence of the Miocene uplift. The Llanbedr fault is a major clue.

(c) Interpretations of the Llanbedr fault

The thick Mesozoic and Palaeogene sediments proved in the Mochras borehole are thrown against Cambrian rocks, exposed little more than a kilometre away, that rise in the Merioneth mountains immediately to the east to heights of 590 m in Moelfre and Llawllech, to 710 m in Rhinog Fawr, to 740 m in Y Llethr, and to 735 m in Diffwys. If the inferred base of the Trias at Mochras, estimated to be at about 3000 m depth, rests on a Lower Palaeozoic floor (a possible intermediate development of Carboniferous rocks being ignored), the apparent throw of the fault, perhaps as a zone of stepped faults, is thus not less than about 3750 m. If the Lower Palaeozoic floor is of Ordovician rocks (see Griffiths, King, and Wilson 1961, pp. 183ff.) not steeply dipping, a thickness of Cambrian rocks of the order of 2500 m should be added on the downthrown side, to increase the apparent additive throw of the fault in the Lower Palaeozoic rocks to 6000 m or more. (See Wood & Woodland in Woodland 1971, fig. 2.)

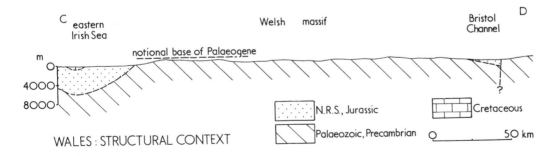

Fig. 96. Section along the line C-D of Fig. 92. The general comments under Fig. 95 are equally applicable. The Cretaceous outlier in the eastern Irish Sea is, after Greenly, conjecturally located. The thickness of Mesozoic rocks in the Bristol Channel may well be much greater than the diagram indicates, and the details of truncation by a fault along the Devon coast are unknown.

Such a calculation is simplistic or naive, but it affords a measure of the magnitude of the contrast between massif and basin. A more sophisticated calculation should be supplemented by assumptions of the staggered growth of the fault in several major pulses. Thus the Palaeogene rocks in the borehole rest directly on Lias, probable Cretaceous and possible Middle and Upper Jurassic rocks being absent in part through Eocene erosion, in part no doubt through early Cretaceous erosion. On opposite sides of the fault the intensity of erosion during any periods of uplift and exposure need not have been commensurate if repeated fault movement accompanied the strains of uplift; and while on the west rocks as young as uppermost Lias were preserved, on the east Jurassic and Triassic might well have been removed by Cretaceous erosion, Cretaceous, Jurassic, and Triassic by Eocene erosion, Palaeogene, Cretaceous, Jurassic, and Triassic by Neogene erosion. A measure of the throw of the fault depends therefore on the notional levels of the datum horizons on the massif: in the Palaeogene rocks it was not less than 1350 m. An interpretation of Bouguer anomalies given by Blundell, Davey, and Graves (1971, fig. 12, R–R[1]) shows the fault to be multiple in a zone of three fractures, growth being both pre-Palaeogene and post-Palaeogene so that within the fracture zone the Palaeogene rocks overstep the Mesozoic to rest directly on Lower Palaeozoic: correspondingly, the Cambrian rocks of the Merioneth mountains were also directly overlain by Palaeogene, all Mesozoic stripped. (See Fig. 97.)

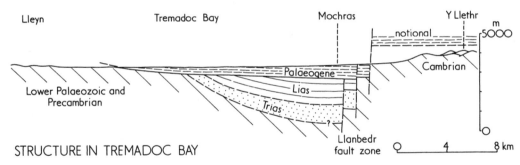

Fig. 97. Section from Lleyn to Ardudwy. The details of overstep by the Palaeogene rocks are probably more complex than the diagram shows, but the substantive relations are to be accepted. The multiple fracturing in the Llanbedr fault zone, and the evidence of repeated movement, are adapted from Blundell, Davey, & Graves (1971). The notional position of the Palaeogene base above the present-day summits of the Merioneth mountains is no more than a simplest structural relationship: what 'Alpine' warping the base suffered is unknown, but if the effects of the Llanbedr fault are in main features as the diagram suggests, the base rested on Palaeozoic rocks, all Mesozoic rocks removed by pre-Palaeogene erosion. The thickness of the Trias is unknown.

The Llanbedr fault appears to fade rapidly northwards. It may be traced southwards along the western foot of the Cader Idris range (Powell 1956, pl. xix), where, near Tonfanau, the Trias floor, at about 600 m, appears to be at much shallower depth than at Mochras. These are signs of warping in the Mesozoic rocks before faulting, or before the final stages of faulting, took place: it accords with the known sub-Palaeogene warping in Trias and Lias in Tremadoc Bay, and it may also reflect the influence of contemporary movement along the neighbouring (and very ancient) Bala fault. The regional meaning given to the Llanbedr (and similar) faults thus rests in a correlated interplay of folding and fracturing, the block relations of which are as yet very imperfectly known; but together they imply that the geomorphic form of the oldland massif as it emerged in Neogene times was not in faintest reflection that of a tilted or gently warped Cretaceous cover, although it might have retained some undestroyed segments of a Triassic floor, however impossible it might now be to identify them.

(d) Analogy with Ulster

It is perhaps repetitive again to draw attention to the similarities between the inferred evolution of pre-Neogene Wales, on the mainland of which few post-Palaeozoic remnants remain except in Clwyd and Glamorgan, and pre-Neogene Ulster, where representatives of every system are still preserved; but the discordant relations in Ulster at the base of the Trias, at the base of the Hibernian Greensands and the Chalk, at the base of the basalts, and particularly at the base of the Lough Neagh Clays are highly relevant comment on the interpretation to be put on the contrasts between the basin of Cardigan Bay and the massif to the east; and the Miocene deformation that is to be seen in the Lough Neagh syncline, with amplitude of over 1000 m, and in the accompanying faults some with throw of hundreds of metres, is a parallel to the deformation recognisable in the downfaulted suncline of Palaeogene rocks in Tremadoc Bay. Evolution of the Welsh massif in Neogene times began in corrugated country.

V. THE NEOGENE ENVIRONMENT

(a) Eustasy and warping

Evidence in the Welsh landscape of events in Neogene times is mainly erosional: the resulting landforms suggest that the principal erosive agent was the sea, and that at least the later events of the period followed from the progressive emergence of Wales, as more widely of much of the British area, from beneath sea level. An earlier stage in Neogene evolution, after the Miocene deformation of the Palaeogene rocks, was thus, if the Welsh terrain did not continuously remain beneath the sea, a general subsidence antecedent to the emergence. An unwillingness to concede an order of emergence beyond a few hundred metres (for which there is direct evidence) arises partly because of the persistent assumption of a Chalk cover, partly because of the absence of confirmatory signs in a stratigraphy of associated marine sediments; but even in southern England, beyond the limit of glaciation, where sediments (not yet consolidated) are still preserved, the residual outcrops are minute, many of them having been found only by sustained search; and in glaciated ground, which includes the whole of Wales, there can be little expectation of finding any Neogene remnants undisturbed or undispersed.

The abundant quartz pebbles widely scattered on the surface of the Castlemartin peninsula, especially on the platform around Flimston at a height of about 50 m (Dixon 1921, p. 169), may be of 'Pliocene' age, although with them are to be found pebbles of Old Red Sandstone and of Carboniferous grits derived from the north

and presumably ice-carried. Otherwise, Neogene deposits are not certainly known in Wales. They are, on the other hand, developed perhaps to a thickness of 200 m over a
and presumably ice-carried. Otherwise, Neogene deposits are not certainly known in Wales. They are, on the other hand, developed perhaps to a thickness of 200 m over a wide submarine tract running south from the southern Irish Sea through St. George's Channel towards the Celtic Sea, in the interpretation put upon the geophysical evidence by Blundell, Davey, and Graves (1971, pp. 350, 365, 368): the precise age of these rocks is unknown, but in appearing locally to rest transgressively, Palaeogene overstepped, on a Mesozoic and perhaps older foundation, they imply discontinuous sedimentation in the basin and an oscillatory floor that emerged temporarily with the Miocene movements.

The outcrops of late-Pliocene gravels at Orleigh Court near Bideford in Devon and near Camborne in Cornwall rest on the '430-foot' platform. The Lenham Beds of Kent, also of late-Pliocene (Diestian, Gedgravian) age, rise above 180 m on the North Downs, and similar sands and gravels, presumably also Diestian, are found widely scattered on the South Downs and on the flanks of the London basin at platform heights ranging between 200 m and 150 m. The Gedgravian Coralline Crag descends to and below sea level on the East Anglian coast in indication of the Neogene down-warping of the North Sea basin (see Wooldridge & Linton 1955, fig. 11; Walsh and others 1972, fig. 6). When these fragmentary signs of a Neogene base—but a base probably not everywhere of the same age— are used as a datum, it is evident that both gentle folding (perhaps in response to a widening Atlantic and and eastward migration of the east-Atlantic continental shelf) and hinterland uplift were instrumental in determining the late-Neogene landforms: the exposed land surface of Wales and southern England may then be looked on as mild swells risin out of the Pliocene sea in complement to the synclinal sags, long-sustained of the southern Irish Sea and the North Sea.

The emergent swells were notched by the sea as they rose, their margins cliffed and retreating at each stage of uplift. The notching was doubtless insignificant if uplift was steadily continuous; but a pulsed uplift, with intervals of still-stand, encouraged the erosion of relatively wide wave-cut benches to form a stepped profile, the step at any one stage being insignificantly warped. Over southern Britain as a whole the swells may then have had the form of exceedingly gentle anticlines; but if folding preceded major uplift, or if there was no wide lateral fluctuation in the loci of the anticlinal crests as the folds emerged, the domed surface would in its profile become dominated by the marine benches, whose form, especially in the crestal regions, would give the impression of simple uplift (compare Brown 1960, fig. 4, p. 48). Moreover, any one wave-cut bench itself sloped seawards from its coastal cliff, at an angle sometimes diverging little from the slope of the dome; and residual deposits on it, although contemporaenous, would neither rest all at the same height nor necessarily belong all to the same facies—and conversely, deposits at the same height but in remnant outliers at a distance from one another might not be of the same age.

In net result, the composite landform characteristic of Wales (and of other oldland cores like Cornubia, the Lake District, the Southern Uplands, the western

Highlands) gives the impression of eustatic uplift; and although the appearance of eustasy was not the product of a world-wide change in sea level, it is effectively an element appropriately incorporated into an explication of the evolving Neogene Wales.

(b) Concordance of landforms

A concordance of geomorphic residuals is a main indication of such eustatic uplift when correlation cannot be direct. The analysis carried out by Hollingworth (1951, pp. 72ff., and pls. x, xi) drew attention to the concordance between North Wales and South Wales especially in summit heights below 600 m. From the concordance at any one approximate level Hollingworth deduced contemporaneity and uplift that was therefore 'eustatic'. In a region where the higher summits are very widely spaced, where all Cenozoic (and Mesozoic) rocks have been removed without leaving a trace of the details of 'Alpine' deformation during Neogene times, and where interpolation in the emptinesses between the isolated summits can be highly conjectural without being easily disproved, a challenge to Hollingworth's statistical conclusions is found in the conflicting statistics used by O. T. Jones (1952, pp. 202ff., and pls. x and xi) to support a correlation between summit levels that shows a steady fall by Cenozoic tilting from North Wales southwards, and then to refute a correlation based on an altimetric concordance of summit levels. While Hollingworth inferred a falling sea-level in stepped pulses, Jones favoured sub-aerial erosion (basically of Triassic origin) as a prime agent in forming the 'high plateau' which he regarded, because of the tilting, to merge southwards with the much more recent 'coastal plateau', marine in origin and untilted. (See Fig. 91.)

Difficulties of integration increase with altitude of the residual surfaces: up to heights of about 1100 feet Hollingworth's steps are not merely individually convincing but as suites correspond closely between North Wales and South Wales. At higher altitudes there is greater uncertainty. The attempt by Brown (1960) to identify the remnants of specific surfaces by direct field observations complemented statistical method, 'objective' in its purpose. Nevertheless, it did not altogether succeed in resolving the several constituents that contribute to the multiple form of upland Wales. His recognition of four major 'plains' (or 'peneplains' or 'plateaux') had as counterpart his regarding them to be of 'subaerial' origin, for each ranges in height through 100 m or more; yet he ascribed the generalised steps between them to 'eustatic' uplift, and considered the changes in regional base-level of erosion to affect more or less uniformly the evolution of each peneplain. Thus to him the 'Summit Plain', now represented only in the residual monadnock peaks between 615 m and 1100 m, was the sub-Mesozoic surface exhumed; the 'High Plateau', 520 m to 580 m in height, rising to 610 m around the summit monadnock peaks, was a product of mid-Cenozoic uplift 'and ran its course in Miocene times'; the 'Middle Peneplain,' 365 m to 490 m in height, was an intermediate stage of Neogene evolution; and the 'Low Peneplain', 215 m to 335 m in height, was the latest of the series formed shortly before an 'early-Pleistocene' sea cut the '600-foot' platform. (See Brown 1960, pp. 103, 169.)

In Brown's synthesis Wales was thus continuously land, at heights above the present '600-foot' platform, from end-Mesozoic times until Pleistocene ice buried it. Moreover, 'peneplanation' (a process the details of which remain undefined) was repeatedly accommodated to changing sea-level, although the altimetric range of any one peneplain much exceeds the intervals between successive plains; and the peneplains retain their identity, their ranges, and their intervals between, whether they are centrally or peripherally located. These are strained relations when in Merioneth the monadnock peaks of Cader Idris, the summits of Y Llethr, Rhinog Fawr, and Diffwys incorporated in the 'Summit Plain', Moelfre and Llawllech in the 'High Plateau', and Moel Goldog, Craig Ddinas, Pen-y-garn, and Esgair Berfa in the 'Middle Peneplain', all lie within a few kilometres of the Llanbedr fault and the depression of Tremadoc Bay. There are also many degraded steps within each peneplain: in the Aberystwyth district, for instance, Rodda (1970) showed that trend-surface analysis offers only marginal preference for separate plains over graded steps.

While Brown's synthesis thus provides no support for O. T. Jones's tilted sub-Triassic floor, it scarcely gives grounds for preferring peneplanation to marine erosion: the echelon of bevelled surfaces is best accommodated in a unitary system if it is regarded as a product of pulsed emergence in post-Alpine Neogene times (see Brown 1957; George 1961, pp. 255, 262). It does not follow that Pliocene or Pleistocene tilting of the platforms was insignificant—the growth of the North-Sea basin and probably of the Cardigan-Bay basin is opposed to that assumption—and precise correlation of the remants of the higher platforms is particularly uncertain if it is based solely on altimetric equivalence.

VI.   NEOGENE PLATFORMS

The reality of the 'coastal plateau' is generally accepted. Its composite nature is manifest in a sequence of stepped platforms rising to a little above 190 m, the lower platforms biting into and locally obliterating the higher, most of them with degraded cliff margins recognisable here and there where they impinge against a hinterland. Their common tripartite division into a graded sequence at '200-foot' intervals is an artefact, and any one of them may span a vertical range of as much as 60 m, presumably due to alternations of slow and fast uplift (see George 1961, pp. 251ff.). Nevertheless, outliers of higher platforms as insular monadnocks above lower platforms are good evidence of the pulses of emergence that give some justification to the emphasis placed on the '200-foot', '400-foot', and '600-foot' platforms, as in Anglesey (Greenly 1919, pp. 779ff.), Pembrokeshire (George 1970, pls. ii, iii), and Gower (George 1938, fig. 3, p. 32).

The so-called '600-foot' platform has been very thoroughly surveyed, as a geomorphic element retaining its consistency with a surprisingly uniform development, around the greater part of the Welsh coast, running deep into Pembrokeshire and the Towy valley for 35 km, into the Teifi valley for 40 km, and into the Vale of Clwyd for 20 km (see Brown 1956; 1960, fig. 31). It is poorly preserved only where the high mountains of Merioneth (in proximity to the Llanbedr fault) fall steeply to the sea, and where the encroachment of lower platforms, in Lleyn and particularly from Anglesey into Arfon, has left only small remnants of the higher platform remaining. The uniformity in general altitude of the platform is strong evidence both of the 'eustatic' uplift that has exposed it (and the platforms below), and of its origin as a waved-planed bench rather than as a subaerial peneplain (compare A. A. Miller 1937).

The so-called '400-foot' platform, which may rise significantly higher, and the '200-foot' platform, which in Gower and Pembrokeshire falls to 50 m and in the Vale of Glamorgan to 30 m, are even more clearly defined, partly because of their relative youth. In the long slopes and minor local steepenings of their composite form they indicate a comparatively gentle rise of the land punctured by pauses not of long duration; and they are most readily distinguished from one another, with a clearly recognisable cliff between, only where a lower element of the sequence, most often at about 60 m, penetrates far into an upper. (See Miller 1939, pp. 35ff.; North 1929, pp. 33ff.; Driscoll 1958; Embleton 1964.) Their good preservation demonstrates their very late Cenozoic origin; but the only direct proof in Wales of their Neogene (or later) age is the truncation of the Flimston pocket of pipe-clays by the '200-foot' platform.

Higher platforms are less readily integrated (Brown 1960, terminal plate). If Brown's four major divisions are accepted as a first approximation, the 'Low Peneplain' is very well developed in south-western Wales, where in long peninsular tongues it forms a dominant element of the landscape on the flanks of the lower Towy and Teifi valleys and in the hinterland of Gwent and the Forest of Dean. Elsewhere it is less spectacular in unitary form, and, imprecisely defined in altitude, it is not always identifiable as a manifestly marine feature of the landscape but may incorporate valley remnants of fluvial or subaerial origin especially where it fingers into the mountains of the Arenig, Hiraethog, and Snowdonian masses, and into the Llyn Tegid embayment. In terrain of graded fall from greater heights, as on the interfluves of the Monnow system or of the coalfield rivers in South Wales, geological structure is a factor in controlling landform, partly offset by discordant minor steps that strongly suggest local cliffing and that support a marine origin (Rice 1957, pp. 359ff.; George 1961, p. 256).

The 'Middle Peneplain' is composed of too great a variety of geomorphic elements, and is too completely broken up, for its original unity to be restored with confidence. In southernmost development it is dominantly a scarp and back-slope feature in Mynydd Epynt and the hills to the south-west, and in the interfluve remnants of much of the South Wales coalfield; but elsewhere—in the hills between Radnor Forest and Clun Forest, widely on the flanks of Drygarn, in the complex of Mynydd Hiraethog—the geological foundation is structurally complicated and is to a far less degree reflected in the geomorphy. Correlation based on an altimetric interval and not subject to other controls may then be suspect and be in part coincidental. It is therefore reassuring that other signs of benching are not lacking. These include bevelled escarpment crests, notably of Mynydd Epynt and Mynydd Bwlch-y-groes in Old Red Sandstone, and of the Rhondda country in Pennant Sandstone; notched dip-slopes in the Brecon Beacons and the Black Mountains; and minor monadnocks standing above regional levels within the altimetric range of the 'peneplain' (Brown 1960, pl. 4; Thomas 1959, pp. 74ff.). These hints of marine planing do not conform closely in height, and point to staggered uplift only generalised in 'Middle Peneplain'. At the lower limit of the 'peneplain' Mynydd Preseli and the Mynydd Llanybyther range (Brown 1960, pl. 2 and fig. 7) stand as impressive monadnocks above the 'Low Peneplain' and strengthen a local inference of sharp fall in sea level rather than in subaerial base-level.

The 'High Plateau' is distinguished from the 'Middle Peneplain' less on geomorphic than on altimetric criteria. As such, it is greatly fragmented, its residuals being very widely scattered and very small in area, again of a variety of geomorphic kinds—escarpments in the Fanau, the Brecon Beacons, and the Black Mountains, in Mynydd Llangynidr and Mynydd Llangattwg, and in Craig-y-Llyn, moel summits in structurally complex terrain in the Drygarn country and south of Plynlimon, in Rhyd Hywel, around Lake Vyrnwy, and in the Longmynd. The dip-slopes of both the Beacons and the Black Mountains show repeated signs of notching (compare Brown 1960, pl. 22) and occasional summits, like Plynlimon and Drygarn, stand out as cliffed monadnocks over the peneplain (see Brown 1960, pl. 24, and figs. 12, 20), the details, not altogether degraded, repeatedly suggesting the work of the sea.

The only unity possessed by Brown's 'Monadnock Group' is the height of the summits above about 600 m, which range up to 1076 m in Yr Wyddfa. The crests of the monadnocks are so isolated that they can be integrated in no compelling ways, and to link the highest summits as a 'Summit Peneplain' is possible only on gratuitous assumptions of warping in domed structures having an amplitude of 300 m in 8 km in Snowdonia. On a different but little less conjectural basis of integration (Fig. 98) it is possible to regard the 'Summit Peneplain' as composite, with a lower member (itself multiple) forming a shelf below the highest Snowdonian peaks and then taking its place in the echelon of platforms built upwards from the 'coastal plateau'. The major problem the high summits present lies in attempts to reconstruct a surface that incorporates such widely removed (and presumably little degraded) remnants as the Snowdonian cluster (Yr Wyddfa at 1076 m, a dozen others exceeding 915 m) in the north, the crests of the Old Red escarpment (Pen-y-fan at 886 m, Waun Fach at 811 m, Banau Sir Gaer at 792 m) 130 km farther south,

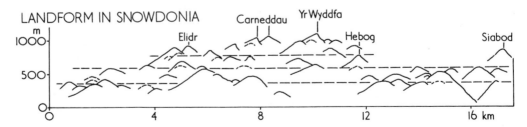

Fig. 98. Interpretation of the 'Summit Peneplain' in North Wales. The distribution of Snowdonian peaks, illustrated in projected profiles along lines running north-west to south-east offers suggestions of the dissection of stepped platforms, not much tilted and therefore Neogene in age, comparable with those seen at lower levels and implying wave-bench erosion.

with heights in intermediate ground of 855 m in Arenig Fawr, 890 m in Cader Idris, 740 m in Plynlimon, and 660 m in Radnor Forest (see p. 366).

If a restoration of a landscape of little-deformed stepped platforms is valid, and if the marine configuration of the platforms of the 'coastal plateau' may be extrapolated upwards, the conclusion is reinforced that there is no discernible place left for the recognition of an ancient Triassic 'peneplain' in any of the present major landforms. Nor, except in the crudest generality, can even the Vale of Glamorgan be looked at as in part an exhumed early Mesozoic oldland: the elements of the Glamorgan coast newly appearing by the erosion of the Trias or Lias cover are seen to be highly local and incidental; none of the residual Mesozoic rocks (through which the Palaeozoic cores emerge) are to be found at heights above the Neogene '600-foot' platform; and the restored profile of the sub-Mesozoic floor (see for instance Trueman 1922, fig. 68; George 1970, fig. 35) is irregularly undulating, archipelagic in the Mesozoic seas, having no semblance to a Triassic peneplain that can with any justification be extended northwards. In vaguer terms, if the Triassic peneplain is supposed to be identifiable with any part of the 'High Plateau' in North Wales, it cannot also be represented in the 'Middle Peneplain' or in the 'Low Peneplain', to which a different and more recent origin must be ascribed. Neogene landforms thus confirm Palaeogene rock-suites in relegating Triassic and Cretaceous events to insignificance in the Cenozoic evolution of wales.

VII. NEOGENE DRAINAGE

(a) Superimposition

The discordant relations of the major primary rivers to the geological structure have been central to most syntheses of Welsh drainage evolution from Lake's day onwards, superimposition from a Chalk cover being a guiding theme. With the Chalk now removed as a controlling agent, another epigenetic surface, post-'Alpine', must be looked for; but the evidence of discordance is no less strong for being Neogene in origin.

The geomorphic coherence of selected self-contained regions leaves little doubt of the incongruence of river courses and structure, as Strahan amply demonstrated in the rivers of the South Wales coalfield, which cross indifferently the major Hercynian folds and some of the faults (sometimes at an acute angle). The Monnow system, draining the Black Mountains and as the integrated Monnow itself continuing into the lowest reach of the Wye, similarly transects the Vale-of-Neath disturbance, the plunging Usk anticline, and the Dean Forest syncline, before entering the Severn estuary at Chepstow. The greater part of the drainage of South Wales, from the crest of the Old Red escarpment to the sea, has a convincing unity in its superimposed relationships with the Hercynian structures. (See A. A. Miller 1935; Clarke 1936; R. O. Jones 1939.)

In Caledonian terrain the Towy, the Wye, and the upper Severn, their sources in Drygarn and Plynlimon, provide equally spectacular evidence (George 1942; Rice 1957a, 1957b); and in North Wales the monadnock masses are foci of stream deployment south-eastwards and southwards across the major structures of the

Harlech–Derwen anticline, the Bala fault, and the Central Wales synclinorium, in a reconstruction that implies a drainage system more or less radial (O. T. Jones 1952, pp. 212ff.; Brown 1957, fig. 7; George 1961, fig. 33).

Wherever the signs have not been mutilated or obliterated by glacial action, all the primary rivers provide clear evidence of rejuvenation, in a variety of geomorphic features especially knick points, valley-wall benches, and abandoned cols. Graded profiles, of which those illustrated by O. T. Jones (1924, pls. xliv, xlv) in the Towy are the prototypes, may in many instances be extrapolated more or less convincingly to suggest a pulsed uplift accordant with the pulsed uplift of the coastal platforms; and they strengthen the evidence of a general uniformity of 'eustatic' uplift from South to North Wales (A. A. Miller 1935, fig. 3; George 1942, pl. ii; Embleton 1960, fig. 6). Usually the incised valleys and gorges resulting from rejuvenation eat back far into the hinterland, and signs of base-levels at less than 300 m are to be seen in long profiles that commence within a few kilometres of the source, as in the uppermost Severn, the Wye, the Towy, the Twrch, and the Taf-fechan. Valley forms as they now appear are thus both relatively mature in the headward encroachment on the interfluves, and geologically very young: in their lower reaches, they imply grading to coastal platforms perhaps as young as Pleistocene.

There are, however, abandoned cols in the heart of some of the highest mountain masses that appear to hint at a much more ancient drainage not readily reconstructed in altimetric range, notably in Snowdonia—Rhydgoch at the source of the Ffrancon, Pen-y-pass at the source of the Peris, Pitts Head at the source of the Gwyrfai. The rivers that formerly broke through the Old Red escarpment of the Fanau, the Brecon Beacons, and the Black Mountains to incise similar cols prompted O. T. Jones (1952, p. 211) to project their profiles to sources in mid-Wales that were perhaps 500 m to 600 m above the present summits of Plynlimon and Drygarn in cover rocks (Triassic and Jurassic) capped by Chalk. He would, no doubt, have applied a similar hypothesis to Snowdonia.

The hypothesis has its difficulties: particularly, it is at variance with a regional inference of Neogene stepped platforms, and with Neogene stratigraphical concepts. The present major streams rising in mid-Wales provide at their sources no evidence whatever of having cut down from higher levels, being scarcely incised to even a few metres below the surrounding ground. Thus Bwlch Bryn-rhudd above Cwn Crai, whose floor at 370 m is shadowed by Fan Gihirych at 725 m, may well have been the course of a combined Towy and Irfon-Cilieni flowing into the Tawe; and an early Wye, before capture along its present route between Glasbury and Whitney, may have flowed through the Pen-y-genffordd col at 323 m, between summits of 608 m and 645 m in the Black Mountains, to follow the water-gap by Crickhowell and Abergavenny to the plain of Gwent. But to make these assumptions is to imply that the Usk between Sennibridge and Crickhowell was not an integrated stream until the Ffos Tarw stage was initiated with base-level at only about 270 m above the present (George 1942, pp. 105 ff.).

Most of the cols of the Old Red escarpment are more difficult to explain, both because there are no obvious origins for them to the north, and because the Epynt scarp, bevelled at about 470 m, and the hill mass running south from Radnor Forest, are unbroken by streams flowing from the Drygarn-Plynlimon ground except by the Wye, the Irfon-Cilieni, and the Towy-Usk. Thus there are no ancestral sources in mid-Wales of the Brân and the Yscir on' Epynt, the Twrch and the Giedd and the Taf-fechan of the Fans–Beacons, the Grwyne and the Honddu and the Olchon of the Black Mountains. Still less are there sources of the Snowdonian cols outside Snowdonia, where (in the absence of more plausible syntheses) the cols are perhaps the accidental products of headward erosion by opposed streams of steep gradient, some draining north-westwards to the Menai Strait, some south-eastwards to the Glaslyn and Gwrhyd valleys (which demand their own explanation)—unless a wholly different explanation is to be sought in the effects of transfluent ice.

It may then be that, on Neogene emergence of the Welsh mountains, the high summits—Snowdonia, the Merioneth mass, the Cader Idris range, Plynlimon-Drygarn, the Old Red escarpment—were mainly independent centres of stream deployment, the asymmetry in the primary drainage pattern, which consists of long gentle rivers to the east and the south, of short steeply falling rivers to Cardigan Bay and Arfon, being ascribed to an asymmetry of geological structure, fault-controlled along the Llanbedr line and possibly farther south, and along the north-west front of Snowdonia.

(b) *Secondary adjustment*

The evolution of superimposed drainage systems is characterised by differential erosion, structurally controlled, seen in river diversion and piracy of a variety of kinds, of which abandoned cols are the typical signs. An ensuing reticulate pattern is widespread in Wales. It is very well illustrated in the growth of the Tawe and Nedd rivers (R. O. Jones 1931, 1939), integrated by the extension of strike streams some on the shales of the Lower and Middle Coal Measures some along west-north-westerly faults (D. G. Jones & Owen 1956, p. 249), the two main rivers along belts of strong disturbance. Similarly, the Towy as a unified river, and the neighbouring Cothi, were the product of the beheading of streams flowing south from mid-Wales, in 'subsequent' adjustment to two major anticlinal belts (George 1942, p. 121). An early Teifi probably had its source on the south flank of Cader Idris, its fragmentation being assisted by the erosive ease with which piratical streams—the lower Dyfi, the Rheidol, the Ystwyth—could cut down their valleys along faults (O. T. Jones & Pugh 1935, p. 296). For 50 km from Llyn Tegid to the sea at Towyn the Bala fault determines the drainage in contributing to the diversion of the headwaters of the Dee, the Tanat, the Vyrnwy, and perhaps the Dyfi.

The superimposed primary river pattern was Neogene in origin. Its modification by piratical 'subsequents' was to be expected. What is surprising is the very late occurrence of piracy. The evidence of long profiles, stepped in rejuvenated segments, demonstrates for instance that the upper Rheidol continued southwards across the present Ystwyth valley at Yspytty Ystwyth to form a headwater of the Teifi approximately when the '600-foot' platform was being planed, and that the Clywedog-Severn was beheaded by the steeply graded Twymyn-Dyfi, and the upper Towy diverted from the Usk by the lower Towy, at about the same time; that the lowest Nedd linked with the reach upstream only with the emergence of the '400-foot' platform; and that the Towy and the Tawe became integrated in their lowest reaches only with uplift of the '200-foot' platform.

The strong implication is that the Welsh massif, and its erosional dissection, are deceptively old. They are in fact not merely post-'Alpine' but late Neogene, and in the 'coastal plateau' perhaps as young as Pleistocene. The 'maturity' of much of the landscape is also deceptive, being a combination of rapid marine planing and rapid subaerial and fluviatile degradation; and there is a perversity in recognising that the oldest geomorphic elements—the Snowdonian mountains, the steep descents to the sea in Merioneth, the Wye in its reach above Glasbury—have a more rugged and unfinished look than the much younger Anglesey and south-Pembrokeshire platforms, the Towy valley below Llandovery, the Vale of Glamorgan. The speed with which the Neogene landscape was sculptured is a measure both of the efficacy of erosional processes, and of the transformations that may take place in very short geological intervals; and a recognition of the speed disposes of conservative prejudice in the interpretation of landscape evolution.

VIII.   QUATERNARY SEA-LEVELS

The glacial events of the Pleistocene being outside the present conspectus, reference need be made only to the continuity from Neogene times that is seen in Quaternary fluctuations in sea level.

The precise dating of the 'coastal plateau' is uncertain, its elements, or at least the lowest of them, perhaps continuing into the Pleistocene if correlation with the East-Anglian post-Gedgravian crags is allowed. Later signs of sea-level changes are two-fold: the 'raised beaches', and the drowned valleys and peats. The beaches, and the platforms on which they rest, appear all to be Pleistocene; but there is disagreement on exact age, and disagreement on correlation of platform and associated beach deposits. It is becoming conventional, although the convention remains unproved, to regard the platforms as ancient 'Middle and Early Pleistocene' (West 1969, p. 258), and the deposits resting on them as much more recent. Thus everywhere—in Gower, south Pembrokeshire, Lleyn, Anglesey—a beach platform, erosionally very well defined as a notch in many kinds of Palaeozoic rock, is to be seen underlying all the local drifts, and in that stratigraphical relationship is certainly 'old' in Pleistocene terms. It lies at a variable level, according to the accident of exposure, up to about 10 m above high-water mark, and despite later oscillations it may be regarded as indicating a sea-level, 'pre-Glacial' in a local sense, not greatly differing from the present sea-level.

The fauna, including the soft-nosed rhino, found in the beach deposits of Gower helps to give a relative age apparently much younger, the several members (an earlier *Patella* Beach, a later *Neritoides* Beach) being referred variously to various of the later Pleistocene stages, sometimes both to the Hoxnian, sometimes both to the Eemian, sometimes one to the Hoxnian one to the Eemian (see West 1969, p. 258; and in Lewis 1970: Bowen, p. 214; John, p. 260; Stephens, Tab. 11.1).

The lowering of sea-level by 'eustatic' relative movement as land-ice accumulated during periods of intense glaciation is shown by the repeated evidence of drowned valley-mouths and their deposits. Off the Vale of Glamorgan a former Severn and its Welsh tributaries descended to levels of the order of 35 m below present sea-level (Anderson and Blundell 1965); and along the coast of Cardigan Bay indirect geophysical evidence suggests a similar drowning at the mouths of the Dyfi, the Dysynni, the Mawddach, and the Glaslyn, with perhaps a descent of the floor (not certainly under a cover of sediments all post-Pliocene) to 90 m offshore from Traeth Mawr and the Dysynni estuary (Blundell, Griffiths and King 1969).

The precise age of the down-cutting (which in some of the valleys was intensified by glacial gouging) is unknown; but post-Glacial recovery, and drowning by the Flandrian sea, are marked by the submerged peats and associated sediments of the coastal flats, to be seen in most sheltered bays from the Severn estuary to Liverpool Bay. Locally the sediments exceed 24 m in thickness and indicate a corresponding rise in sea level of not less magnitude; and they span some 9,000 years of Recent times (see Godwin 1940; George 1970, p. 133). They are complemented by the alluvium that chokes most of the Welsh valleys; but otherwise the effect of the Glacial drowning on the landscape was negligible, except in the formation of rias. The continuing Flandrian subsidence, perhaps still not ended, implies that the present-day shore platform, over a kilometre wide in places, and very perfectly planed in hard rock, is much older than it appears: patches of drift resting on it suggest that it is (in local terms) pre-Glacial, and that in immediately post-Glacial times it was some tens of metres in height above contemporary sea level, a prominent member of the 'coastal plateau'.

IX. REFERENCES

AL-SHAIKH, Z. D. 1969. Geophysical results from Caernarvon and Tremadoc bays. *Nature*, **224**, 897–99.

ANDERSON, J. G. C., and BLUNDELL, C. R. K. 1965. The sub-drift rock surface and buried channels of the Cardiff district. *Proc. Geol. Ass.*, **76**, 367–78.

ARCHER, A. A. and ELLIOTT, R. W. 1965. The occurrence of olivine-dolerite dykes near Llanrwst, North Wales. *Bull. geol. Surv. G.B.*, **23**, 145–52.

AUDLEY-CHARLES, M. G. 1970. Stratigraphical correlation of the Triassic rocks of the British Isles. *Quart. J. geol. Soc. Lond.*, **126**, 19–48.

BACON, M., and McQUILLIN, R. 1972. Refraction seismic surveys in the north Irish Sea. *J. geol. Soc.*, **128**, 613–22.

BLUNDELL, D. J., DAVEY, F. J., and GRAVES, L. J. 1971. Geophysical surveys over the south Irish Sea and Nymphe Bank. *J. geol. Soc.*, 127, 339–75.

——, GRIFFITHS, D. H., and KING, R. F. 1969. Geophysical investigations of buried river valleys around Cardigan Bay. *Geol. Jl*, **6**, 161–80.

——, KING, R. F., and WILSON, C. D. V. 1964. Seismic investigations of the rocks beneath the northern part of Cardigan Bay, Wales. *Quart. J. geol. Soc. Lond.*, **120**, 35–50.

BOTT, M. H. P., and WATTS, A. B. 1971. Deep structure of the continental margin adjacent to the British Isles. *Rep. Inst. geol. Sci.*, **70/14**, 89–110.

BROWN, E. H. 1952. The River Ystwyth, Cardiganshire: a geomorphological study. *Proc. Geol. Ass.*, **63**, 244–69.

——. 1956. The 600-foot platform in Wales. *Proc. 8th gen. Ass. 17th intern. geogr. Union.*, 304–12.

——. 1957. The physique of Wales. *Geogr. Jl*, **126**, 318–34.

——. 1960. *The relief and drainage of Wales.* Cardiff.

BULLERWELL, W. (editor). 1965. Aeromagnetic map of Great Britain: Sheet 2 (England and Wales). *Inst. geol. Sci.*

——. and McQUILLIN, R. 1969. Preliminary report on a seismic reflection survey in the southern Irish Sea. *Rep. Inst. geol. Sci.*, **69/2.**

CLARKE, B. B. 1936. The post-Cretaceous geomorphology of the Black Mountains. *Proc. Birmingham nat. Hist. Phil. Soc.*, **16,** 155–73.

DEWEY, H. 1918. The origin of some land-forms in Caernarvonshire. *Geol. Mag.*, (6)**5,** 145–57.

DIXON, E. E. L. 1921. The country around Pembroke and Tenby. *Mem. geol. Surv. Gt. Britain.*

DODSON, M. H, and LONG, L. E. 1962. Age of Lundy granite, Bristol Channel. *Nature*, **195,** 975–6.

DONOVAN, D. T. (editor). 1968. *The geology of shelf seas.* Edinburgh and London.

——, LLOYD, A. J., and STRIDE, A. H. 1971. Geology of the British Channel. *Proc. geol. Soc.*, **1664,** 294–5.

—— and Others. 1961. Geology of the floor of the Bristol Channel. *Nature*, **189,** 51–2.

DRISCOLL, E. M. 1958. The denudation chronology of the Vale of Glamorgan. *Trans. Inst. Brit. Geogr.*, **25,** 45–57.

EDEN, R. A., WRIGHT, J. E., and BULLERWELL, W. 1971. The solid geology of the east Atlantic continental margin adjacent to the British Isles. *Rep. Inst. geol. Sci.*, **70/14,** 111–28.

EMBLETON, C. 1957. Some stages in the drainage evolution of part of north-east Wales. *Trans. Inst. Brit. Geogr.*, **23,** 19–35.

——. 1964. The planation surfaces of Arfon and adjacent parts of Anglesey. *Trans. Inst. Brit. Geogr.*, **35.**

FEARNSIDES, W. G. 1910. North and central Wales. In Monckton, H. W., and Herries, R. S., (editors): *Geology in the field*, 786–825. London.

FITCH, F. J., and others. Isotopic age determinations on rocks from Wales and the Welsh borders. In Wood, A. (editor): *The Precambrian and Lower Palaeozoic rocks of Wales*, Cardiff. 23–45.

GEORGE, T. N. 1938. Shoreline evolution in the Swansea district. *Proc. Swansea sci. and Field Nat. Soc.*, **2,** 23–48.

——. 1942. The development of the Towy and upper Usk drainage pattern. *Quart. J. geol. Soc. Lond.*, **98,** 89–137.

——. 1955. British Tertiary landscape evolution. *Sci. Progr.*, **43,** 291–307.

——. 1961. North Wales. *Geol. Surv. Mus.*

——. 1961. The Welsh landscape. *Sci. Progr.*, **49,** 242–64.

——. 1962. Tectonics and palaeogeography in southern England. *Sci. Progr.*, **50,** 192–217.

——. 1966. Geomorphic evolution in Hebridean Scotland. *Scott. J. Geol.*, **2,** 1–34.

——. 1967. Landform and structure in Ulster. *Scott. J. Geol.*, 3, 413–48.

——. 1970. South Wales. *Inst. geol Sci.*

GODWIN, H. 1940. A Boreal transgression of the sea in Swansea Bay. *New Phytol.*, **39,** 308–21.

GREENLY, E. 1919. The geology of Anglesey. *Mem. geol. Surv. Gt. Brit.*, 2 vols.

——. 1938. The age of the mountains of Snowdonia. *Quart. J. geol. Soc. Lond.*, **94,** 117–24.

——. 1942. The later igneous rocks of Arfon. *Geol. Mag.*, **79,** 328–31.

GRIFFITHS, D. H., KING, R. F., and WILSON, C. D. V. 1961. Geophysical investigations in Tremadoc Bay, North Wales. *Quart. J. geol. Soc. Lond.*, **117,** 171–91.

Inst. geol. Sci. 1972. 1:2,500,000 map of the sub-Pleistocene geology of the British Isles and the adjacent continental shelf.

JONES, D. G. and OWEN, T. R. 1957. The rock succession and geological structure of the Pyrddin, Sychryd, and Upper Cynon valleys, South Wales. *Proc. Geol. Ass.*, **67,** 232–50.

JONES, O. T. 1924. The Upper Towy drainage system. *Quart. J. geol. Soc. Lond.*, **80,** 568–609.

——. 1930. Some episodes in the geological history of the Bristol Channel region of Glamorgan. *Rep. Brit. Assoc.*, 57–82.

——. 1952. The drainage system of Wales and the adjacent regions. *Quart. J. geol. Soc. Lond.*, **107,** 201–25.

——. 1956. The geological evolution of Wales and the adjacent regions. *Quart. J. geol. Soc. Lond.*, **111,** 323–51.

—— and PUGH, W. J. 1935. The geology of the district around Machynlleth and Aberystwyth. *Proc. Geol. Ass.*, **66,** 247–300.

JONES, R. O. 1931. The development of the Tawe drainage. *Proc. Geol. Ass.*, **44**, 305–21.
——. 1939. The evolution of the Neath-Tawe drainage. *Proc. Geol. Ass.*, **50**, 530–66.
LAKE, P. 1900. Bala Lake and the river system of North Wales. *Geol. Mag.*, (4)7, 204–15, 241–5.
LEWIS, C. A. (editor). 1970. *The glaciation of Wales and adjoining regions*. London.
LINTON, D. L. 1951. Midland drainage. *Advanc. Sci.*, **7**, 449–56.
LLOYD, A. J. 1963. Upper Jurassic rocks beneath the Bristol Channel. *Nature*, **198**, 375–6.
MATLEY, C. A. 1913. The geology of Bardsey Island. *Quart. J. geol. Soc. Lond.*, **69**, 514–33.
——. 1928. The Pre-Cambrian complex and associated rocks of south-western Lleyn. *Quart. J. geol. Soc. Lond.*, 84, 440–504.
MILLER, A. A. 1935. The entrenched meanders of the Herefordshire Wye. *Geogr. Jl.*, **85**, 160–78.
——. 1937. The 600-foot platform in Carmarthenshire and Pembrokeshire. *Geogr. Jl*, **90**, 148–59.
——. 1939. Preglacial erosion surfaces around the Irish Sea basin. *Proc. Yorks. geol. Soc.*, **24**, 31–59.
MILLER, J. A., and FITCH, F. J. 1962. Age of the Lundy granites. *Nature*, **195**, 553–5.
NORTH, F. J. 1929. The evolution of the Bristol Channel. *Nat. Mus. Wales*.
OWEN, T. R. 1967. 'From the south'. *Proc. Geol. Ass.*, **78**, 595–99.
——. 1971. The structural evolution of the British Channel. *Proc. geol. Soc.*, **1664**, 289–94.
POWELL, D. W. 1956. Gravity and magnetic anomalies in North Wales. *Quart. J. geol. Soc. Lond.*, **111**, 375–97.
RICE, B. J. 1957a. The erosional history of the Upper Wye basin. *Geogr. Jl*, **123**, 357–70.
——. 1957b. The drainage pattern and upland surfaces of south-central Wales. *Scott. geogr. Mag.*, 73, 111–22.
RODDA, J. C. 1970. Trend-surface analysis trial for the planation surfaces of north Cardiganshire. *Trans. Inst. Brit. Geog.*, **50**, 107–13.
STRAHAN, A. 1902. On the origin of the river system of South Wales and its connection with that of the Severn and the Thames. *Quart. J. geol. Soc. Lond.*, **58**, 207–25.
THOMAS, T. M. 1959. The geomorphology of Brecknock. *Brycheiniog*, **3**, 55–136.
TRUEMAN, A. E. 1922. The Liassic rocks of Glamorgan. *Proc. Geol. Ass.*, **33**, 245–84.
WALSH, P. T., BOULTER, M. C., IJTABA, M., and URBANI, D. M. 1972. The Preservation of the Neogene Barrington formation of the southern Pennines and its bearing on the evolution of upland Britain. *J. geol. Soc.*, **128**, 519–59.
——, and BROWN, E. H. 1971. Solution subsidence outliers containing probable Tertiary sediment in north-east Wales. *Geol. Jl*, **7**, 299–320.
WEST, R. D. 1969. *Pleistocene geology and biology*. London.
WILKINSON, H. R., and GREGORY, S. 1956. Aspects of the evolution of the drainage pattern of north-east Wales. *Lpool. and Manchr. Geol. J.*, **1**, 543–58.
WILLIAMS, D. 1924. On two olivine- dolerite dykes in Snowdonia. *Proc. Lpool geol. Soc.*, **14**, 38–47.
WILSON, C. D. V. 1959. Geophysical investigations in the Vale of Clwyd. *Lpool and Manchr geol. Jl*, **2**, 253–70.
WOBBER, F. J. 1966. A study of the depositional area of the Glamorgan Lias. *Proc. Geol. Soc.*, **77**, 127–37.
WOOD, A. and WOODLAND, A. W. 1968. Borehole at Mochras, west of Llanbedr, Merioneth. *Nature*, **219**, 1352–5.
WOODLAND, A. W. (editor). 1971. The Llanbedr (Mochras Farm) borehole. *Rep. Inst. geol-Sci.*, **71/18**.
WOOLDRIDGE, S. W. and LINTON, D. L. 1955. *Structure, surfaces and drainage in south-east England*. London.
WRIGHT, J. E., HULL, J. H., McQUILLIN, R., and ARNOLD, S. E. 1971. Irish Sea investigations 1969–70. *Rep. Inst. geol. Sci.*, **71/19**.

THE QUATERNARY OF WALES

D. Q. Bowen

I. INTRODUCTION

'At Pont-aber-glass-lyn, 100 yards below the bridge, on the right bank of the river, and 20 feet above the road, see a good example of the furrows, fluting and striae on rounded and polished surfaces of the rock, which Agassiz refers to the action of glaciers. See many similar effects on the left, or south-west side of the pass of Llanberis'.

William Buckland.

(Visitors' book, Goat Hotel, Beddgelert, North Wales, 16th October 1841).

SHORTLY after the above was written, and before the year was out, Buckland (1842), newly influenced by Agassiz, had communicated evidence for former glaciation in Snowdonia to the Geological Society of London. In this, he provided the answer for the erratic boulders of Llanberis Pass which had been discussed by Edward Llwyd a century and a half earlier (Davies 1969). Almost simultaneously Darwin (1842) came to similar conclusions when he discussed the former activity of glaciers in Caernarfonshire, but Murchison (1843), in alternative view, failed to find evidence for glaciation in Snowdonia. Indeed, he was, in 1862 (Agassiz 1885), among the last to embrace the 'glacial theory'.

Buckland's conversion from his belief in an universal deluge as expounded in his *Requiquae Diluvianae* (1823), was, however, incomplete, for he continued to regard the shelly drifts of the Vale of Clwyd and Moel Tryfan as marine. The Moel Tryfan drift, discovered by Trimmer (1831) and at the time regarded as valuable support for the thesis that such material was indeed diluvium, was to figure prominently in the delayed complete acceptance of the 'glacial theory'. Lyell (1863) after inspection of the exposure was fully convinced of its marine origin, while in the second edition of Ramsay's (1881) memoir on north Wales marine submergence to 610 m above sea level was postulated. Even after general opinion had moved away from submergence to full acceptance of the glacial origin of such shelly drift, especially after Croll's (1870) paper on the shelly deposits of Caithness, lingering doubts were expressed. Hull (1910) still believed in the marine origin of the Tryfan sands, while T. I. Pocock as recently as 1938 suggested that some shelly drifts in the Cheshire-Shropshire lowland were marine.

In distribution the Pleistocene deposits of Wales are best known, and most extensively exposed, along its three coastlines where together with important exposures in the northern borderlands, evidence for subdividing the succession lies. Inland, with the notable exception of much of south Wales mapped by Strahan and his colleagues, parts of north Wales, and the borderlands, the details of distribution are still not fully known. It is only slowly that gaps are being filled, recent important additions including the Harlech Dome (Foster 1968), Arenig (B. Rowlands 1971), Montgomeryshire (M. Brown 1971) and mid-Wales (Potts 1968).

The unconformity at the base of the Quaternary succession, or sub-drift surface is known with varying degrees of accuracy in different parts. Not surprisingly it is well known in the coastal lowlands of denser population where dock and harbour installation, and in south Wales coal mining, enabled its configuration to be determined. The course and depth of buried channels in north Wales (Whittow 1965), Cardigan Bay (Blundell *et al.*, 1969. Allen 1960) South Wales (Anderson 1968), are reasonably well known, most river channels being traced to depths of about 30 m or so, with exceptions in the case, for example of the Dysynni and Dyfi, where excavation to –90 m or so, probably includes overdeepening by ice-action.

Inland knowledge of the sub-drift surface is patchy. It is best known in areas of coal mining and in the Cheshire and northern part of the Shropshire lowland where Wills (1912) and Howells (1956) showed that buried channels bear little relationship to present day streams. Elsewhere, apart from isolated work, chiefly of a geophysical nature, for example on the Rheidol-Teifi buried valley (Coster & Gerrard 1947), and on the lower Teifi (Francis 1964), it is inadequately known.

One of the earliest subdivisions of the Welsh Pleistocene was by Ramsay (1852) who recognised three stages in north Wales: (1) an episode of glaciation when striations and roches moutonneés were formed, followed by (2) marine submergence and deposition of the shelly drifts, and finally (3) an episode of valley glaciation when moraines were formed. This was fore-runner to the tripartite characterisation of the drifts of Lancashire and Cheshire proposed by Hull (1864), subsequently adopted by the Geological Survey and used for over 100 years as a system of time-stratigraphic division, despite objection and alternative mono-glacial view by Binney (1848) and Carvill Lewis (1894). Farther west Jehu (1904, 1909) erected similar tripartite schemes for north Pembrokeshire and Llyn, Williams (1927) for Cardiganshire, and Griffiths (1940) for parts of south and west Wales.

H. Carvill Lewis (1894) attempted to indicate the maximum extent of ice in Wales (not of 'newer drift' ice, as was subsequently supposed by some), but was handicapped at that time by dearth of information on southern districts. Differentiation of 'older' and 'newer drifts' by W. B. Wright (1914) suggested that Wales was ice-covered during the 'newer' glaciation, but he later accepted Charlesworth's (1929) version of the extent of the 'newer' ice sheets (Wright 1937). Charlesworth believed that the 'newer drift' glaciation was of Early Magdalenian age (Cresswellian) because its deposits sealed Aurignacian implements in the Cae Gwyn caves of the Vale of Clwyd. In south Wales T. N. George (1932, 1933) amplified the stratigraphic relations between 'older' and 'newer drifts', and suggested that the *Patella* raised beach antedated both glaciations.

Unlike previous authors, Mitchell (1960) proposed time-stratigraphic division of the Welsh Pleistocene starting with the Cromerian Interglacial. In alternative view (e.g. Bowen 1970), the Welsh Pleistocene succession has been incorporated into time-stratigraphic division of Upper Pleistocene age. This reflects contemporary controversy between the 'Irish School', and more traditional and orthodox view (e.g. Shotton 1962).

The difficulties of erecting time-stratigraphic divisions in Quaternary deposits is well known (Flint, 1971, Morrison, 1968) and in Britain a scheme of subdivision based on climatic change has been proposed (Shotton and West, 1969), the belief being that the alternating cold and temperate stages have time-stratigraphic validity. Such a belief is partly re-inforced in respect of the last, or Devensian, cold stage for it is subdivided in terms of radiocarbon years: the base of the Middle Devensian, Upper Devensian and Flandrian being drawn respectively at 50,000, 26,000 and 10,000 years B.P. These divisions are believed to correspond broadly with climatically significant events, a belief supported by the climatic fluctuations revealed through isotopic analysis of a core through the Greenland ice sheet at Camp Century (Dansgaard *et al.*, 1969). The nomenclature proposed by the Stratigraphy Committee of the Geological Society (Shotton & West, 1969) is adopted in the following account, although locally named lithostratigraphic units are specified in the correlation tables.

II. REGIONAL PLEISTOCENE STRATIGRAPHY

(a) The North-East

Drift antedating the Main Irish Sea Glaciation

Glacial deposits older than the Main Irish Sea drifts (Wills, 1937) are represented by the lower boulder clay of Macclesfield (Evans *et al.*, 1968), the basal clay of the Burland bore-hole (Poole & Whiteman, 1966), and erratics incorporated into younger sediments such as the gravels at Four Ashes (Shotton, 1967b).

Interglacial deposits are known from five localities. At Bontnewydd, in the Elwy Valley, Denbighshire, cave excavations over some years recovered what is regarded as a typical Ipswichian mammalian assemblage (Sutcliffe, 1960). It included *Hippopotamus amphibius* (Linné), *Elephas antiquus* (Falconer), *Dicerorhinus hemitoechus* (*leptorhinus*) (Falconer), and *Hyaena crocuta* (Goldfuss), (Neaverson, 1942).

Vegetational evidence for what is probably the same interglacial is afforded by organic deposits at Four Ashes, Tryssul, and Seisdon. At Four Ashes an organic silt resting on bedrock, overlain by fluviatile gravels of Devensian age, contained plant debris and pollen indicating contemporary temperate deciduous woodland conditions (A. V. Morgan 1970). A radiocarbon date on a wood sample gave > 45,000 years BP (Shotton *et al.*, 1971). At Tryssul, on the Shropshire-Staffordshire border, silts and clays lying in small basins on the Trysull sands and gravels ('older drift'), and overlain by till of the Main Irish Sea (Devensian) Glaciation, have yielded pollen which included *Pinus*, *Betula*, *Quercus*, *Alnus*, *Corylus*, *Salix*, and *Ulmus* (A. V. Morgan 1970). Several thousand opercula of the freshwater gastropod *Bithynia tentaculata* (Linné) were collected and subsequently gave radio-carbon dates of > 25,000 and > 34,000 years BP (Shotton *et al.*, 1970). At nearby Seisdon plant debris from grey silt gave a radiocarbon date of > 44,000 years BP (Shotton *et al.*, 1971). This agrees with the evidence of fauna and flora that the deposit is interglacial. The stratigraphical position of the interglacial deposits, their flora, gastropod, and insect fauna of temperate aspect, suggest that they represent the Ipswichian Interglacial.

The stratigraphic status of silts recovered from the Burland borehole (Poole and Whiteman 1966) is uncertain. They have yielded a temperate insect fauna and a radiocarbon date of $>$ 38,100 years BP (Shotton *et al.*, 1969), thus could be Ipswichian.

Underlying Main Irish Sea deposits in eastern Cheshire is the Chelford (Congleton) Sands Formation. It was formerly thought to be fluvioglacial and part of the Middle Sands Formation (Simpson & West 1958), but is now regarded as a low-angled alluvial fan complex derived from the east (Boulton & Worsley 1965) Evans *et al.*, 1968. Worsley 1970). Intraformational ice-wedge pseudomorphs indicate deposition over a permafrost table (Worsley 1966a), while, between periodic flooding, ventifacts were fashioned on the alluvial surfaces (Thompson & Worsley 1967). The petrography of the sands has been studied by Harrison (Evans *et al.*).

STAGE		THE NORTH EAST : Cheshire, Denbigh, Flint, Montg. Salop, Staffs.	THE NORTH WEST : Anglesey, Caerns, Merioneth.
FLANDRIAN	Zones IV-VIII	Alluvium. Colluvium. Marine alluvium. Bryn Carrog coastline of the Vale of Clwyd. Dune sand. Calcareous tufa at Caerwys and Ysceifiog. Peat at Bettisfield, Fenn's Moss C14 AD 1295, 1210. Whixall Moss C14 50 BC to 1275 BC.	Alluvium. Colluvium. Dune sand, Newborough Warren. Marine alluvium. Basin peat and lake deposits at Cwm Idwal, Llyn Dwythwch, Nant Ffrancon, Glanllynau and Cors Goch. Raised beach (Main Postglacial) of Anglesey, Llyn and Bardsey Island.
DEVENSIAN — Late Glacial Zones I - III		Upland solifluction deposits in Montgomeryshire. Church Stretton valley: head and fan deposits younger than 11,048, 11,000, 12,135, and 13,555 BP. Pingo ramparts near Crewe and Llangurig.	Cirque morainic & protalus drifts of Snowdon, Y Carneddau, Y Glynderau, Arenig Fach, Cader Idris. Head. Stratified screes. Upper lake clays, interstadial lake muds (C14 12,050 BP), lower lake clays at N. Ffrancon & L. Dwythwch. Glanllynau mosses C14 14,000 BP.
DEVENSIAN — Upper Devensian		Welsh Ice — Welsh Readvance Tills; ? Tremeirchion CF 18,00 BP; Tills of the Berwyn, Severn valley & south Salop. ? Tremeirchion CF 18,000 BP / Irish Sea Ice — Ellesmere morainic complex = Uffington (Worcester) Terrace; Main Irish Sea Glaciation (Stockport Formation Wolverhampton till)	Irish Sea Ice — Upper till at Porth Nefyn; Zone of weathering, permafrost and head; Trevor and Porth Neigwl Irish Sea tills. Moel Tryfan drift. / Welsh Ice — Llanystumdwy, Clynnog & Mochras tills. Sarn Badrig moraine; Criccieth and Llangelynin tills. Rhinog till, Arenig till, Drysgol till
DEVENSIAN — Lower & Middle Devensian		Ffynon Beuno and Cae Gwyn caves Aurignacian implements; Chelford Mud; Llandudno head. / Chelford Sands Form. / Four Ashes Gravel / MAIN TERRACE	Redeposited (solifluction) glacial drift at Llwyngwril & Porth Neigwl; Head deposits at Criccieth Castle & Red Wharf Bay
IPSWICHIAN		Bontnewydd Temperate Fauna. Tryssul silts. Four Ashes organic beds. ? Burland silts.	Porth Oer and Red Wharf Bay Raised Beaches.
PRE-IPSWICHIAN		? Macclesfield lower boulder clay; ? Burland clay	Raised beach & Lr. & Middle Devensian erratics Glaciation.

Table 10.   Correlation table of Quaternary units in North East Wales and the borderlands and North West Wales.

At Farm Wood Quarry, Chelford, pollen analytical investigations of an organic horizon within the Chelford Sands indicated forest conditions with *Betula sp.*, *Pinus sylvestris* and *Picea abies* (the last two with tree stumps in position of growth), and *Salix* (Simpson and West 1957). The appropriate climatic conditions were subsequently confirmed by the analysis of a prolific beatle fauna (Coope 1959), while a radiocarbon date of 60,000 years BP (57,000 years BP in Simpson & West) indicated correlation with the floristically similar Brörup interstadial of Denmark.

The Four Ashes gravels, which underlie the Wolverhampton till, consist mainly of local Bunter material with some erratics, and include several organic lenses. Radiocarbon dating of these gave the following dates: 30,655, 36,340, 42,530,

(Shotton 1967b), (Shotton *et al.*, 1968), 43,500 (Shotton *et al.*, 1970), 38,000, 30,500 and 40,000 years BP (Shotton *et al.*, 1971). The contemporary insect faunas indicate fluctuating temperature conditions (A. Morgan 1970), which as the dates show obtained during the 'Upton Warren interstadial complex' (Coope & Sands 1966). While the gravels beneath the Main Irish Sea till are thus Middle Devensian, those below, resting on the temperate organic silt, must be lower Devensian.

Other deposits, of possibly Lower and Middle Devensian in age, consist of head, found below glacial drifts of the Upper Devensian, e.g. on the Little Orme, Llandudno (Hall 1870) and near Shrewsbury (Whitehead 1961).

Devensian Glacial Deposits

For over 100 years officers of what is presently the Institute of Geological Sciences systematically mapped the glacial drifts, and delimited the respective provinces of the coeval Welsh and Irish Sea ice sheets. The former, characterised by Lower Palaeozoic erratics from as far afield as Arenig, are blue and yellow grey in colour, except in the Vale of Clwyd where the local Trias imparts a red hue. Irish Sea drifts consist predominantly of sand, with less frequent till, and are characteristically red in colour with erratics from the Lake District, Scotland and the Isle of Man. They also contain shells and shell fragments dredged up from the sea floor which include Boreal, Celtic and Lusitanian species (Jeffreys in Pocock 1906). As they antedate the glaciation, interglacial as well as interstadial species occur. Of some importance is the discovery of *Tellina tenuis* (da Costa), which points to an Ipswichian age for part of the fauna (Thompson and Worsley 1966).

Although there is evidence along the north Wales coast to show that Welsh ice was present prior to that from the Irish Sea, the latter was sufficiently powerful to penetrate the Vale of Clwyd, to deposit shelly drift up to almost 305 m on Halkyn Mountain (Strahan 1886 , up to 300 m near Wrexham, and over 305 m on Gloppa Hill near Oswestry (Wedd *et al.*, 1929). The coeval Welsh ice effected complete glaciation of the uplands as B. M. Rowlands (1970) has shown in the Arenig country, Travis in the Berwyns (1944), Brown (1971) in Montgomeryshire, and P. H. Rowlands (1966) and Brown (1971) in the south Shropshire hill country. P. H. Rowlands, however, argued that some higher summits formed nunataks, but Brown favoured a greater ice cover which deposited the tills around Marshbrook, deposits which Rowlands and Shotton (*in press*) regard as older.

While the distribution of glacial drift is well known in outline opinion on its subdivision has never been unanimous. Binney (1848) first outlined the monoglacial view by interpreting the multi-till and fluvioglacial sequences as the product of one advance and retreat. He was supported by Carvill Lewis (1894), but Hull (1864) in alternative explanation classified the drifts in tripartite division of Lower Boulder Clay, Middle Sands, and Upper Boulder Clay, a scheme to be adopted by the Geological Survey and used for almost a hundred years. Hull's scheme thus recognised two glacial advances, with the Middle Sands indicating a period of retreat. The use of these lithostratigraphic units as time-stratigraphic divisions is illustrated by Poole & Whiteman's (1966) correlation of the Lower Boulder Clay and Upper Boulder Clay with Würm I and Würm II of Europe respectively.

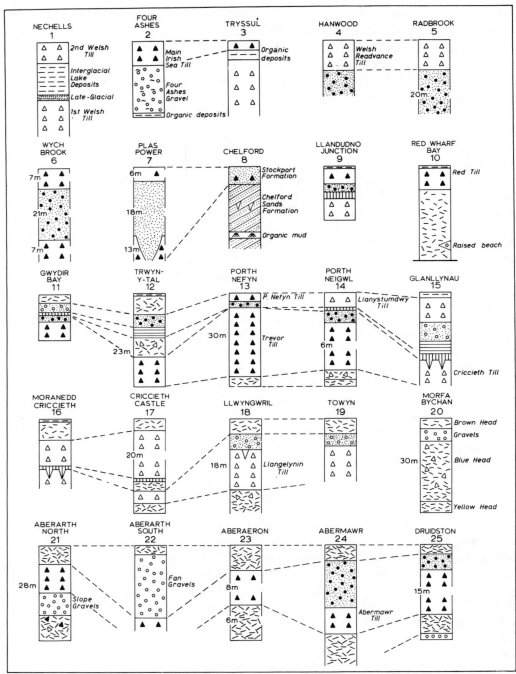

Fig. 99.   The Pleistocene succession at sites in Wales and the borderlands. Many of these are generalised and except where indicated most columns are of the order of 7 to 9 m; thicknesses of individual units, however, may vary considerably in lateral extent.

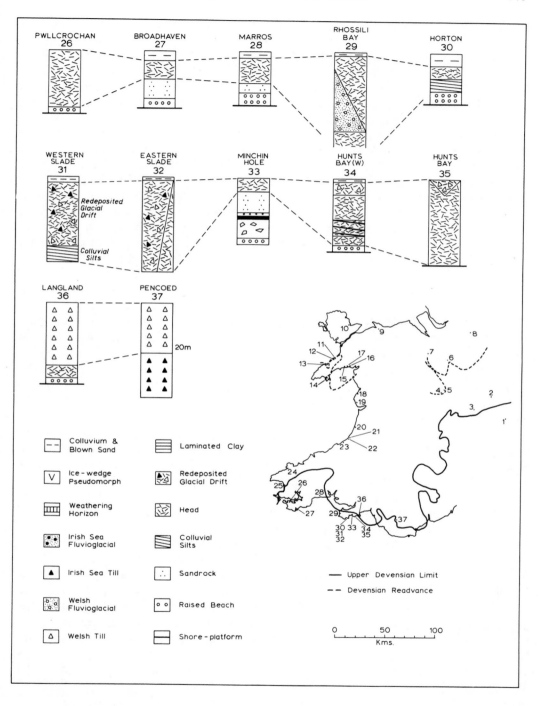

On the other hand the complexity of depositional sequences produced by contemporary glaciers have been used as analogues in interpreting the frequently chaotic arrangement of till, sand and gravel in the Cheshire-Shropshire lowland exposures. In the Macclesfield district the Gawsworth Sand (part of the Middle Sands of earlier authors) is thought to be a deposit partly of subglacial origin deposited by the glacier which was responsible for the Upper Boulder Clay during the last glaciation of the area (Evans *et al.*, 1968). This explanation is more acceptable than that of Poole & Whiteman (1961, 1966) who argued that the Middle Sands consisted of outwash formed during retreat of the Lower Boulder Clay ice sheet, and the considerable topography associated with it survived the Upper Boulder Clay Glaciation because of their frozen state (Taylor 1958). Moreover, the finely detailed morainic and fluvioglacial landforms of, for example, the Wrexham-Ellesmere-Whitchurch-Bar Hill morainic complec (Yates & Mosely 1967) are not consistent with the Poole & Whiteman hypothesis. Worsley (1970) has commented that multi-till exposures interbedded with sand units, as in the Wrexham delta terrace, show the difficulties of enforcing the tripartite scheme, while Boulton (*in press*) has suggested that sequences of till and sands forming at the margins of contemporary glaciers, notably in Vestspitsbergen, are appropriate analogues for those of the Cheshire-Shropshire lowland.

Wills (1950) placed his Main Irish Sea Glaciation, terminating at the celebrated Wolverhampton Line of himself and Whitehead, in the last Glaciation, and suggested that two subsequent stages occurred. These were the Little Welsh Glaciation (Welch Readvance of Whitehead, in Pocock *et al.*, 1938) coeval with Irish Sea advance to the Ellesmere morainic belt, and finally an advance of Scottish Readvance age penetrating the north Wales coastal belt and sealing the Aurignacian implements in the Vale of Clwyd caves.

In discussion with Simpson (1959) Poole suggested that the Upper Boulder Clay of the Cheshire-Shropshire lowland was Riss (Wolstonian) in age, although Simpson believed it post-dated the Chelford Sands, with their included radiocarbon date of 57,000 years BP, thought at that time to represent the Middle Sands of the tripartite sequence. Shotton & Strachan (1959) maintained that the maximum Irish Sea advance occurred shortly before 42,000 years BP, radiocarbon dates of 41,000 and 41,900 years BP having been obtained from a terrace of the River Salwarpe at Upton Warren which was correlated with the Main Terrace of the Severn, the outwash train of the Main Irish Sea Glaciation (Coope *et al.*, 1961). A similar date from the base of the Avon No. 2 Terrace at Fladbury, a terrace correlated with the Severn Main Terrace, was apparent confirmation of this view.

The tripartite scheme was upheld again by Poole & Whiteman (1961) who equated the Main Irish Sea advance of Wills (1937) with the Upper Boulder Clay Glaciation, but unlike Wills who had terminated this advance along the Wolverhampton line, Poole & Whiteman contended that the Upper Boulder Clay extended southwards to the Birmingham district. They argued that the large erratics in the Wolverhampton area did not indicate extent of glaciation but merely of northern erratics in the Irish Sea Ice Sheet, which was coeval with a Welsh ice sheet south of that line. In the Birmingham area Pickering (1957) suggested that the local upper boulder clay was Saalian (Wolstonian), a view presumably taken into account by Poole when he made a similar claim for the Upper Boulder Clay of Cheshire (Simpson 1959).

Further work by the Geological Survey in the Stockport and Knutsford district (Taylor, Price and Trotter 1963), again upholding the tripartite notion, held that the Lower Boulder Clay of that area was the Main Irish Sea Glaciation of Wills (1937) and the Upper Boulder Clay was a readvance of the same ice sheet.

Apparent confirmation that two separate boulder clay formations, of time-stratigraphic significance, did exist was provided by Boulton & Worsley (1965) who determined depths of leaching on the glacial drifts north and south of the Ellesmere morainic complex. Their results showed that tills south of that line had been leached to greater depths, which was taken to imply a time distinction. A radio-carbon date of 28,000 years BP from two specimens of *Nucella lapillus* (Linné) recovered from glacial drift at Sandiway (Boulton & Worsley 1965) suggested that drifts north of the Ellesmere moraine were late Weichselian (Devensian).

Poole & Whiteman (1966) subsequently added to their previous veiws by suggesting that the Lower Boulder Clay was probably Würm I in age, and that the Upper Boulder Clay of the Cheshire-Shropshire lowland as well as the Birmingham district is Würm II in age. Shotton (1967a), however, maintained that the Main Irish Sea Glaciation occurred during the Early Würm, terminating at the Wolver-hampton Line, and, accepted Boulton & Worsley's (1965) argument, that the Main Würm maximum limit was the Ellesmere moraine. At Chelford he suggested that the Early Würm horizon was locally lost in the unconformity between the Chelford Sands and the Gawsworth Sands (Stockport Formation of Worsley (1967).

Later that year Shotton (1967b) published radiocarbon dates obtained from organic horizons in the Four Ashes Gravels, a fluviatile deposit overlain by the Wolverhampton Till of the Main Irish Sea Glaciation. This showed clearly that the Main Irish Sea Glaciation was Upper Devensian in age and that the subsequent readvance to the Ellesmere moraine was therefore younger. The radiocarbon dates from Upton Warren and Fladbury, which previously figured so prominently, were thus seen to be maximum dates for terrace aggradation which spanned the Middle and Upper Devensian (see Lower Severn) (Table 11).

The authors of the Macclesfield memoir (Evans *et al.*, 1968) had already anticipated such a possibility. In it Evans suggested that the Main Irish Sea Glacia-tion represented the Main Weichselian (Upper Devensian) Glaciation, and that previous to that only residual patches of boulder clay, greatly dissected, and called Lower Boulder Clay, manifestly older than the Congleton (Chelford) Sands Formation, indicated earlier glaciation.

In summary, with the exception of residual patches of Lower Boulder Clay antedating the Chelford Sands Formation, it would appear that the glacial drifts of the Cheshire-Shropshire lowland are Upper Devensian in age. The outer limit of this glaciation, although disputed by Poole (1968), is the Wolverhampton Line, from which the Main Terrace of the Severn represents the contemporary outwash train. Subsequently a readvance occurred, the Welsh Readvance in the Shrewsbury

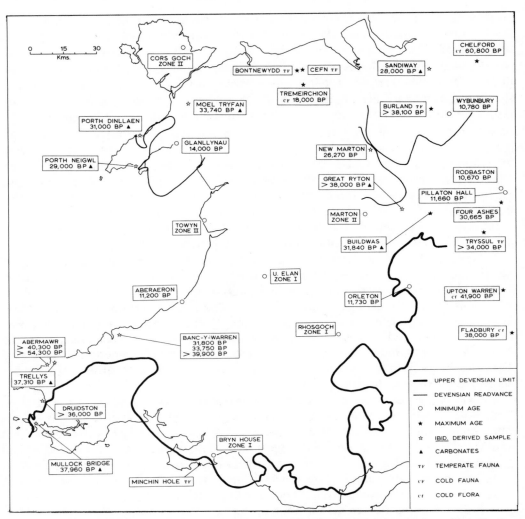

Fig. 100.   Upper Devensian glacial limits in Wales and the borderlands, together with indications of maximum and minimum time-stratigraphic age determinations. The proximity in time of these to the glacial event varies according to the local availability of organic materials. Some considerable doubt attaches to the validity of dates obtained from carbonates (text). The minimum age at Glanllynau should read 14,468 yrs BP; dating of shell fragments from Mullock Bridge has produced another date of 38,160 yrs BP (Shotton & Williams 1971). Derived samples were all recovered from glacial deposits.

district (Whitehead in Pocock 1938) where the Uffington (Worcester) Terrace of the Severn is deemed to be its contemporary outwash, and the Ellesmere Readvance, of Irish Sea Ice, farther north (Fig. 100). The youngest date from the Four Ashes gravels of 30,000 years BP shows that the Main Irish Sea Glaciation occurred sometime afterwards, but the exact significance of the radiocarbon date of 18,000 years BP from a carpal bone of *Mammuthus primigenius* (Blumenbach) recovered from the Tremeirchion caves of the Vale of Clwyd is uncertain (B. M. Rowlands 1971). The caves were sealed by Irish Sea till, but whether this was deposited by the Main Irish Sea advance, Ellesmere advance, or even by the Scottish Readvance as suggested by Wills (1950) is not clear. It is possible that the date provides a maximum age for the Main Irish Sea Glaciation for dates of 18,500 and 18,240 years BP from Dimlington in Yorkshire (Penny *et al.*, 1969) antedate the Last Glaciation of eastern England.

The lithostratigraphic units of Lower Boulder Clay, Middle Sands and Upper Boulder Clay, while useful locally, cannot be used in time-stratigraphic division; further subdivision of the glacial drifts must be based on other criteria, such as biostratigraphy.

Minimum age determinations have been obtained from radiocarbon dating of organic deposits in kettle holes (Shotton & Strachan 1959) or by pollen dating of the same materials (e.g. Birks 1965). These all indicate Devensian Late-glacial dates (Fig. 100).

During deglaciation it is clear from laminated clays and deltaic gravels that glacial lakes, probably highly ephemeral in nature and restricted in extent, occurred. Wills (1924) and Poole & Whiteman (1961, 1966), the latter despite opposition from some colleagues (Earp in Poole & Whiteman 1961) suggested that extensive lakes accumulated between the retreating ice front and the emerging topography to the south. Notable amongst these was Lake Buildwas, overspill from which was said to have initiated the Ironbridge gorge (Wills 1924), along which the gravels of the Main Terrace were transported. This and Lake Newport coalesced to form Lake Lapworth, but M. Brown (1971) doubts that such a water body existed in the Shrewsbury-Church Stretton-Buildwas area, and in an alternative explanation suggested that the ice decayed in place.

After deglaciation the glacial drifts were subject to cryoturbation in the form of large scale involutions and ice-wedge penetration. North of Wolverhampton considerable areas of ice-wedge polygonal patterned ground occur (A. V. Morgan 1971), while in Cheshire Worsley (1966b) has described late Pleistocene ice-wedge casts.

Whenever the form of the ground and bedrock allowed head formation occurred. In the Church Stretton valley radiocarbon dating (table 10) has shown that much of the head formed during the time equivalent of Pollen Zone III (P. Rowlands 1966).

(b) *The North-West* (Table 10)

The lowlands

Antedating the oldest glacial deposits of Anglesey and Llyn are shore-platforms which can be separated into an upper level at 7·5 m and a lower one at 3 m O.D. respectively. Lying on these at Porth Oer, 7·5 m, and Red Wharf Bay, 3 m, are raised beaches, the oldest Pleistocene deposits in north Wales, although the erratics contained in the former were probably derived from pre-existing drifts. It does not follow that the platforms and beaches are the same age, for at Red Wharf Bay they are separated, in part, by as cree deposit (Whittow & Ball 1970).

Representing the interval between the raised beaches and the onset of glaciation are head deposits. These consist of local scree as well as soliflucted local tills (Synge 1964. Saunders 1968a). The latter, like the raised beach erratics, are all that remains of formerly coherent deposits.

Llyn: Jehu (1909) established a tripartite scheme for the glacial deposits of western Caernarfonshire consisting of: (1) Lower Boulder Clay, (2) Middle Sands and Gravels, (3) Upper Boulder Clay. Units (1) and (3) indicate separate glaciations, the former being more extensive than the latter, while the middle unit (2) was deemed to indicate complete deglaciation and lacustrine deposition. Many years later Synge (1964) restricted the 'upper boulder clay' to the area lying north of his Clynnog-Brynchir moraine. He suggested that the extent to which the drifts were weathered and disturbed by cyroturbation south of that moraine was sufficiently great to warrant the insertion of a full interglacial between the two boulder clay formations. By comparing the succession with that of eastern Ireland he argued that they were of Wolstonian and Devensian age respectively. More extensive work by Saunders, Simpkins, Whittow and Ball, however, has now shown that the upper till (3) is found outside Synge's limit over wide areas, and that interstadial rather than interglacial conditions occurred between the two glacial episodes.

During the lower boulder clay glaciation Irish Sea and Welsh ice sheets were in contact along a zone between Abersoch and Nefyn. Irish Sea till, the Trevor till (Simpkins 1968), is characteristically purple-grey, shelly, calcareous and clay rich, full of northern erratics from the Southern Uplands and Firth of Clyde (Jehu 1909), while the corresponding Welsh (Criccieth) till (Simpkins 1968) consists of erratics from the Nantlle slate belt, the pattern of ice-movement being confirmed by till fabric analyses (Saunders 1968b).

Shell fragments from the Trevor till at Porth Neigwl and shelly outwash at Porth Dinllaen were radiocarbon dated to 29,000 and 31,800 years BP respectively (Saunders 1968c). If these are reliable they indicate an Upper Devensian age for the glaciation.

The Middle Sands and Gravels of Jehu include fluvioglacial deposits of the retreating lower boulder clay ice-sheet as well as proglacial outwash and laminated clays of the advancing upper boulder clay ice sheet (Plate 5). Separating the two formations, however, is a zone of weathering with evidence of permafrost as well as head deposits (Fig. 99). The horizon of weathering is finely divided material from the surface of the lower till at Criccieth (Simpkins 1968), and similarly on the surface of the Trevor till (Saunders 1968a). Associated with that horizon at Criccieth and Glanllynau (Plate 5) are ice-wedge pseudomorphs which penetrate the Criccieth till and are infilled with the weathered material. Synge (1970a) has questioned the existence of such a hiatus. He argued that the lower till was deposited by ice moving in a slightly different direction than when it deposited the upper till. The horizon of weathering was suggested to be an iron-pan effect, and the wedges as load structures.

The upper boulder clay consists of three facies: (1) Northern facies such as the Porth Nefyn till, with erratics from Anglesey. (2) The Clynnog till of north-west Llyn with erratics from Snowdonia (Simpkins 1968). (3) The Llanystumdwy till (Simpkins 1968) with Ordovician shales, purple slates and ophitic dolerites from southern Snowdonia.

Further data adduced in support of readvance are the glacial tectonics at Dinas Dinlle, and a radiocarbon date of 16,830 years BP for 'organic material' collected by Foster (1968) in gravels west of Bryncir. Examinations of similar material, however, revealed no evidence for its organic nature; it contained no pollen, spores, or recognisable cellular material (Simpkins 1968).

During deglaciation terraces of sand and gravel accumulated in the Bodfaen-Cors Geirch area which Matley (1936) regarded as lake shorelines, as did Saunders (1968b). But Whittow & Ball (1970) in alternative view regard them as kame terraces.

Minor disagreement holds on the extent of upper boulder clay, but its overall outcrop is continued seawards as a thickening of the submarine drift cover (Al-Shaikh 1970), thus confirming the drift limit of Mitchell (1960) and Synge (1964), although they, unlike more recent workers, chose to regard it as the limit of Devensian ice instead of a readvance feature.

Subsequent to deglaciation head deposits accumulated, the drift surfaces were frost-heaved, and sedimentation commenced in newly formed kettle holes, notably in southern Llyn where at Glanllynnau pollen analysis of such sediments by Simpkins (1968) recognised Pollen Zones I, II and III. Radiocarbon dating of the base of Zone I, defined as commencing with the rise in *Rumex*, gave 12,050 years BP, while the Zone I/II boundary, marked by *Rumex* decline and rise of *Betula*, gave 11,300 years BP. Subsequently G. R. Coope discovered a moss layer in the solifluction clays below the organic deposits which was dated as 14,468 years BP (Shotton *et al.*, 1971). This indicates a minimum age for the Llanystumdwy till. The latter lies well outside the Devensian limit of Mitchell & Synge, but as it is separated from the underlying Criccieth till (Plate 5) by indications of interstadial conditions, it seems likely that the basal till is also Devensian.

Investigation of the pleniglacial solifluction clays of the Glanynnau kettle hole by Coope (Coope *et al.*, 1971. Coope 1970) showed that they were characterised by an arctic assemblage of Coleoptera. The radiocarbon date of 14,468 years BP was obtained from a moss layer embedded in grey silt 200 cm below the All 	eröd horizon. During the latter part of Zone I this cold fauna was replaced by a highly thermophilous assemblage of Coleoptera appearing first 10–15 cm below the horizon dated at 12,050 years BP. These indicated average July temperatures slightly above 17 degrees centigrade. Supporting work elsewhere (Coope *et al.*, 1971) has confirmed that a thermal maximum occurred between 13,000 and 12,000 years ago, but this is not indicated in the pollen record (except as a time lag phenomenon) as thermophilous trees were unable to reach this country in the time available. When they did arrive, during the Alleröd, the climatic optimum of this warm phase had already passed. The Zone II fauna is thus characterised by less thermophilous assemblages of Coleoptera.

Anglesey and Arfon: A tripartite succession of (1) Lower Boulder Clay, (2) Middle Sands and Gravels, (3) Upper Boulder Clay was established by Greenly (1919), who also noted esker gravels on the upper till, and 'moraines' on Holyhead Mountain. The latter are now regarded as block fields of periglacial origin

(Whittow & Ball 1970). Striations, erratic trains and the alignment of ice-moulded forms such as drumlins, showed that the island was glaciated from north-east to south-west. Coeval Welsh ice extended into the area parallel to Menai Strait for about 5 km into the island, bringing with it many erratics, notably the porphyritic felsites of the Bala volcanics.

Greenly was able to sustain the validity of his tripartite scheme only with difficulty as the superposition of upper on lower boulder clay occurred at only 14 sites on the east coast, where a compacted blue till is overlain by an unconsolidated red till, both of which contain erratics from the north and north-east, and are sometimes separated by sand and gravel up to 7 m thick. But over the remainder of the island the drift is characterised by local rocks, with northern erratics and heavy minerals (Smithson 1953) being chiefly subordinate. Greenly explained the absence of two discrete tills by suggesting that the second glaciation had removed all trace of the lower boulder clay, a suggestion endorsed by Hopley (1963) who argued that a second advance had moulded earlier deposits to fashion the drumlin fields of the island. Whittow & Ball (1970) also agreed with this view and suggested that the deposits of Anglesey are the same age as the upper boulder clay of Llyn, but differed from Greenly in recognising the red till of the east coast as yet a further advance, their Liverpool Bay phase. It must be stressed, however, that no stratigraphical evidence exists to support more than one advance. The blue and red tills of the east coast could readily be deposited from different layers within the same ice sheet, a proposition endorsed by their respective compaction characteristics. Pollen analytical investigations at Cors Goch (Seddon 1958) showed that silty organic muds lying on the glacial drift are Late-glacial Pollen Zones II and III in age.

The glacial drifts of Arfon were derived mainly from Snowdonia. At the Pen-y-bryn brickworks, near Caernarfon, two glacial units separated by a horizon of weathering occur (Whittow & Ball 1970); this is the only exposure which sustains the tripartite view.

The Merioneth coastlands: Unpublished work by David Morris (pers. comm. 1970) on the succession south of Harlech has revealed marked similarities to that farther south (see South West). A basal head complex, including redeposited glacial drift, and which is penetrated by ice-wedge pseudomorphs, is overlain by the Llangelynin till between Tonfannau and Llwyngwril (Fig. 99). This, and its fluvio-glacial component, are both cryoturbated and penetrated by ice-wedge casts. It contains erratics from Cader Idris and the Arans, and Morris argues that the southerly movement of ice involved in its deposition was the result of pressure by an Irish Sea ice sheet to the west. It would seem, therefore, that the Llangelynin till is the same age as those at Criccieth and Trevor. (Table 10).

Numerous pockets of red clay, rich in kaolinite, within the till are deemed to have derived from older glacial deposits. But such material occurs in the Tertiary beds of the Mochras borehole (C. O'Sullivan pers. comm.) so that its provenance may be local and Tertiary.

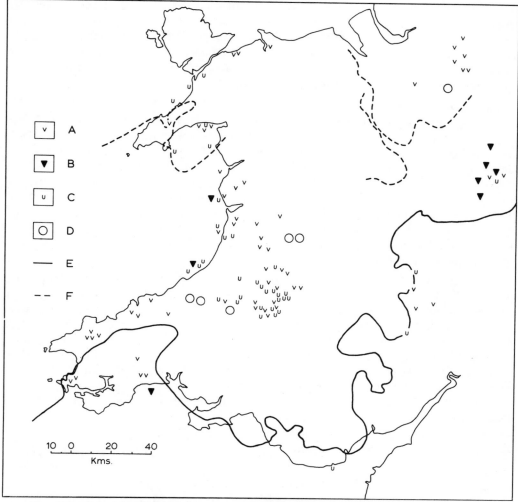

Fig. 101. Selected periglacial features and their relationship to ice limits. Key: A—ice-wedge pseudomorph. B—ice-wedge polygonal patterned ground. C—involution. D—remains of pingos. E— Upper Devensian Ice limit. F—Limit of Devensian Readvance. With the exception of periglacial structures at Llandre, west Carms. and at Barry Island, the features indicated all post-date local glacial deposits, and are Devensian Late Glacial in age.

At Towyn, till deposited by ice which moved down the Dysynni valley is over-lain by peat deposits of Pollen Zone II (A. Wood pers. comm.). The olive grey till at Mochras, Harlech, forming the upper part of Sarn Badrig, is thought to be younger than the Llangelynin till (Morris pers. comm.). Onshore, the moraine of Sarn Badrig follows the bedrock wall of Traeth bach, but subsequently leaves it at an angle of 45 degrees. Morris takes this as an indication that any pressure from Irish Sea ice to the west was lacking at that time. Foster (1970a) takes the view that Sarn Badrig represents a medial moraine deposited between Irish Sea and Welsh ice sheets, but failed, however, to take cognizance of the Llanystumdwy till at Criccieth which was deposited by Welsh ice at a time when Irish Sea ice failed to cross Llyn. As the Mochras till is coextensive with the Llanystumdwy till it shows that Sarn Badrig represents the limit of Welsh ice from the Glaslyn and Ffestiniog valleys during the upper boulder clay glaciation of Llyn. Geophysical evidence has confirmed the morainic structure of Sarn Badrig (Whittington 1971 pers. comm.).

The uplands

Traces of only the most recent glaciation occur in the uplands, although the presence of Arenig felsites in older drift near Birmingham show that the Merioneth ice cap functioned on more than one occasion. During the most recent glaciation, which on the evidence of the coastal succession, must be Devensian, the ice cap distributed flows east and west of an ice-shed which ran from north to south just east of Arenig Fawr and Rhobell Fawr, but east of Aran Fawddwy (Foster 1968). The tills of the Harlech Dome (Table 10) reflect, in their changing erratic content, depositional activity of different layers within the ice. The upper ones crossed the Rhinogs to enter Tremadoc Bay, while the lower ones followed the Vale of Trawsfynydd and Afon Eden (Foster 1970b).

Foster's conclusions, which echo those of Greenly (1919), show that the greatest thickness of ice in north Wales lay over the area around Trawsfynydd, a proposition fully supported by Rowlands (1970) in the Arenig country to the east where he showed that an easterly extension of the Merioneth ice cap over-rode the lesser peripheral masses of north-east Wales and was sufficiently powerful for a time to fend off Irish Sea ice in the Vale of Clwyd. Recessional moraines of this glaciation were recognised at Pwll Glas and Bryneglwys, while the cirque glaciation of Arenig Fach was thought to have occurred during the Devensian Late-glacial. This was a time of widespread solifluction, and when the large sorted stone-stripes of Rhinog Fawr formed (Ball and Goodier 1968).

South of the Bala-Mawddach line, Foster (1968) showed that Merioneth ice deposited till up to 427 m on Aran Fawddwy, and escaped through cols carrying felsites to the tills of the Dyfi estuary (Jones & Pugh 1935). In the upper Dyfi basal tills are overlain by multi-phase periglacial deposits (A. Machin 1969, pers. comm.).

Nourishing its own ice, but subordinate to the Merioneth ice cap, was Cader Idris. Volcanic erratics occur at Llwyngwril (Reade 1898) north Cardiganshire (Keeping 1882) and farther south (Williams 1927). Watson (1968) has described nivation cirque forms below the main Cader Idris cirques, which presumably formed during the Late-glacial.

The glacial drifts of Snowdonia are imperfectly known but it is clear that local ice masses were subordinate to the Merioneth ice cap. In the marginal uplands Whittow & Ball (1970) described an 'intermediate moraines' phase, correlated with the Liverpool Bay phase of Anglesey and accorded the status of a readvance. But the most celebrated drifts of north Wales are the shelly sands and gravels on Moel Tryfan (Trimmer 1831) at nearly 400 m, and which Ramsay (1881) used to support his concept of a great marine submergence. Their glacial origin was recognised before the turn of the century and recently Foster (1970a) obtained a radiocarbon date of 34,740 years BP from shell material contained in the drift. If this date is correct it shows that the Irish Sea Ice sheet overwhelmed the northern uplands during the Upper Devensian. As such, this event finds a ready parallel in the north east.

Pollen analytical investigations by Godwin (1955) and Seddon (1957, 1961) have revealed the sequence of Devensian Late Glacial events. Most of the Snowdonian cirques (Table 10) have inner and outer moraines. It is only outsidet he former that Devensian Late-glacial deposits occur, which suggests that they post-date the Alleröd climatic amelioration of Pollen Zone II, and formed during the deterioration of Pollen Zone III. The outer, earlier, moraines formed during Zone I. At Cors Geuallt, Capel Curig, Crabtree (1966) found pollen evidence, in lake sediments, for the Bölling oscillation (Pollen Zone Ib).

Towards the close of the Pleistocene periglacial drifts accumulated throughout the region, while stratified scree deposits probably formed during the Late-glacial (Ball 1966). Even today active frost polygons occur between Carnedd Llewelyn and Foel Grach (Tallis & Kershaw 1959, 1970).

(c) The Lower Severn (Table 11)

This region lies outside the limit of Upper Devensian Glaciation, although it is linked to the Main Irish Sea advance by means of the Severn Main Terrace. Wills (1937) recognised two earlier glaciations which he called the 1st Welsh and 2nd

STAGE		SOUTH WALES : Carmarthenshire & Glamorgan		THE LOWER SEVERN : Gloucs. & Worcs.		
Flandrian	IV - VIII	Alluvium. Colluvium (temperate fauna at Nash Point). Dune sands, Merthyr Mawr. Marine alluvium, Swansea Bay, Burry estuary, Cardiff-Chepstow. Organic deposits at Craig-y-llyn, Bryn House, Cwmllynfell, & Llanllwch.		Alluvium. Marine alluvium. Colluvium. Peat deposits.		
Upper Devensian	Zones I - III	Cirque morainic & protalus drifts of the Carms Fans, Craig-y-llyn scarp, Graig Fawr, Graig Fach & Cwm Saerbren. Organic deposits at Cwmllynfell, Bryn House. Head and scree deposits. Pingo ramparts at Ffald-y-brenin. Buried peats at Margam.				
Upper Devensian		Extra-glacial areas	Glaciated areas	Power House Terrace	Avon No.1 Terrace	
Upper Devensian		Cave deposits with head and cold fauna : Gower & Coygan.	Rhossili head. Rhondda head.	Worcester Terrace		
Upper Devensian		Head deposits at Marros, Ragwen Point, Horton, Port Eynon Bay, Hunts Bay, & Cefn Hengoed.	Brecknock till at Langland Glamorgan till in Rhondda valleys C. Wales till in Tywi valley & Ammanford-Mynydd Bettws.	Main Irish Sea Glaciation		
Upper Devensian				MAIN		
Middle Devensian		Redeposited (solifluction) glacial drift at Western & Eastern Slade and Hunts Bay.		TERRACE Oolitic (CF)	Avon No.2 Terrace	
Lower Devensian		Colluvial silts at Horton, Hunts Bay, W. & E. Slade & Stormy Down.	Head between Langland & Mumbles.	Permian breccia & Malvernian tjacle gravels		
Lower Devensian				Kidderminster Terrace	Avon No.3 Terrace surface	
Lower Devensian					Avon No.4 Terrace surface	
Ipswichian		Neritoides beach Minchin Hole breccia (temp. fauna) Patella beach, Gower	Newton Down terra rossa. Parent material of Ston Easton soil series. Marros & Ragwen Point beaches.	Gravels of No.3 Terrace		
Pre-Ipswichian		West Gower mixed Welsh & Irish Sea till and outwash Western Carms Irish Sea drifts Pencoed Irish Sea till Ewenny Irish Sea tills		WOLSTONIAN STAGE	2ND WELSH GLACIATION	Bushley Green. Avon No Woolridge Terrace 5.
Pre-Ipswichian				HOXNIAN STAGE	Nechells lake deposits	
Pre-Ipswichian				ANGLIAN STAGE	1ST WELSH GLACIATION	

Table 11. Correlation table of Quaternary units in South Wales and the Lower Severn area.

Welsh Glaciations respectively. Their deposits are characterised by Arenig felsites, notably between Walsall and Birmingham, erratics from the Berwyn, Denbighshire Moors, the Wrekin, as well as by more local Bunter pebbles.

At Nechells, Birmingham, the older drifts are separated by interglacial lake deposits and peaty silts with plant remains. Pollen analytical investigation of these (Duigan 1956, Kelly 1964) revealed that the vegetational succession compares with that at Hoxne, and hence permit correlation with the Hoxnian Stage. It follows that the underlying drifts (1st Welsh) could be Anglian, and the upper till (2nd Welsh) could be Wolstonian. But Poole (1968) maintained that the Upper Boulder Clay of the Birmingham area is Devensian in age, and to be correlated with that at Four Ashes. The Nechells deposits thus afford a valuable basis for relating the Severn, and much of the Welsh, succession, to that of East Anglia.

Deposits of the 1st Welsh are thin and are restricted to the lower parts of the Severn basin. It is possible that they are the same age as the Plateau drift of the Oxford region. Their hill-top and inter-stream distributions shows that considerable dissection occurred after their formation. Indeed they bear no apparent relationship to the drainage lines of the present time.

Unlike the older deposits, those of the 2nd Welsh, although highly dissected and cut into by present day valleys, can occasionally be related to distinct drainage lines and even examples of glacial diversion (Wills 1937, 1950). 2nd Welsh ice advanced as afar as Gloucester, and pressed against the Cotswold scarp as far as the vicinity of Moreton. Hey (1958) has shown that Welsh ice was at least 92 m thick against the eastern side of the Malvern Hills, and that the Woolridge gravels between Gloucester and Tewkesbury, and within the Leadon valley between 60 m and 86 m containing Malvernian, Silurian and Bunter pebbles, as well as flints, represents the outwash of this ice. Its eastern margin impounded glacial Lake Harrison (Shotton 1953), which on retreat of the ice drained into the Severn, initiated the present Avon, which thus reversed its previous course. During the 2nd Welsh Glaciation Irish Sea ice made its presence felt, but its erratics are limited and restricted to a very small area.

During deglaciation the terrace systems of the Severn and Avon were initiated, and deposited in valleys below the level of the older glacial drifts. The Bushley Green Terrace and Avon Terrace No. 5 consist of outwash from the decaying Welsh and Eastern glaciers respectively (Bishop 1958) and only occur downstream of Stratford on the Avon and Tewkesbury on the Severn. The later terraces have been described by Wills (1938) for the Severn, and Tomlinson (1925, 1935) and Shotton (1953, 1968) for the Avon.

The Kidderminster Terrace of the Severn and Avon Terraces No. 3 and 4 have long been regarded as interglacial and correlated with the Ipswichian. This, however, is only partly correct for a major time break occurs within the gravels. The gravels of No. 3 and No. 4 terraces are part of one aggradation which increases in age downwards (Tomlinson 1925, Shotton 1953), all of which antedates the formation of the terrace flats. The temperate mammalian fauna of *Hippopotamus*, *Elephas antiquus* (Falconer), and the temperate non-marine molluscs *Belgrandia marginata*

(Michaud) and *Unio littoralis* (Cuvier), come from the lower gravels beneath the No. 3 Terrace feature (Tomlinson 1925, 1935). While *Unio littoralis* (Cuvier) and *Corbicula fluminalis* (Müller) were found at the base of the No. 4 Terrace gravels in the Stour, a tributary of the Avon (Tomlinson 1925). These gravels formed under temperate conditions and correlation with the Ipswichian is accepted. But the upper gravels of No. 4 Terrace are younger, and were deposited under colder conditions, for they have yielded *Mammuthus primigenius* (Blumenbach) and show signs of cryoturbation (Shotton 1968). It is clear, as Shotton has shown, that the No. 4 gravels, and the terrace flats of both No. 4 and No. 3 Terraces, must be placed in the Devensian cold stage. As both terraces are older than the basal deposits of the Main and Avon No. 2 Terraces, radiocarbon dated respectively to 41,900 and 38,000 years BP, they must be placed in the Lower Devensian. It follows that the Kidderminster must also be placed in the Lower Devensian, a conclusion not in conflict with the frequently coarse nature of its constituents.

Contrary to the above, Wills (1937, 1938) believed that No. 3 Terrace was later than No. 2, the gravels of both forming one aggradation increasing in age downwards and terminating with interglacial conditions. As No. 2 Terrace is related to the Main Irish Sea Glaciation, by means of the Main Terrace, he argued that interglacial conditions occurred after this glaciation, but prior to the Welsh Readvance. The views of Tomlinson and Shotton have now supplanted this scheme.

The Kidderminster Terrace goes up the Stour and not the Severn above Bewdley, showing that the latter did not exist at that time. Following the erosion which fashioned the Kidderminster and Avon No. 4 and No. 3 Terraces, aggradation of the Main Terrace and Avon No. 2 Terrace commenced, the latter being separated from Avon No. 3 by a rock step. Although these terraces were, until recently, related to the Main Irish Sea Glaciation (Shotton 1967a, Coope *et al.*, 1961), the aggradation is now known to have spanned the Middle Devensian, culminating during the Upper Devensian when the glaciation occurred (Shotton 1967b). (Table 11).

While the Main Terrace in the Severn Valley has not yielded a fauna, prolific fossil remains were recovered from one of its tributary terraces at Upton Warren on the Salwarpe. The fauna and flora were described by Coope, Shotton and Strachan (1961), the former including *Mammuthus primigenius* (Blumenbach), *Tichorhinus antiquitatis* (Blumenbach), *Bison priscus* (Bojanus), *Rangifer tarandus* (Linné), and *Equus caballus* (Linné); and the latter including plant remains and pollen. The above, together with the evidence from a large insect fauna point to an arctic climate. A sample of peat gave a radiocarbon date of 41,000 years BP.

Confirmation of this was obtained at Fladbury from the base of Avon No. 2 Terrace. Here Coope (1962) described a beetle fauna indicating a subarctic climate together with *Mammathus primigenius* (Blumenbach) and *Tichorhinus antiquitatis* (Blumenbach). A radiocarbon date gave 38,000 years BP.

Related to the Main and Avon No. 2 Terraces respectively are extensive spreads of taele gravels (Table 11). In the Avon and lower Severn are Oolitic gravels (which contained a cold fauna of mammoth and woolly rhino), Malvernian, Permian breccia and Bunter solifluction gravels. Large-scale ice-wedge polygonal patterned ground was described on the surface of Avon No. 2 terrace (Shotton 1960).

The Main Terrace occurs between the Severn estuary and Bridgnorth, where it passes into the Worfe valley to merge with Main Irish Sea Glaciation sands and gravels (Wills 1938), but also occurs upstream of Bewdley as far as the Ironbridge gorge, to the fashioning of which it has been realted (Wills 1924, 1938). Its constituents are gravels of Irish Sea provenance in contrast to those of older terraces. The Main Irish Sea Glaciation, whose outwash train the Main Terrace represents, occurred after 30,000 years BP, possibly even after 18,000 years BP (North East), so that the total aggradation of the terrace gravels spanned a considerable time. Radiocarbon dates previously mentioned from Upton Warren and Fladbury show that aggradation commenced during the Middle Devensian. Fluviatile gravels of the same age have been discovered in many Midland valleys, notably in that of the Tame, where contemporary insect faunas show that 32,000 years ago the mean July temperature was depressed to about 4°C.

Younger than the Main Terrace, the Worcester Terrace (Avon No. 1) occurs on both sides of the Ironbridge gorge and is related to the Welsh Readvance. The still younger Power House Terrace (Wills 1938), which demonstrates aggradation after incision from the Worcester Terrace level, may be related to a further advance, and has been tentatively related to Peake's (1961) Llay Readvance; but uncertainty attaches both to the correlation and reality of readvance.

The buried channel of the Severn and its Flandrian fill have been discussed by Beckinsale and Richardson (1964), while Anderson (1968) has described the buried channel systems south of Upton on the Severn.

STAGE		THE SOUTH-WEST : Cards. and Pembs.			THE SOUTH EAST : Brecknock, Hereford, Mon., & Radnorshire.	
Flandrian	IV – VIII	Alluvium. Marine alluvium. Dune sands. Freshwater West. Coastal peats at Ynys las, Clarach & Freshwater West. Peat deposits at Tregaron, Cors Fochno, Upper Elan Valley & Plynlimon.			Alluvium. Iron Age alluvial gravels. Marine alluvium at Llanwern post 2660 BP. Basin and lake muds, clays & peats at Rhosgoch, Mynydd Illtyd and Waen ddu.	
Upper Devensian	Zones I – III	Head. Stratified screes. Alluvial fan gravels at Aberarth and Llanon. Redeposited (solifluction) glacial drifts in upland and coastal Cards. Basin clay and organic deposits at Aberaeron and Upper Elan valley. Cirque morainic & protalus drifts of Plynlimon and Cwm Ystwyth.			Head. Stratified screes. Redeposited (solifluction) glacial drifts of upland Radnorshire. Block fields. Cirque morainic & protalus drifts of Brecon Beacons. Basin deposits at Rhosgoch, M. Illtyd and Waen ddu.	
		Extra-glacial areas	Irish Sea Ice area	Welsh Ice area	Glaciated areas	Extra-glacial areas
		Cave deposits with head and cold fauna in South Pembs. Head at West Angle Bay, Pwllcrochan and in Milford Haven, Broadhaven, Manorbier, Swanlake, Lydstep, Caldy Island.	Abermawr rubble drift Llechryd clays Banc-y-warren & Cemmaes sands and gravels. Aberaeron, Newquay, Gwbert Abermawr, Druidston tills.	Morfa bychan head Elan valley, Tregaron Tywi valley outwash. Elan valley, Pentrecwrt, Morfa bychan, Llanon tills.	Orleton outwash Clun & Radnor Forest tills Lower Usk valley glaciation Herefordshire newer drifts	Risca and Pontypool Head deposits.
Middle Devensian			Abermawr, Porth Clais, Druidston, Aberarth head.	Lowest head at Morfa Bychan.		
Lower Devensian					Caerleon Terrace	
Ipswichian		? West Angle peat and estuarine clay Milford Haven estuarine clays Raised beaches at : Poppit, Porth Clais, Druidstone, Broadhaven, Caldy Island.				
Pre-ipswichian		Older Irish Sea drifts of South Pembrokeshire ? Whitesands Bay giant erratic. West Angle till			Herefordshire older drifts. St. Julians Terrace. Tredunnock- Llangibby Terrace (= ? Bushley Green) Afon Lwyd drift. Henllys Brook drift. Usk No.2 Terrace.	

Table 12.   Correlation table of Quaternary units in South East and South West Wales.

(d) The South-East (Table 12)

The Older Drifts

Glacial drifts antedating those of the Devensian are unknown in Radnorshire and Breconshire (Lewis 1970b), unless they are represented by the highest deposits in Radnor and Clun Forests (Dwerryhouse & Miller 1930, Luckman 1966), but this is thought to be unlikely by Brown (1971).

In Herefordshire, outside the Devensian limit (Fig. 100) are remanié drifts found capping hill-tops and inter-stream areas. These are characterised by Welsh erratics from the west and have been described by Grindley (1954), Pocock (1940), Dwerryhouse and Miller (1930) and Burnham (1964). Recently Luckman (1970) presented a synthesis of previous work. Exact dating of these drifts is not possible, although their highly dissected nature makes it probable that they antedate the Ipswichian.

Older drift deposits in the lower Usk are not as extensive as those of Herefordshire. The inter-stream drifts of the lower Afon Lwyd, and between the Henllys Brook and lower Ebbw Valley may be the same age as Will's's (1937) 2nd Welsh Glaciation (Williams 1968a, Squirrell & Downing 1969). The above authors argued that these drifts are related to the Tredunnock or Llangibby Terrace of the Usk (Usk No. 2 of Welch & Trotter 1961), an outwash terrace, which was correlated on altitudinal grounds with the Bushley Green Terrace of the Severn. Williams showed that this pre-Devensian glaciation occupied shallower valley floors than the present, and named it the Upper Usk Maximum.

The St. Julians Terrace, at about 30 to 34 m (Williams 1968a), may also represent a valley train, but lack of exposures make this uncertain. Correlation of the four Wye terraces has not been attempted (Welch & Trotter 1961). The Caerleon Terrace, 14 to 18 m, again recognised by Williams, Squirell and Downing, has been correlated with the Kidderminster Terrace of the Severn on altitudinal grounds and deemed to be of Ipswichian age. Squirell & Downing (1969) also regard the terraces between 10 and 15 m at Pencarn and Maesglas, south-west of Newport, as the same age on similar grounds. This correlation, however, is open to question. Not only have temperate conditions not been demonstrated, but now that the Kidderminster Terrace is regarded as Lower Devensian it seems reasonable to suppose that the Caerleon Terrace may also be this age. The region where the Pencarn and Maesglas terraces are found was covered by Devensian ice (Bowen 1970a), hence they cannot be older than that episode.

The Newer Drifts

Although no unequivocal traces of the Ipswichian occur in the four counties under consideration, the limit of Devensian glaciation on the other hand is remarkably clear (Fig. 100). In the Monmouthshire coalfield valleys it is approximately that indicated by Charlesworth (1929) and Bowen (1970a). But in the Usk Valley Lewis's (1970b) portrayal required considerable modification as a result of mapping by the I.G.S. (Trotter & Welch 1961). In this region Williams (1968a) recognised two advances after the Caerleon stage. An earlier one, suggested to be Middle Devensian, called the Vale of Gwent Maximum, and represented by deposits beyond the morainic drift of Welch & Trotter (1961) with a valley train called the Llanbadoc terrace. This occurs in the present valley, in contrast with the older drifts. The subsequent advance reached Kemys Commander downstream of which the contemporary valley train passes below the present alluvium of the Usk to form part of the buried gravels of the Newport district investigated by Williams (1968b), who devised a scheme of correlation with the sequence postulated by Beckinsale & Richardson (1964) for the lower Severn. Stratigraphic evidence for two glacial advances after the Caerleon Terrace stage is lacking and in Table 12 the view is taken that only one advance occurred.

In Herefordshire the extent of the Wye glacier and the central Wales ice sheet is taken as the limit of continuous drift of an undissected nature. This forms a striking contrast with the highly dissected drifts east of such a line (Luckman 1970. Burnham 1964).

Inside the outer limit of Devensian glaciation (fig. 100) signs of recent ice occupation are ubiquitous. In Herefordshire and Radnorshire Luckman (1966) described valley moraines, kettle holes and kame terraces, while Cross (1966, 1968) showed that ice was 243 m thick in the Wigmore and Presteigne basins where extensive lacustrine deposits formed during deglaciation. At the same time several examples of glacial drainage diversion occurred (Dwerryhouse & Miller 1930). The Orleton moraine is pitted with kettle holes along its entire length between Kington and Orleton, while near the latter a radiocarbon date of 11,730 years BP (Shotton et al., 1969) indicates a minimum date for glaciation. This date may, however, be too 'young' as the sample was collected from outwash deposits where the risk of contamination by modern radiocarbon was high.

Lewis (1970b) and Potts (1968) described glacial deposits in the Builth-Llanwrtyd lowland, the latter commenting particularly on the extensive outwash deposits south-west of Llanwrtyd itself. The moraines and outwash terraces of the Wye Valley were described by Pocock (1940). Erratics from this area were used by Dwerryhouse & Miller (1930) to show the direction of ice movement eastwards and by Howard (1904) in Mynydd Eppynt to show a southerly component of flow. At Rhosgoch Common, Radnorshire, Bartley (1960) showed, through pollen analytical investigations, that immediately after deglaciation lake sediments formed in a hollow on morainic debris. He demonstrated the presence of Pollen Zones I, II and III, Pollen Zone I fixing a close minimum age for the glaciation.

West of Rhayader and Llanwrtyd, in the uplands proper, Potts (1968) mapped extensive areas of till and outwash in all the major valleys, which he deemed to be Upper Devensian in age. From the marginal distribution of considerable spreads of outwash gravels at Rhayader, Tregaron, upstream of Llandovery and Llanwrtyd, he argued that the ice stood at these positions during the time equivalent of Pollen Zone Ia, which Penny (1964) has suggested was when the Scottish Readvance occurred farther north. During the Late-glacial much till was re-deposited, through solifluction, along steep slopes, and head and stratified screes formed, while on suitable lithologic types block fields and tors formed (Potts 1971). The stratified screes, as well as some of the redeposited drift sequences, commonly show a roughly three-fold succession. Potts interpreted this in terms of the three Late-glacial Pollen

Zones, the thicker deposits of Pollen Zone II indicating milder conditions when a greater number of freeze-thaw cycles occurred. Ice-wedge pseudomorphs show that permafrost conditions obtained during Pollen Zone III.

Severe periglaciation occurred in the Brecon Beacons after deglaciation (Lewis 1970b). The cirques were occupied by small glaciers and inert snow patches during the time equivalent of Pollen Zones I and III. So spectacular is the periglacial inheritance that at one time Lewis (1966) suggested that they had been ice-free during the Devensian. Pollen analytical investigations at Mynydd Illtyd (J. Moore 1970) and Waen ddu (Trotman 1964) have confirmed the appropriate climatic conditions for Lewis's (1970c) subsequent reconstruction. Only deposits of Pollen Zone IV occur within the inner blocky moraine at Cwm Cerrig Gleisiad (Moore 1970) which shows clear analogy with Snowdonia.

(e) South Wales (Table 11)

The 'older drifts'

The oldest Pleistocene deposits known in the area are the so-called 'older drifts' which are glacial in origin. Their stratigraphic relations are shown between Pencoed and Llanharan where they are overlain by 'newer' glacial deposits. It was long held that the *Patella* beach of Gower (George 1932) antedated the 'older drift' of South Wales (Zeuner 1959. Mitchell 1960), but it has now been shown to post-date that glaciation both in South Wales (Bowen 1970a) as well as in South West England (Bowen 1969a, 1969b), erratic bearing drifts overlying the raised beaches in such areas being re-cycled glacial deposits now incorporated in slope, solifluction and alluvial drifts.

East of Swansea Bay and in western Carmarthenshire the older glacial drifts have been greatly dissected, and coherent deposits of till are only to be found around Pencoed and at Ewenny. Strahan and Cantrill (1904) showed that an Irish Sea ice sheet invaded Glamorgan carrying erratic boulders as far east as Cowbridge and possibly beyond, although they noted that the so-called erratic flints in the Ely Valley drifts only appear where the Lias begins to contain chert. At Pencoed they described deep red and purple tills containing igneous erratics, while shell fragments were found in nearby sands. Recent opencast operations have shown that these deposits are widespread and are overlain by Welsh drifts, a relationship in conflict with Woodland and Evans (1964) who held that the Irish Sea and Welsh drifts are mutually exclusive. The morainic deposits of the latter around Llanilid are of 'newer drift' age (Bowen 1970a). At Ewenny exposures of shelly calcareous till may still be seen, while transient exposures have revealed a multi-till sequence.

Unlike Strahan and his colleagues, Griffiths (1940) had no doubt that the Vale of Glamorgan was glaciated and traced erratics, as well as the characteristic Irish Sea suite of heavy minerals as far east as Cardiff. Later Crampton (1966) was to suggest that Lias cobbles found in the Vale were remanié glacial deposits, and in addition he traced the distinctive heavy minerals east of Cardiff. Reports of boulder clay in the Bristol District (Hawkins & Kellaway 1971) add confirmation to the above.

In western Carmarthenshire, as well as Pembrokeshire, erratics in the 'older drift' which consists largely of sands and gravels, the finer tills having been removed by erosion, as well as large erratic blocks, show that Irish Sea Ice crossed the area from north-west to south-east (Strahan *et al.*, 1914. Griffiths 1940).

It is only in west Gower that unequivocal older drifts of Welsh provenance occur. But they are mixed with materials derived from the Irish Sea (George 1933a). In contrast to the drifts, described above, these are little dissected and bear a subdued topography. George (1933a) argued that the mixture of Irish Sea and Welsh erratics showed a commingling of the two ice sheets. But Bowen (1970a) suggested that the contrast in the degree of drift dissection could represent an age distinction, the 'older drift' of west Gower representing readvance of Welsh ice incorporating Irish Sea erratics into its deposits.

The 'older drift-newer drift' Interglacial

The *Patella* beach of Gower was at one time regarded as having accumulated in cold conditions (George 1932), but now, together with the raised beaches of Carmarthen Bay (Bowen 1970a), it is accepted as interglacial. In Gower the *Patella* beach lies on shore-platforms from just above O.D. to over 15 m, but most frequently at 10 m or so. As well as local pebbles it contains igneous erratics from farther afield, and whereas these were formerly accounted for by ice-rafting, they are now regarded as having derived from the destruction of 'older drifts'. At Marros and Ragwen Point the beaches consist of local material and are overlain by sandrock and sand. A further beach, the *Neritoides* beach was identified by George (1932) in Gower where at Minchin Hole cave it is separated from the *Patella* beach by breccia containing a temperate fauna (George 1932). (Plate 6).

The fauna of the Patella beach includes: *Patella vulgata* (Linné), *Littorina littorea* (Linné), *Littorina rudis* (Maton), *Neritoides obtusata* (Linné), and *Purpura lapillus* (Linné), but the assemblages does not permit palaeoecological or chronological inferences to be made (George 1932). At Minchin (Mitchin) Hole cave, Gower, a mammalian fauna which included *Dicerorhinus hemitoechus* (Falconer), was recovered from the cave breccia between the *Patella* and *Neritoides* beaches, as well as from the latter beach (George 1932). *D. hemitoechus* (Falconer) was considered by Hinton (George 1932) of some value for correlation purposes, and he correlated the Minchin Hole breccia with the Middle Terrace brickearth of the lower Thames at Ilford. This was subsequently confirmed by Sutcliffe's (1960) work on interglacial mammalia in Britain, and on this basis the Minchin Hole breccia, *Neritoides* and *Patella* beaches have been placed in the Ipswichian (Bowen 1970a). Further support for this view derives from the superposition on the raised beaches of west Gower and Carmarthen Bay of slope deposits of one unitary cold stage (Devensian).

Non-marine mollusca of temperate aspect were recorded from blown sand of the raised beach series at Caswell Bay (George 1932), while at Marros, pollen identified from a dune-slack overlying the raised beach indicated late temperate interglacial conditions (Bowen 1970a).

It has been argued that relict soils (palaeosols) found in Glamorgan are

Ipswichian (Bowen 1970a). On the slopes of Newton Down, Porthcawl, Crampton (1966) described a *terra rossa* soil, while *terra fusca* soils have been identified elsewhere. Many of these form the parent material for Flandrian soils such as the Ston Easton Soil Series, but as they have yet to be discovered in stratigraphic succession with other Pleistocene deposits they cannot be accorded the status of geosols (Morrison 1968). But because they outcrop only beyond the outer limit of Devensian glaciation on ground previously traversed by the 'older drift' Irish Sea ice sheet, and their colluviated facies immediately overlies the *Patella* beach, an Ipswichian age is indicated.

The 'newer drifts'

Extra-glacial districts: Immediately overlying the raised beaches are blown sands, commonly in areas where none occurs on the present foreshore, thus suggestive of a regressional origin, a conclusion in accord with the pollen evidence at Marros; such deposits span the change over from temperate to colder conditions at the beginning of the Devensian.

Tiddeman (1901) and later George (1933a) described red loams associated with the raised beaches of Gower, which Ball (1960), after examination of the exposure on Worms Head pronounced to be *terra fusca* soils. Subsequently Bowen (1965, 1970a) showed that they were not *in situ* and were colluvial deposits which included the eroded relics of interglacial soils. By analogy with the *limon rouges* of the Mediterranean he called them colluvial silts, a genetic term, for in practice grain size varies between wide limits. Several facies occur, some with erratics incorporated from the 'older drift', others with sub-rounded or sub-angular limestone blocks formerly interpreted at Horton as head (Wirtz 1953). Indeed, much of the so-called 'lower head' around the Irish Sea basin is colluvial, and did not form in periglacial conditions. Investigation of the clay mineralogy of the colluvial silts north of Port-Eynon Bay showed a preponderance of illite, but with subordinate quantities of kaolinite (Bowen and Dobson). The latter indicates derivation and contamination from other than interglacial sources.

The overlying head consists of two principal facies (Fig. 99): first, angular and sub-angular frost riven fragments of local rock, occasionally incorporating 'older drift' erratics, and second, redeposited 'older drift' (Bowen 1965). Recognition of the latter as a re-cycled glacial deposit was crucial to the establishment of the correct sequence in the area. The critical exposures at Western and Eastern Slade were described by Bowen (1971a).

The thermoclastic head, in lateral variation as cave sediments, has yielded a prolific mammalian fauna of cold aspect. Unfortunately most of the faunal lists (Allen & Rutter 1944, 1947) are useless as they are not related to the cave successions. At Cathole Cave Cresswellian implements were recovered from the head (McBurney 1959).

In summary, the coastal slope deposits, periglacial or otherwise, of west Gower and Carmarthen Bay (as well as S. Pembs. and S. W. England), show clearly that the last local glaciation antedated the *Patella* and related raised beaches.

Glaciated districts: Head deposits of Lower and Middle Devensian age lying between the *Patella* beach and Upper Devensian glacial drifts may only be seen between Langland and Mumbles, where they lie inside Charlesworth's (1929) outer limit of 'newer drift' glaciation, a limit since modified in Gower and Carmarthenshire (Bowen 1970a). Three drift provinces associated with the Upper Devensian Welsh ice-sheet (Bowen 1970a). These are those of the Glamorgan drift, composed of Coal Measures material and lacking Devonian erratics (David 1883), the Brecknockshire drift containing abundant Devonian erratics, and the Central Wales drift with erratics from formations older than the Devonian.

Several instances of basal ice-sheds have been recognised. Of these the most important is on Mynydd Sylen, north of Llanelli. Basal ice followed the Gwendraeth Fawr from north-east to south-west, but was crossed by upper ice layers with a north to south gradient (Strahan & Cantrill 1907). This demonstrates the existence farther north of a not inconsiderable Central Wales ice-sheet.

It has been suggested that boulder strewn uplands, such as Mynydd Eglwysilian (Squirell & Downing 1969), and elsewhere in the coalfield (Woodland & Evans 1964), were glaciated during a previous advance as the upper limit of coherent till lies below the summits. However, stratigraphic confirmation of two advances thus postulated is lacking. Indeed in Carmarthenshire central Wales ice carried Silurian boulders, in analogous role, to over 300 m on Mynydd Bettws where they are associated with extensive till.

A minimum age for this glaciation is indicated by the Pollen Zone I deposits found in a kettle hole at Bryn House (Singleton), Swansea (Trotman 1964). During the Late Glacial head continued to form at appropriate sites, e.g. at the foot of Rhossili Down, while in the Maesteg and Pontypridd area it is frequently difficult to distinguish from till. Although confirmation by pollen analysis is lacking it seems likely that the Carmarthenshire Fan and Rhondda Fawr cirques were occupied by small glaciers and snow patches at this time; morainic and protalus drifts occur in all of these.

Near Ffald-y-brennin, north Carmarthenshire, north of Llansadwrnen, and around Llanpumpsaint and Pont-ar-sais, Bowen (1972a) has described the remains of pingos. In the last named area well over a hundred individual pingo mounds are indicated from their fossil ramparts. As such they demonstrate the occurrence of permafrost in those areas at a time subsequent to deglaciation.

(f) The South-West (Table 12)

Pembrokeshire

The 'older drifts': These are only found in south Pembrokeshire outside the limit of Devensian glaciation (fig. 100), although scattered erratics in the raised beach and overlying head of north Pembrokeshire derive from deposits of the same age. The older glaciation was effected by Irish Sea ice which crossed the country from north-west to south-east, as revealed by the several indicator fans (Griffiths 1940). In common with older drifts of other areas they are confined to hill-top and interstream areas, thus demonstrating their considerable dissection, and hence antiquity.

The absence of glacial drift overlying the raised beaches of south Pembrokeshire shows, like Carmarthen Bay and west Gower, that the 'older drift' glaciation antedates the raised beach interglacial. Indeed, at West Angle Bay raised beach deposits overlie boulder clay (Dixon 1921), while the large erratic block partly sealed by raised beach at Whitesands Bay (John 1971) may also be the same age.

The 'older drift-newer drift' interglacial: This is represented by marine deposits consisting of raised beaches and estuarine clays, as well as the West Angle peat. The beaches lie on shore-platforms up to 10 m O.D. in the north (John 1965a), and from below high-water-mark to about 7 m above it in the south, although within Milford Haven they are lower, between 2·5 and 3 m above high-water-mark (Dixon 1921).

Foreign erratics as well as local pebbles occur within the raised beaches which are well exposed at Poppit (John 1970a), Porth Clais (Leach 1911, John 1970b), Freshwater West (Dixon 1921) and Caldey Island (Dixon 1921). In Milford Haven the raised beach is sometimes replaced by estuarine clay (Dixon 1921) which is now known to contain foraminifera (Kaill 1971 pers. comm.).

At West Angle Bay, Dixon (1921) described a succession consisting of basal gravelly drift, with igneous rocks, which he thought might be boulder clay, overlain by sand and gravel typical of the raised beach of south Pembrokeshire. The beach unit was overlain by loams, with plant remains, and finally by red sand and gravel.

Recently, mechanically excavated exposures have revealed the following succession:

> Sandy gravel (head) 4 m
> Marine alluvium 2·4 m (base at 6·7 m O.D.)
> Peat 0·3 m
> White sandy gravel 0·75 m
> Raised beach sand and shingle 0·6 m (base at 5·3 m O.D.)
> Weathered Irish Sea till (base not seen) 1·5 m

Pollen analysis of the peat and marine alluvium has revealed contemporary temperate deciduous woodland (Field 1968). Unfortunately the vegetational succession does not compare with any obvious interglacial, but on regional stratigraphic grounds it is unlikely to be anything other than Ipswichian. The upper sandy gravel, described as till by John (1970a), is a facies of the head formation which is found along adjacent coastlines both within Milford Haven, and on the south coast.

Cave sediments of Ipswichian age occur at Eel Point, Caldy, and Blackrock Quarry, Tenby, where bones of *Hippopotamus* and *Hyaena* (Dixon 1921) were recovered.

The 'newer drifts' South of the outer limit of Upper Devensian glaciation (fig. 100) one unitary head formation, overlying the raised beaches, represents the whole of Devensian time. It may contain scattered erratics derived from 'older drifts', while at Blackrock Quarry and Hoyle's Mouth, Tenby a cold fauna, including

---

* Radiocarbon dating of *Alnus* tree stumps in position of growth on the surface of the marine alluvium immediately below the Head unit gave a date of > 35,500 years BP (Birmingham 327), thus confirming the interglacial status of the beds above the till and beneath the head.

Mammuthus primigenius (Blumenbach) and *Tichorhinus antiquitatis* (Blumenbach) was recovered from it (Dixon 1921). At Little Hoyle Cave, Tenby, Cresswellian artifacts from the head confirm its Devensian age (McBurney 1959).

In north Pembrokeshire the Lower and Middle Devensian is represented by head and other slope deposits overlying the raised beach. In turn these are overlain by glacial drift of the Upper Devensian glaciation. Jehu (1904) proposed a tripartite arrangement for the glacial succession of this area consisting of: (1) Lower Boulder Clay, (2) Middle Sands and Gravels, (3) Upper Boulder Clay. The lower boulder clay, or till, is a stiff purple shelly calcareous deposit containing erratics from the north. It is overlain by outwash (2). The upper boulder clay or 'rubble drift' of Jehu, however, was shown to be head by John (1965a). This unit frequently incorporates erratic material within more localised scree. Such redeposition of the glacial drift also occurs in other localities, such as the western end of the Dale Valley, and as such reflects local site characteristics notably of steep adjacent slopes. On the other hand, Synge (1970a) interprets Jehu's 'rubble drift' unit at Abermawr, as glacial. His recognition of a 'soliflucted drift' unit within the basal head at that exposure means that according to his interpretation three separate glacial episodes are recorded there. This conflicts with the monoglacial view of John (1970) and Bowen (1971b), although both admit that erratic pebbles in the lower head were derived from an 'older drift' glaciation.

The extent of Upper Devensian ice along the coast is indicated by those exposures revealing raised beach antedating glacial drift. Druidston Haven marks the most southern instance of this stratigraphic control, and as such, is consistent with the drift distribution inland, for north of the Roch-Trefgarne anticlinal upland glacial drift is extensive, while to the south it is sparse and highly dissected. Moreover, the outwash terraces (valley trains) of the Western Cleddau and Camrose Brook commence near the ice limit shown on figure 100. Although the stratigraphic control of the raised beach is lacking, it seems likely that the ice pressed against the coastline south of Druidston Haven (Bowen 1971b) for on the Marloes peninsula small valleys are infilled with calcareous till, on which lie hummocky sands and gravels. The western side of the Dale peninsula also carries outcrops of coherent glacial drift, while outwash filled the Mullock Bridge Valley.

In alternative explanation John (1968) suggested that Upper Devensian glaciation covered Pembrokeshire and extended south to the Isles of Scilly. He subsequently modified this view, placing his new limit between Milford Haven and the south coast of the county (John 1971). These views are discussed in the synthesis but neither are adopted in figure 100.

Charlesworth (1929) was the first to suggest that ice of the Last Glaciation terminated in Pembrokeshire, but restricted its influence to the north of the county where he identified ice-marginal overflow channels in the Fishguard district, and end-morainic deposits between Newport and Cardigan on the Creigiau Cemmaes upland. The channels were subsequently shown to be subglacial (Bowen & Gregory 1965), and the 'morainic' deposits to be fluvioglacial, consisting largely of stratified ice-contact drift (Gregory & Bowen 1966). These demonstrated that the ice covered the Fishguard area, and as such are consistent with the stratigraphic evidence.

Such a re-interpretation also brought into question the existence or otherwise of Charlesworth's (1929) and Jones' (1965) ice-damned lakes. Laminated clays at Llechryd (Bowen 1967) and deltas at Pentrecwrt (Jones 1965) show that water bodies did exist, but in view of the new model of ice-wastage, were probably ephemeral and restricted in extent.

Several attempts have been made to obtain a maximum radiocarbon date, or series of dates, for the glaciation (John 1965b. John & Ellis-Gryfydd 1970). Dates from shell and wood fragments found in the impervious Irish Sea till have shown that the samples are older than dating by the radiocarbon method allows: > 36,300 to > 54,300 years BP (John 1970). On the other hand most of the dates on shell fragments and 'organic' materials found in sands and gravels have given finite dates, 31,800 to 38,160 years BP (John 1970). Carbonate material in silicate debris behaves as an open system and dates on such materials are almost certainly too 'young' as a result of contamination by younger C14. Moreover the Pleistocene age of the organic samples has not always been beyond doubt (Boulton 1968). For these, and other reasons, the C14 dates on derived samples from south Cardiganshire and Pembrokeshire have been criticised and thought unreliable by Bowen (1966, 1971b and 1972a), Shotton (1967a) and Boulton (1968): they are discussed further in the synthesis.

Cardiganshire.

Coastal Cardiganshire. The Coastal Pleistocene deposits were described by Keeping (1882), Reade (1898) and Williams (1927) who made an attempt to rationalise the succession throughout the county and at the same time recognising a tripartite arrangement of the glacial drifts. More recently Mitchell (1962) and Watson (1965–70) have re-interpreted the results of some of the earlier work.

Despite the apparent complexity of the coastal successions the general sequence is not unlike Pembrokeshire. The 'pre-glacial' shore-platform and raised beach appear to be absent unless represented respectively by the rock outcrop noted by Williams at Aberarth and the gravels at Newquay mentioned by Synge (1963) and Mitchell (1962). Both antedate the drift platforms which bury the basal fossil cliff at Newquay, between Aberaeron and Aberarth and between Morfa and Aberystwyth.

The general sequence is: (1) Basal periglacial and other slope deposits including redeposited glacial drift. (2) Glacial drift, including tills from both Irish Sea and Welsh sources, as well as the associated outwash. (3) Upper periglacial deposits.

Deposits at the base of the coastal succession (1) include slope gravels, scree, solifluction clays, silts, and glacial drift in secondary positions. This now includes Mitchell's (1960) 'Newquay boulder clay' (Watson and Watson 1967).

Both Welsh and Irish Sea glacial deposits outcrop along the coast, Williams (1927) lower boulder clay, Irish Sea in provenance, was re-named the Fremington Till by Mitchell (1962). This distinctive purple clay-rich shelly calcareous till is well exposed at Aberaeron, Newquay and Gwbert. In lithostratigraphic correlation it is the equivalent of those exposed at Abermawr and Druidston Haven farther south, and it shows that Irish Sea ice impinged on the coastline from Llanrhystud southwards. At that locality, Llansantffraid and Llanon, drift of the coeval Welsh ice sheet is exposed as a gravelly deposit which includes some Irish Sea erratics. At Llansantffraid Mitchell (1960, 1962) argued that a red clay overlying the till was an interglacial soil, but examination by pedologists failed to reveql any attribute normally associated with a soil. The material is here compared with that described by Morris farther north in Merioneth, and is manifestly glacial in origin.

At Morfa Bychan, south of Aberystwyth, Watson & Watson (1967) suggested that the blue head (Fig. 99) was composed of periglacial material, the bedding characteristics and fabric of the deposit being unlike that of glacial drift, but that in its lower horizons it might include glacial material ('older drift') which had been redeposited by solifluction. The deposit consists almost entirely of local material together with rare foreign erratic pebbles. The shape characteristics of many pebbles or clasts, however, is diagnostic of glacial transportation (Flint 1971), and many of them bear deep striations. Wood (1959) maintained that it consisted of Welsh boulder clay redeposited by solifluction 'at the end of the last glacial period', a conclusion not inconsistent with the considerable relief immediately to the rear of the drift platform. The thickness of the basal yellow head compares with other head deposits in Pembrokeshire and south Wales in topographically similar situations which accumulated prior to Upper Devensian Glaciation.

The lack of extensive constructional topography on the glacial drifts has led many to believe that they belong to the older drifts of Britain. But this absence is not altogether surprising. Both Irish Sea and Welsh tills are highly argillaceous and hence prone to solifluction. Moreover when they contain a large proportion of gravel sized material they form features such as the kettle-holed moraine at Llanon, the hummocky drift at Llwyncelyn, and the spectacular kamiform deposits, consisting of fluvioglacial gravels overlying basal sands, at Banc-y-Warren (Williams 1927). Overlying the latter unit at Llechryd are laminated clays, which formed in Lake Teifi. So arresting are the constructional deposits of Banc-y-Warren, that Wirtz (1953) inserted an advance of Devensian ice across Cardigan Bay from Llyn to impinge on the Teifi Estuary, leaving central Cardigan Bay ice free. In so doing he was clearly unaware of comparable features south of Cardigan at heights up to 180 m O.D. (Gregory & Bowen 1966). Moreover his reconstruction is contrary to all the stratigraphic evidence.

Several different deposits are included in the upper periglacial complex (3). Williams's (1927) misgivings on the supposed glacial origin of his upper boulder clay have been confirmed and it is now regarded as head (Watson & Watson 1967). Alluvial fan gravels occur at Llansantffraid (Mitchell 1962) and Aberarth (Watson 1970), while redeposited glacial drift occurs in appropriate topographic positions.

A feature of the glacial drift is the degree to which disturbance by cryoturbation has occurred. Large involutions are found between Llansantffraid and Morfa (Watson 1965), together with vertical stones (Watson & Watson 1971) and ice-wedge pseudomorphs. Coarse glacial gravels are greatly weathered, a feature which together with their cryoturbation led to Wirtz (1953) placing them in the Riss glaciation, a practice followed by Mitchell (1960) and Watson (1970).

Inland Cardiganshire

Until recently the drift geology of this region was almost unknown. Keeping (1882) and Williams (1927) had mentioned scattered exposures but now more is known due to the mapping of Watson (1970) and Potts (1968).

Two opposing schools of thought have emerged on the nature and significance of the drifts. Mitchell (1960) and Watson (1970) believe that they are principally periglacial, and that the area was ice-free during the Devensian. On the other hand Potts (1968, 1971) maintains that the same deposits are glacial, re-cycled to greater or lesser extent by late solifluction, and that the area was glaciated by a late Devensian ice sheet. He derives support from work in adjacent areas. Foster (1968) in the Harlech Dome, B. Rowlands (1970) in Arenig, M. Brown (1971) in Montgomeryshire, and the implications manifest in the Welsh borderland (Luckman 1970) and South Wales (Bowen 1970a), all point to the view of Devensian glaciation being the most likely.

At the centre of controversy is the ubiquitous blue-grey stoney clay of the uplands deemed to be head by Watson (1970) but till and re-cycled till by Potts (1971). Watson's view is held for a combination of reasons: the clay is characterised by strong preferred orientation downslope; it consists largely of angular and sub-angular rock fragments though rounded and sub-rounded ones do occur; it is crudely bedded parallel to local slope, horizons of dirty gravels silts and sands occur; the drift terrace which it forms is usually thicker when the rear backslope is highest; and, such terraces occur principally at the foot of slopes facing north-west to east.

On the other hand, Potts (1968, 1971) maintains that sedimentological and morphometric parameters support a glacial rather than periglacial origin for the deposit. He noted that rounded pebbles are frequently deeply striated, and in some areas the clays are apparently genetically related to spreads of fluvioglacial outwash gravels. Statistical tests showed that there was no preferred orientation for the drift terraces, and he argued that when bedded this indicated redeposition, downslope, of original glacial drift by solifluction. Moreover, that the high relief of much of the area, the general character of the country rock, and the fact that cryonival processes are still active today, lends colour to the proposition that redeposition of glacial drift in the past would have occurred soon after deglaciation (cf. with A. Wood's view on Morfa bychan).

Both agree on the nature and origin of the widespread stratified screes typically developed on shales and thinly beded mudstones. Permafrost obtained locally when these were deposited.

In the Teifi valley, partly infilled by glacial drift, controversy is not so much concerned with the origin of that material, than with its age. On the one hand Watson (1970) believes the drifts are Wolstonian (Riss), while on the other hand Potts (1968) and Bowen (1967) believe they are Devensian. The undissected nature of the drift, fresh constructional landforms, including kettle holes, lends a measure of support to the second view. O. T. Jones (1965) believed that they were deposited during the Last Glaciation.

Near Cwrtnewydd, on the right bank side of the Teifi, Watson (1971) described the remains of pingos. A further concentration occur near Talgarreg (Watson pers. comm.). Collectively they demonstrate the former presence of permafrost, probably during the Late-glacial.

III. PLEISTOCENE SYNTHESIS

In the Pleistocene of Wales and the borderlands three base lines, more or less accurately defined, permit subdivision and correlation of the succession by working backwards in time in the Lyellian manner. These are, first, the organic deposits of Pollen Zones I, II and III of the Devensian Late-glacial; second, the organic deposits of the Middle Devensian or 'Upton Warren interstadial complex'; and third, the coastal raised beaches. Of these the first two have been calibrated by radiocarbon dating.

Pollen zonation of organic deposits enabled Godwin (1955) and Seddon (1962) to show that during the time equivalent of Pollen Zone III cirque glaciers and snow patches occupied the Snowdonia cirques. Similar conditions were inferred for the Brecon Beacons (Lewis 1970c), while the same probably holds for other upland cirques also.

Devensian Late-glacial deposits attain greatest significance when infilling kettle holes, for as such they demonstrate a minimum age for the underlying glacial drifts. It is not unreasonable to assume that consequent upon the formation of the kettle hole sedimentation followed without major time break. Pollen analytical investigation of such sediments not only affords a record of climatic amelioration after deglaciation, but also of a means of dating.

In many kettle holes basal solifluction clays underlie the Late Glacial organic deposits, but a radiocarbon date on mosses found in those at Glanllynau (14,468 years BP) shows that the interval of time after deglaciation and prior to vegetational colonisation (12,050 years BP at Glanllynau) was not excessive. Radiocarbon and pollen zonational minimum ages for Upper Devensian Glaciation are shown on figure 99.

The sedimentary infill of the vast majority of kettle holes remain unexamined, but their preservation alone demonstrates the youthfulness of the glacial drift on which they lie for such landforms older than the Devensian are unknown in Britain (West 1967). This is not surprising in view of the great interval of erosion manifest between 'older' and 'newer drift' glaciations in the Midlands and East Anglia, to say nothing of the maritime west such as in Pembrokeshire where 'older drift' has all but been removed. Kettle holes as significant landforms occur at Mathry, north Pembrokeshire, on the Llanon moraine, Cardiganshire, and in the Teifi valley outside the Tregaron moraine.

Fluviatile gravels of the Middle, and probably Lower Devensian also, at Four Ashes were critical evidence for the demonstration that Upper Devensian Irish Sea Ice reached the Wolverhampton Line (Shotton 1967b). The youngest radiocarbon date from that exposure of 30,655 years BP showed that the overlying Wolverhampton till was Upper Devensian. Other Lower and Middle Devensian deposits underlying glacial drift of the Main Irish Sea Glaciation include the Chelford Sands Formation, with its included organic horizon dated as 60,000 years BP. In a similar, though indirect way, the radiocarbon dates for the basal aggradational deposits of the Main Terrace at Upton Warren (41,900 years BP) and Avon No. 2 Terrace at Fladbury (38,000 years BP) and Brandon (32,270 years BP) also provide maximum radiometric ages for Upper Devensian glaciation.

Glaciation of the borderlands by Welsh Ice, coeval with that of the Main Itish Sea Glaciation north of the Wolverhampton Line, confirms the presence of a considerable ice-cap farther west, as well as enabling the periglacial inheritance of upland central Wales to be viewed in correct time perspective (Potts 1971). Greenly's (1919) expectation that the thickest part of that ice-cap overlay Trawsfynydd was fully confirmed by Foster's (1968) work in the Harlech Dome, B. Rowlands' (1970) work in Arenig, M. Brown's (1971) work in Montgomeryshire, and that of Potts (1968) in mid-Wales.

The foregoing, in support for a concept of extensive Upper Devensian Glaciation, is fully consistent with the distribution of 'newer' as contrasted with 'older' drifts. The former, unlike the latter, mantle interfluve, hill-slope and valley floor alike, and are relatively little dissected. These several lines of mutually confirming data, together with a further datum, that of the coastal raised beaches of Ipswichian age (below), combine to indicate the extent of Upper Devensian glaciation (fig. 2).

It is somewhat more extensive, though still approximating to, the reconstruction of Charlesworth (1929), but differs in several fundamental respects to that of Mitchell (1960); these are discussed below. John's successive reconstructions to the Isles of Scilly (John 1968), and to Pembrokeshire south of Milford Haven (John 1971) are manifestly erroneous: the former in failure to appreciate the homotaxial nature of Quaternary sequences, and the latter because the Pleistocene succession on either side of his line is identical (fig. 99). The argument that absence of rebound shorelines in west and south Wales mitigates against the existence of a Devensian ice cover in those parts is invalid (Synge (1970a), for the Bardsey raised beach lies at 6 m O.D. (Matley 1913); there are now detectable signs of rebound in south Wales (Bowen 1970a. Anderson 1968); moreover in Cardigan Bay such features were probably fashioned on drift cliffs before they receded to their present position.

Not only is the extensive nature of Upper Devensian Glaciation now firmly demonstrated, but so also is a subsequent readvance in north-west Wales and in the north-east borderlands. In both areas the ground within the readvance limit is diversified by constructional landforms subjectively assessed as marginally fresher in form than those of the Upper Devensian maximum. Farther south, it is possible that a reduced central Wales ice-cap (Potts 1968) was time-equivalent to readvance in the areas mentioned.

A limited number of radiocarbon dates allow something to be said for Upper Devensian Glaciation and Readvance in general terms. Minimum dates have already been discussed. Maximum radiocarbon dates are those from Four Ashes, and the several determinations from shells, shell fragments and other organic material carried by the ice and incorporated into its drifts. The latter, however, cannot be accepted uncritically. Nor can it be argued that because they fall into the required time-range for demonstrating Upper Devensian Glaciation, i.e. between 37,960 and 28,000 years BP the suggestion being that contamination by modern radiocarbon in approximately equal percentage on infinitely aged material to produce dates is highly unlikely, that they are valid. For this is precisely the range where the data of Shotton (1967a) shows a whole range of contamination values by modern radiocarbon would produce the observed results. The fact remains, therefore, that the only fully reliable dates are those from Four Ashes, and possibly from Tremeirchion.

The Tremeirchion date (B. Rowlands 1971) of 18,000 years BP may provide a solitary maximum date for the Devensian Readvance. Or is it a closer maximum date for the Upper Devensian Maximum? Comparison with Dimlington, Yorks, suggests that it may be the latter. But a recent date from the Isle of Man by G. R. Coope of 18,900 years BP (Shotton et al., 1971) shows that the island was ice free by that date. As this was obtained from mosses capable of subaquatic photosynthesis a 'hard water' effect must be allowed for. This means that the correct date is certainly younger and is by no means inconsistent with the maximal date obtained from Tremeirchion although at first sight the two appear incompatible.

It would appear, therefore, that the Upper Devensian Maximum occurred sometime after 30,655 years BP, probably after about 23,000 years BP in comparison with northern Europe (Penny 1964), to be followed after deglaciation by readvance to Llyn, Sarn Badrig and the Ellesmere morainic complex and districts between Chirk and Shrewsbury sometime after 18,000 years BP. The oldest minimum age determination for the readvance is 14,468 years BP from Llyn. Many more dates are required to test the above hypothesis, but what is interesting is that western Britain is, at last, falling into step with events in northern Europe.

The alternative view of restricted Devensian glaciation has occupied such a prominent position in the last decade that some discussion of it is merited. Mitchell (1960), Synge (1963) and Watson (1970) would restrict Devensian ice so as to leave Cardigan Bay and the adjacent Welsh upland ice free, except perhaps for the possibility of a lobe of Irish Sea Ice crossing the bay to impinge upon the Teifi estuary (Wirtz 1953). This general reconstruction, criticised by Charlesworth (1963), is held for four principal reasons.

First, the glacial drifts of the area are deemed to be 'older' as they are 'weathered' and subject to considerable cryoturbation as involutions, ice wedges and vertically inclined stones (Watson 1965, Watson & Watson 1971). Degree of weathering, however, is a highly arbitrary criterion, as work in the Cheshire-Shropshire lowland has shown (Boulton & Worsley 1965 cf. with Shotton 1967b). Moreover instances of deposits demonstrably younger than Pollen Zone III and weathered to a corresponding if not greater extent are known from the Isle of Man, an acceptable environmental analogue for west Wales. Cryoturbation cannot be used to establish age distinctions for Pleistocene deposits. The climatic significance of large scale involutions is unknown (Pewe 1968), but till, younger than 30,655 years BP, around Wolverhampton displays large scale involutions, ice-wedge pseudomorphs as well as bearing on its surface large scale ice-wedge polygonal patterned ground (A. V.

Morgan 1970). Farther north, in Cheshire, several generations of ice-wedges occur in Devensian glacial deposits (Worsely 1966b). Indeed this argument is no longer valid in Ireland as periglacial structures have now been extensively demonstrated in the Sperrin Mountains (Colhoun 1971).

Second, the ice-free hypothesis has been championed on account of the 'periglacial landscape' of mid-Wales (Watson 1968): Such a proposition in substitution for stratigraphic demonstration lacks validity. Moreover the perceived 'periglacial' character of the landscape has not won wide support (Potts 1971). While an undoubted periglacial legacy occurs, locally spectacular, it is to be seen as a final, comparatively swift ornamentation superimposed on a fundamentally glaciated landscape (Wood 1959. Potts 1968. 1971. Thomas 1970), much of which probably occurred during the readvance farther north. Since the western part of Wales was ice free some time before 14,468 years ago the area was subject to a periglacial regime for almost 5,000 years, if not more (Isle of Man C14 date), prior to the opening of the Flandrian. It is not without significance that in present day Scandinavia Rapp (1960) has shown that periglacial processes operate exceedingly swiftly and with great intensity and efficacy.

Third, the Bryncir-Clynnog moraine is deemed to represent the outer limit of Devensian glaciation (Mitchell 1960. Synge 1964), thus allowing glacial deposits farther south to be regarded as 'older', just as the South Irish End Moraine is so defined. While constructional landforms of glacial drift are locally spectacular and ubiquitously prominent within such limits they are by no means absent farther south. In any event morphological arguments are subordinate to stratigraphic demonstration which now upholds the view of extensive glaciation at the time in question.

Fourth, the fundamental basis of the case lies in correlation of the Pleistocene succession of south-east Ireland with that of Wales. This is considered later.

Coastal raised beaches, on whose age much controversy has occurred, provide a final datum line in north-west Wales, Pembrokeshire, Carmarthen Bay and Gower, as well as permitting correlation still farther afield. While it is likely that the coastline has been re-occupied by successive interglacial sea levels (Bowen 1969b), in theory beaches of different ages will be differentiated by their overlying successions. In practice, however, only deposits of one cold stage overlie them in Anglesey and Llyn, Pembrokeshire, Marros and Gower.

In north Pembrokeshire and east Gower glacial drift of Upper Devensian age is separated from raised beach exposures by a limited thickness of head and other slope deposits. By analogy with the Cheshire-Shropshire lowland, where periglacial deposits of Lower and Middle Devensian age underlie Upper Devensian glacial drift, the above mentioned deposits are also considered to have accumulated during the earlier part of the Devensian cold stage (John 1970, Bowen 1970a). On lithostratigraphic grounds it follows that the underlying raised beach is most logically placed in the Ipswichian interglacial.

Outside the area of Upper Devensian glaciation (fig. 100) the coastal raised beaches are overlain by slope deposits of one unitary cold stage which do not lend themselves to time-stratigraphic subdivision. The so called 'lower head' of Wales and South West England consists of colluvium and cliff talus representing the onset of colder conditions during which thermoclastic scree formed. Any weathered material in the coastal successions now lies in secondary position (Bowen 1969a and b).

The temperate mammalian fauna of Minchin Hole, Gower, and its correlation, confirms the Ipswichian age of the *Patella* and *Neritoides* beaches, as does the field relations and facies variation of interglacial relict soils in Glamorgan. Moreover it is to be noted that a raised beach of the Last Interglacial is found at approximately the same height throughout the world. At Selsey and Wretton such beaches have been dated by pollen analysis (Sparkes & West 1970. West & Sparks 1960).

Mitchell & Orme (1967) suggested that two raised beaches of Hoxnian and Ipswichian age respectively, separated by head, occurred in the Isles of Scilly. As the principal beach of South West England is regarded by them as Hoxnian, its separation, by 'giant' erratics, from the shore-platform on which it lies shows that the platform is Cromerian (Mitchell 1960) or earlier (Stephens 1966).

In Gower, both beaches at Minchin Hole are Ipswichian, the intervening cave breccia being interglacial, as its fauna shows, and not periglacial, corresponding to the Wolstonian cold stage. Further evidence of an Ipswichian twin sea level occurs at West Angle Bay, Ragwen Point and Horton (Bowen 1970a), as well as from the Vale of Gordano (ap Simon & Donovan 1956) (Table 13), while it is known in South East England (Hollin 1965), on the north-east Atlantic littoral from the Channel Islands to North Wales (Bowen 1972b), as well as in the eastern U.S.A. (Hollin 1970).

At many localities in South Wales and Southern Ireland upper angular layers of the beach have been truncated in a way reminiscent of marine planation. As such it probably indicates the second Ipswichian level, for in Jersey the most ubiquitous raised beach is separated from the shore platform by talus and water-deposited loess (Elhai 1963). Apart from Minchin Hole and West Angle Bay demonstration that the raised beaches and their subjacent platforms are not contemporary is found at Red Wharf Bay (fig. 99), and at Whitesands Bay, St. David's where a 'giant' erratic is sealed on the platform by raised beach (John 1971). But 'giant' erratics (Mitchell 1960, Stephens 1966) cannot be critical in any scheme of chronology until their mode of emplacement on the platform is known.

It was long thought that the raised beaches antedated the 'older drift' glaciation, but is it now known that this relationship is reversed (Bowen 1970a). Particularly critical for this realisation was recognition of the 'older drift', in sequences overlying the raised beaches, as a re-cycled glacial deposit through solifluctional and alluvial processes. Thus was the long standing problem of the 'older drift of South Wales', and its correlation with the Little Eastern of East Anglia or Warthe of Germany (Zeuner 1959) resolved, and the difficulties inherent through the Last Interglacial beach appearing to antedate two separate glaciations in south Wales finally removed (Bowen 19710a).

In papers read to Section C at the Dundee British Association, 1968, and the Ussher Society 1969, Bowen argued that this relationship also held in south-west

PLATE 5

The succession at Glanllynnau or Afon Wen (SH 456373), near Criccieth, Caernarfonshire, as exposed in 1967. It consists of: (1) the basal Criccieth Till (see text) to the level of the viewer's head. (2) Weathered surface of the Criccieth Till, with the finely divided particles infilling the ice-wedge pseudomorph (cast) penetrating the till immediately left of the viewer. (3) Laminated stoneless clays, silts and sands. (4) Proglacial outwash gravels. (5) The Llanystumwyd Till (see text), the upper surface of which is disturbed by cryoturbation. (6) Recent soil. Note figure on right for scale.

PLATE 6

The succession at the entrance to Minchin Hole Cave, Gower.

    1: *Patella* beach, 0.51m. thick, lying at 9.4m. O.D.

    2: Cave breccia (temperate fauna), 2m.

    3: Cave earth, 114mm.

    4: Coarse, loose, (*Neritoides*) sand, 0.46m. thick, lying at 12.2m. O.D.

    5: Sandrock (aeolianite), 2.31m.

    6: Head, 1.22m.

    The large boulder left of units 1 and 2 enables the present exposure to be related to that in George, T.N. (1932).

PLATE 7

Drift cliff at Marros Sands, at the head of Carmarthen Bay showing: a basal blue shale head (below hammer), overlain by a brown sandstone head, and colluvium and recent soil. Ice-crack pseudomorphs penetrate the brown head. Length of hammer is 30cms.

PLATE 8

Ice-crack polygons exposed on the foreshore at Marros Sands, Carmarthen Bay. These may be seen in section in the adjacent drift cliff (plate 3). Length of hammer is 30cms.

STAGE	SOUTH WALES*		SOUTH WEST ENGLAND
DEVENSIAN	HEAD	Cover sand. Coastal head deposits. Redeposited 'older drift'. Colluvial silts	HEAD	Coastal head deposits. Croyde Bay, Fremington Quay, Isles of Scilly redeposited glacial drift. Croyde, Brean Down Colluvial silts.
IPSWICHIAN	Neritoides beach West Angle Marine Alluvium. ——— ——— ——— Minchin Hole breccia. West Angle peat. ——— ——— ——— Patella Beach		Porth Seal (Scilly) Beach. Vale of Gordano upper beach and marine clay. ——— ——— ——— Vale of Gordano Red Clay. Scilly 'lower head' (in part). ——— ——— ——— Chad Girt (Scilly) Beach. Vale of Gordano lower beach.
PRE-IPSWICHIAN	'Older drift' of South Wales.		Fremington till. Isles of Scilly outwash gravels. Kenn gravels. 'Giant' erratics.

* i.e. the extra-glacial area with respect to Devensian Glaciation.

Table 13. A correlation between the Pleistocene succession of South Wales and that of South West England with particular reference to interglacial marine and related sediments.

England, where glacial material lies in secondary position above the beach at Croyde Bay, Trebetherick and the Isles of Scilly (Table 13) (Bowen 1969a, 1969b). At no locality outside the limit of Upper Devensian Glaciation does glacial drift in primary position overlie the raised beaches.

In the lower Severn region the 'older drifts' are severely dissected and bear very little relationship to present topography. By analogy, the 'older drifts' of Wales also occur beyond the outer limit of Upper Devensian Glaciation, a conclusion supported by comparison with the North East where only dubious fragments of pre-Chelfordian glacial material remain. From the remaining outcrops, however, it is clear that Wales was completely glaciated during an 'older drift' glaciation. By allowing for the measure of controversy over the age of the Upper boulder clay in the Birmingham district (cf. Pickering 1957 and Poole 1968), all that can reasonably be said is that the 'older drift' glaciation of Wales antedated the raised beaches and hence the Ipswichian. (Note, however, the adoption of Pickering's view in Table 11).

In comparison with the Midlands, Welsh borderlands and South Wales, the 'older drifts' of Ireland are more extensive inland, and in coastal situations form drift cliffs of considerable extent, a situation which does not obtain in Wales. Workers from Ireland have long regarded the Welsh raised beaches as Hoxnian, the coastal glacial deposits of Cardigan Bay as Wolstonian, and the Bryncir-Clynnog moraine as the Devensian maximum. The cornerstones of the Irish succession on which such a correlation is based are the Gortian Interglacial deposits (Watts 1967) correlated with the Hoxnian Interglacial of England, and the identification of the South Irish End Moraine as the outer limit of Devensian glaciation (Mitchell 1960).

Now that the Ardcavan deposit, Co. Wexford (Mitchell 1960) is no longer regarded as interglacial (Watts 1970), the complete absence of Ipswichian deposits in Ireland is puzzling. The Shortalstown marine clay (Mitchell & Colhoun 1971) is the only interglacial deposit with serious claim to an Ipswichian age. Gortian deposits, therefore, represent the only unequivocal and discrete interglacial known in Ireland: these are crucial for two reasons. First, because pollen bearing slits within dune slacks overlying raised beaches have been correlated with this interglacial at Newtown, Co. Waterford (Watts 1959) and between Fenit and Spa, Co. Kerry (Mitchell 1970), thus dating the beaches as Gortian (Hoxnian). And second, because Gortian deposits, notably at Kilbeg (Watts 1959) have been used to date overlying glacial drifts as Eastern General (Wolstonian).

At Newtown, however, the pollen spectra simply record the end of an interglacial cycle; the same is true between Fenit and Spa, but with the addition of admittedly derived temperate pollen genera (Mitchell 1970), in assemblage strongly reminiscent of the later part of the Gortian cycle. Pollen dating the raised beaches is thus not unequivocal.

The Kilbeg site has been used to provide a maximum age, i.e. Gortian (Hoxnian) for overlying glacial drift. A minimum age was arrived at on the basis that the drifts lie outside the South Irish End Moraine, and hence must be Eastern General (Wolstonian) (Watts 1959). Such a combination of stratigraphic and geomorphic parameters is open to serious doubt.

It would appear that the present unsatisfactory situation, outlined above, has achieved time-stratigraphic significance far in excess of its intrinsic worth. In alternative interpretation three views in particular do not seem to have been given equal status.

(i) Could it be that the Irish Gortian deposits are to be correlated with the Ipswichian Interglacial and not the Hoxnian? This possibility was considered by Watts (1970) but rejected on the grounds that the botanical evidence points to the equivalence of the Gortian and Hoxnian cycles, a conclusion supported by Turner (1970) who, however, pointed out that the unifying feature of Gortian, Hoxnian and Holstein (continental) pollen diagrams is the strong development of *Abies* during the Late Temperate Zone, and that many continental workers had misgivings about correlation between Hoxnian and Holsteinian. Despite the confidence of the palaeobotanist it remains nevertheless a correlation based on long-distance tele-correlation for the regional variation of both Hoxnian and Ipswichian vegetational successions has yet to be determined over most of England, and over all of Wales (West 1968).

(ii) If, on the other hand, the Gortian deposits are to be correlated with the Hoxnian Stage, then in Ireland they only demonstrate a maximum age for overlying glacial drifts. The only organic deposits overlying the post-Gortian glacial deposits are Devensian. Indeed in some areas, for example, Baggotstown, Kildromin and Burren, Gortian Interglacial deposits are covered only by a thin layer of till of the Last Glaciation (Devensian). All that distinguishes such sites from others, as at Kilbeg, is that the latter is covered by till lying outside the outer limit of the Last Glaciation as placed along the 'South Irish End Moraine'. Stratigraphically, therefore, the existence of a post-Gortian (Hoxnian), pre-Ipswichian Glaciation cannot be established in Ireland. Nor can it be until Ipswichian deposits are found. Geomorphic arguments, involving the freshness of morainic features, are notoriously unreliable, as the example of the Cheshire-Shropshire lowlands illustrate. It follows that the glacial drifts outside the 'South Irish End-Moraine' could well be Devensian in age. Substitution of Devensian drift for Eastern General (Wolstonian) drift overlying raised beaches in coastal sections could, therefore, bring the Irish sequence into harmony with that of Wales and England. It is a well known fact that the drift distribution of Ireland inland from the coast is far from being satisfactorily known (Synge 1970b).

(iii) The third alternative is a compromise view. On homotaxial grounds the successions on either side of the Irish-Celtic Sea may not be comparable. In explanation for such a possibility the Ipswichian shoreline may have run parallel to the contemporary one but at some distance seawards of it. Subsequently the Flandrian transgression removed all trace of it in Ireland and is even now only partly successful in exhuming a Hoxnian shore-zone. The low tidal range of Irish coasts bordering on St. George's Channel may, therefore, have uniformitarian implication.

When reviewed in terms of the stratigraphic succession of Wales, it is the second alternative (b), which is deemed most likely on economy of hypothesis at the present time. Moreover, it is noted that one of the most pressing problems of the British Quaternary today is the need to demonstrate unequivocally the existence of a post-Hoxnian, pre-Ipswichian glaciation (Gipping/Wolstonian/EasternGeneral Glaciation) during that cold stage defined on palaeobotanical grounds.

Particularly damaging to the Irish view as extended to Wales is that such a scheme would require drifts of two cold stages, to say nothing of the two temperate stages involved, to be present in the Welsh succession. Any such proposition is contrary to all the known evidence in Wales (Fig. 99).

Work in the last decade has shown that the Irish chronology is inapplicable in Wales. Pre-Ipswichian events are recorded by 'older' glacial drifts, with indication of multiple glaciation at Nechells. Only one interglacial may be recognised with any degree of confidence, the Ipswichian, during which the coastal raised beaches, locally of composite nature, were deposited. The Devensian cold stage involved extensive late glaciation followed by readvance. Prior to glaciation periglacial deposits accumulated throughout Wales and the borders.

The implications for Ireland are considerable. With the extent of Devensian ice being determined from the distribution of the raised beach antedating glacial drift, it is clear that Ireland was almost completely glaciated, an interesting reflection of Wright's (1914) early views. Such proposition cannot be refuted on unequivocal stratigraphic grounds. A corollary to this is that the South Irish End Moraine is to be regarded as a readvance feature, thus being correlated with the Sarn Badrig and Ellesmere/Welsh Readvance positions in Wales and the borderlands.

The Pleistocene succession in Wales (Tables 10 to 12) is thus based primarily on lithostratigraphy, though not without biostratigraphic support, and can be correlated most readily with the Midlands, where the Middle Pleistocene and part of the Upper Pleistocene (West 1967) has been subdivided on biostratigraphic evidence, and the Devensian cold stage calibrated by radiocarbon dating as far back as that method allows. 'Irish Pleistocene studies are still in the stage of working hypothesis' (Charlesworth 1963), but so too are those of Wales, the essential difference being ready correlation with, and proximity to, the Midlands of England for the latter.

IV.   THE FLANDRIAN

Subdivision of both the Devensian Late-glacial and the Flandrian (Holocene/ Post-Glacial) is based on pollen zonation. Late Glacial Pollen Zones I and III contain mostly non arboreal and an absence of tree genera, e.g. at Rhosgoch, Radnorshire (Bartley 1960). During the amelioration of Zone II (Alleröd) the proportion of arboreal pollen rose. Continuity with the Pleistocene is obtained from sediments with evidence of the vegetational succession across the Zone III–IV boundary, which is the base of the Flandrian.

Non-marine deposits

The Flandrian vegetational succession in Wales and England records the development of temperate deciduous forest and its subsequent clearance by man (Godwin 1960). After Zone VIIb (Table 13) regional variation is more pronounced, as a result of man's activities (Turner 1965).

Raised bog and blanket bogs are common in Wales. The stratigraphy of the Tregaron raised bog was studied by Godwin & Mitchell (1938). In common with similar bogs elsewhere it showed a development of sphagnum peat after early Flandrian lake sediments and fen peats. Several recurrence surfaces have been demonstrated and radiocarbon dated, including the Grenz Horizont (Turner 1964). Turner's studies at Tregaron and the Whixall and Fenns Mosses of Shropshire enabled her to demonstrate the agricultural activities not only of Bronze Age people at Tregaron, but clearance of the Landnam type until at least the opening of the Iron Age. At Tregaron this was followed by a long period of pastoral farming when Celtic tribes occupied the area. This ended at 1182 AD (C14) when predominantly arable farming is indicated by the cereal pollens. It is known that Cistercian monks settled at nearby Strata Florida Abbey in 1164 AD, and they grew barley and oats on their granges. At Borth Bog and Plynlimmon P. Moore (1968) was able to demonstrate, by pollen analytical investigations, historical events up to the activities of the Forestry Commission in the 1920's. Welsh blanket bogs started to form during Pollen Zone VIIa, and are to be attributed to the rise in precipitation at that time.

Except locally (e.g. Crampton 1965) most alluvial deposits await chronologic differentiation; studies of alluvial stratigraphy have hardly begun in this country.

This is also true of colluvial deposits, although colluvium at Nash Point, Glamorgan has been dated by means of its included non-marine mollusca (Bowen 1970a) as post-Flandrian climatic optimum. The same probably holds for the hill-washes of the South Wales Coalfield (Woodland & Evans 1964).

Tufa deposits occur at several localities in Flintshire, notably at Caerwys (Wedd & King 1924) where it contains a microlithic industry. Skeleton remains at that site, thought at one time to be Mesolithic, have been radiocarbon dated as 2100 years BP (Barker *et al.*, 1971) thus demonstrating its intrusive origin.

In west Wales pollen analytical analysis of samples obtained from soil profiles has enabled the general pattern of pedogenesis to be related to the Flandrian pollen zones (Smith & Taylor 1969) as well as the activity of Bronze Age man and his successors.

Marine and related sediments

Abundant evidence for Flandrian sea level change occurs along the Welsh coastline: submerged forest, submerged peat beds, raised beaches, marine alluvium, and alternations of fresh-water and brackish water deposits. Many of these were investigated before, or during the earliest years of pollen analysis, and certainly long before radiocarbon dating. It is, therefore, a tribute to the earlier workers, notably Strahan (1896) at the Barry docks excavation, together with Clement Reid, that their descriptions have been successfully incorporated into later investigations with more powerful tools.

Godwin (1940) showed that the Flandrian transgression occurred in Swansea Bay, affecting levels above—16 m OD during the later part of the Boreal Period (Zone VI). This was interrupted by stages of halt or minor regression indicated by peat beds. He showed that throughout south Wales that peat beds at the same level are approximately the same age, and was able to integrate the previous work of Strahan (1896), Hyde (1936), George (1936), von Post (1933) and George & Griffiths (1938) with the results of the pollen analytical investigations.

Subsequent pollen analysis and radiocarbon dating of submerged peat beds at Margam and Port Talbot have confirmed Godwin's (1940) results. At Port Talbot the main eustatic rise in sea level did not affect levels of—18 m OD until well into the Flandrian. On an irregular surface of glacial drift fresh-water organic muds and peats formed from Pollen Zone II, C14: 11,980 and 11,260 years BP (Godwin & Willis 1961) to Pollen Zone IV (later Boreal, C14: 8,970 years BP when they were submerged by the rise in sea level (Table 14).

In Britain generally, the main eustatic rise in sea level ended about 5,500 years BP, and the terminal dates from Margam of 6184 years BP at —3 m, Ynys Las of 6026 years BP at —0·61 m, and Freshwater West of 5960 years BP near low tide, approximate to this. At Margam there followed a period free from marine transgression until at least 3402 years BP (Godwin & Willis 1964) when a minor transgression, probably indicated by Gibson's record of *Scrobicularia* clay at Port Talbot overlying a peat between —0·3 and 0·61 m, occurred (Strahan 1907). Similar marine beds at comparable heights have been reported from several sites. The maximum height of the marine limit in south Wales according to Bowen (1970a) is 4·5 m, and

Stage	C 14 BP	Culture	Zone	Period	Sea-level (in metres O.D.)	Representative Archaeological Sites
F L A N D R I A N		Medieval	VIII	SUB-ATLANTIC		Pen Dinas Mynydd Bychan
	2000	Rom-B.			post 2,660 s.l. rise to 4.3m, Llanwern	
		Iron Age				
	3000	Bronze	VIIb	SUB-BOREAL	s.l. at -2.3, Margam earliest blown sand, Merthyr Mawr.	Llyn Fawr Hoard Ynysgwenant Bromfield
	4000	Beaker				
		Neolithic				Bryn celli ddu Bryn yr hen bobl Pentre Ifan Tinkinswood
	5000					
	6000		VIIa	ATLANTIC	marine transgression ends Y. Las. P. Talbot. Freshwater W.	
	7000	Mesolithic			marine transgression	Red loam at Nana's Cave, Caldey Prestatyn Burry Holms
	8000		VI	BOREAL		
	9000		V		s.l. below -18m at P. Talbot.	
	10,000		IV	Pre-Boreal		
UPPER DEVENSIAN	10,800		III			Cresswellian sand at Nana's cave Hoyle's Mouth Cathole Cave
	12,000	Cresswellian	II	Allerod		
			I			
					s.l. -130m	Tremeirchion mammouth (18,000 BP) 'Red Lady' of Paviland (18,460 BP)
	26,000	UPPER PALAEOLITHIC Aurignacian				Aurignacian occupation of Paviland, Cae Gwyn, Ffynon Beuno.
MIDDLE DEVENSIAN	50,000					
LOWER DEVENSIAN		Mousterian				Mousterian occupation of Bontnewydd and Coygan caves
	70,000					
IPSWICHIAN & EARLIER		LR. PAL.			s.l. +15m. (Ipswichian)	Paviland Levallois flake Acheulian hand axes of Pen y lan, and Coygan (Hoxnian ? or later ?)

Table 14.   Late Quaternary Stages, cultures, pollen zones, sea level change and representative archaeological sites in Wales. Data for sea level change lies consistently above synthetic curves for world wide movements thus indicating a measure of isostatic recovery for Wales.

the average height of the Flandrian infill determined by Anderson (1968) in south Wales and up the Severn estuary to Upton, is 5·3 m. At Llanwern, a peat underlying marine clay which extends to 4·4 m, was radiocarbon dated as 2660 years BP (Godwin & Willis 1964). This suggests that the peat is the same age as the Iron Age peat described by von-Post at Backpill (Godwin 1940). It also shows that the Llanwern transgression compares with Godwin's (1960) Romano-British transgression in Somerset.

Unlike Wales no large-scale dock excavations were made in the north. This has resulted in a comparative dearth of knowledge of the buried peat beds. The only detailed investigations have been those of McMillan in Anglesey (1949) and Bibby (1940) near Rhyl. A solitary C14 date from Cheshire shows that at Moreton a marine transgression took place, burying peat beds, after 3695 years By (Godwin & Willis 1964).

The 25 foot (c. 7·5 m) or Neolithic beach of northern Britain formed when the decelerating rise of sea level was matched by isostatic uplift, and was subsequently uplifted when the sea level rise ceased. In Scotland the Main Post-Glacial beach lies at about 14 m near Fort William, in Lancashire at 5·2 m, and at much the same height in north Wales. It forms the Bryn Carrog coastline of B. Rowlands (1955) in the Vale of Clwyd, where cliffs were cut into boulder clay, while in Anglesey several raised beach exposures occur (Greenly 1919, Hopley 1963), sometimes re-occupying the 'pre-glacial' shore-platform (Whittow 1965). They occur up to 5·5 m and contain a molluscan fauna including *Patella vulgata* (Linné) and *Cardium edule* (Linné). Fewer instances occur in Llyn; at Porth Neigwl it occurs at 3·6 m (Whittow 1965). On Bardsey Island raised beach shingle occurs up to 6 m (Mateley 1913). It is entirely absent in Cardigan Bay, possibly through erosion of the drift terraces, but near Barmouth George (1933b) suggested that the pebble ridge west of Ynys-y-Brawd, may be a raised beach relic. Farther south the beach must pass below sea level, but George (1932) suggested that it might be represented by the Heather-slade beach in Gower at 0·15 m O.D. (Bowen 1970a). Considerable uncertainty attaches to this deposit: it has been placed in the Hoxnian, Ipswichian and Flandrian interglacials in turn, but may be simply a facies of the redeposited glacial drift of Gower.

Extensive sand dune areas are found at Newborough Warren, Anglesey, in Cardigan Bay, Freshwater West and Merthyr Mawr. At the last named Higgins (1933) investigations showed that sand was accumulating during the Bronze age (table 14), and had advanced inland by the early Iron Age. North (1929) has shown that during the 13th century further encroachment took place.

The investigation of recent Flandrian and contemporary marine sediments has only recently commenced. The Cardigan Bay area has been (Caston 1965, Moore 1968, Jones 1971), and is still being investigated by a team from Aberystwyth, while the Bristol Channel area is under investigation from teams at Bristol and Swansea.

V. ARCHAEOLOGY

Lower Palaeolithic implements are known from only three localities in Wales. A quartzite Acheilian hand-axe was picked up on an allotment at Pen-y-lan, Cardiff

(McBurney 1965), while two other similar ones were found at Coygan cave, Carmarthenshire (Houlder & Manning 1966). In the borderland, at Claverly, east Shropshire, it was possible to show that a similar axe came from Middle Devensian gravels (Harrison and Shotton 1970). The other find consists of a Levallois flake, found at Paviland cave (Grimes 1951), which McBurney (1965) suggested came from the Ipswichian raised beach. It is, however, difficult to place these finds in any chronological sequence (Table 14).

Mousterian flake tools of the Upper Palaeolithic have been recorded from Coygan cave, and Bontnewydd cave, Denbighshire. The latter is the northernmost record of penetration by this culture from the south-west. The sparse population of Wales at this time, the Lower Devensian (Table 14), was probably due to the lack of flint, from which implements were fashioned.

Paviland cave, Gower, ranks among the most important Aurignacian sites ever found in Britain (Sollas 1913). Unfortunately the succession was none too clear as a result of previous excavation, notably by Buckland (1823) who discovered the skeleton of the 'Red Lady', and Cresswellian implements are mixed with earlier types. But at Cae Gwyn and Ffynon Beuno, in the Cale of Vlwyd (Garrod 1926), the succession shows clearly that the Aurignacian implements antedate Irish Sea glacial drift.

Cresswellian implements have not been found in north Wales. But they have been found at several sites in the south: King Arthur's cave in the Wye valley, Hoyle's Mouth, Tenby, Prior Farm Cave, Monkton, Pembrokeshire, Nana's Cave, Caldy (Lacille & Grimes 1955) and Cathole Cave, Gower, where a bone sewing needle provided a link with the Magdalenian (McBurney 1959).

Two radiocarbon dates are available for the Palaeolithic. The skeleton of the 'Red Lady' of Paviland, the remains of a ceremonially buried youth, gave a date of 18,460 years BP (Oakley 1968). This accords with some of the younger implements recovered from the cave. The contemporary environment was discussed by Bowen (1970b).

Using specimens collected by Hicks nearly 90 years ago from the Tremeirchion caves, and which presumably came from the red cave-earth layer (Garrod 1926), a successful dates was obtained from a mammoth's carpal bone of 18,000 years BP (Rowlands 1971). It is presumed that the cave was occupied by *Homo sapiens* at this time.

The Mesolithic cultures of Wales are essentially survivals from the Palaeolithic, and are characterised by the appearance of microliths. Several sites have been explored: Paviland, Nana's Cave, Caldy, Burry Holms, Nab Head, Pencaer, Craig-y-llyn, Aberdaron, Pencilan Head, and Prestatyn. Many sites were submerged by the Flandrian transgression. At Freshwater West a microlithic industry found in submerged forest beds was radiocarbon dated to 5960 years BP (Godwin & Willis 1964).

Neolithic man's activities as stock raiser and cultivator are revealed in the pollen record. This period in Wales has been reviewed by Daniel (1963), Grimes (1965), and Savory (1963). Settlement was pronounced in the western areas of

Anglesey, Llyn and Pembrokeshire, while inland concentrations occurred in the Black Mountains, upland Carmarthenshire, Conway valley and upper Dee. Petrological identification of hand-axes (Shotton 1959), manufactured in several 'factories', such as Graig Lwyd, is still continuing. Of these, the most celebrated is the spotted dolerite from Preseli. Several varieties of cromlechau or megaliths occur. These are discussed by Grimes (1965), and include, in addition to those on Table 4, Sweyne's Houses, Rhossili, and Arthur's Stone, Reynoldston. The radiocarbon date of 5000 years BP from hazel nut shells found in the Coygan camp is the first Neolithic date for Wales. (Callow, W. J. & Hassal 1968).

Savory (1965) recently reviewed the Bronze Age in Wales, which includes the transitional beaker cultures. At Ynysgwent, Denbighshire, a C14 date of 3423 years BP (Shotton *et al.*, 1969) was obtained from a burial site containing beakers. At Bromfield, Shropshire, several C14 dates between 3940 and 2712 years BP were obtained for a Middle Bronze Age cemetery (Shotton *et al. ibid.*).

A tool of unexpected power was that of combining radiocarbon dating and pollen analysis. In this way Bronze Age man's forest clearing and agricultural activities have been dated at Llanllwch, Carmarthen to 3230 and 3178 years BP (Thomas 1966), at Whixall Moss to 2307 and 3238 years BP (Turner 1964), and at Tregaron to 2669 and 2879 years BP (Turner 1964). The numerous stone circles built during this period have been described by Grimes (1963).

Although Iron Age Wales lacks 'museum objects', almost 350 hill forts are known, a concentration unequalled in southern Britain except by Cornwall (Hogg 1965). Of these the most notable are Pen Dinas, Aberystwyth, and Craig-y-dinas; Mynydd Bychan, Glamorgan, is a fortified farm. Some of the earliest iron objects to be discovered were from the Llyn Fawr Hoard, dated to about the 6th century BC.

VI.   REFERENCES

AGGASSIZ, E. C. 1885, *Louis Agassiz*. London.

ALLEN, A. 1960. Seismic refraction investigations of the preglacial valley of the River Teifi near Cardigan, *Geol. Mag.*, **97,** 276–82.

ALLEN, A. E., and RUTTER, J. G. 1944 and 1947. A survey of the Gower caves with an account of recent excavations. Pt. 1 (1944), Pt. 2 (1947), *Proc. Swansea Sci. Fld. Nat. Soc.*, **2,** 221–46 and 263–90.

AL-SHAIK, Z. D. 1970. *Geophysical investigation in the northern part of Cardigan Bay*, Unpublished Ph.D. thesis, University of Wales.

American Commission on Stratigraphic Nomenclature, 1961. Code of Stratigraphic Nomenclature. *Bull. Am. Ass. Petrol. Geol.* **45,** 645–65.

ANDERSON, J. G. C. 1968. The concealed rock surface and overlying deposits of the Severn Valley and estuary from Upton to Neath. *Proc. S. Wales Inst. Engrs.* **83,** 27–47.

AP SIMON, A. M., and DONOVAN, D. T. 1956. Marine Pleistocene deposits of the Vale of Gordano. *Proc. Spel. Soc. Bristol Univ.* **7,** 130–36.

BALL, D. F. 1960. Relic-soil on Limestone in South Wales. *Nature Lond.* **187,** 4736, 497–98.

——. 1966. Late-glacial scree in Wales. *Biul. peryglac,* **15,** 151–63.

—— and GOODIER, R. 1968. Large sorted stone stripes in the Rhinog Mountains, North Wales. *Geogr. Annlr.*, **50, A1,** 54–9.

BARKER, H., BURLEIGH, R. and MEEKES, N. 1971. British Museum Natural Radiocarbon Measurements, VII. *Radiocarbon,* 13.

BARTLEY, D. D. 1960. Rhosgoch Common, Radnorshire: Stratigraphy and pollen analysis. *New Phytol.*, **59,** 238–62.

BECKINSALE, R. P., and RICHARDSON, L. 1964. Recent findings on the physical development of the lower Severn valley. *Geog. J.*, **130**, 87–105.

BIBBY, H. C. 1940. The submerged forests at Rhyl and Abergele, North Wales. Data for the study of post-glacial history III. *New Phytol.* **39**, 220–25.

BINNEY, E. W. 1848. Sketch of the drift deposits of Manchester and its neighbourhood. *Mem. Manchr. lit. phil. Soc. Ser.*, **2, 8**, 195–234.

BIRKS, H. J. B. 1965. Late-glacial deposits at Bagmere, Cheshire and Chat Moss, Lancashire. *New Phytol.*, **64**, 270–85.

BISHOP, W. W. 1958. The Pleistocene geology and geomorphology of three gaps in the Midland Jurassic escarpment. *Phil. Trans. R. Soc.*, **B, 241**, 255–305.

BLUNDELL, D. J., GRIFFITHS, D. H. and KING, R. F. 1969. Geophysical investigations of buried river valleys around Cardigan Bay. *Geol. J.*, **6**, 161–80.

BOWEN, D. Q. 1965. *Contributions to the geomorphology of central south Wales*, unpublished Ph.D. thesis, University of London.

——. 1967. On the supposed ice-dammed lakes of South Wales. *Trans. Cardiff. Nat. Soc.*, **93** (1964–1966), 4–17.

——. 1969a. A re-evaluation of the coastal Pleistocene succession in south-west Britain. *Res. des Com. VIII INQUA Congr.*, Paris 183.

——. 1969b. The Pleistocene history of the Bristol Channel. *Proc. Ussher Soc.*, **2(2)**, 86.

——. 1970a. South-east and central South Wales, in, Lewis, C. A. (Ed.). *The Glaciations of Wales and adjoining regions*, 197–227, London.

——. 1970b. The Palaeoenvironment of the 'Red Lady' of Paviland. *Antiquity*, XLIV, 134–36.

——. 1971a. The Quaternary succession of South Gower, in Basset, D. A. and M. G. (Eds.). *Geological Excursions in South Wales and the Forest of Dean*, 135–42.

——. 1971b. The Pleistocene succession and related landforms in north Pembrokeshire and south Cardiganshire, *ibid.* 260–66.

——. 1972a. Further comment on the 'Red Lady' of Paviland and related matters. *Antiquity*, XLVI.

——. 1972b. Lithostratigraphy supporting Hollin's test for Wilson's theory of ice ages in the north-east Atlantic littoral (in press).

——. 1972c. Remains of pingos in Carmarthenshire (in press).

—— and DOBSON, M. R. (in preparation).

—— and GREGORY, K. J. 1965. A glacial drainage system near Fishguard, Pembrokeshire. *Proc. Geol. Ass.*, **74**, 275–82.

BOULTON, G. S. *Journ. Geol. Soc. Lond.* (in press).

——. 1968. A Middle Würm Interstadial in South-West Wales. *Geol. Mag.*, **105**, 190–91.

—— and WORSLEY, P. 1965. Late Weichselian glaciation of the Cheshire-Shropshire Basin. *Nature, Lond.* **207**, 704–6.

BROWN, M. J. F. 1971. *Glacial geomorphology of Montgomeryshire and West Shropshire.* Unpublished Ph.D. thesis, University of London.

BUCKLAND, W. 1823. *Reliquiae Diluvianae.*

——. 1842. On the glacia-diluvial phaenomena in Snowdonia and the adjacent parts of North Wales. *Proc. Geol. Soc. Lond.*, **3**, 579–84.

BURNHAM, C. P. 1964. The soils of Herefordshire. *Trans. Woolhope Nat. Fld. Club*, **38**, 27–35.

CALLOW, W. J., and HESSALL, G. I. 1968. Nat. Physical Lab. Natural Radiocarbon Measurements. *Radiocarbon*, 10.

CASTON, V. N. D. 1966. *A study of the recent sediments and sedimentation in Tremadoc Bay, North Wales*, unpublished Ph.D. thesis, Univ. of Wales.

CHARLESWORTH, J. K. 1929. The South Wales end-moraine. *Quart. J. geol. Soc. Lond.* **85**, 335–58.

——. 1963. Some observations on the Irish Pleistocene. *Proc. R. Ir. Acad.*, **B, 62**, 295–322.

COLHOUN, E. A. 1971. Late Weichselian periglacial phenomena of the Sperrin Mountains, Northern Ireland. *Proc. R. Irish Acad.* **B, 71**, 53–71.

—— and MITCHELL, G. F. 1971. Interglacial Marine Formation and Late glacial Freshwater Formation in Shortalstown Townland, Co. Wexford. *Proc. R. Irish Acad.*, **B, 71,** 211–45.

COOPE, G. R. 1959. A Late Pleistocene insect fauna from Chelford, Cheshire. *Proc. R. Soc.,* **B, 151,** 70–86.

——. 1962. A Pleistocene Coleopterous fauna with arctic affinities from Fladbury, Worcestershire. *Quart. J. geol. Soc. Lond.*, **118,** 103–23.

——, SHOTTON, F. W., and STRACHAN, I. 1961. A late Pleistocene fauna and flora from Upton Warren, Worcestershire. *Phil. Trans. R. Soc.*, **B, 244,** 379–417.

—— and SANDS, C. H. S. 1966. Insect faunas of the last glaciation from the Tame Valley, Warwichshire. *Proc. R. Soc.*, **B, 165,** 380–412.

——. 1970. Climatic interpretations of Late Weichselian Coleoptera from the British Isles. *Rev. Geographie Phys. et Geol. Dyn.*, **12,** 149–155.

——, MORGAN, ANNE and OSBORNE, P. J. 1971. Fossil Coleoptera as Indicators of Climatic Fluctuations during the Last Glaciation in Britain. *Palaeogeog. Palaeoclim. Palaeoecol.*, **10,** 87–101.

COSTER, H. P., and GERRARD, J. A. F. 1947. A seismic investigation of the history of the River Rheidol in Cardiganshire. *Geol. Mag.*, **84,** 360–68.

CRABTREE, K. 1966. *Late Quaternary deposits near Capel Curig, North Wales.* Unpublished Ph.D. thesis University of Bristol.

CRAMPTON, C. B. 1966. Certain effects of glacial events in the Vale of Glamorgan, South Wales. *J. Glaciol.*, **6,** 261–6.

CROLL, J. 1870. The boulder clay of Caithness a product of land ice. *Geol. Mag.*, **7,** 200–14.

CROSS, P. 1966. *The glacial geology of the Wigmore and Presteigne Basins and some adjacent areas.* Unpublished M.Sc. thesis, University of London.

——. 1968. Some aspects of the glacial geomorphology of the Wigmore and Presteigne districts. *Trans. Woolhope Nat. Fld. Club*, **39,** 198–220.

DANIEL, G. 1963. The Personality of Wales, in, Foster, I. L. and Alcock, L. *Culture and Environment*, London, 7–24.

DANSGAARD, W., JOHNSEN, S. J. and MOLLER, J. 1969. One thousand centuries of climatic record from Camp Century on the Greenland Ice Sheet. *Science*, **166,** 377–81.

DARWIN, C. 1842. Notes on the effects by the ancient glaciers of Caernarvonshire etc. *Phil. Mag.*, **21,** 180–88.

DAVID, J. W. E. 1883. On the evidence of glacial action in south Brecknockshire and east Glamorgan. *Quart. J. Geol. Soc. Lond.*, **39,** 39–54.

DAVIES, G. L. 1969. *The Earth in Decay*, London.

DIXON, E. E. L. 1921. *The Geology of the South Wales Coalfield. Pt. XIII. The Country around Pembroke and Tenby. Mem. geol. Surv. Gt. Britain.*

DUIGAN, S. L. 1956. Pollen-Analysis of the Nechells Interglacial deposits, Birmingham. *Quart. J. geol. Soc. Lond.*, **112,** 373–91.

DWERRYHOUSE, A. R., and MILLER, A. A. 1930. The glaciation of Clun Forest, Radnor Forest and some adjoining districts. *Quart. J. geol. Soc. Lond.*, **86,** 96–129.

ELHAI, H. 1963. *La Normandie Occidentale: Entre La Seine et le Golfe Normand-Breton.* Bordeaux.

EVANS, W. B., WILSON, A. A., TAYLOR, B. J. and PRICE, D. 1968. *Geology of the country around Macclesfield, Congleton, Crewe and Middlewich,* Memoir Geol. Surv. Gt .Britain.

FALCONER, H. 1868. *Palaeontological Memoirs and Notes of the late Hugh Falconer.* (C. Murchison).

FIELD, E. 1968. Unpublished dissertation, University of Durham (Botany Dept.).

FLINT, R. F. 1971. *Glacial and Quaternary Geology,* New York.

FOSTER, H. D. 1968. *The Glaciation of the Harlech Dome.* Unpublished Ph.D. thesis, University of London.

——. 1970a. Sarn Badrig, a submarine moraine in Cardigan Bay, North Wales. *Z. Geomorph.* 473–86.

——. 1970b. Establishing the age and geomorphological significance of sorted stripes in the Rhinog Mountains, North Wales. *Geogr. Annlr.* **52A,** 96–102.

FRANCIS, T. J. G. 1964. A seismic refraction survey of the preglacial Teifi valley near Cenarth. *Geol. Mag.*, **101**, 108–12.

GARROD, D. A. E. 1926. *The Upper Palaeolithic Age in Britain*, Oxford.

GEORGE, T. N. 1932. The Quaternary Beaches of Gower. *Proc. Geol. Ass.*, **43**, 291–324.

——. 1933a. The glacial deposits of Gower. *Geol. Mag.*, **70**, 208–32.

——. 1933b. The submerged forest near Barmouth. *Proc. Swansea Sci. Fld. Nat. Soc.*, **1**, 187–91.

——. 1936. The geology of the Swansea main drainage excavations. *Proc. Swansea Sci. Fld. Nat. Soc.*, **1**, 331–60.

—— and GRIFFITHS, J. C. 1938. The superficial deposits at the mouth of the River Tawe. *Proc. Swansea Sci. Fld. Nat. Soc.*, **2**, 23–48.

GODWIN, H. 1940. A Boreal transgression of the sea in Swansea Bay; data for the study of post-glacial history, VI. *New Phytol.*, **39**, 308–21.

——. 1955. Vegetation history at Cwm Idwal: a Welsh plant refuge. *Svensk. Bot. Tids.*, **49**, 35–43.

——. 1960. Radiocarbon dating and Quarternary history in Britain. *Proc. R. Soc.*, **13**, **153**, 287–320.

—— and MITCHELL, G. F. 1938. Stratigraphy and development of two raised bogs near Tregaron, Cardiganshire. *New Phytol.*, **37**, 425–54.

—— and WILLIS, E. H. 1961. Cambridge University Natural Radiocarbon Measurements III, *Radiocarbon*, **3**, 60.

—— and WILLIS, E. H. 1964. Cambridge University Natural Radiocarbon Measurements VI. *Radiocarbon*, **6**, 123–25.

GREGORY, K. J., and BOWEN, D. Q. 1966. Fluvioglacial deposits between Newport, Pembs., and Cardigan, in, Price, R. J. (Ed.) Deglaciation. *Occ. Pubn. Br. Geomorph. Res. Gp.*, **2**, 25–28.

GRIFFITHS, J. C. 1940. *The glacial deposits west of the Taff*. Unpublished Ph.D. thesis, University of London.

GRIMES, W. F. 1951. *The Prehistory of Wales*, Cardiff.

——. 1963. The stone circles and related monuments of Wales, in Foster, I. Ll. and Alcock, L. (Eds.) *Culture and Environment*, London. 93–152.

——. 1965. Neolithic Wales, in Foster, I. Ll. and Daniel G. (Eds.) *Prehistoric and Early Wales*, London, 35–70.

GREENLY, E. 1919. The Geology of Anglesey. *Mem. Geol. Surv. Gt. Britain*.

GRINDLEY, H. E. 1954. The Wye Glacier, in, *Herefordshire*, Woolhope Club Centenary Vol. Chapter 3.

GREIG, D. C., WRIGHT, J. E., HAINS, B. A. and MITCHELL, G. H. 1968. Geology of the country around Church Stretton, Craven Arms, Wenlock Edge and Brown Clee. *Mem. Geol. Surv. Gt. Britain*.

HALL, H. F. 1870. On the glacial and post-glacial deposits in the neighbourhood of Llandudno. *Geol. Mag.*, **7**, 509–13.

HAWKINS, A. B., and KELLAWAY, G. A. 1971. Field meeting at Bristol and Bath with special reference to new evidence of glaciation. *Proc. Geol. Ass.*, **82**, 267–92.

HEY, R. W. 1957. High level gravels in and near the lower Severn valley. *Geol. Mag.*, **95**, 161–68.

HIGGINS, L. S. 1933. Coastal changes in South Wales—the excavation of an old beach (Merthyr Mawr Warren). *Geol. Mag.*, **70**, 541–49.

HOGG, A. H. A. 1965. Early Iron Age Wales, in Foster, I. Ll. and Daniel, G. (Eds.), *Prehistoric and Early Wales*, London, 109–50.

HOLLIN, J. T. 1965. Wilson's theory of ice-ages. *Nature Lond.*, **208**, 12–16.

——. 1970. Antarctic Ice Surges. *Antarctic J. U.S.A.*, **5**, 155–56.

HOPLEY, D. 1963. *The coastal geomorphology of Anglesey*. Unpublished M. Sc. thesis, University of Manchester.

HOWELL, F. T. 1965. *Some aspects of the sub-drift surface of some parts of north west England*. Unpublished Ph.D. thesis Univ. of Manchester.

HOULDER, C. H., and MANNING, W. H. 1966. *South Wales*. Regional Archaeologies. London.

HOWARD, F. J. 1904. Notes on glacial action in Brecknockshire and adjoining districts. *Trans. Cardiff Nat. Soc.*, **36** (1903), 92–108.

HULL, E. 1864. Geology of the country around Oldham, including Manchester and its suburbs. *Mem. Geol. Surv. Gt. Br.* H.M.S.O.

HYDE, H. A. 1936. On a peat bed at the East Moors, Cardiff. *Trans. Cardiff Nat. Soc.*, **67** (1934), 39–48.

JEHU, T. J. 1904. The glacial deposits of northern Pembrokeshire. *Trans. R. Soc. Edinb.*, **41**, 53–87.

——. 1909. The glacial deposits of western Caernarvonshire. *Trans. R. Soc. Edinb.*, **47**, 17–56.

JOHN, B. S. 1965a. *Aspects of the glaciation and superficial deposits of Pembrokeshire.* Unpublished D.Phil. thesis, University of Oxford.

——. 1965b. A possible Main Würm glaciation in west Pembrokeshire. *Nature Lond.*, **207**, 4997, 622–23.

——. 1968. A reassessment of glacial limits in the southern Irish Sea basin. *Inst. Br. Geogr. Newsletter*, **5**, 28.

——. 1970a. Pembrokeshire, in, Lewis C. A. (Ed.) 1970.

——. 1970b. The Pleistocene drift succession at Porth Clais, Pembrokeshire. *Geol. Mag.*, 107.

——. 1971. Glaciation and the West Wales landscape. *Nature in Wales*, 12.

—— and ELLIS-GRUFFYDD, I. D. 1970. Weichelian stratigraphy and radiocarbon dating in south Wales. *Geol. en Mijnb.*, **49**, 285–96.

JONES, A. S. G. 1971. A textural study of marine sediments in a portion of Cardigan Bay (Wales). *J. Sed. Pet.*, **41**, 505–16.

JONES, O. T. 1965. The glacial and post-glacial history of the lower Teifi Valley. *Quart. J. geol. Soc. Lond.*, **121**, 247–81.

—— and PUGH, W. J. 1935. The geology of the district around Machynlleth and Aberystwyth. *Proc. Geol. Ass.*, **46**, 247–300.

KEEPING, W. 1882. The glacial geology of central Wales. *Geol. Mag.*, **19**, 251–

KELLY, M. R. 1964. The Middle Pleistocene of north Birmingham. *Phil. Trans. R. Soc.*, **B. 247**, 533–92.

LACAILLE, A. D., and GRIMES, W. F. 1955. The Prehistory of Caldey. *Arch. Camb.*, **104**, 85–165.

LEACH, A. L. 1911. On the relation of the glacial drift to the raised beach near Porth Clais, St. David's. *Geol .Mag.*, **8**, 462–66.

——. 1918. Flint-working sites on the submerged land (Submerged Forest) bordering the Pembrokeshire coast. *Proc. Geol. Ass.*, **29**, 46–47.

LEWIS, C. A. 1966. The Breconshire end-moraine. *Nature, Lond.*, **212**, 1159–61.

——, (Editor). 1970a. *The glaciations of Wales and adjoining regions*, London.

——. 1970b. The Upper Wye and Usk Regions, in Lewis, C. A. (Ed.) 1970a.

——. 1970c. The glaciations of the Brecknock Beacons. *Brycheiniog*, **14**, 97–120.

LEWIS, H. CARVILL, 1894. *Papers and notes on the glacial geology of Great Britain and Ireland*, London.

LUCKMAN, B. 1966. *Some aspects of the geomorphology of the Lugg and Arrow Valleys.* Unpublished M.A. thesis, University of Manchester.

——. 1970. The Hereford Basin, in, Lewis, C. A. (1970a).

McBURNEY, C. B. M. 1959. Report on the first season's field work on British Upper Palaeolithic cave deposits. *Proc. Prehist. Soc.*, **25**, 260–69.

——. 1965. The Old Stone Age in Wales, in Foster, I. Ll. and Davies, G. *Prehistoric and Early Wales*, London.

McMILLAN, N. F. 1949. Notes on post-glacial clays in Anglesey. *Proc. Lpool. Geol. Soc.*, **20**, 106–10.

MACNEY, D. and BURNHAM, C. P. 1966. *The Soils of the Church Stretton district of Shropshire.* Mem. Soil Surv. Eng. and Wales, H.M.S.O.

MATLEY, C. A. 1913. The geology of Bardsey Island. *Q. Jl. Geol. Soc. Lond.*, **69**, 514–33.

——. 1936. A 50-foot coastal terrace and other late-glacial phenomenon in the Lleyn peninsula. *Proc. Geol. Ass.*, **43**, 222–33.

MITCHELL, G. F. 1960. The Pleistocene history of the Irish Sea. *Advmt. Sci. Lond.*, **17**, 313–25.

——. 1962. Summer Field Meeting in Wales and Ireland. *Proc. Geol. Ass.*, **73**, 197–214.

——. 1970. The Quaternary deposits between Fenit and Spa on the north shore of Tralee Bay, Co. Kerry. *Proc. R. Irish Acad.*, **70, B**, 131–63.

—— and ORME, A. R. 1967. The Pleistocene deposits of the Isles of Scilly. *Quart. J. geol. Soc. Lond.*, **123**, 59–92.

MORISON, R. B. 1968. Means of Time-Stratigraphic Division and Long Distance Correlation of Quaternary Successions, in Morrison, R. B. and Wright, H. E. 1968. *Means of Correlation of Quaternary Successions.* 1–114, (Vol. 8, VII INQUA Congr.) Salt Lake City.

MOORE, J. J. 1970. In Lewis, C. A. 1970b.

MOORE, J. R. 1968. Recent sedimentation in north Cardigan Bay, Wales. *British Museum (Nat. Hist.) Bull. Miner.*, 2.

MOORE, P. D. 1968. Human influence upon Vegetational History in North Cardiganshire. *Nature, Lond.*, **217**, 5133, 1006–9.

MORGAN, A. V. 1971. *The glacial geology of the area north of Wolverhampton.* Unpublished Ph.D. thesis, University of Birmingham.

MORGAN, ANNE. 1971. Unpublished Ph.D. thesis, University of Birmingham.

MURCHISON, R. I. 1843. Anniversary address of the President. *Proc. Geol. Soc. Lond.*, **4**, 65–151.

NEAVERSON, E. 1942. A summary of the records of Pleistocene and post-glacial Mammalia from North Wales and Merseyside. *Proc. Lpool. Geol. Soc.*, **18** (1940–43), 70–85.

NORTH, F. J. 1929. *The evolution of the Bristol Channel.* Cardiff (2nd Ed., 1955).

OAKLEY, K. P. 1968. The date of the 'Red Lady' of Paviland. *Antiquity*, **42**, 306–7.

PEAKE, D. S. 1961. Glacial changes in the Alyn river system and their significance in the glaciology of the North Welsh Border. *Quart J. geol. Soc. Lond.*, **117**, 335–66.

PENNY, L. F. 1964. A review of the Last Glaciation in Britain. *Proc. Yorks. Geol. Soc.*, **34**, 387–411.

——, COOPE, G. R., and CATT, J. A. 1969. Age and insect fauna of the Dimlington Silts, East Yorkshire. *Nature, Lond.*, **244**, 65–67.

PEWE, T. L. 1969. *The Periglacial Environment: past and present.* Montral.

PICKERING, R. 1957. The Pleistocene geology of the South Birmingham area. *Quart. J. geol. Soc. Lond.*, **113**, 223–37.

POCOCK, T. L. 1906. *The geology of the country around Macclesfield, Congleton, Crewe and Middlewich. Mem. geol. Surv. Gt. Britain.*

——. 1938. Glacial deposits between North Wales and the Pennine range. *Zeit. Gletscherk. Eiszeit. forsch Gesch. Klimas*, **26**, 52–69.

——. 1940. Glacial drift and river terraces of the Hereford Wye, *Z. Gletscherk. Eiszeit forsch Gesch. Klimas*, **27**, 98–117.

——, WHITEHEAD, T. H., WEDD, C. B. and ROBERTSON, T. 1938. *Geology of the country around Shrewsbury. Mem. geol. Surv. Gt. Britain.*

—— and WHITEHEAD, T. H. 1948. *The Welsh Borderland British Regional Geology*, London,

POOLE, E. G. 1968. Age of the Upper Boulder Clay glaciation in the Midlands. *Nature, Lond.* **217**, 1137–8.

—— and WHITEMAN, A. J. 1961. The glacial drifts of the southern part of the Shropshire Cheshire basin. *Quart. J. geol. Soc. Lond.*, **117**, 91–130.

—— and WHITEMAN, A. J. 1966. *Geology of the country around Nantwich and Whitchurch. Mem. geol. Surv. Gt. Britain.*

POTTS, A. S. 1968. *The glacial and periglacial geomorphology of Central Wales.* Unpublished Ph.D. thesis, University of Wales.

——. 1971. Fossil cryonival features in Central Wales. *Geogr. Annlr.* **53, A**, 39–51.

RAMSAY, A. C. 1852. On the superficial accumulations and surface markings of North Wales. *Quart. J. Geol. Soc. Lond.*, **8**, 371–76.

——. 1881. *The Geology of North Wales*, (2nd Ed.). Mem. Geol. Surv. Gt. Britain.

RAPP, A. 1960. Recent developments of mountain slopes in Kärkevagge and surroundings, northern Sweden. *Geogr. Annal.*, 42.

READE, T. M. 1898. Notes on the drift of the Mid-Wales coast. *Proc. Lpool Geol. Soc.*, **7**, 410.

REED, A. H. and SHOTTON, F. W. 1970. A Quartzite Hand-Axe from near Claverley, East Shropshire. *Proc. Birmham. Nat. Hist. Soc.*, **21**, 243–49.

ROWLANDS, B. M. 1955. *The glacial and post-glacial evolution of the landforms of the Vale of Clwyd.* Unpublished M.A. thesis, University of Liverpool.

——. 1970. *The glaciation of the Arenig region.* Unpublished Ph.D. thesis, University of Liverpool.

——. 1971. Radiocarbon Evidence of the Age of an Irish Sea Glaciation in the Vale of Clwyd. *Nature, Lond.*, **230**, 9–11.

ROWLANDS, P. H. 1966. *Pleistocene Stratigraphy and Palynological investigation in the Marton, Church Stoke and Church Stretton Valleys, West of Shropshire.* Unpublished Ph.D. thesis, University of Birmingham.

—— and SHOTTON, F. W. 1971. Pleistocene deposits of Church Stretton (Shropshire) and its neighbourhood. *Jl. Geol. Soc.*, **127**, 599–622.

SAUNDERS, G. E. 1968a. A reappraisal of Glacial Drainage Phenomena in the Lleyn Peninsula, *Proc. Geol. Ass.*, **79**, 305–24.

——. 1968b. A fabric analysis of the ground moraine deposits of the Lleyn Peninsula of South West Caernarvonshire. *J. Geol.*, **6**, 105–18.

SAUNDERS, G. E. 1968c. Glaciation of possible Scottish Readvance age in North West Wales. *Nature, Lond.* **218**, 5136, 76–8.

SAVORY, H. 1963. The Personality of the Southern Marches of Wales in the Neolithic and Early Bronze Age, in Foster, I. L. and Daniel, G. (Eds.), *Culture and Environment*, London.

SEDDON, B. 1957. Late-glacial cwm glaciers in Wales. *J. Glaciol.*, **3**, 94–9.

——. 1958. *Geology and Vegetation of the Late Quaternary period in North Wales.* Unpublished Ph.D. Cambridge University.

——. 1962. Late-glacial deposits at Llyn Dwythwch and Nant Ffrancon, Caernarvonshire. *Phil. Trans. R. Soc. Lond.*, **B**, **244**, 459–81.

SHOTTON, F. W. 1953. The Pleistocene deposits of the area between Coventry, Rugby and Leamington and their bearing upon the totographic development of the Midlands. *Phil. Trans. R. Soc. Lond.*, **B**, **237**, 209–60.

——. 1959. *Proc, Prehist. Soc.*, **25**, 141–43.

——. 1962. The physical background of Britain in the Pleistocene. *Advanc. Sc.*, **19**, 193–206.

——. 1960. Large-scale patterned ground in the valley of the Worcestershire Avon. *Geol. Mag.*, **97**, 404–8.

——. 1967a. The problems and contributions of methods of absolute dating within the Pleistocene period. *Quart. J. geol. Soc. Lond.*, **122**, 357–83.

——. 1967b. Age of the Irish Sea glaciation in the Midlands. *Nature, Lond.*, **215**, 1366.

——. 1968. The Pleistocene succession around Brandon, Warwickshire. *Phil. Trans. R. Soc.*, **B**, **254**, 387–400.

—— and STRACHAN, I. 1959. The investigation of a peat moor at Rodbaston, Penkridge, Staffordshire. *Quart. J. geol. Soc. Lond.*, **155**, 1–16.

——, BLUNDELL, D. J., and WILLIAMS, R. E. G. 1968. Birmingham University Radiocarbon Dates II. *Radiocarbon*, **10**, 200–6.

——, —— and WILLIAMS, R. E. G. 1969. Birmingham University Radiocarbon Dates III, *ibid.*, **11**, 263–70.

——, —— and WILLIAMS, R. E. G. 1970. Birmingham University Radiocarbon Dates IV, *ibid.*, **12**, 395–99.

—— and WILLIAMS, R. E. G. 1971. Birmingham University Radiocarbon Dates V, *ibid.* **13** (2), 141–56.

—— and WEST, R. G. 1969. In Recommendations on Stratigraphical Usage. *Proc. Geol. Soc. Lond.* 1656.

SIMPKINS, K. 1968. *Aspects of the Quaternary history of Central Caernarvonshire.* Unpublished Ph.D. thesis, University of Reading.

SIMPSON, I. M., and WEST, R. G. 1958. On the stratigraphy and palaeobotany of a Late-Pleistocene organic deposit at Chelford, Cheshire. *New Phytol.*, **57**, 239–50.

——. 1959. The Pleistocene succession in the Stockport and South Manchester area. *Quart. J.*

geol. Soc. Lond., **115**, 107–21.

SMITH, R. T. and TAYLOR, J. A. 1969. The Post-Glacial development of vegetation and soils in North Cardiganshire. *Trans. Inst. Br. Geogr.*, **48**, 75–96.

SMITHSON, F. 1953. The micro-mineralogy of North Wales soils. *J. Soil. Sci.*, **4**, 194–209.

SOLLAS, W. J. 1913. Paviland cave: an Aurignacian station in Wales. *J. R. Anthrop. Inst.*, **43**, 325.

SPARKS, B. W. and WEST, R. G. 1970. Late Pleistocene deposits at Wretton, Norfolk. *Phil. Trans. R. Soc.*, **B, 258**, 1–30.

SQUIRELL, H. C. and DOWNING, R. A. 1969. *Geology of the South Wales Coalfield. Part 1. The Country around Newport (Mon.), Mem. geol. Surv. Gt. Britain.*

STEPHENS, N. 1966. Some Pleistocene deposits in north Devon. *Biul. Peryglac.*, **15**, 103–14.

STRAHAN, A. 1886. On the glaciation of south Lancashire, Cheshire and the Welsh Border. *Quart. J. geol. Soc. Lond.*, **42**, 369–90.

——. 1896. On submerged land-surfaces at Barry, Glamorgan, with notes on the fauna and flora by Clement Reid. *Quart. J. geol. Soc. Lond.*, **52**, 474–89.

—— and CANTRILL, T. C. 1904. *The Geology of the South Wales Coalfield. Part VI. The Country around Bridgend, Mem. geol. Surv. Gt. Britain.*

—— et al. 1907. *Ibid. Part VIII. The Country around Swansea.*

——, CANTRILL, T. C., DIXON, E. E. L., THOMAS, H. H. and JONES, O. T. 1914. *Ibid. Part XI. The country around Haverfordwest. Mem. geol. Surv. Gt. Britain.*

SUTCLIFFE, A. J. 1957. *Cave Fauna and Cave Sediments.* Unpublished Ph.D. thesis. University of London.

——. 1960. Joint Mitnor Cave, Buckfastleigh. *Trans. Torquay Nat. Hist. Soc.*, **13** (1958–59), 1–26.

SYNGE, F. M. 1963. A correlation between the drifts of south-east Ireland and those of Wales. *Ir. Geogr.*, **4**, 360–66.

——. 1964. The glacial succession in west Caernarvonshire. *Proc. Geol. Ass.*, **75**, 431–44.

——. 1970a. The Pleistocene Period in Wales, in Lewis C. A. (1970a) Ed. 315–50.

——. 1970b. The Irish Quaternary: Current Views 1969, in Stephens, N. and Glasscock R. E. (Eds.). *Irish Geographical Essays*, 34–48.

TALLIS, J. H., and KERSHAW, K. A. 1959. Stability of stone polygons in North Wales. *Nature Lond.*, **183**, 485–6.

TAYLOR, B. J. 1958. Cemented shear-planes in the Pleistocene Middle Sands of Lancashire and Cheshire. *Proc. Yorks. Geol. Soc.*, **31**, 359–65.

TAYLOR, B. J., PRICE, R. H., and TROTTER, F. M. 1963. *The Geology of the Country around Stockport and Knutsford, Mem. geol. Surv. Gt. Britain.*

THOMAS, K. W. 1965. The stratigraphy and pollen analysis of a raised bog at Llanllwch, near Carmarthen. *New Phyt.*, **64**, 101–17.

THOMAS, T. M. 1970. The imprint of structural grain on the micro-relief of the Welsh uplands. *Geol. J.*, **7**, 69–100.

TIDDEMAN, R. H. 1901. On the age of the raised beach of Southern Britain, as seen in Gower. *Geol. Mag.*, **7**, 441–43.

THOMPSON, D. B., and WORSLEY. 1966. A Late Pleistocene Molluscan fauna from the drifts of Cheshire Plain. *Geol. J.*, **5**, 197–207.

—— and WORSLEY, P. 1967. Periods of ventifact formation in the Permo-Triassic and Quaternary of the North East Cheshire Basin. *Mercian Geol.*, **2**, 279–98.

TOMLINSON, M. 1925. River Terraces of the Lower Valley of the Warwickshire Avon. *Quart. J. geol. Soc. Lond.*, **81**, 137–69.

——. 1935. The superficial deposits of the Country north of Stratford-on-Avon. *Quart. J. geol. Soc. Lond.*, **91**, 423–62.

TRAVIS, C. B. 1944. The glacial history of the Berwyn Hills, North Wales. *Proc. Lpool geol. Soc.*, **19** (1943–47), 14–28.

TRIMMER, J. 1831. On the diluvial deposits of Caernarvonshire between the Snowdon chain and the Menai Strait. *Proc. geol. Soc.*, **1**, 331–32.

TROTMAN, D. 1964. *Data for Late-Glacial and Post-Glacial history in South Wales.* Unpublished Ph.D. thesis, University of Wales.

TURNER, C. 1970. The Middle Pleistocene deposits at Marks Tay, Essex. *Phil. Trans. R. Soc.*, **B, 257**, 374–437.

TURNER, J. 1964. The anthropogenic factor in vegetational history. I. Tregaron and Whixall Mosses. *New Phytol.*, **63**, 73–90.

TURNER, J. 1965. A contribution to the history of forest clearance. *Proc. R. Soc.*, **B, 161**, 343–54.

WATSON, E. 1965. Periglacial structures in the Aberystwyth region. *Proc. Geol. Ass.*, **76**, 443–62.

——. 1966. Two nivation cirques near Aberystwyth, Wales. *Biul. peryglac.*, **15**, 79–101.

——. 1968. The periglacial landscape in the Aberystwyth region, in Bowen, E.G. *et al.*, (Eds.). *Geography at Aberystwyth*. Cardiff.

——. 1970. The Cardigan Bay area, Lewis, C. A. (Ed.) 1970a. 125–45.

——. 1971. Remains of pingos in Wales and the Isle of Man. *Geol. J.*, **7**, 381–92.

—— and WATSON, S. 1967. The periglacial origin of the drifts at Morfa-bychan, near Aberystwyth. *Geol. J.*, **5**, 419–40.

—— and ——. 1971. Vertical stones and analogous structures. *Geogr. Annal.*, **53**, A, 107–14.

WATTS, W. A. 1959. The interglacial deposits at Kilbeg and Newtown, Co. Waterford. *Proc. R. Ir. Acad.*, **60B**, 79–134.

——. 1967. Interglacial deposits in Kildromin Townland near Herbertstown, Co. Limerick. *Proc. R. Irish Acad.*, **65, B**, 339–48.

——. 1970. Tertiary and Interglacial Floras in Ireland, in Stephens, N. and Glasscock, R. E. (Eds.). *Irish Geographical Essays*, 17–33.

WEDD, C. B., SMITH, B., KING, W. B. R., and WRAY, D. A. 1929. *The Geology of the country around Oswestry. Mem. geol. Surv. Gt. Britain.*

—— and KING, W. B. R. 1924. *The Geology of the country around Flint, Howarden and Caergwrle, Mem. geol. Surv. Gt. Britain.*

WELCH, F. B. A. and TROTTER, F. M. 1961. *The Geology of the country around Monmouth and Chepstow. Mem. geol. Surv. Gt. Britain.*

WEST, R. G. 1967. The Quaternary of the British Isles, in Rankama, K. (Ed.). *The Quaternary* Vol. 2. New York.

——. 1968. *Pleistocene Geology and Biology*. London.

—— and SPARKS, B. W. 1960. Coastal interglacial deposits of the English Channel. *Phil. Trans. R. Soc.*, **B, 243**, 95–133.

WHITTOW, J. B. 1965. The Interglacial and Post-Glacial Strandlines of North Wales, in Whittow, J. B. and Wood, P. D. (Eds.). *Essays in Geography for Austin Miller*. 94–117. Reading.

WHITTOW, J. B., and BALL, D. F. 1970. North-west Wales, in Lewis, C. A. (Ed) 1970a, 21–58.

WHITEHEAD, T. H. In Poole and Whiteman (1961).

WILLIAMS, G. J. 1968a. *Contributions to the Pleistocene Geomorphology of the Middle and Lower Pleistocene Geomorphology of the Middle and Lower Usk*. Unpublished Ph.D. thesis, University of Wales.

——. 1968b. The Buried Channel and Superficial Deposits of the Lower Usk, and their correlation with similar features in the Lower Severn. *Proc. Geol. Ass.*, **79**, 325–48.

WILLIAMS, K. E. 1927. The glacial drifts of western Cardiganshire. *Geol. Mag.*, **64**, 205–27

WILLS, L. J. 1912. Late-Glacial and post-Glacial changes in the lower Dee valley. *Quart. J. geol. Soc. Lond.*, **68**, 180–98.

——. 1924. The development of the Severn valley in the neighbourhood of Ironbridge and Bridgnorth. *Quart. J. geol. Soc. Lond.*, **80**, 274–311.

——. 1937. The Pleistocene history of the West Midlands. *Rep. Br. Ass. Advt. Sc. London*, 71–94.

——. 1938. The Pleistocene development of the Severn. *Quart. J. geol. Soc. Lond.*, **94**, 161–242.

——. 1950. *The Palaeogeography of the Midlands* (2nd Ed.). Liverpool.

WIRTZ, D. 1953. Zur Stratigraphie des Pleistocäns in Westen der Britischen Inseln. *Neues Jb. Geol. Palaont. Abh.*, **96**, 267–303.

WOLSTEDT, P. 1967. The Quaternary of Germany, in Rankama, K. (Ed.). *The Quaternary* Vol. 2. New York.

WOOD, A. 1959. The erosional history of the cliffs around Aberystwyth. *Lpool Manchr geol. J.*, **2**, 271–

WOODLAND, A. W. and EVANS, W. B. 1964. *The Geology of the South Wales Coalfield. Part IV. The country around Pontypridd and Maesteg. Mem. geol. Surv. Gt. Britain.*

WORSLEY, P. 1967. Problems in naming the Pleistocene deposits of the North-East Cheshire Plain. *Mercian Geol.*, **2,** 51–55.

——. 1966a. Fossil frost wedge polygons at Congleton, Cheshire, England. *Geogr. Annlr.*, **48A,** 211–219.

——. 1966b. Some Weichselian fossil frost wedges from East Cheshire. *Mercian Geol.*, **1,** 357065.

——. 1970. The Cheshire-Shropshire Lowlands, in, Lewis, C. A. (Ed.). 1967a, 83–106.

WRIGHT, W. B. 1914. *The Quaternary Ice Age*. London (2nd Ed. 1937).

YATES, E. M., and MOSELEY, F. 1967. A contribution to the glacial geomorphology of the Cheshire Plain. *Trans. Inst. Br. Geogr.*, **42,** 107–25.

ZEUNER, F. E. 1959. *The Pleistocene Period* (2nd Ed.). London.